CONSTRUCTION PLANNING, EQUIPMENT, AND METHODS

McGraw-Hill Series in Construction Engineering and Project Management

Raymond E. Levitt, *Consulting Editor*

Barrie and Paulson: *Professional Construction Management*
Jervis and Levin: *Construction Law: Principles and Practice*
Koerner: *Construction and Geotechnical Methods in Foundation Engineering*
Levitt and Samelson: *Construction Safety Management*
Oglesby, Parker, and Howell: *Productivity Improvement in Construction*
Peurifoy and Ledbetter: *Construction Planning, Equipment, and Methods*
Peurifoy and Oberlander: *Estimating Construction Costs*
Shuttleworth: *Mechanical and Electrical Systems for Construction*

CONSTRUCTION PLANNING, EQUIPMENT, AND METHODS

Fourth Edition

R. L. Peurifoy, P.E.

Consulting Engineer
Bryan, Texas

W. B. Ledbetter, P.E.

Professor of Civil Engineering
Texas A & M University

McGraw-Hill Publishing Company

New York St. Louis San Francisco Auckland Bogotá Caracas
Hamburg Lisbon London Madrid Mexico Milan
Montreal New Delhi Oklahoma City Paris San Juan
São Paulo Singapore Sydney Tokyo Toronto

This book was set in Times Roman
by Intercontinental Photocomposition Limited.
The editors were Kiran Verma and J. W. Maisel;
the production supervisor was Diane Renda.
New drawings were done by ECL Art.
Halliday Lithograph Corporation was printer and binder.

CONSTRUCTION PLANNING, EQUIPMENT, AND METHODS

567891011 HDHD 99876543210

ISBN 0-07-049763-X

Library of Congress Cataloging in Publication Data

Peurifoy, R. L. (Robert Leroy),
Construction planning, equipment, and methods.

(McGraw-Hill series in construction engineering and
project management)
Includes bibliographies and index.
1. Building. I. Ledbetter, William Burl. II. Title.
III. Series.
TH145.P45 1985 624 84-17118
ISBN 0-07-049763-X

CONTENTS

PREFACE

Since publication of the third edition of this book improvements in planning, methods, and equipment have proved beneficial to the construction industry. The authors of this book have incorporated these improvements in the belief that they will better serve the industry and increase the value of the book to the reader.

New material includes expanded and updated coverage of construction planning to include time value of money, discounted present worth analysis, rate of return analysis, precedence diagramming, and PERT; soil stabilization and compaction; operation analyses; rock drills and methods; tunnel mucking; pressure grouting methods and materials; methods of determining the supporting capacities of load bearing piles; and methods and materials used in concrete construction.

Illustrations of equipment and methods are included as an aid to the reader in understanding the printed material more fully.

Many of the tables show value in both U.S. Customary (also called English) and metric (SI) units. Appendixes are included to assist the reader in converting values from U.S. Customary to metric units. Also, other tables listing conversion factors of interest to the reader are included.

The number of publications and sources of information listed under References at the close of many of the chapters has been increased by including more recent publications.

Names and addresses of the manufacturers of construction equipment illustrated and described in the book are given as an aid to the readers who may wish to contact the manufacturers.

The use of generic masculine pronouns generally has been retained in text references to individuals whose gender is not otherwise established. It should be emphasized that this has been done solely for succinctness of expression and such references are intended to apply equally to men and women.

The authors are deeply grateful to the many persons who have given generous assistance in obtaining much of the information appearing in the book.

Comments from readers will be welcomed.

R. L. Peurifoy
W. B. Ledbetter

LIST OF ABBREVIATIONS AND SYMBOLS

ACI	American Concrete Institute
AOA	Activity on arrow
AON	Activity on node
ASTM	American Society for Testing and Materials
AGC	Associated General Contractors of America, Inc.
bbl	barrel
bhp	brake horsepower
bm	bank measure, volume of earth prior to loosening
°*C*	Celsius temperature
cfm	cubic feet per minute
const	construction
cpm	cycles per minute
cps	cycles per second (hertz)
cu ft	cubic foot
cu m	cubic meter
cu yd	cubic yard
cwt	100 pounds
deg	degree
est	estimated
°*F*	Fahrenheit temperature
f.o.b.	free on board
fpm	feet per minute
fps	feet per second
ft	foot
ft-lb	foot-pound
gal	gallon
gpm	gallons per minute

hp	horsepower
hr	hour
in.	inch
kW	kilowatts
lb	pound
lin ft	linear foot
M	1,000
m	meter
m^3	cubic meter
max	maximum
M fbm	1,000 feet board measure of lumber
min	minute, minimum
mm	millimeter
mph	miles per hour
op	operation
PERT	Program evaluation review technique
plf	pounds per linear foot
psf	pounds per square foot
psi	pounds per square inch
rpm	revolutions per minute
sec	second
sq ft	square foot
sq in.	square inch
square	100 square feet of area
sq yd	square yard
std	standard
tph	tons per hour
wk	week
yd	yard
yr	year

CHAPTER
ONE

INTRODUCTION

THE PURPOSE OF THIS BOOK

The efforts of an engineer or architect, who designs a project, and the constructor, who builds the project, are directed toward the same goal, namely, the creation of something which will serve the purpose for which it is built in a satisfactory manner. Construction is the ultimate objective of a design. It is hoped that this book will assist the reader in more fully understanding the total construction process, from inception of the idea through startup. It is hoped that the material presented in the book will illustrate how the application of engineering fundamentals and analyses to construction activities may reveal methods of better controlling quality, schedule (time), and costs of a construction project. Engineering for construction begins long before a contractor moves onto the construction site. In fact, the total process of engineered construction may be thought of as consisting of six major elements, as shown in Fig. 1-1. Although shown as a linear process from conception through startup, the process elements often overlap. *Project management* may be defined as the overall control of this total process to optimize the three major attributes of the process: *quality, schedule, and cost.* Often overlooked, or ignored, is the fact that the degree of influence over these three attributes is *not* constant throughout the construction process. In reality, the greatest influence can be exerted during the project definition phase, and this influence *rapidly* decreases as the process continues (Fig. 1-2). Therefore, to effectuate significant opti-

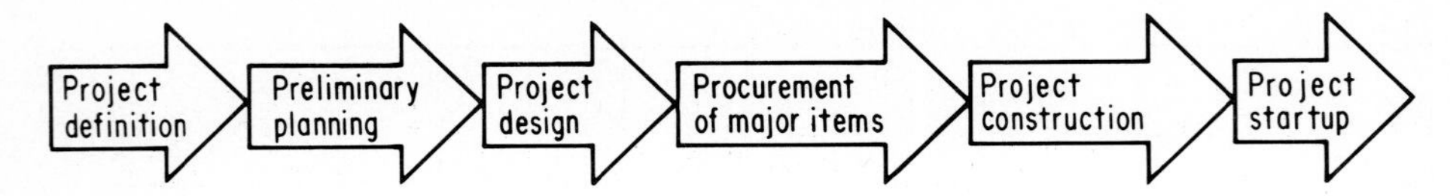

Figure 1-1 The total process of engineered construction.

mization of quality, schedule, and cost, project management is perhaps needed the most during the processes of project definition and preliminary planning. It is during the execution of these early phases that the major decisions are made concerning overall project size and complexity, project location, time constraints, desired level of quality, etc. Proper management of the planning and decision process is, therefore, extremely important.

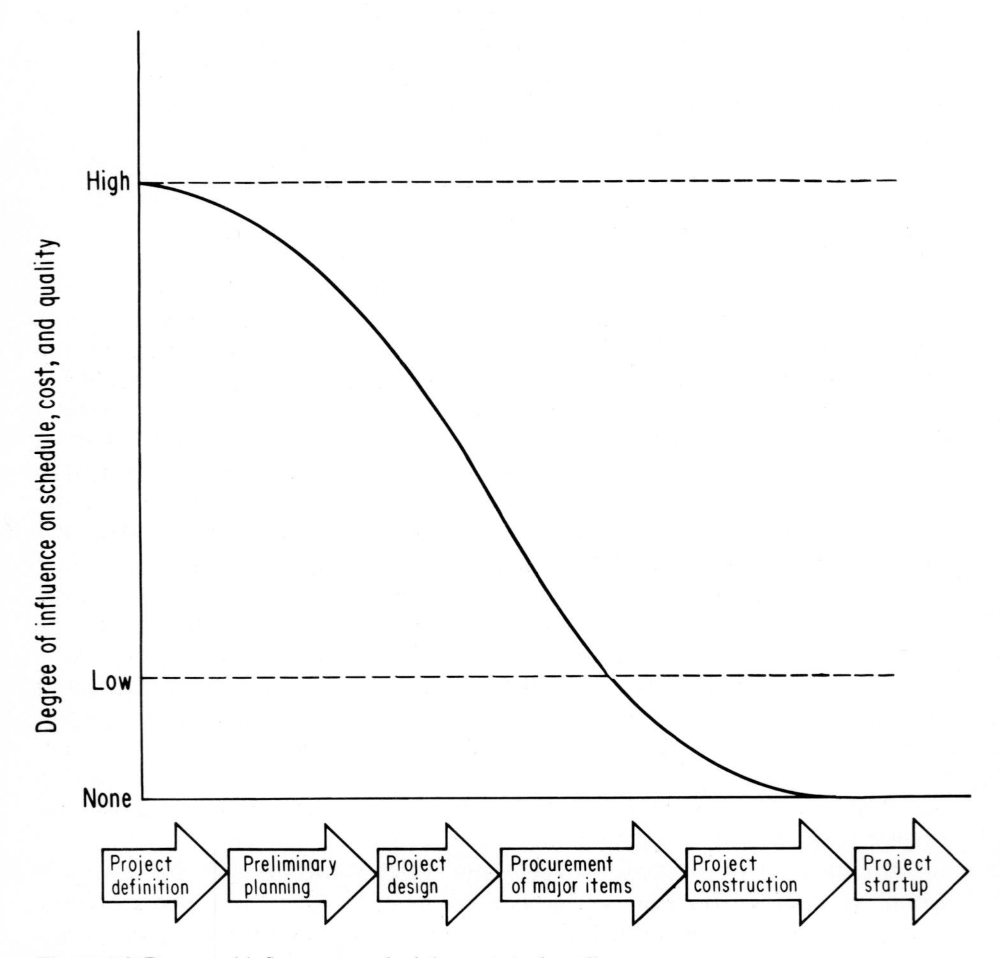

Figure 1-2 Degree of influence on schedule, cost, and quality.

THE ENGINEER AND CONSTRUCTION

When the prospective owner of a project under consideration recognizes a need for the project, he usually employs an engineer to make a study to determine the feasibility of the project. If the study indicates that it is justified, an engineer will be engaged to prepare the plans and specifications and usually to supervise the construction of the project. It is the duty of the engineer to design that project which will most nearly satisfy the needs of the owner at the lowest practicable cost. The engineer should study every major item to determine if it is possible to reduce the cost without unduly reducing the desired level of quality or prolonging the time of construction. It may be possible to change the design, modify the requirements for construction, or revise portions of the specifications in such a manner that the cost of the project will be reduced without sacrificing quality or schedule. An engineer who practices this philosophy is rendering a real service to the client. To accomplish this effectively, the engineer should be reasonably familiar with construction materials, methods, and costs if he or she is to design a project that is to be constructed at the lowest practicable cost.

THE CONSTRUCTION INDUSTRY

Although the construction industry is essentially a service industry, whose responsibility is to convert plans and specifications into a finished product, it is exceedingly complex and highly individual in character. The impact of construction on the economy of the United States is considerable. According to the *Business Roundtable*, construction in the United States amounted to over $300 billion in 1979, which is considerably more than reported by the federal government. Whatever the actual amount, it is spent by numerous owners to produce capital plant improvements in which the contractor generally assumes the responsibility for delivery of the completed facility at a specified time and cost. In so doing he accepts legal, financial, and managerial obligations. Under the stimulus of increasing demand for its services, the construction industry has expanded and is expanding in geographical scope and technological dimensions. There are more than 900,000 companies in the United States, ranging in size from small proprietorships with one or two employees to huge design/construct firms employing many thousands of employees and handling work in the billions of dollars.

CONSTRUCTION CONTRACTS

An understanding of construction contracts is essential for the proper operation of a construction project, and the engineer contributes an important service in developing the contract. The subject is exceedingly complex and beyond the scope of this book to cover in all details. However, there are fundamental concepts and definitions which must be understood, and these are presented in the following paragraphs. For more complete information on construction contracts there are a number of good texts—two of which are listed at the close of this chapter. Although there are many types of construction contracts, they all must contain four attributes to be valid. First there must be an *agreement* between the parties involved. Such an agreement involves *offer* (for example, the signed bid by a contractor proposing to construct a project constitutes an offer) and *acceptance* (for example, when the owner notifies the winning proposer). Second there must be *consideration.* In the case of a construction contract, if a contractor promises to build an addition to a home without being compensated and then changes his mind, he generally cannot be forced to build the addition because there was no *consideration.* Third, there must be *capacity.* This means that both parties must be of sufficient age to enter into a contract and must be mentally aware of what they are doing. Finally, for a contract to be valid, it must be *legal.* Obviously, a contract between two parties in which one agrees to commit an illegal act cannot be enforced. Of the many types of construction contracts, they generally will fall into one or more of the following three types.

Lump-sum contract The terms of this contract provide that the owner will pay to the contractor an agreed sum of money for the completion of a project conforming to the plans and specifications furnished by the engineer or the architect. It is common practice for the owner to pay the contractor a portion of this money at specified intervals, such as monthly, with the amount of each payment depending on the value of the work completed during the prior period of time, or according to some other schedule. Under the terms of this contract, a contractor may earn a profit (if he prepared a good estimate and stayed within budget) or he may sustain a loss (if his actual costs exceed his estimate). This type of contract generally requires the engineer or architect to have plans and specifications in sufficient detail for a contractor to prepare a detailed cost estimate. This is the preferred type of contract for the construction of buildings, homes, and the like because the owner obtains the benefits of competitive bidding and knows what the project will cost before he enters into a contract with a contractor.

Unit-price contract The terms of this type of contract provide that the owner will pay to the contractor an agreed amount of money for each unit of work completed in a project. The units of work may be any items whose quantities can be determined, such as cubic yards of earth, lineal feet of pipe, square yards of concrete pavement in-place, etc. Payments are usually made by the owner to the contractor at specified intervals during the construction of the project, with the amount of each payment depending upon work actually completed during the prior period of time. This type of contract also requires complete plans and specifications and is the preferred type of contract when the actual quantities in-place are not known with certainty beforehand. For example, the exact amounts of soil and rock to be excavated may not be known until the contractor actually performs the excavation. The owner, by requiring this type of contract, obtains the benefits of competitive bidding without having contractors bidding higher to cover the unknown quantities involved. Under the terms of this contract, the contractor may earn a profit or may incur a loss, depending upon the accuracy of his estimate.

Cost-plus-fee contract The terms of this type of contract provide that the owner will reimburse the contractor for all costs specified to construct the project, including all labor costs, material costs, equipment usage costs, subcontractor costs, and job supervision costs. In addition the owner agrees to pay the contractor an additional fee, which is essentially a management fee, to reimburse the contractor for the costs incurred at his head office resulting from the construction of the project. Items which are usually included in the fee are such costs as rent, taxes, insurance, interest on borrowed money for the project, and main office supervision and control costs, to name a few. Finally, the fee will include some expected profit for the contractor, as that is the primary reason for the contractor being in business. Whether or not the contractor actually makes a profit depends upon how accurately he estimated his other costs which make up the remainder of the fee. Under this type of contract the contractor usually takes the least risk and, therefore, has the least incentive to keep costs down. It is used primarily in situations where the scope of work cannot be well defined ahead of construction or where the state of the art for the particular construction is not well known. To exercise some control and give some incentive to the contractor to hold costs down, there are many, many variations to this type of contract, including cost plus a percentage of cost, cost plus a fixed fee, and cost plus a sliding fee, all with guaranteed maximums or with incentives for holding down costs. These types of contracts are complicated, and the reader is referred to the references at the close of this chapter for more detailed information.

PERFORMANCE BOND

Contractors frequently are required to furnish a performance bond for each project. This bond, which is not to be confused with insurance, is a three-party instrument in which a bonding company (termed surety) guarantees (or bonds) to the owner that the project will be built by the contractor in accordance with the contract. If, for any reason, the contractor becomes unwilling or unable to complete the contract, the surety will take steps to engage another contractor to complete the contract or take such other steps as will be satisfactory to the owner. The cost of a performance bond depends upon the size of the project, but will generally be in the range of 1 percent of the total project cost, provided the contractor has a good reputation and can get a performance bond.

CONTRACTOR TYPES

Contractors tend to specialize somewhat in various types of work. Although there are no clear-cut lines separating the many fields of construction, they may roughly be divided into *residential, building-commercial, industrial, highway-heavy*, and *specialty*. The specialty contractors are very numerous and include those involved with pipeline, power, transmission line, steel erection, railroad, offshore, pile driving, concrete pumping, and on and on. The reasons for specialization are complexity and capital investment. Few contractors are large enough to have the necessary expertise and the large inventory of expensive equipment to engage in all types of construction. The subject of owning and operating costs of construction equipment is discussed in Chap. 3.

CONSTRUCTION ECONOMY AND THE DESIGN ENGINEER

The cost of a project is determined by the requirements of the contract documents. Prior to completing the final design, the engineer should give careful consideration to the method and equipment which may be used to construct the project. Requirements which increase the cost without producing commensurate benefits should be eliminated. The decisions of the engineer should be based on a sound knowledge of the construction methods and equipment to be employed.

The budget for a project may be divided into six or more items: materials, labor, equipment, subcontracts, overhead, and profit. The design engineer has a strong influence over the costs of the first five of these items. If the engineer specifies materials which must be transported over long distances, or specifies excessive testing, or does not allow substitution of equal-quality materials, the costs will be higher than necessary. Other costly engineering practices include requiring many one-of-a-kind items which cannot be mass-produced, the use of nonstandard materials or techniques when not required, and establishing standards of quality that are higher than necessary.

The following list indicates methods which an engineer may use to reduce the costs of construction:

1. Design concrete structures with as many duplicate members as practical in order to permit the reuse of forms without rebuilding.
2. Simplify the design of the structure where possible.
3. Design for the use of cost-saving equipment and methods.
4. Eliminate unnecessary special construction requirements.
5. Design to reduce the required labor to a minimum.
6. Specify a quality of workmanship that is consistent with the quality of the project.
7. Furnish adequate foundation information where possible.

8. Refrain from requiring the contractor to assume the responsibility for information that should be furnished by the engineer or for adequacy of design.
9, Use local materials when they are satisfactory.
10. Write simple, straightforward specifications which clearly state what is expected. Define the results expected, but within reason permit the contractor to select the methods of accomplishing the results.
11. When possible, use standardized specifications, ones with which the contractors are familar.
12. Hold prebidding conferences with contractors in order to eliminate uncertainties and to reduce change orders to a minimum.
13. Use inspectors who have sufficient judgment and experience to understand the project and have authority to make decisions.

Other examples, illustrating methods of effecting economy in construction, will be found in succeeding chapters of this book.

CONSTRUCTION ECONOMY AND THE CONTRACTOR

One desirable characteristic of a successful contractor is a degree of dissatisfaction over the plans and methods under consideration for constructing a project. Complacency in members of the construction industry will not contribute toward developing new equipment, new methods, or new construction planning, all of which are desirable for continuing improvements in the products of the industry at lower costs. A contractor who does not keep informed on new equipment and methods will soon discover that his competitors are underbidding him. It is hoped that the analyses and examples presented in this book will impress on the reader the value of carefully studying each project in order to select the methods and equipment that will produce the greatest construction economy.

Suggestions for possible reductions in construction costs by the contractor include, but are not limited to, the following:

1. Prebidding studies of the project and the site to determine the effect of:
   a. Topography
   b. Geology
   c. Climate
   d. Sources of material
   e. Access to the project
   f. Housing facilities if required
   g. Storage facilities if required
   h. Labor supply
   i. Local services

2. The use of alternate construction equipment, having higher capacities, higher efficiencies, higher speeds, more maneuverability, and lower operating costs.
3. The payment of a bonus to the key personnel for production in excess of a specified rate.
4. The use of radios as a means of communication between the headquarters office and key personnel on projects covering large areas.
5. The practice of holding periodic conferences with key personnel to discuss plans, procedures, and results. Such conferences should produce better morale among the staff members and should result in better coordination among the various operations.
6. The adoption of realistic safety practices on a project as a means of reducing accidents.
7. Consideration of the desirability of subcontracting specialized operations to other contractors who can do the work more economically than the general contractor.
8. Consideration of the desirability of improving shop and service facilities for better maintenance of construction equipment.
9. The use of methods-improvement techniques, such as time-lapse photography.

VALUE ENGINEERING IN CONSTRUCTION

This is a formalized application of a specialized branch of engineering whose objective is to effect overall economy in the total-life-cycle cost of an engineered project, generally with special emphasis on the cost of constructing the project. A government agency, a corporation, or a private owner of a project to be designed and constructed normally uses the services of an architect and/or engineer to perform the necessary investigations to design the project most suitable to the objectives of the owner and presumably at the lowest practicable overall cost. Following completion of the plans and specifications, the project is advertised and a contract is awarded to the successful bidder. Alternatively, the project may be awarded on a negotiated basis. The terms of the contract provide that the contractor may use his own staff or employ consultants to make a study of the design, specifications, materials, and methods of construction to determine if any of these items can be modified to permit the construction of the project at a cost less than the amount of the contract *without reducing the quality or usefulness* of the project and without unduly delaying the completion of the project.

Value engineering may be applied to a project in one of two stages. The first stage is during or immediately after completion of the plans and specifications, and prior to release to potential contractors. If economies can be effected in the design or use of materials, etc., these savings will be realized by the owner. These savings, if any, will be reduced by the cost of the value engineering study.

The second state is after the award of the contract. The study is made by the contractor concerning savings which can be effected through modifications in the contract. The contractor is invited to submit to the owner detailed statements describing the modifications, with estimates showing the anticipated reductions in costs. If the owner, usually after consultation with his architect and/or engineer, approves the modifications, the resulting net savings are shared by the contractor and the owner on a preagreed basis.

MAKING A VALUE ENGINEERING STUDY

A value engineering study should be made by persons who are experienced in the design and/or the construction methods and materials involved. As a beginning point, value studies should initially concentrate on those items that represent the larger costs in the project and those that are repetitive. Value engineering studies have been shown to be very cost-effective. For example, the Environmental Protection Agency estimated that value engineering applied during the design phase of waste treatment facilities can potentially reduce costs, not only for construction but over the life cycles of the projects as well [1].

OBJECTIONS TO VALUE ENGINEERING STUDIES

Although numerous studies have demonstrated that reductions in costs are possible, there may be some objections by contractors to such studies and any recommended modification. Possible objections include the following:

1. How would it affect the project with regard to cost?
2. How would it affect the project with regard to time?
3. How would it affect the owner-contractor relationship?
4. How would it affect the architect-contractor or engineer-contractor relationship?
5. What is the probability of the modification being approved?

EXAMPLES OF VALUE ENGINEERING STUDIES

The examples presented below illustrate how value studies reduced (or might have reduced) the costs of the projects to which they apply. The examples are included to illustrate the possible or realized advantages of value studies. Their inclusion is not intended to imply that each recommendation had merit.

Example 1-1 The design of a building required the installation of a number of sliding doors to be fabricated from an expensive panelling material available in one stock size, namely, 4 ft 0 in. wide by 8 ft 0 in. long. The specified sizes of the doors were 2 ft 7 in. wide by 4 ft 1 in. high, equal to an area of 11 sq ft, to be cut from a panel whose area was 32 sq ft. The excess material, amounting to 21 sq ft for each panel, would be wasted because it could not be used elsewhere in the project. Prior to awarding a contract for the construction of the building, the owner, who was familiar with construction practices, directed that the sizes of the doors be modified to 4 ft 0 in. by 4 ft 0 in. In addition to reducing the cost, the modification improved the function of the doors.

Example 1-2 A contractor was awarded a contract to construct a concrete taxiway, 75 ft wide by approximately 1,500 ft long, for an existing airport. The specifications required that the taxiway be constructed in five strips, each 15 ft wide, with dowels of 2-in.-diameter steel pipe installed at 18-in. spacings in the joint between adjacent strips. Prior to starting construction, the contractor offered to reduce the amount of the contract by $25,000 if he were permitted to lay three strips, each 25 ft wide. The decision in analyzing this alternate method should be based on determining if the purpose of the runway would be better served by five strips 15 ft wide or three strips 25 ft wide.

Example 1-3 It was estimated that a radio facility designed for the U.S. Department of Defense, Bureau of Yards and Docks, would cost $16,845,620 as originally designed. When the design was subjected to a value engineering analysis by a consultant, it was determined that certain modifications could be made, without affecting the operating function of the facility, which reduced the cost to $14,999,521. The fee paid for the investigation was $600. Thus the net saving was $1,840,099 [2].

Example 1-4 Following the completion of a combination flood control and hydroelectric dam, whose cost was approximately $76,000,000, the contractor released an article listing six modifications, which, if they had been adopted, would have reduced the cost of the project by an estimated $7,200,000 [3].

The suggested modifications in methods and materials, together with the estimated savings, are listed and briefly described below.

Care and diversion of the river If a cofferdam lower than the one specified had been permitted (with the contractor to assume the risk of any costs resulting from the overtopping of the dam) and combined with a provision for flood waters to be diverted through a portion of the dam, the cost of this item could have been reduced by an estimated $500,000.

Height differential limited to 20 ft This limitation on height differential reportedly forced a shutdown of concreting operations for 3 months during the second-stage conversion and retarded concreting progress for 4 months thereafter. This requirement reportedly delayed the completion of the project at least 6 months, resulting in an increase in costs of at least $500,000.

Concrete lifts limited to 30 in. This limitation, which was applied to a portion of the concrete placed in the dam, increased the cost of cleanup and curing, with a resulting increase of at least $300,000 in the cost of the project.

Refrigeration The requirement that concrete be cooled before, during, and after placement resulted in an additional estimated cost of $1,400,000.

Five-day limit between successive lifts This requirement was estimated to have delayed the completion of the project by 3 months, at an additional cost of at least $250,000.

Aggregate The requirement that the aggregate for the project be produced from a quarry 7 miles from the dam, instead of permitting the use of natural sand and aggregate from a nearby quarry, as originally contemplated, increased the cost of the project by at least $4,250,000.

Example 1-5 A reinforced concrete bridge to be constructed in Hawaii was designed to be built on falsework. Because the superstructure was 150 ft (45.7 m) above the ground, the contractor chose to build it by cantilevering out from the piers, thus eliminating the need for falsework. The savings resulting from this change in construction plans, amounting to about $400,000, were shared equally by the contractor and the owner [4].

Persons interested in pursuing this subject further may wish to obtain one or more of the publications listed in the bibliography at the end of this chapter.

THE TIME VALUE OF MONEY

Today almost everyone is aware of the fact that money has a time value. One dollar today is worth more than one dollar tomorrow. This fact is vividly reinforced when the monthly charge bills are examined. Failure to pay the bill promptly results in an added charge being imposed. This added charge amounts to rent on the money that is owed, and is termed *interest.* Interest, usually expressed as a percentage of the amount owed, becomes due and payable at the close of each period of time involved in the statement of the bill. For example, if $1,000.00 is borrowed at 14% interest, then $0.14 \times 1{,}000$, or $140.00, in interest is owed on the *principal* of $1,000.00 after 1 year. If the borrower pays back the total amount owed after 1 year, he will pay $1,140.00. If he does not pay back any of the amount owed after 1 year, then normally the interest owed, but not paid, is considered as additional "principal" and thus the interest is *compounded.* Then, after 2 years he will owe $\$1{,}140.00 + 0.14 \times 1{,}140.00$, or $1,299.60. If your credit is good and you have borrowed the $1,000.00 from the bank, the banker normally does not care whether you pay him $1,140.00 after 1 year or $1,299.60 after 2 years. To him, the three values ($1,000, $1,140, and $1,299.60) are *equivalent.* In other words, $1,000 today is equivalent to $1,140 one year from today, which is equivalent to $1,299.60 two years from today. The above three values are obviously not equal, but they are "equivalent." Note that the concept of equivalence involves time and a specified rate of interest. The three values above are only equivalent for an interest rate of 14%, and then only at specific times. *Equivalence* means that one sum or series differs from another only by the accrued, accumulated interest at rate i for n periods of time.

Note that in the preceding example the principal amount was multiplied by an interest rate to obtain the amount of interest due. To generalize this concept the following symbols will be used:

P = a present single amount of money
F = a future single amount of money, after n periods of time
i = the rate of interest per interest period (usually 1 year)
n = the number of periods of time (usually years)

Equations for single payments. To calculate the future value F of a single payment P after n periods at an interest rate i, the following calculation would be made:

At the end of the first period: $F_1 = P + Pi$
At the end of the second period: $F_2 = P + Pi + (P + Pi)i = P(1 + i)^2$
At the end of the nth period: $F_n = P(1 + i)^n$

Or, in general, the future single amount of a present single amount is

$$F = P(1+i)^n \tag{1-1}$$

Note that F is related to P by a factor which depends only upon i and n. This factor, termed the *single payment compound amount factor* (SPCAF), makes F *equivalent* to P. This factor may be expressed in a functional form as

$$(1+i)^n = (F/P,i,n)$$

and Eq. (1-1) can be expressed as

$$F = P(F/P,i,n) \tag{1-1A}$$

If a future amount F is given, the present amount P can be calculated by transposing the equation to

$$P = \frac{F}{(1+i)^n} \tag{1-2}$$

or in functional form as

$$P = F(P/F,i,n) \tag{1-2A}$$

The factor $1/(1+i)^n$ is known as the *present worth compound amount factor* (PWCAF). To compute either P or F, one need only solve Eq. (1-1) or Eq. (1-2). To aid in the solving of these equations, tables of values for the two functions for typical values of i and n are given in Appendix A. Following are examples using these relations.

Example 1-6 A contractor wishes to set up a revolving line of credit at the bank to handle her cash flow during the construction of a project. She believes she needs to borrow \$12,000 with which to set up the account, and she can obtain the money at 1.45% per month. If she pays back the loan and accumulated interest after 8 months, how much will she have to pay back?

To solve, use Eq. (1-1).

$$F = 12{,}000(1+0.0145)^8 = 12{,}000(1.122061) = 13{,}464.73 = \underline{\$13{,}465}$$

The amount of interest she paid was $\underline{\$1{,}465.}$

Example 1-7 A construction company wants to set aside enough money today in an interest-bearing account to have \$100,000 five years from now for the purchase of a replacement piece of equipment. If the company can receive 12% interest on its investment, how much should be set aside now to accrue the \$100,000 five years from now?

To solve, use Eq. 1-2:

$$P = \frac{100{,}000}{(1+0.12)^5} = \frac{100{,}000}{1.762342} = \$56{,}742.68 = \underline{\$56{,}740}$$

An alternate solution is to use the tables in Appendix A. Here the functional form of the equation is used [Eq. (1-2A)]:

$$P = 100{,}000(P/F,12,5) = 100{,}000(0.56743) = \$56{,}743 = 56{,}740$$

In the preceding examples and explanation, equivalent single payments now and in the future were equated. Four parameters were involved: P, F, i, and n. *Given any three, the fourth can easily be calculated.*

Formulas for a uniform series of payments. Often, payments or receipts occur at regular intervals, and such uniform values can be handled by use of additional functions. First let us define another symbol. Let

A = uniform *end-of-period* payments or receipts continuing for a duration of n periods

If this uniform amount A is invested at the end of each period for n periods at a rate of interest i per period, then the total equivalent amount F at the end of the n periods will be

$$F = A[(1+i)^{n-1} + (1+i)^{n-2} + \cdots + (1+i) + 1]$$

By multiplying both sides of the equation by $(1+i)$ and subtracting the result from the original equation, the result will be

$$Fi = A(1+i)^n - 1$$

which can be rearranged to

$$F = A\left[\frac{(1+i)^n - 1}{i}\right] \tag{1-3}$$

This relationship can also be expressed in functional form as

$$F = A(F/A,i,n) \tag{1-3A}$$

The relationship $(1+i)^n - 1)/i$ is sometimes known as the *uniform series, compound amount factor* (USCAF).

The relationship can be rearranged to yield

$$A = F\left[\frac{i}{(1+i)^n - 1}\right] \tag{1-4}$$

which can be expressed in functional form as

$$A = F(A/F,i,n) \tag{1-4A}$$

The relationship $i/[(1+i)^n - 1]$ is known as the *uniform series sinking fund factor* (USSFF) because it determines the uniform end-of-period investment A that must be made in order to provide an amount F at the end of n periods.

In order to determine the equivalent uniform periodic series required to replace a present value of P, simply substitute Eq. (1-2) for F into Eq. (1-4). The resulting equation is

$$P = A\left[\frac{(1+i)^n - 1}{i(1+i)^n}\right] \tag{1-5}$$

In functional form the equation is

$$P = A(P/A,i,n) \tag{1-5A}$$

This relationship is known as the *uniform series present worth factor* (USPWF).

By inverting Eq. (1-5), we can obtain the equivalent uniform series of end-of-period values A from a present value P. The resulting equation is

$$A = P\left[\frac{i(1+i)^n}{(1+i)^n - 1}\right] \tag{1-6}$$

In functional form the equation is

$$A = P(A/P,i,n) \tag{1-6A}$$

This relationship is often called the *uniform series capital recovery factor* (USCRF).

To aid in the calculations using these relationships, values for F/A, A/F, P/A, and A/P can also be found in Appendix A for typical values of i and n. For each relationship, knowing any three of the four parameters will allow you to calculate the fourth.

As an aid toward the understanding of the above six equivalence relationships, appropriate cash flow diagrams can be drawn. *Cash flow diagrams* are drawings where the horizontal line represents time and vertical arrows represent cash flows at specific times. The cash flow diagrams for each

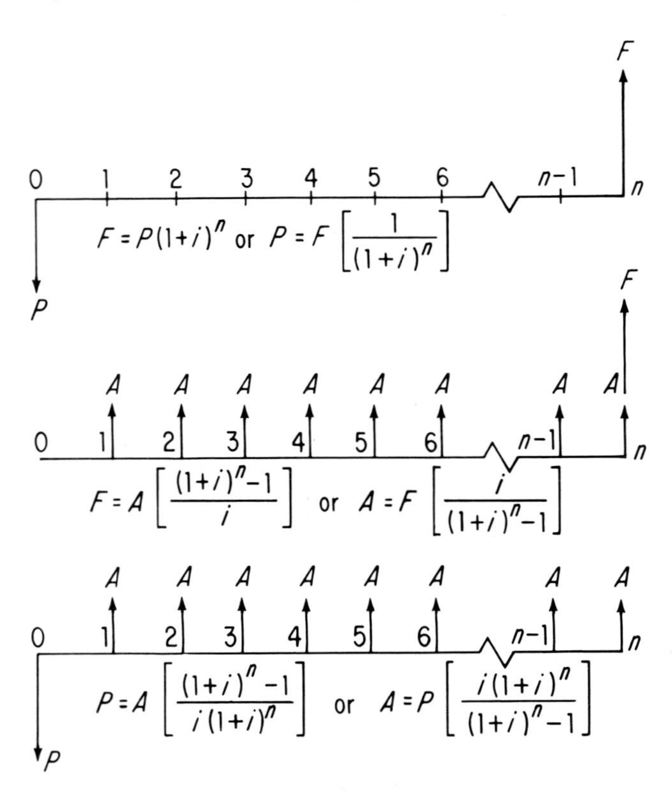

Figure 1-3 Cash flow diagrams.

relationship are shown in Fig. 1-3. These relationships form the basis for many complicated engineering economy studies involving the time value of money, and there are many excellent texts on the subject (see references at the close of this chapter for further reading on this subject).

Most complicated engineering economy problems can be broken down into parts where the above six relationships can be utilized. The following examples illustrate how this can be done.

Example 1-8 A piece of construction equipment costs $45,000 to purchase. Fuel, oil, and minor maintenance are estimated to cost $12.34 for each hour the equipment is used. The tires cost $3,200 to replace (estimated to occur every 2,800 hr of use), and major repairs of $6,000 are expected after 4,200 hr of use. The piece of equipment is expected to last for 8,400 hr, after which it will have an estimated salvage value of 0.1 the purchase price. How much should the owner of the equipment charge, per hour of use, if he expects to use the piece of equipment about 1,400 hr per year? Assume an interest rate of 15%.

To solve, a good first step is to draw a cash flow diagram (Fig. 1-4). Note that the cash disbursements are shown going down on the diagram.

$$n = \frac{8,400}{1,400} = 6 \text{ years}$$

$$A_1 = -45,000(A/P,15,6) = -45,000(0.26424) = -11,890.80$$
$$A_2 = -12.34(1,400) = -17,276.00$$
$$A_3 = -3,200(A/F,15,2) = -3,200(0.46512) = -1,488.38$$

Note that this is analogous to

$$-3,200((P/F,15,2) + (P/F,15,4) + (P/F,15,6))(A/P,15,6)$$

$$A_4 = -6,000(P/F,15,3)(A/P,15,6)$$
$$= -6,000(0.65752)(0.26424) = -1,042,46$$
$$A_5 = +(4,500 + 3,200)(A/F,15,6) = +7,700(0.11424) = +879.65$$
$$A_T = \text{the total annual cost} = -\$30,817.99 = -\$30,820$$

The hourly cost would then be

$$\frac{30,817.99}{1,400} = \$22.01/\text{hr}$$

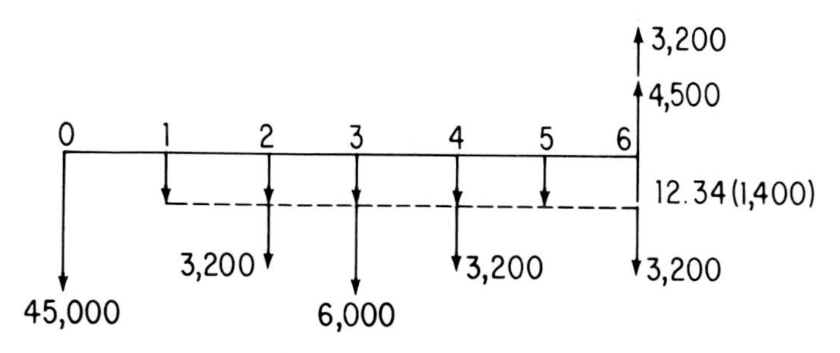

Figure 1-4 Cash flow diagram.

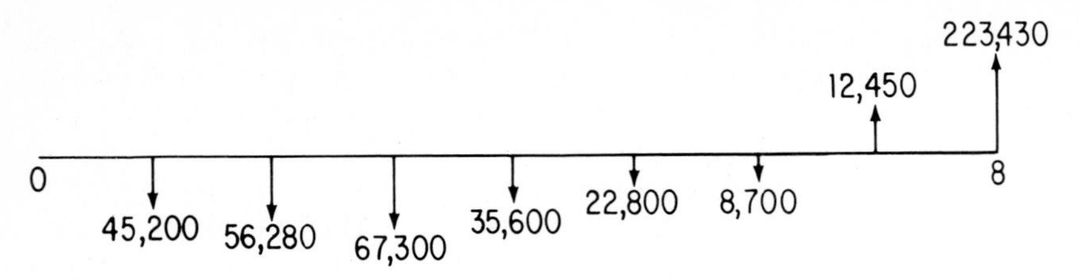

Figure 1-5 Cash flow diagram for Example 1-9.

Example 1-9 A contractor calculates that he will have the following cash flow overdrafts on a construction project. These overdrafts are due to the delays in receiving progress payments from the owner and from retainage being held by the owner until completion of the project. If the contractor's cost of money is 2.0 percent per month, how much total interest will he have to pay as a result of these cash flow overdrafts?

Month 1	−$45,200
Month 2	−56,280
Month 3	−67,300
Month 4	−35,600
Month 5	−22,800
Month 6	− 8,700
Month 7	+12,450
Month 8	+223,430

The cash flow diagram for this problem is shown in Fig. 1-5. To solve, calculate the future worth (F) of the cash flow, as the total amount of money involved (excluding time value) equals zero.

$$F = -45{,}200(F/P,2,7) - 56{,}280(F/P,2,6) - 67{,}300(F/P,2,5) - 35{,}600(F/P,2,4)$$
$$-22{,}800(F/P,2,3) - 8{,}700(F/P,2,2) + 12{,}450(F/P,2,1) + 223{,}430$$
$$F = -45{,}200(1.1486) - 56{,}280(1.1261) - 67{,}300(1.1040) - 35{,}600(1.0824)$$
$$-22{,}800(1.0612) - 8{,}700(1.0404) + 12{,}450(1.0200) + 223{,}430$$
$$F = -\$25{,}244$$

Note: This problem points out the potential significant costs of overdraft cash flows during construction, especially if interest rates are high. In this example, 2 percent per month is equivalent to $(1 + .02)^{12} - 1$ or 26.8 percent per year. If the contractor could borrow money for 1 percent per month (12.7 percent per year), his total interest would reduce to $12,339.

DISCOUNTED PRESENT WORTH ANALYSIS

Often in engineering economic studies, as well as in general financial analyses, a discounted present worth analysis is made of each alternative under consideration. This type of analysis involves calculating the *equivalent* present worth or present value of all the dollar amounts involved in the alternative to determine its present worth. This present worth is *discounted* at a predetermined rate of interest, often termed the *minimum attractive rate of return* (MARR or i^*). The MARR is usually equal to the current rate of interest for borrowed capital plus an additional rate for such factors as risk, uncertainty, contingencies, and the like. The following example illustrates the use of discounted present worth to evaluate three mutually exclusive alternatives.

Example 1-10 The Ace-in-the-Hole Construction Company is considering three methods of acquiring company pickups for use by field engineers. The alternatives are:

A. Purchase the pickups for $7,200 each and sell after 4 years for an estimated $1,200 each.

B. Lease the pickups for 4 years for $2,250 per year paid in advance at the beginning of each year. The contractor pays all operating and maintenance costs on the pickups.

C. Purchase the pickups on special time payments with $750 down now and $2,700 per year at the end of each year for 3 years. Assume the pickups will be sold after 4 years for $1,200 each.

If the contractor's MARR is 15 percent, which alternative should he choose? The cash flow diagrams for these three alternatives are given in Fig. 1-6. To solve calculate the net present worth (NPW) of each alternative at 15 percent and select the least costly alternative.

$$\text{NPW}_A = -7{,}200 + 1{,}200(P/F,15,4) = -7{,}200 + 686.1 = -\$6{,}514$$

$$\text{NPW}_B = -2{,}250 - 2{,}250(P/A,15,3) = -2{,}250 - 5{,}137 = -\$7{,}387$$

$$\text{NPW}_C = -750 - 2{,}700(P/A,15,3) + 1{,}200(P/F,15,4) = -\$7{,}742$$

The least costly alternative is A.

The foregoing example was simplified in two respects. One, the amount of calculations required was quite small. Two, all three alternatives involved the same lives (4 years in the example). Problems involving more data and calculations are no different in approach than

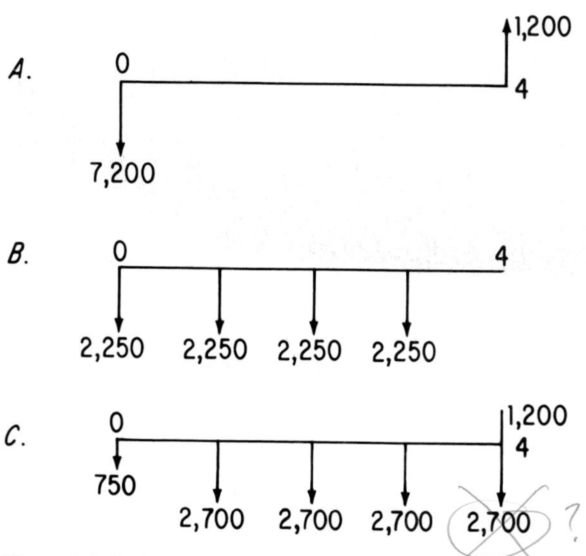

Figure 1-6 Cash flow diagrams for Example 1-10.

the example cited. But, when alternatives involve different lives, the analysis becomes more difficult. Obviously, if you are comparing one alternative with a life of 5 years with an alternative with a life of 10 years, their respective discounted present worths are not directly comparable. How do you handle this situation? There are two approaches generally used. They are:

Approach 1: Truncate (cut off) the longer-lived alternative(s) to equal the shorter lived alternative and assume a salvage value for the unused portion of the longer-lived alternative(s). Then make the comparison on the basis of equal lives.

Approach 2: Assume equal replacement conditions (costs and incomes) for each alternative and compute the discounted present worth on the basis of the least common multiple of lives for all alternatives.

The following example illustrates these two approaches.

Example 1-11 A contractor is considering the purchase of either a new track-type tractor for \$73,570, which has a 6-year life with an estimated net annual income of \$26,000 and a salvage value of \$8,000, or a used track-type tractor for \$24,680, with an estimated life of 3 years and no salvage value and an estimated net annual income of \$12,000. If the contractor's MARR is 20%, which tractor, if any, should she choose?

The cash flow diagrams for these two alternatives are shown in Fig. 1-7. Note the unequal lives. In order to use approach 1, a suitable salvage value must be assumed for the new tractor after 3 years. For this example, let us assume a 3-year salvage value for the new tractor of \$30,000. Now the cash flow diagram for the new factor looks like that shown in Fig. 1-8. The discounted present works of each alternative are

$$\text{NPW}_{\text{new}} = -73{,}570 + 26{,}000(P/A{,}20{,}3) + 30{,}000(P/F{,}20{,}3) = -\$1{,}443$$

$$\text{NPW}_{\text{old}} = -24{,}680 + 12{,}000(P/A{,}20{,}3) = +\$597$$

Using approach 1, purchasing the used tractor is the better alternative. Using approach 2, the cash flow diagram for the use tractor is shown in Fig. 1-9. The discounted present worths of each alternative are

$$\text{NPW}_{\text{new}} = -73{,}570 + 26{,}000(P/A{,}20{,}6) + 8{,}000(P/F{,}20{,}6) = +\$15{,}570$$

$$\text{NPW}_{\text{old}} = -24{,}680 + 12{,}000(P/A{,}20{,}6) - 24{,}680(P/F{,}20{,}3) = +\$944$$

Using approach 2, purchasing the new tractor is the better alternative. Neither of the

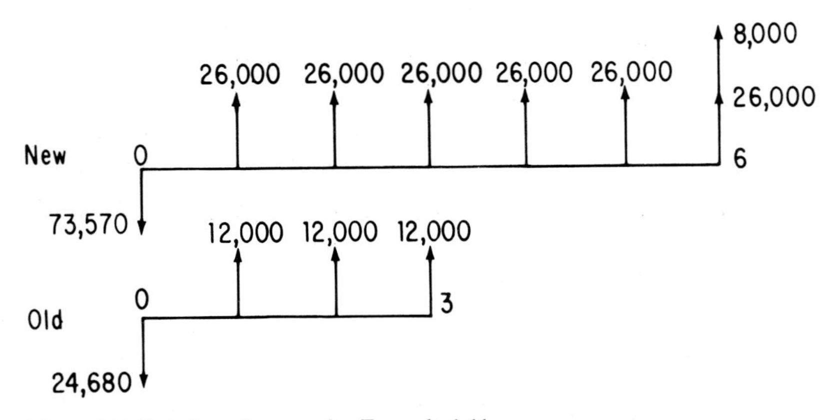

Figure 1-7 Cash flow diagrams for Example 1-11.

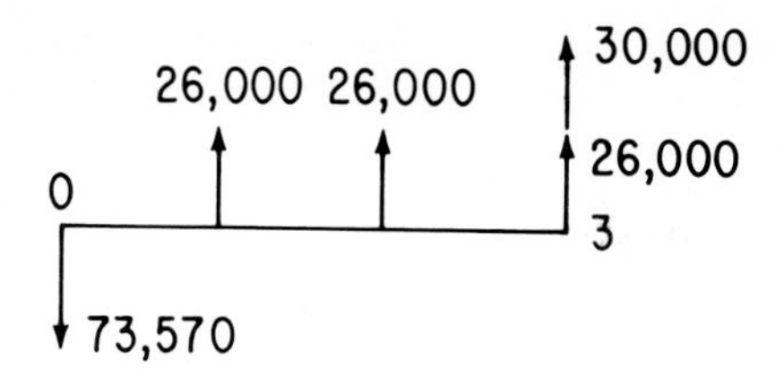

Figure 1-8 Cash flow diagram for Example 1-11.

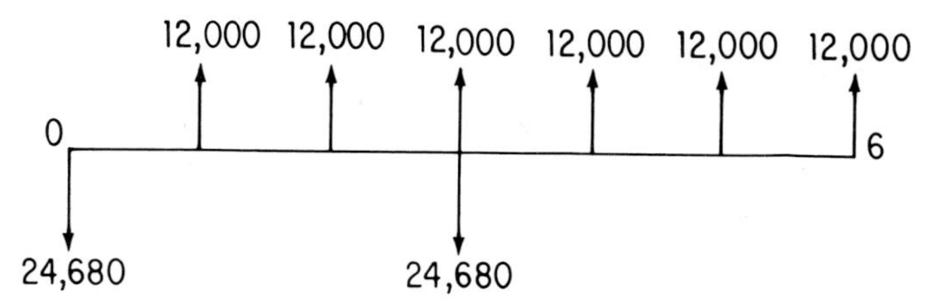

Figure 1-9 Cash flow diagram for Example 1-11.

approaches discussed is entirely satisfactory, and they can yield different solutions (as shown in the example). Which solution, if either, is correct? The answer is "whichever one best fits the situation." To assume you can receive a high (or low) salvage value by selling a piece of equipment before the end of its useful life may be very erroneous. On the other hand, to assume an equal replacement cost for a used piece of equipment 3 years in the future may be equally erroneous. To further complicate matters, you can assume inflation (or deflation) on replacement of the shorter-lived alternatives and use different values for the replacement conditions. But, in all approaches you are making assumptions and your results are only as good as your assumptions. The real answer is to use the most *reasonable* approach to each specific problem.

RATE OF RETURN ANALYSIS

Often, the use of discounted present worth analysis raises more questions than it answers, because the assumptions involved in its use are very visible to anyone reviewing the analysis. Interestingly, if the *rate of return* analysis is used, many people will accept the results without considering the assumptions involved, even though the resulting answer will be the same. There is something about knowing the anticipated *rate of return* of an investment that permits a decision to be made with more "perceived" confidence.

The *rate of return* of a proposed investment is that interest rate which makes the discounted present worth of the investment equal to zero. To calculate the *rate of return*, simply set up the equation and solve for i. The following example illustrates the procedure.

Example 1-11 (same example) A contractor is considering the purchase of either a new track-type tractor for $73,570, which has a 6-year life with an estimated net annual income of

\$26,000 and a salvage value of \$8,000, or a used track-type tractor for \$24,680, with an estimated life of 3 years and no salvage value and an estimated net annual income of \$12,000. Which tractor, if either, should she choose?

The cash flow diagrams are given in Fig. 1-7, except that in this case no MARR is given. Note again the unequal lives. The resulting equations are

$$\text{NPW}_{\text{new}} = -73{,}570 + 26{,}000(P/A,i,6) = 0$$

$$(P/A,i,6) = \frac{73{,}570}{26{,}000} = 2.82962 = \frac{(1+i)^6 - 1}{i(1+i)^6}$$

This equation can be solved directly if you have a calculator that will perform the necessary calculations for you, or you can use the interest tables, selecting values of $(P/A,i,3)$ just higher and just lower than the value needed, and then interpolate to find the value of i which makes the equation equal to zero.

The resulting value of i is

$$i_{\text{new}} = 28.0\%$$

$$\text{NPW}_{\text{old}} = -24{,}680 + 12{,}000(P/A,i,3) = 0$$

$$(P/A,i,3) = \frac{24{,}680}{12{,}000} = 2.05667 = \frac{(1+i)^3 - 1}{i(1+i)^3}$$

$$i_{\text{old}} = 21.5\%$$

The use of this approach suffers from the same problem as the discounted present worth analysis in that it ignores the different lives of the two alternatives. In this case, while the rates of return for each alternative are correct, what happens during the second 3-year period using the old tractor? If equal replacement conditions are assumed, then the rate of return for the old tractor does not change and the conclusion reached is that the new tractor yields a higher rate of return.

However, if we assume the salvage value of the new tractor is \$30,000 after 3 years and compare the rates of return on the basis of 3-year lives, the resulting equation for the new tractor is

$$\text{NPW}_{\text{new}} = -73{,}570 + 26{,}000(P/A,i,3) + 30{,}000(P/F,i,3) = 0$$

This equation can be solved as before, and the result is

$$i_{\text{new}} = 18.9\%$$

Using the values for i of 21.5 percent and 18.8 percent for the old and the new tractor, respectively, you can see that the old tractor will yield a better rate of return. Again, the answer depends upon the assumptions used. Note that a decision as to which tractor to purchase is still not made. The only thing you know is their respective rates of return. Before a decision can be reached, you must know your MARR. If we use the 3-year analysis period, if the MARR is 20 percent, then the choice would be to purchase the used tractor. But if the minimum attractive rate of return is 30 percent, then the choice would be to choose *neither*! When you analyze alternatives involving positive rates of return, the ***do-nothing*** alternative is always a possibility.

If the previous example is analyzed further, a curious result can occur. If the MARR were 15 percent, which alternative should be selected? They both *exceed* the MARR. But, since the old tractor yields a higher MARR, shouldn't it be selected? To answer this question, determine each alternative's net present worth at 18 percent.

$$\text{NPW}_{\text{new}} = -73{,}570 + 26{,}000(P/A,15,3) + 30{,}000(P/F,15,3) = \$5{,}519$$

$$\text{NPW}_{\text{old}} = -24{,}680 + 12{,}000(P/A,15,3) = \$2{,}719$$

According to our net present worth analysis, the new tractor yields a higher value *for a*

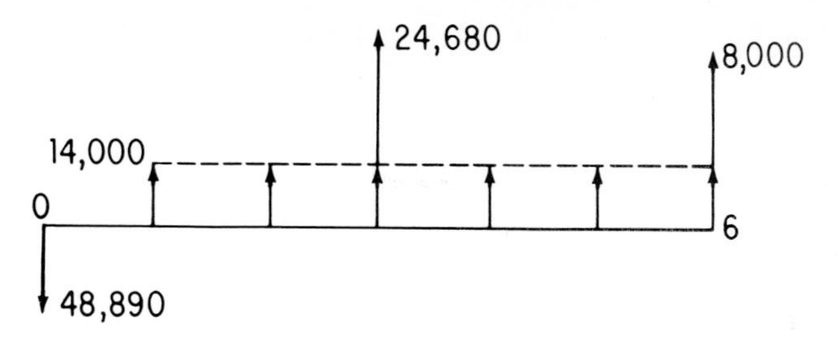

Figure 1-10 Incremental cash flow diagram for two tractors (Example 1-11).

MARR of 15 percent. How can this be? Intuitively, it would seem that the alternative with the higher rate of return would yield the higher NPW regardless of the assumed interest rate. That is not true! In this example there is a hidden consideration that strongly influences the result. Notice that the initial investments are *not the same.* We are looking at initial investments of $73,570 and $24,680. When we examine the rate of return of each of these alternatives, we have ignored their respective differences in initial cash flows. Therefore, we can obtain misleading results through such an analysis.

To handle unequal initial investments, *an incremental rate of return analysis* (IROR) is required. Essentially, the question that must be answered is this: "For alternatives that have a satisfactory rate of return (ROR), what is the IROR of the difference in the cash flows of the alternatives?" To make this analysis, first arrange the alternatives in ascending order of initial cash flow. Then compare alternatives, two by two, alternatively rejecting the alternative with the lower IROR.

For our example of the two tractors, the incremental cash flow diagram is given in Fig. 1-10. The cash flow equation is

$$\text{NPW}_{\text{new-old}} = -48{,}890 + 14{,}000(P/A,i,6) + 24{,}680(P/F,i,3) + 8{,}000(P/F,i,6) = 0$$

Solving for i (IROR) yields

$$i = 30.9\%$$

Now the picture is clear. While the initial investment of $24,680 for the old tractor will yield a ROR of 21.5 percent, the incremental increase in initial investment of $48,890 (by purchasing the new tractor) will yield an incremental rate of return (IROR) of 30.9 percent. Now that all the rates of return are known, a decision can be reached *which is dependent upon the MARR*. For a MARR of 20 percent, the ROR of the new tractor is too low and therefore the old tractor is chosen. For a MARR of 15 percent, both alternatives exceed it and we have to examine the IROR. In this case the IROR is higher than the MARR, so we should choose the new tractor. Another way to look at the problem is, for the MARR of 15 percent a $73,570 investment earning 18.9 percent is better than a $24,680 investment earning 21.5 percent if the incremental investment of $48,890 can only earn the MARR of 15 percent.

FURTHER WORK IN THE TIME VALUE OF MONEY

In this chapter, only the basic information on the subject of the time value of money has been given. The subject is complex, and whole texts have been devoted to its proper presentation. The reader can appreciate the pitfalls and complexity by noting the different answers that can be obtained depending on what assumptions are made and how the problem is approached. For further information, the reader is referred to the many excellent texts on the subject of engineering economy. A few are listed at the close of this chapter.

PROBLEMS

1-1 Solve the following problems with the rate of interest equal to 12 percent compounded anually.:

(*a*) If $10,000 is invested today, what will be its value after 10 years?

(*b*) How long will it take for invested money to double in value?

(*c*) If $1,000 is invested today, and an equal amount invested each year for 8 years (nine payments), what will be the value after 8 years? After 10 years?

(*d*) If $12,000 is needed at the close of 5 years, what equal annual payment must be made starting today and ending at the close of 4 years (five payments)?

1-2 A friend offers to double your money in 5 years if you will invest in a venture of his. What annual rate of interest will you receive if you invest and your friend's prediction is correct? If it takes 8 years to double your money, what rate of interest will you receive?

1-3 A contractor is considering the following three alternatives:

A. Purchase a new microcomputer system for $11,500. The system is expected to last for 8 years with a salvage value of $1,000.

B. Lease a new microcomputer system for $2,400 per year, payable in advance. It should last 8 years.

C. Purchase a used microcomputer system for $7,200. It is expected to last 4 years with essentially no salvage value.

(*a*) For a MARR of 15 percent, which alternative should be selected?

(*b*) For a MARR of 20 percent, which alternative should be selected?

(*c*) What is the IROR between alternatives A and B?

Note: Assume equal replacement conditions for this problem.

1-4 Same as Prob. 1-3, except assume that the new computer can be sold after 4 years for $5,000.

1-5 Same as Prob. 1-3, except that the new computer and leased computer should each last 9 years and the used computer is expected to last 3 years. Assume equal replacement conditions.

1-6 Discuss the importance of not ignoring the unequal lives in Prob. 1-3.

REFERENCES

Value engineering

1. Dell'isola, A. J.: *Value Engineering in the Construction Industry*, 3d ed., Van Nostrand, New York, 1982, 364 p.
2. U.S. Department of Transportation, Federal Highway Administration: *Value Engineering for Highways*, U.S. Government Printing Office, Washington, D.C., 1980.
3. Zimmerman, L. W., and G. D. Hart: *Value Engineering, A Practical Approach to Owners, Designers and Contractors*, Van Nostrand, New York, 1982.
4. Barrie, D. S., and M. Gordon: The Professional Contruction Management Team Discovers Value Engineering, *Journal of the Construction Division, ASCE*, vol. 103, no. CO3, September 1977.

Engineering economy

5. Collier, Courtland, and W. B. Ledbetter: *Engineering Cost Analysis*, Harper & Row, New York, 1982.
6. Grant, Eugene L., W. Grant Ireson, and R. S. Leavenworth: *Principles of Engineering Economy*, 6th ed., Ronald, New York, 1976.
7. Taylor, George A.: *Managerial and Engineering Economy*, 3d ed., Van Nostrand, New York, 1980.

CHAPTER

TWO

PROJECT PLANNING AND MANAGEMENT

GENERAL INFORMATION

This chapter deals with the planning that is necessary both prior to and during the actual construction on an engineered project. Such planning is necessary in order to construct the project within cost and on time. Items which need to be adequately planned include:

1. The identification of specific activities of work required and the inter-relationships between those items (precedence relationships)
2. The proper sequencing of the specific activities of work so as to complete the project in the optimum amount of time
3. The time for delivery of materials and installed equipment
4. The types, quantities, and duration of construction plant and equipment
5. The classification and numbers of workers needed and the periods of time they will be needed
6. The amount and timing of financial assistance that is needed

Ideally, a contractor should complete some of this planning prior to bidding a project, since such planning will frequently reveal a number of factors which significantly affect the cost of the project. Of course such planning is in itself costly and often very time-consuming. But the potential benefits of such planning are enormous, and every effort should be made to do as much preplanning before bidding as is practicable.

CONSTRUCTION ACTIVITIES

Most construction projects are divided into specific activities of work, each with a specific objective and length of time to accomplish. Each activity generally has a specific beginning and ending point and may require a specific piece of equipment or a specific trade classification. For example, the construction of a reinforced concrete retaining wall might be divided into such activities as:

1. Lay out site
2. Excavate earth (machine)
3. Excavate earth (hand)
4. Build and erect forms
5. Place reinforcing steel
6. Place concrete
7. Cure concrete
8. Remove forms
9. Finish concrete surface
10. Clean up site

Once the specific activities of work are defined, the project planner should then determine the quantity of work involved in each activity using some easily measured quantity. For example, the placement of concrete could involve the total number of cubic yards of concrete involved in the activity (or the number of square yards of concrete surface area if the activity were a concrete slab on grade). The building of forms would normally be measured by the hundred square feet of concrete surface area. Then he should estimate the probable rate at which the work will be performed, usually assuming typical productivity rates from the contractor's experience. From this information, the probable time required to complete each activity can be calculated. Resource allocation (men and machines), as well as any requirements for installed equipment, can be assigned to each activity. Once the activities have been identified, the project planner must sequence them *properly* (for example, you cannot place concrete until after the forms have been erected). From this information it is then possible to estimate the time necessary to complete the project and the amount of resources that will be required. At this point, normally the project planner has been assuming unlimited resources and probably has not considered such factors as delays due to weather and the like. So the plan thus far conceived may not be entirely realistic, but remember it has been formulated before bidding on the project in order to see how the plan might impinge on (1) the ability of the contractor to accomplish the project and (2) the resulting costs of the project. The following section describes the methodology used in sequencing activities of work and describing their interrelationships.

PROJECT NETWORK ANALYSIS (CPM)

This method of project planning involves the identification of specific activities, their durations, and their interrelationships, as described in the preceding section. There are two types of networks in general use: (1) the activity-on-arrow (AOA) type, commonly called *arrow diagramming* and (2) the activity-on-node (AON) type, commonly called *precedence diagramming*. Each of these types, both generally termed *critical path methods* (CPM), uses the same information in a slightly different form, and both types will be presented.

ACTIVITY-ON-ARROW DIAGRAMMING

Using an example of the placement of a mobile home on a permanent site, Table 2-1 lists the activities, their probable durations, and precedence relationships. The units of time involved should be convenient for the degree of accuracy desired consistent with keeping the level of detail reasonable. Units such as days, weeks, and months are generally used. Once the units are selected, usual practice is to use integer quantities only, rounding off all durations to the next higher integer value. In this example, work days are chosen. Note that each activity is identified by both a symbol and a numbering sequence which describes the from-to relationship for the activity. To clearly express the interrelationships between activities, the "from" node *must* be a lower number than the "to" node. In that way the direction of the arrow is clearly established. Figure 2-1 illustrates the AOA diagram for this example problem. Note that the figure depicts the construction sequence and interrelationships between the activities. The following paragraphs explain the construction of this type of diagram and its various parts.

Definitions Because terms and symbols are used in analyzing a project and constructing the arrow diagram, it is necessary to define these items.

Table 2-1 List of activities, durations, and precedences for a mobile home installation project

Activity		Events			Activities which	
Code	Description	From	To	Duration, days	Precede	Follow
A	Site layout	1	2	1	–	B, C
B	Slab excavation	2	3	4	A	D
C	Slab forms	3	5	1	B	G
D	Place blocks	2	4	1	A	E, F
E	Rough plumbing	4	6	2	C	H
F	Rough electrical	4	8	2	C	K
G	Place concrete	5	7	1	D	I, J
H	Place home	6	8	1	E	K
I	Remove forms	7	9	2	G	X
J	Cure concrete	7	10	7	G	L
K	Hookup home	8	11	4	F, H	L
L	Cleanup	10	11	2	J, K, X	–
X	Dummy	9	10	0	I	L

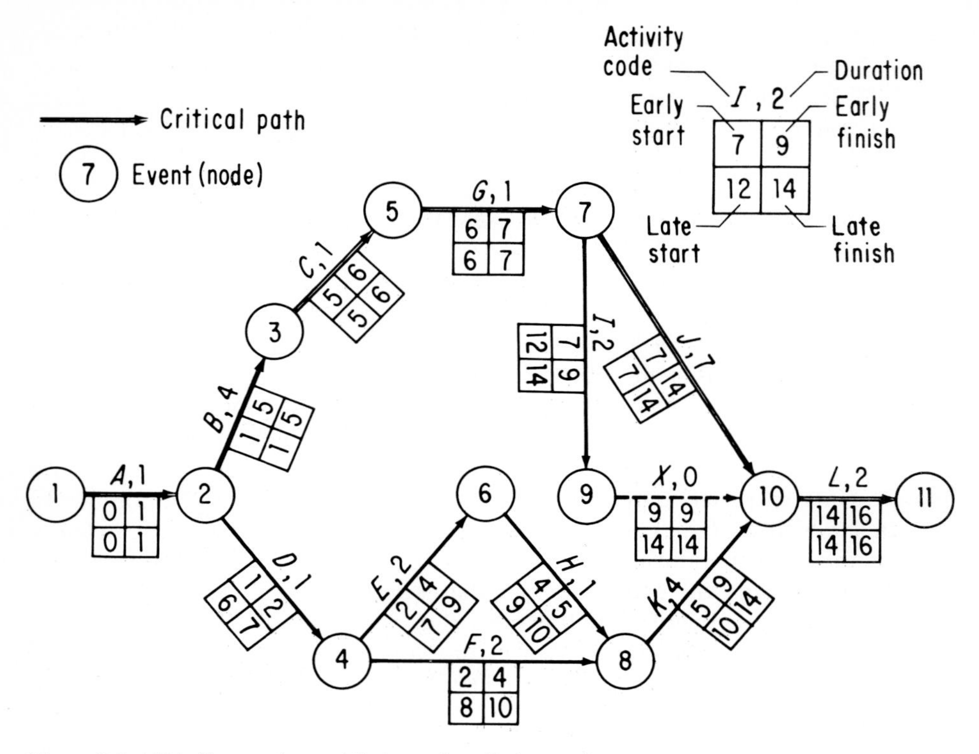

Figure 2-1 AOA diagram for mobile home installation project.

Activity A specific job or task that has to be performed. Normally time is required to complete an activity. For example, "cure concrete" is an activity in the above example.

Event The start or completion of an activity. It requires no time in itself and is usually indicated on the AOA diagram by a number enclosed in a circle. The event is sometimes referred to as a *node*.

Arrow A line drawn to represent each activity in a network, joining two events (the start and the finish of an activity). The arrow is usually designated by two numbers, one at the tail (the "from" event) and one at the head (the "to" event). To avoid confusion, the "from" number should *always* be less than the "to" number. Generally, the length of the arrow has no relation to the duration of the activity which it represents.

AOA network This is the arrow diagram drawn to portray the proper relationships between the activities in a project. It is common practice to start time with zero, and to start the first arrow or arrows at the left end of the network and proceed to the right.

Dummy This is an artificial activity, usually represented on the diagram by a dotted line, to describe the proper relationship between activities. For example, activity X in Fig. 2-1 is a dummy activity and indicates that activity L cannot start until both I and J are completed. It is included to avoid having two different activities possess the same from-to numbers. A dummy activity does not require any time.

Early start (ES) This is the earliest time an activity can start.

Duration (D) The estimated time to perform an activity. For example, it is estimated that the concrete must be cured for 7 days.

Early finish (EF) This is the earliest time an activity can be finished. It is equal to the early start time plus the duration of the activity: EF = ES + D.

Late start (LS) This is the latest time an activity can be started without delaying the completion of the project: LS = LF − D.

Late finish (LF) This is the latest time an activity can be finished without delaying the completion of a project: LF = LS + D.

Total float (TF) This is the amount of time that an activity may be delayed without delaying the completion of the project. It is equal to the difference between the late start and early start, or late finish and early finish: TF = LS − ES = LF − EF. In Fig. 2-1, the total float of activity I is 5 days (10 − 5).

Free float (FF) This is the time that the finish of an activity can be delayed without delaying the early start time of any following activity: FF = ES of following activity − EF of this activity.

Critical path This is the *longest* interconnected path of activities through the network. Its length determines the overall duration of the project. All activities on the critical path have zero float times. A project may have more than one critical path.

Critical activity This is an activity on the critical path. It has zero float time, that is, LS − ES = 0 and LF − EF = 0. Activities A, B, C, G, J, and L are critical activities and make up the critical path for the mobile home installation project shown in Fig. 2-1.

The uses of these terms and symbols are illustrated more fully in the examples which follow. Persons who wish more comprehensive information on this subject can find many excellent texts covering critical path networks. A few are listed at the close of this chapter.

Steps in critical path scheduling To apply the critical path method (CPM) of scheduling the construction of a project, the following steps should be used.

1. Separate the project into discrete activities, each with a definite starting point and ending point.
2. Estimate the duration of each activity.
3. Determine the proper sequencing of each activity, including which activities must precede or follow other activities.
4. Draw an AOA network with the activities and events properly interconnected. Where necessary, include dummy activities to clarify the network and avoid redundancy in activity event numbers.

5. Examine the network and optimize, if possible, to eliminate unnecessary dummy activities.
6. Assign numbers to all events, being sure that the "from" number is always less than the "to" number. This establishes the proper direction of the activities.
7. Make a forward pass and a backward pass through the network to establish early start, late start, early finish, and late finish times for all activities.
8. Determine the critical path, or paths, and critical activities.
9. Prepare a table listing all activities, their designations, durations, and ES, LS, EF, and LF times, and their total float. Their free float can also be listed, if desired.

To illustrate these steps, let us go through the preceding example of the placement of a mobile home on a permanent site. Steps 1, 2, and 3 are illustrated in Table 2-1. Note that each activity has a specific duration and the precedence relationship between all the activities is shown in the last two columns of the table. Of course, at this stage the node numbers would not be known, and those two columns would initially be left blank. The dummy activity would also be unknown until the network was constructed. Once this preliminary information is determined, the AOA diagram can be constructed (step 4). Figure 2-1 shows the completed AOA diagram with the dummy included. Remember that initially you might have several dummy activities which had to be included to avoid redundancy, but after optimizing the network (step 5), only one dummy activity would be needed in this case. Step 6 involves assigning numbers to all the events (nodes). Note that the "from" node is always less than the "to" node. Also note that the numbers do not have to be consecutive and any sequence of numbers could be used.

Step 7 is a forward pass calculation. Normally the first activity starts at time zero (rather than time 1). Starting at the left-hand side of the diagram with 0 as the ES time for activity A, add the duration to its ES to obtain its EF time (in this case, time 1 at node 2). Here the diagram splits, and two activities begin at node 2. Continuing the forward pass with activity B, we see that its ES is identical with the EF of its preceding activity (1 in this case). Adding the duration of B to its ES time yields its EF time of 5 $(1 + 4)$. In a similar manner you can progress through the network, calculating ES and EF times for all activities, until you reach activities K and L. Each of these activities has more than one activity preceding it with different EF times. Which one do you choose? The answer is to choose the preceding EF with the *largest* time. Thus, for activity K, its ES time is 5 (rather than 4) because it cannot start until activity H is completed at time 5. The fact that activity F could be finished at time 4 does not mean that activity K can start then, as activity H still has one more day before it is finished. Similarly, activity L cannot start until time 14. Adding the duration of 2 to the ES time of activity L yields an EF time of 16 days. As this is the last activity, you now know the duration of the project (16 days) and you have completed the forward pass. Next comes the backward pass. On the last activity, the EF time becomes the LF time, if you want to finish the project as soon as you can after starting it. The LS time of the last activity is simply its LF time minus its duration (16 – 2), which yield an LS time of 14 for activity L. The LF time of all activities immediately preceding activity L must equal the LS time of activity L (day 14). Working backward, the LF and LS times for preceding activities can be determined until you come to activity G. It has two activities which immediately follow it (activities I and J) with different LS times. The LF time of activity G must be the *smaller* of the two following LS time (7 days rather than 12). In other words, for activity G *not* to delay either activity I or J, it cannot finish any later (LF) than time 7. In a similar manner the remaining LF and LS times can easily be calculated until you are at the beginning of the network.

The next step is to determine the critical path. The critical path is defined as the longest interconnected path through the network. All activities on this path have the same ES and LS times (and similarly they have the same EF and LF times). In the example problem, the critical path is 1-2-3-5-7-10-11 and contains activities A, B, C, G, J, and L. The critical path is denoted by a double line in Fig. 2-1. Note that these activities have no "float" to their durations, which is a definition of a critical activity.

The final step, which is to calculate the floats (total and free) for all activities, can easily be calculated knowing each activity's ES, LS, EF, and LF times and the adjacent activity's times. Placing these values in a table completes step 9 and is illustrated in Table 2-2.

Table 2-2 List of activities, times, and floats for a mobile home installation project

Activity							
Code	Duration	ES	EF	LS	LF	TF	FF
A	1	0	1	0	1	0	0
B	4	1	5	1	5	0	0
C	1	5	6	5	6	0	0
D	1	1	2	6	7	5	0
E	2	2	4	7	9	5	0
F	2	2	4	8	10	6	1
G	1	6	7	6	7	0	0
H	1	4	5	9	10	5	0
I	2	7	9	12	14	5	0
J	7	7	14	7	14	0	0
K	4	5	9	10	14	5	5
L	2	14	16	14	16	0	0
X	0	9	9	14	14	5	0

ACTIVITY-ON-NODE DIAGRAMMING

One problem with the AOA diagram (arrow) is the necessity to include dummy activities. Such activities increase the length of the tables, enlarge the graph, and take time to calculate. They also increase the complexity of the network. Another type of project network can be used, called the activity-on-node (AON) diagram, or the precedence diagram, which overcomes these problems. In the AON diagram the activities are denoted by boxes, and arrows are used *only* to designate the interrelationship between activities. Figure 2-2(*a*) and (*b*) illustrate the differences between the AOA and AON diagrams.

Figure 2-3 is the AON (precedence) diagram for the mobile home installation project. Note the absence of dummy activities and the clear precedence relationship between activities. In this figure neither the size of the box nor the length of the arrows denotes a time interval. The ES, LS, EF, and LF times, as well as the critical path, are clearly depicted in the figure. Calculation of these values and float times is identical to that described earlier for the AOA diagram.

To most users, the AON diagram is simpler to draw, is easier to explain, and presents a clearer picture of the project than the arrow diagram. If that is the case, then why describe the arrow diagram at all? The fact is, since arrow diagramming was developed first, it is still widely used. Further, its use of numbers for nodes (or events) makes it simpler to program on the computer. Both methods have their advocates, and the beginner needs to understand both as both will be encountered in practice.

COMMON ERRORS IN NETWORK DIAGRAMMING

The three most common errors in network diagramming are dangling, looping (cycling), and redundancy (AOA diagrams only). They are depicted in Fig. 2-4. In Fig. 2-4(*a*) activity 10–16 is left "dangling" as it obviously is not the last activity in the network. The looping error in Fig. 2-4(*b*) results when an

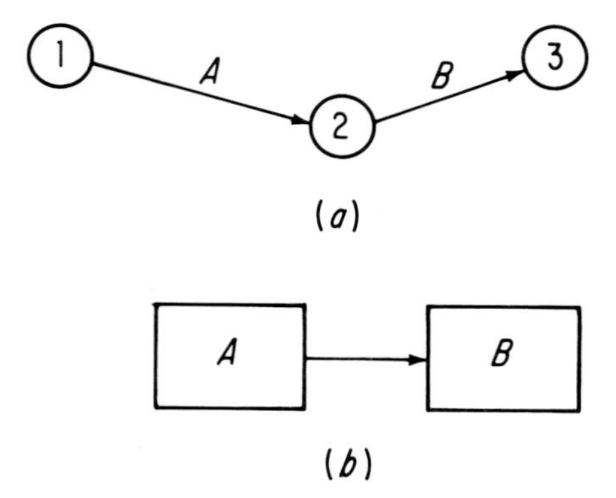

Figure 2-2 (*a*) AOA diagram for two activities. (*b*) AON diagram for two activities.

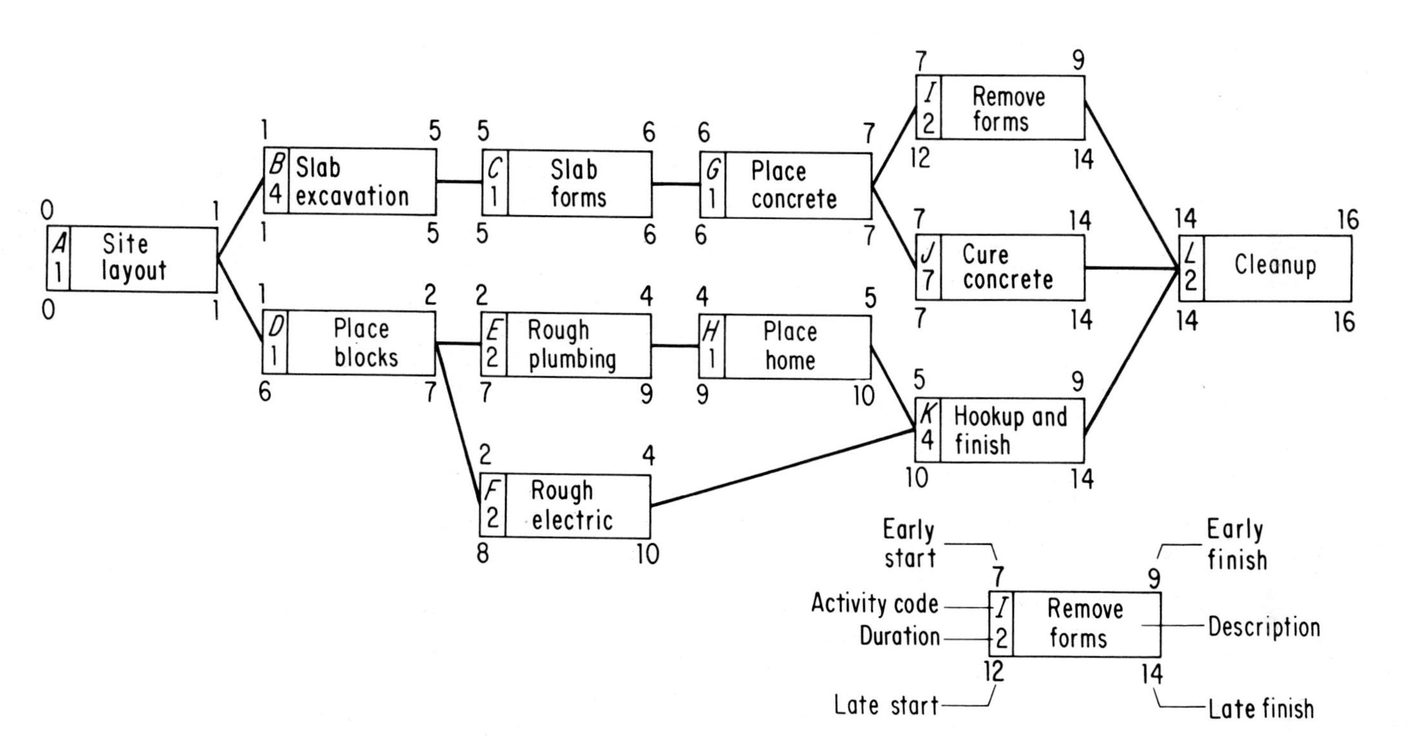

Figure 2-3 AON diagram for mobile home installation project.

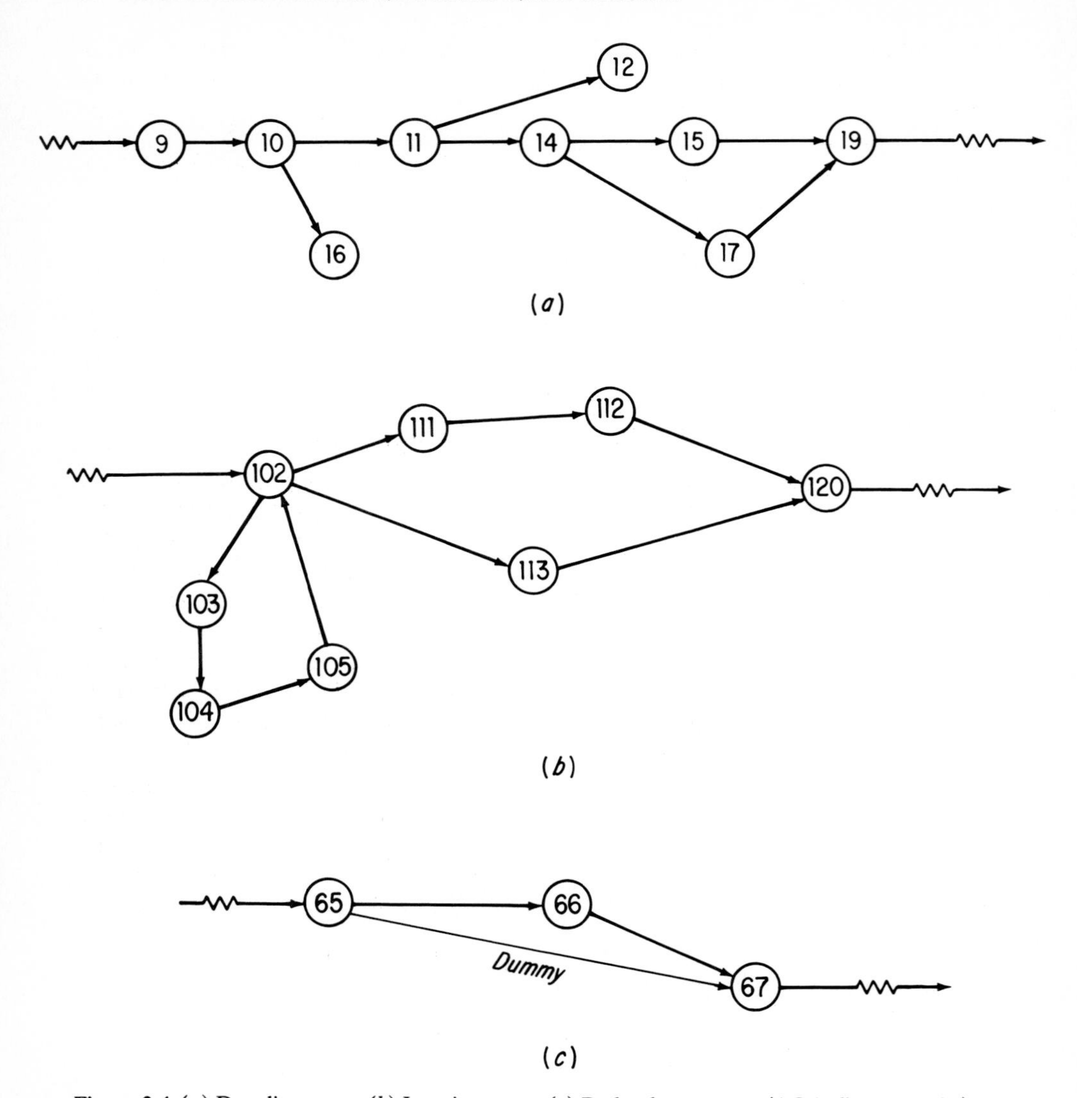

Figure 2-4 (*a*) Dangling error. (*b*) Looping error. (*c*) Redundancy error. (AOA diagram only.)

endless loop is shown in the diagram and you cannot progress beyond it. The dummy activity (65–67) in Fig. 2-4(*c*) is unnecessary as it is not needed for network logic.

ADVANCED NETWORK ANALYSIS TECHNIQUES

While AOA and AON diagrams are valuable planning and management techniques, they have several shortcomings, the handling of which requires modifications and extensions of the basic techniques. One major shortcoming is in the handling of overlapping, or semicontinuous, activities. Another shortcoming is the fact that often the durations of activities are not known with any high degree of certainty. Each of these shortcomings can be partially handled by modifications and extensions to the basic models. These modifications and extensions are presented in the following paragraphs.

Overlapping activities Assume that the construction of a bridge requires six intermediate piers consisting of concrete caps supported by steel piles. All piles will be driven before any caps are started. The construction of the caps might be divided into three activities: the erecting of forms, the placing of reinforcing steel, and the placing of concrete. If this plan is used in preparing the arrow diagram, it will be necessary to show all forms completed as one activity before placing any reinforcing steel, and likewise all reinforcing for all caps must be placed before any concrete is placed. However, the adopted schedule for constructing the caps provides that after the forms for two caps are erected, the reinforcing will be placed in the forms. Then after the reinforcing is placed, the concrete for these two caps will be placed. In the meantime the forms of other caps will be erected, in units of two, using the side forms from the caps previously constructed, followed by placing the reinforcing steel and concrete. A diagram containing only three arrows (or only three activities) cannot represent these activities correctly. One way to handle these overlapping activities would be to divide the three activities into three subactivities each, for a total of nine activities for this part of the diagram.

The sequential information appears in Table 2-3, and the arrow diagram for this portion of the project is illustrated in Fig. 2-5. Note that an appropriate duration must be assigned to each of these nine activities. Obviously the network has been made much more complicated by this modification, as the activities in this particular example have *tripled* in number and, if an arrow

Table 2-3 Sequential information for pier caps

Activity	Activities which immediately precede	Activities which immediately follow
J—drive piles		*K*
K—erect forms for caps 1 and 2	*J*	*L*, *N*
L—place reinforcing for caps 1 and 2	*K*	*M*, *O*
M—place concrete for caps 1 and 2	*L*	*P*
N—erect forms for caps 3 and 4	*J*	*O*, *Q*
O—place reinforcing for caps 3 and 4	*L*, *N*	*P*, *R*
P—place concrete for caps 3 and 4	*M*, *O*	*S*
Q—erect forms for caps 5 and 6	*N*	*R*
R—place reinforcing for caps 5 and 6	*O*, *Q*	*S*
S—place concrete for caps 5 and 6	*P*, *R*	

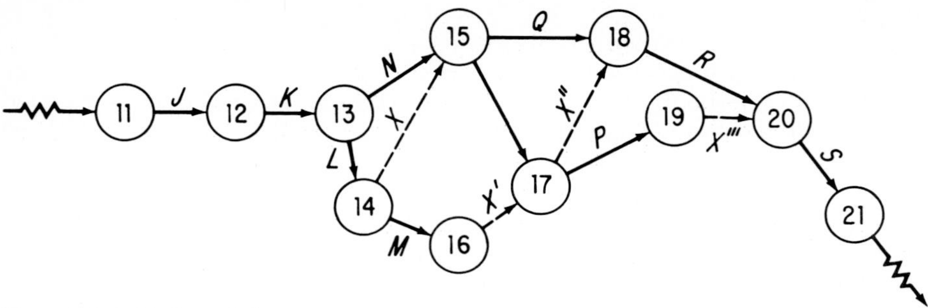

Figure 2-5 AOA (arrow) diagram for pier caps.

diagram is used, four dummy activities are required (X, X′, X″, and X‴ in Fig. 2-5).

Another way to handle this situation is through the use of *lags* and *leads*. Such an extension of the model is more suited to the AON (precedence) method. To illustrate, the pier cap network showing the necessary lags in this case is given in Fig. 2-6. The lag time in starting a subsequent activity is shown in the diagram as a start-start lag, meaning that the start of the placement of reinforcing steel "lags" behind the start of the erection of the forms by at least a predetermined amount of time. This modification of the basic model of the CPM network violates the principle that one activity must be finished before a following activity may be started. But, such a modification permits much more simple and realistic modeling of construction projects.

There are four types of lags and leads which are used. They are depicted in Fig. 2-7. If the lag of the first one (the finish-start or FS) is equal to zero, you have the traditional precedence relationship.

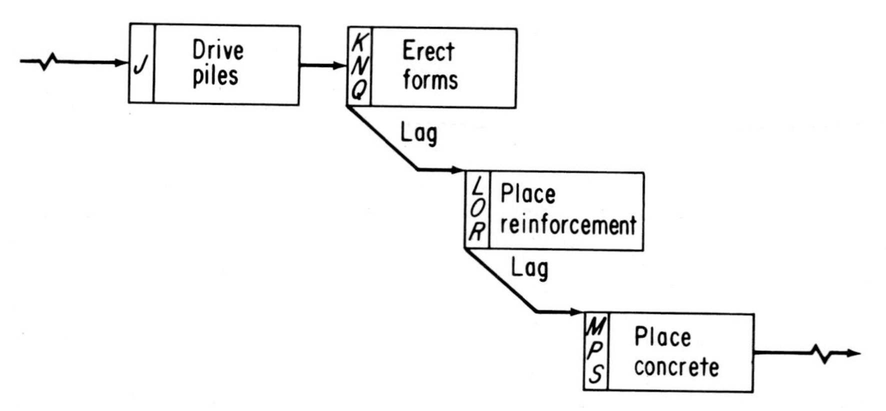

Figure 2-6 AON (precedence) diagram, with lags, for pier caps.

Precedence diagram	Relationship	Definition
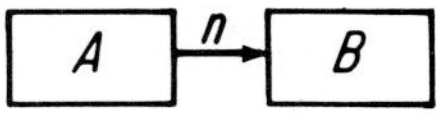 A n B	FS = n	Finish to start. Activity B cannot start until n days after activity A is finished.
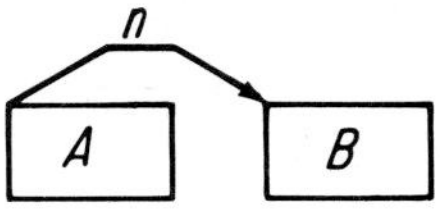 n A B	SS = n	Start to start. Activity B cannot start until n days after activity A is started.
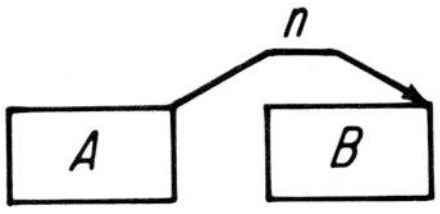 n A B	FF = n	Finish to finish. Activity B cannot finish until n days after activity A is finished.
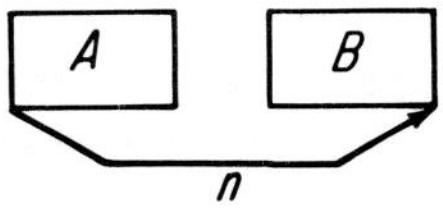 A B n	SF = n	Start to finish. Activity B cannot finish until n days after activity A is started.

Figure 2-7 Lag-lead relationships for AON (precedence) diagrams.

If lags and leads are incorporated in a precedence diagram, the forward and backward passes become more difficult to calculate and the user must be very careful not to become confused and make arithmetical mistakes in calculating ES, EF, LS, and LF times. It is beyond the scope of this text to go into any further detail concerning this technique. For further reading the reader is referred to the references given at the close of this chapter, among which are those by Ahuja and by Weist and Levy.

THE TIME-GRID DIAGRAM METHOD

Whereas the lengths of arrows in Fig. 2-1 do not indicate the durations of the activities, the lengths of the arrows in a time-grid diagram of the AOA method can be drawn to indicate the durations of the activities which they represent, as illustrated in Fig. 2-8. In preparing a time-grid diagram all arrows representing activities and float are drawn horizontally, with each arrow tail starting at the head of the arrow for the immediately preceding activity and then continuing to the right through a path of interrelated activities.

In the diagram vertical lines do not indicate any elapsed time. They simply indicate the precedence of the activities. For example, the broken vertical line

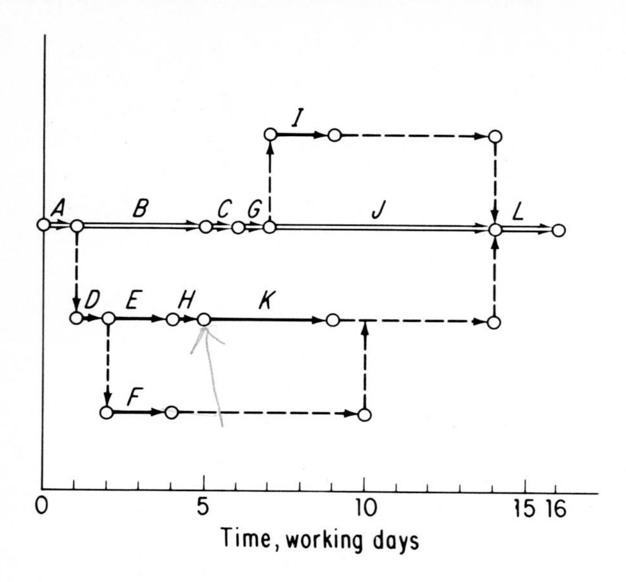

Figure 2-8 Time-grid diagram (AOA) for mobile home demonstration project.

from the head of activity A to the tail of activity C indicates that activity A must be completed before activity C can be started.

Because the float times are represented by broken horizontal lines whose lengths indicate time, it is relatively easy to determine the float time for any activity by inspecting the diagram. Because the arrows are drawn to a time scale, it is possible to show the calendar dates for the activities, which the typical arrow (AOA) diagram does not show. Space in the time schedule is usually provided for working days only, but weekends and holidays could be easily included if desired.

Prior to drawing the time-grid diagram, steps 1, 2, 3, 4, and 7 appearing on pages 27–28 should be completed. Step 7 will indicate those activities lying on the critical path, that is, those with zero float time. Then, when drawing the time-grid diagram, the critical path can be drawn through or near the middle of the diagram, which is usually desirable.

If the name of each activity is written along its arrow and the dates are shown along the bottom of the diagram, the diagram can be a very useful reference during construction.

The diagram in Fig. 2-8 is based on the information listed in Tables 2-1 and 2-2. Note that the differences between the AOA and AON networks become insignificant when presented as a time-grid diagram. In fact, the time-grid diagram may be thought of as a fancy bar chart in which the precedence relationships are clearly identified using vertical broken lines.

Because the diagram shows a time relationship between the activities, it is possible by a visual examination to determine the desirability of shifting the construction schedules for some activities to obtain better distribution of materials, labor, or equipment, or for other reasons. In a similar manner it is possible that altering the schedule for activities, within the periods permitted by float time, may eliminate the need for providing additional equipment on a project for short periods of time only. The subject will be covered later in this chapter under the heading Resource Scheduling.

INTEGRATION OF SCHEDULE AND COST FOR "CRASHING"

When planning the construction of a project by using the critical path method, it is usual practice to select for each activity a rate of progress that will produce the lowest practical direct cost. This progress is based on the delivery of materials, if required, the number of laborers available or the number who can work efficiently, and the number and types of equipment that are available, at a minimum cost, for each activity. After the network diagram is drawn, it is determined that the minimum time required to complete the project is the sum of the durations of those activities lying on the critical path. Such a construction plan is referred to as a *normal* program.

However, it may become desirable for some reason to either reduce or lengthen the total duration of the project. When the normal time is reduced, the project, or a portion of it, is said to be under a *crash* program. If a project is constructed under a crash program, it will be necessary to do some or all of the following:

1. Increase the rates of providing materials.
2. Increase the number of workers, which may reduce their efficiency.
3. Assign the workers to overtime work, which will require premium wage rates.
4. Increase the number of units of equipment assigned to critical activities, which may require the rental of equipment not presently owned, or which may reduce the productivity per unit of equipment.

If some or all of these steps are adopted, there will usually be an increase in the direct cost of the activities involved in the crash program. Similarly, increasing the project duration will often also increase the direct cost of those activities that are increased in duration, especially since equipment ownership costs generally accrue at a constant rate, regardless of the productivity rate of the equipment.

Notice that thus far we have been addressing *direct* costs only (defined here as labor, equipment, materials, and the like directly and solely associated with the particular activities involved). The indirect costs—supervision at the job site, general overhead, job security, etc.—are generally provided as a direct function of project duration. The longer the duration of the project, the higher the indirect costs. An idealized cost schedule for both direct and indirect costs for a project is shown in Fig. 2-9.

If crashing is desired, in order to keep the total increase in cost to a minimum, it is necessary to select for crash operations those activities that will permit the desired reductions in construction time at the least total increase in cost. Because the duration of a project is determined by the activities lying on the critical path, the desired reduction in time should be attained by reducing the durations of one or more critical activities. In order to determine which critical activities to crash first, it will be necessary to determine the *increase* in cost per unit of time reduction associated with each of the critical activities, as well as near critical activities which may become critical as the project duration is reduced. This increase in cost, which is defined as the cost slope, may be expressed as dollars per day, dollars per week, or in other suitable units. While not strictly accurate, the cost slope for a given activity is usually considered to be a straight line with time.

Consider a project whose time-grid diagram is illustrated in Fig. 2-10. Assume that it is desired to reduce the total duration of this project from 36 to 32 working days with a minimum increase in project cost. Table 2-4 shows the duration and total direct cost for each critical activity under normal and crash programs, the possible reduction in duration, and the direct cost slope in dollars per day for reducing the duration of each activity. Because all other

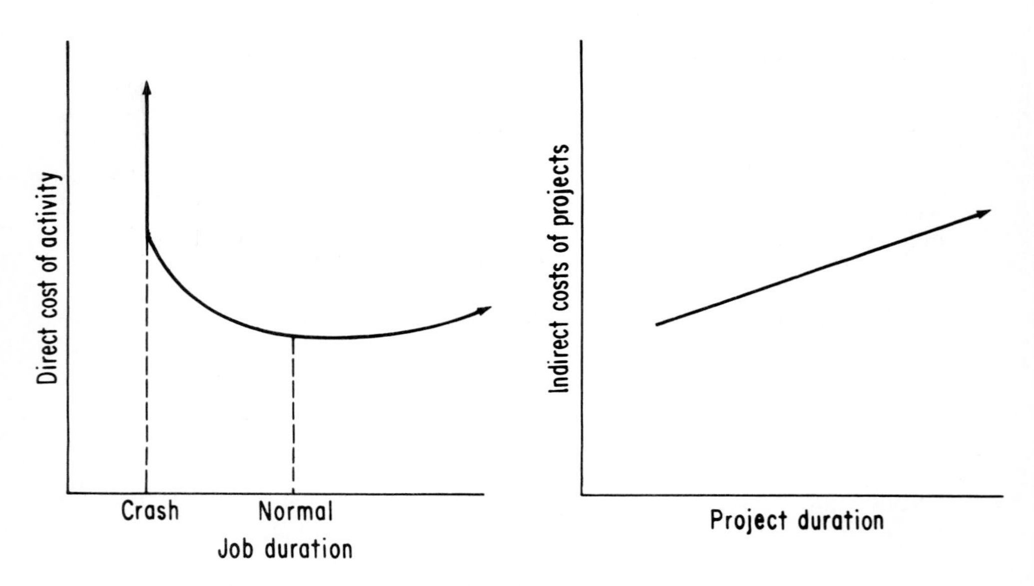

Figure 2-9 Costs associated with an idealized project.

activities except Q and R have adequate float, their crash costs may be disregarded and they can be constructed under normal conditions.

An examination of Table 2-4 reveals that reductions in the following activities should be considered, since they have the lowest cost slopes.

Activity	Possible reduction, days	Direct cost slope, dollars per day
D	2	$116
H	2	106
P	2	95
Q	1	138

It appears that the desired reduction can be obtained from activities H and P at the least increase in total direct cost. However, Fig. 2-10 reveals that activity Q has only 1 day of float. If the duration of P is reduced only 1 day, there is no effect on Q. But, if the duration of P is reduced 2 days, it will be necessary to reduce the duration of Q by 1 day, at an extra cost of $138, for a total cost for P and Q equal to $95 + $138 = $233. Thus the reductions in durations for a minimum increase in direct cost should be:

Reduce D 1 day at $116 = $116
Reduce H 2 days at $106 = $212
Reduce P 1 day at $95 = $95
Total reduction of 4 days = $423 increase in direct cost

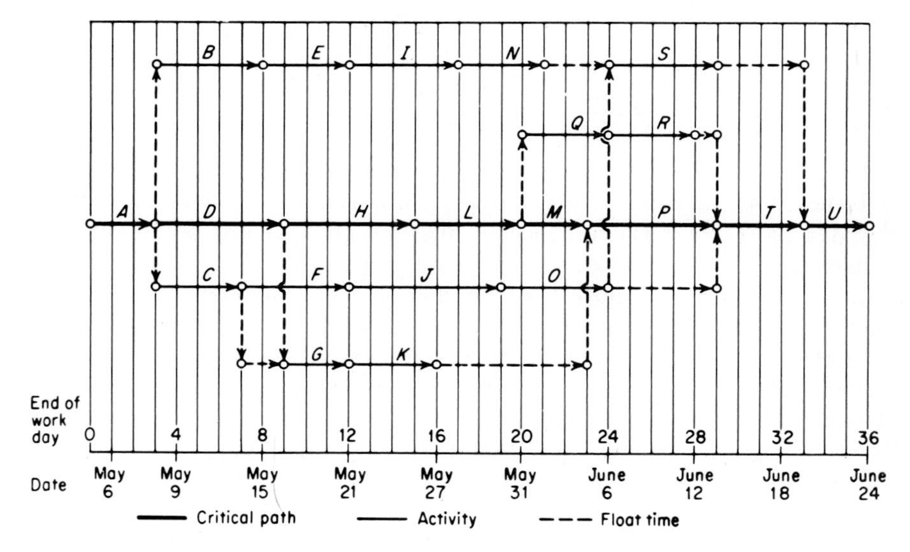

Figure 2-10 Example time-Grid AOA diagram.

Table 2-4 Determining the cost slope for activities under a crash program

	Normal		Crash			
Activity	Duration, days	Total direct cost	Duration, days	Total direct cost	Possible reduction, days	Direct cost slope, dollars per day
A	3	$ 876	2	$ 1,164	1	$288
D	6	16,454	4	16,686	2	116
H	6	14,231	4	14,443	2	106
L	5	8,592	4	8,744	1	152
M	3	6,490	3	6,490	0	0
P	6	18,670	4	18,860	2	95
T	4	12,836	3	13,264	1	428
U	3	944	2	1,168	1	224
Q	4	3,848	1	3,986	1	138
R	4	7,614	1	7,814	1	200

Because activities C, F, J, O, and S have a float of 4 days, the reduction (crashing) of 4 days in total project duration will cause these activities to lie on a critical path also.

The previous example dealt with direct costs only, as these costs will increase as the project duration is crashed (shortened). The indirect costs, as shown in Fig. 2-9, will generally decrease as the project duration is crashed. Therefore, the overall increase in total project cost from selective crashing may be slightly less than the calculated increase in direct cost.

COMPUTERIZED SCHEDULING

Computers, like calculators, continue to become more and more sophisticated while at the same time less costly. Today, microcomputers with more than 64,000 bytes of internal random access memory (RAM) are being sold for about the same price as a good, self-correcting typewriter. They are becoming more and more "user friendly," which means that persons not versed in computer science can manipulate and use them with ease. The technology is literally exploding, with new products and increased capability being introduced almost daily. In the field of project planning and management, more and more companies are using the computer to schedule, plan, and control projects of all sizes. While it is certainly no substitute for decision making, it can quickly provide vast amounts of information to the planner, including large quantities of calculations under various assumptions likely to be encountered, to enable him to make better-informed engineering decisions.

There are a number of very sophisticated, yet easy to use, computer software programs (which are the instructions placed into the RAM of the computer to make the necessary calculations based on data entered by the user in easy-to-follow steps) available in the areas of cost and schedule planning and control. One only has to examine almost any issue of such publications as *Engineering News Record*, *Civil Engineering*, and *Professional Engineer* and specialty publications such as *Construction Computer Applications News*, to name only a few, and there will be numerous advertisements from companies offering computer software programs in this area. Considering the savings in manpower, reduction in error, and increased capability to plan, control, and update schedules with a minimum of effort, these software programs have been shown to be quite cost-effective and often pay for themselves long before completing the first project in which they are used.

The computer overcomes one of the major shortcomings to using CPM—that of updating and revising the network schedule. While many companies recognize the value of *initially* preparing a CPM schedule before commencing with the construction of a complex project (and many owners insist on contractors providing such an initial plan), the CPM schedule has then often been "shelved" and the more easily updated bar charts used to control the progress of work. The reason for this was the large effort required to periodically update the network, especially if there were several hundred activities which all had to be changed manually when the schedule along the critical path changed. The computer has changed all that! It will update and produce a revised schedule literally with "the speed of light," without error (if the input to the revision is correct). Then it can prepare lists of selected activities, in almost any desired form, including a list of only those activities which are behind schedule (with their revised ES, LS, EF, LF, and float—negative in this case—times), for any particular phase of the project.

There is no question that the computer is revolutionizing the project planning and management process, enabling the planner to play "what if" games such as "what happens to the schedule and cost if additional needed equipment is rented or leased," or "what happens to the schedule and cost if critical delivery of major equipment items are delayed?" Many of the software companies offering these programs will help set them up, adapt them to the particular needs of the user, and quickly train the project planners in their use. Most of them are designed to be used on particular computer systems, so the planner must study his or her needs and capabilities to choose the most useful system.

RESOURCE SCHEDULING

If construction on a project is to proceed efficiently and at the scheduled rates, it is necessary to know accurately the types and quantities of resources that will be needed and the times they will be needed. For example, consider the mobile home installation project presented in Table 2-1 and Fig. 2-1. Let us assume the following laborers are required for each activity:

Code	Description	Duration, days	Number of laborers
A	Site layout	1	1
B	Slab excavation	4	4
C	Slab forms	1	4
D	Place blocks	1	3
E	Rough plumbing	2	2
F	Rough electrical	2	2
G	Place concrete	1	6
H	Place home	1	2
I	Remove forms	2	3
J	Cure concrete	7	1
K	Hookup	4	1
L	Cleanup	2	4

One way to look at resource scheduling is to construct a resource allocation chart for the early and late start schedules. Figure 2-11 contains the early start time-grid diagram and resource allocation charts for both the early start and the late start schedules. Two things are immediately apparent from an examination of the charts. One is that the demand for laborers fluctuates from day to day and, if the late start schedule is followed, during day 7 a total of nine laborers will be needed. Further examination reveals that, if the contractor desires to have as uniform a demand for laborers as practicable, he will follow a slightly altered schedule from the early start, probably delaying activity C for a few days, not doing any other activities at the same time as activity G, and thus getting by with a maximum of only six laborers during any given day. Another observation which can be made from this example is that there are a number of possible schedules which can be followed to allocate resources. Thus resource scheduling can be very complex, especially if resources are limited (as they almost always are).

There are two types of resource scheduling problems. They are:

1. *Resource leveling.* This involves the scheduling of resources to smooth out the peaks and valleys in resource use, within the constraint of project duration.
2. *Resource allocation.* This involves the allocation of available resources to project activities to determine the shortest project duration consistent with fixed resource limits.

Resource leveling and resource allocation programs have been written for projects, and they generally are of the "heuristic" type (which means simply "rule of thumb") in which certain rules are followed because they make sense most of the time. Resource leveling or allocation is often accomplished by constructing manually (or with the computer) resource allocation charts similar to the two in Fig. 2-11 and then adjusting the schedule by use of lags and leads to smooth and/or reduce the maximum amount of the resource needed. For complex projects this can become exceedingly tedious, but fortunately there are some very good heuristic programs in such texts as those by Weist and Levy and by Ahuja (listed at the close of this chapter) and there are some excellent computer software programs which can be used effectively.

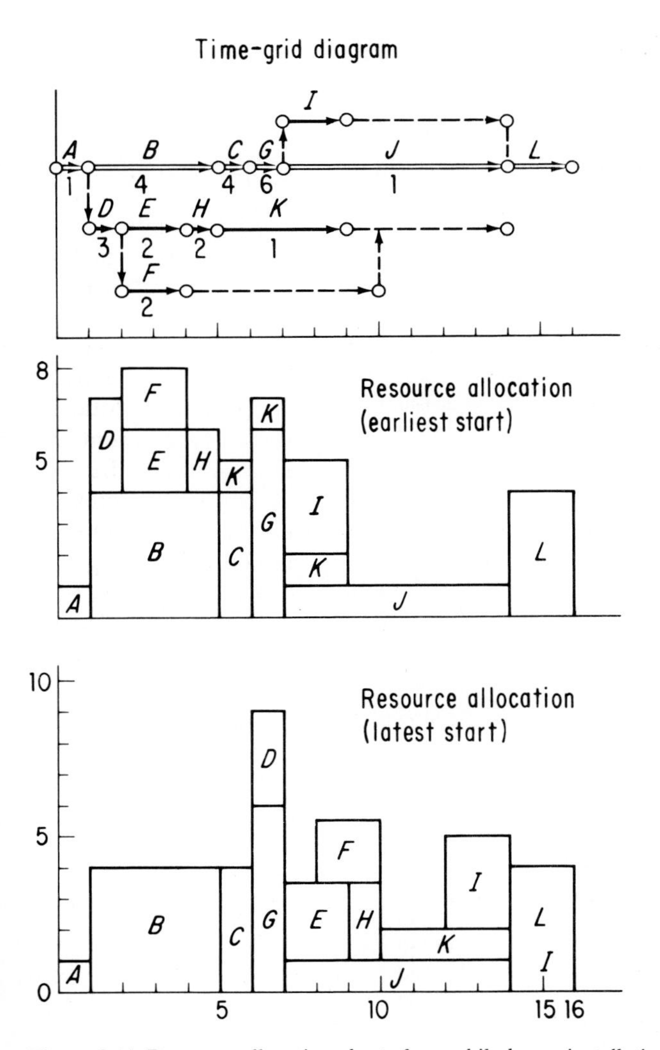

Figure 2-11 Resource allocation charts for mobile home installation project.

CONSTRUCTION PROJECT CASH FLOW

A construction schedule may be used to estimate the amount of funds that a contractor must provide in financing a project during construction. Most construction contracts specify that the owner will pay to the contractor a stated percent of the value of work completed during each month. The payment for work completed during a month is usually made by the tenth of the following month. Upon completion of the project, the retained funds, often 10 percent of the contract value of the work, is paid to the contractor. An analysis of the construction schedule will indicate the approximate expenditures and receipts

Table 2-5 Form for estimating expenditures during construction

Weeks after starting	Activities under construction	Expenditure per week	Cumulative expenditures
1	*A*	$ 5,680	$ 5,680
2	*B*	1,540	7,220
3	*B*	1,540	8,760
4	*B*	1,540	10,300
5	*C, D*	4,780	15,080
6	*C, D*	4,780	19,860
7	*C, D*	4,780	24,640
8	*C, D*	4,780	29,420
9	*C, D*	4,780	34,200
10	*D*	3,240	37,440
11	*E, F*	13,540	50,980
12	*E, F*	13,540	64,520
13	*E, F*	13,540	78,060
14	*E, F*	13,540	91,600
15	*E, F*	13,540	105,140
16	*E, F*	13,540	118,680
17	*G*	10,300	128,980
18	*G*	10,300	139,280
19	*G*	10,300	149,580
20	*G*	10,300	159,880
21	*G*	10,300	170,180
22	*G*	10,300	180,480
23	*H*	55,500	235,980
24	*H*	55,500	291,480
25	*H*	55,500	346,980
26	*H*	55,500	402,480
27	*H*	55,500	457,980
28	*H*	55,500	513,480
29	*H*	55,500	568,980
30	*H*	55,500	624,480
31	*I*	1,200	625,680
32	*J*	1,860	627,540

through any desired dates. The excess of expenditures over receipts indicates the amount of financing which the contractor must provide from sources other than the owner.

The estimated expenditures are determined as illustrated in Table 2-5. The amounts shown are the costs of materials, equipment, labor, and general overhead. Although a contractor does not pay an outside party for the use of the equipment, unless it is rented, he must make monthly payments, or other periodic payments, on the equipment until its cost is liquidated. Thus it is proper to include the cost of owning and operating equipment as an expenditure.

Table 2-6 illustrates a form that may be used to estimate the receipts from

Table 2-6 Estimated receipts during construction

Month	Activities under construction	Weeks under construction	Units completed per week	Unit price received during construction	End-of-period receipts	Total period receipts	Cumulative receipts
April	*A*	1.0	1	$ 0	$ 0	0	
	B	2.4	6	270.00	3,888	$ 3,888	$ 3,888
May	*B*, *C*	4.6	6	270.00	7,452		
	D	4.0	1	3,240.00	12,960	20,412	24,300
June	*C*	1.0	6	270.00	1,620		
	D	2.0	1	3,240.00	6,480		
	E	2.0	1	3,240.00	6,480		
	F	2.0	14,367	0.72	20,688	35,268	59,568
July	*E*	4.0	1	3,240.00	12,960		
	F	4.0	14,367	0.72	41,377		
	G	0.6	14,367	0.72	6,207	60,544	120,112
August	*G*	4.4	14,367	0.72	45,515	45,515	165,627
September	*G*	1.0	14,367	0.72	10,344		
	H	3.2	12,674	4.41	178,854	189,198	354,825
October	*H*	4.6	12,674	4.41	257,105	257,105	611,930
November	*H*	0.2	12,674	4.41	11,179		
	I	1.0	1	0.00	0		
	J	1.0	1	0.00	0	11,179	623,109
Amount retained							69,234
Total amount of contract							$692,343

* Amount payable by the tenth of the following month.

the owner of the project. The prices received during construction are 90 percent of the contract prices for the respective items. The receipts for a given month are payable to the contractor by the tenth of the following month.

Table 2-7 illustrates a form that may be used to determine the estimated expenditures and receipts for the end of each month during construction. At the end of July the estimated cumulative expenditures amount to $124,860. At this time the cumulative receipts, shown for the end of June and payable by the tenth of July, amount to $59,568. Thus there is a difference of $124,860 − $59,568 = $65,292 which the contractor may have to provide from another source for 10 days.

Information contained in this table should assist a contractor when he discusses with his bank a schedule of financial assistance that he may need from the bank.

Table 2-7 Estimated expenditures and receipts during construction

Month	Activities under construction	Weeks under construction	Expenditures per week	Expenditures for month	Cumulative expenditures	Total receipts for month	Cumulative receipts*
April	*A*	1.0	$ 5,680	$ 5,680			
	B	2.4	1,540	3,696	$ 9,376	$ 3,888	$ 3,888
May	*B, C*	4.6	1,540	7,084			
	D	4.0	3,240	17,960	29,420	20,412	24,300
June	*C*	1.0	1,540	1,540			
	D, E	4.0	3,240	12,960			
	F	2.0	10,300	20,600	64,520	35,268	59,568
July	*E*	4.0	3,240	12,960			
	F, G	4.6	10,300	47,380	124,860	60,544	120,112
August	*G*	4.4	10,300	45,320	170,180	45,515	165,627
September	*G*	1.0	10,300	10,300			
	H	3.2	55,500	177,600	358,080	189,198	354,825
October	*H*	4.6	55,500	255,300	613,380	257,105	611,930
November	*H*	0.2	55,500	11,100			
	I	1.0	1,200	1,200			
	J	1.0	1,860	1,860	627,540	11,179	623,109
Amount retained							69,234
Total amount of contract							$692,343

* Amount payable by the tenth of the month.

JOB LAYOUT

One of the first duties of a superintendent when he assumes the responsibility of starting construction is to prepare a job layout for the project. On this layout he will draw to scale the area available for offices, warehouses, storage of materials, equipment, and earth, and constructing forms and fabricating reinforcing steel. In preparing the job layout, the superintendent should endeaver to arrange all areas to reduce the time consumed in carrying materials from storage areas to the project. Materials that are similar in use should be stored close together, where possible. The general office and warehouse should be located near the main entrance in order that persons visiting the project for business purposes will not have to travel around the construction areas to reach the office. This should reduce the danger of injuries to visitors and the confusion that frequently is associated with the presence of strangers around a project. If the general warehouse is near the entrance, it will facilitate the delivery of material to be stored in the warehouse. However, if a warehouse is needed to store heavy materials, such as machines that will be incorporated into the project, it may be desirable to consider using additional warehouses, located nearer the project.

Figure 2-12 illustrates a job layout for a multistoried reinforced-concrete frame building. The contractor is fortunate in having adequate area for easy storage of all materials at the job site. This is not commonly the case for

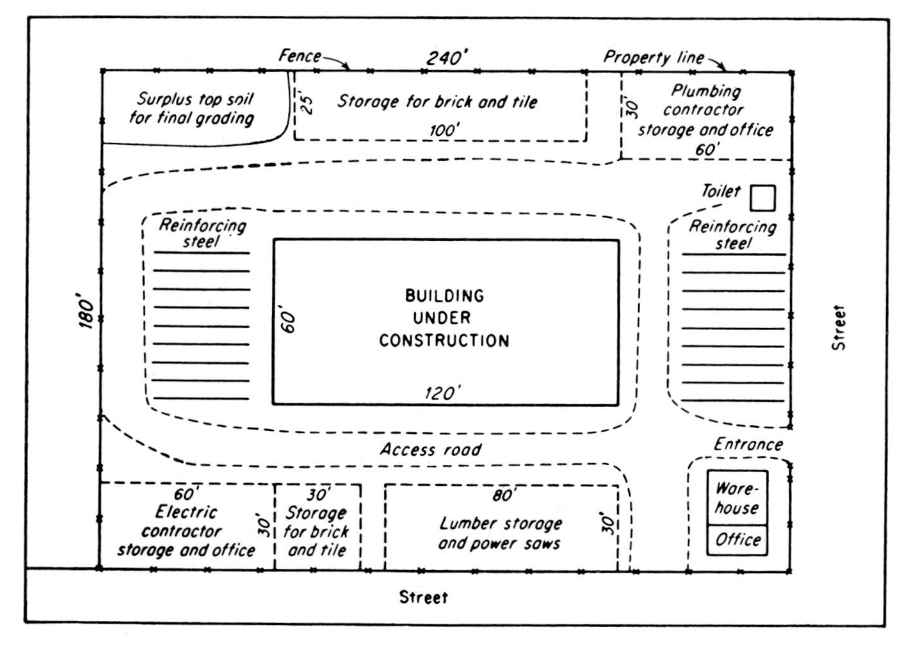

Figure 2-12 Job layout for multistoried reinforced-concrete building.

buildings erected in congested cities, where storage areas at the job site are limited or nonexistent. If area is not available at the job site, the contractor must obtain storage area as near the site as possible.

PROJECT CONTROL DURING CONSTRUCTION

At specified intervals, usually weekly or monthly, reports should be submitted by the project superintendent to the headquarters office showing the actual progress on each activity during the appropriate time interval or through the effective date of the report. If the progress on one or more activities or on the entire project is behind schedule, such information will be known early enough to take corrective steps.

SUPPLEMENTARY MATERIAL

The material presented in the section is supplementary to the project network analysis covered in Chap. 2 and may be omitted as it is not referenced or used elsewhere in the book. It contains a brief presentation of a very powerful planning tool which is beginning to be utilized by the construction industry.

PERT Program evaluation and review technique, or PERT, was developed to overcome the difficulty associated with planning projects for which durations of the specific activities could not be estimated reliably. This technique was developed in the aerospace industry to better manage the development of complex missile systems, and more recently it has been applied to the construction industry. PERT, like traditional CPM, assumes that specific activities and their precedence relationships can be clearly defined for a particular project. PERT makes use of the powerful mathematical tools of probability and statistics, starting out with a fairly complex probability density function called the beta distribution. This distribution handles variable data which are "bounded" in that values exist only within a certain range. If you ask a knowledgeable construction field superintendent how long it might take to erect the structural steel frame for a small building, he may say "10 days." When queried further, you might find out that, if everything goes just right, he believes the erection may take only "5 days." Further, if everything goes wrong, the erection procedure may take as long as "12 days." Now you have *three* possible durations for this activity. Using traditional CPM techniques you would have to pick one of the three estimates for use (probably his best guess of 10 days). But what about the other two estimates of 5 and 12 days? Those estimates provide valuable information to the project planner and can be used with PERT. It can be shown that, using this complex beta distribution, a very good estimate of the expected, or average, duration (t_e) can be determined using the following simple equation:

$$t_e = \frac{O + 4M + P}{6} \tag{2-1}$$

where t_e = expected duration (average)
O = optimistic duration
M = most probable duration
P = pessimistic duration

For the example of the structural steel erection, the expected duration would be:

$$\frac{5 + 4(10) + 12}{6} = 9.5 \text{ days}$$

What does the value of 9.5 days mean? The beta distribution says this value is the "expected" or "average" value, which means that there is a 50 percent probability that the duration of this activity will be *greater than* 9.5 days, and a 50 percent probability that the duration of the activity will be *less than* 9.5 days. Remember, we are dealing with probabilities—in this case an activity duration between 5 and 12 days. Our knowledgeable superintendent really said there was essentially a 0 percent probability of finishing in less than 5 days and a 100 percent probability of finishing *within* 12 days. The resulting probability density function is sketched in Fig. 2-13.

Note that the expected and most probable durations are not identical (although they are close in this case). Also note that the distribution is bounded by the pessimistic and optimistic estimates of the duration, with the most probable duration always in between. The expected duration (t_e) splits the area under the probability distribution function in half. One other point should be made here. By définition, the area under a probability distribution function must equal 1.0 (or 100 percent), hence we can talk about probabilities once we have defined the shape of the distribution function.

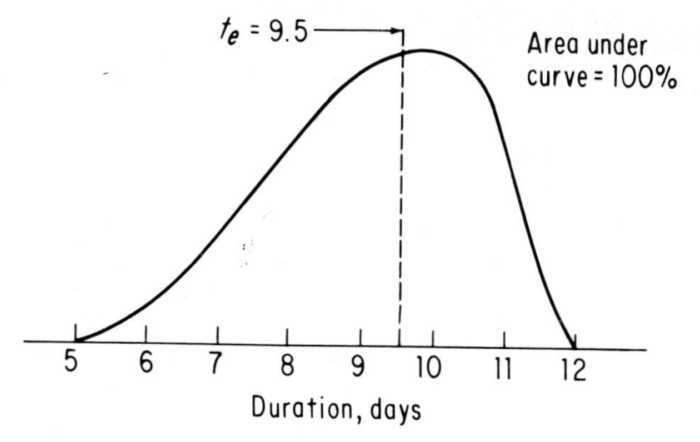

Figure 2-13 Probability density function for structural steel erection.

Although it is valuable to know the expected, or average, duration, it is equally valuable to have some easily understood (and standardized) measure of the spread, or dispersion, of values permissible for each activity's duration. One obvious measure of the spread would be the range R, which is simply the difference between the pessimistic and optimistic durations, or

$$R = P - O \tag{2-2}$$

In our example, the range would be $12 - 5 = 7$ days.

Another, often more useful, measure of the dispersion is the variance, which is the average squared difference of all numbers from their expected or average value. The square root of the variance is called the standard deviation and has the same units as the durations. Using the beta distribution, the variance (S^2) can be closely approximated by the following simple equation:

$$S^2 = \left(\frac{P - O}{6}\right)^2 \tag{2-3}$$

In our example, the variance would be:

$$S^2 = \left(\frac{12 - 5}{6}\right)^2 = 1.36$$

The standard deviation (S) is the square root of the variance, or

$$S = \frac{P - O}{6} \tag{2-4}$$

And the standard deviation of the example activity would be:

$$S = \frac{12 - 5}{6} = 1.17$$

It should be noted here that the figures for the expected value and the standard deviation are carried to the nearest 0.1 day, which is contrary to traditional CPM calculations. Also, the variance is usually carried to the nearest 0.01 unit. The reason for this will become clear in the following discussion.

Of particular importance to the planner is the overall expected project duration (T_e) and its standard deviation (S_T). The overall expected project duration is simply the sum of the expected activity durations *along the critical path,* just the same as with traditional CPM methods. The overall project standard deviation is not so simple. Logically, it would seem unreasonable to assume *every* activity along a critical path could go to either its optimistic or pessimistic values, so summing either of these values along the critical path would yield extremely unrealistic estimates of T_e (either too optimistic or pessimistic). By turning again to the power of probability and statistics, we see that the overall project variance (S_T^2) can be reliably estimated by summing the variances of the individual activities along the critical path, that is,

$$S_T^2 = \sum S^2 \text{ along the critical path} \tag{2-5}$$

Since the standard deviation is the square root of the variance, the overall project standard deviation is *not* the sum of the individual critical activity standard deviations, but rather the square root of the sum of the individual critical activity variances. Thus:

$$S_T = \sqrt{\sum S^2} \text{ along the critical path} \tag{2-6}$$

The following example illustrates the PERT technique.

Example PERT problem A network using the AOA method is shown in Fig. 2-14. Table 2-8 lists the activities, their events (nodes), the three estimates for their durations, the expected durations, the variances, and the standard deviations.

Using the expected values for each of the durations, the critical path is 1-2-3-5-8 and the critical activities are A, C, E, and G. The expected project duration (T_e) is 36.0 days, as shown on Fig. 2-14. Also shown on the figure are the ES, LS, EF, and LF times, from which the free and total floats can be easily calculated. If we used the most probable durations instead of the expected durations, the overall project duration (T_e) would be $6 + 12 + 11 + 4 = 33$ days, which is 1 day less than the expected duration of the overall project. Which value is the more nearly correct? The answer is *neither*. One duration is the most probable duration, and the other is the expected duration. The important point that *can* be stated is that we know there is a 50 percent probability that the duration will be equal to or less than 36 days (and an equal probability that the duration will be equal to or greater than 36 days). We do not know (yet) the probability associated with completing the project in less than the most probable duration of 33 days. Therefore the most useful duration, and the one that is almost always used in practice, is the overall expected project duration.

The overall project variance (S_T^2) is the sum of the individual activity

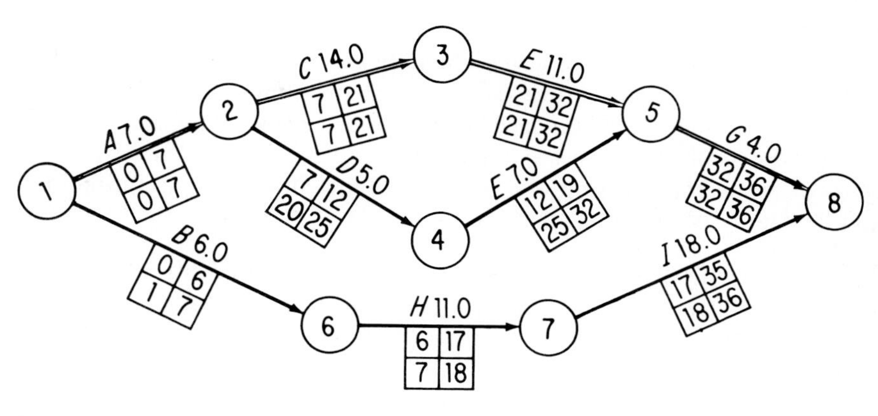

Figure 2-14 AOA diagram for example PERT problem.

Table 2-8 Activities, events, duration estimates, expected durations, variances, and standard deviations for PERT example problem

Activity			Durations, days					
Code	From	To	Optimistic	Most probable	Pessimistic	Expected duration, days	Variance	Standard deviation
A	1	2	3	6	15	7.0	4.00	2.0
B	1	6	2	5	14	6.0	4.00	2.0
C	2	3	6	12	30	14.0	16.00	4.0
D	2	4	2	5	8	5.0	1.00	1.0
E	3	5	5	11	17	11.0	4.00	2.0
F	4	5	3	6	15	7.0	4.00	2.0
G	5	8	1	4	7	4.0	1.00	1.0
H	6	7	3	8	31	11.0	21.78	4.7
I	7	8	3	17	37	18.0	32.11	5.7

variances along the critical path [Eq. (2-5)]. In our example,

$$S_T^2 = 4.00 + 16.00 + 4.00 + 1.00 = 25.00$$

and the overall project standard deviation (S_T) is [from Eq. (2-6)]

$$S_T = \sqrt{25.00} = 5.0 \text{ days}$$

Once the overall expected project duration (T_e) and standard deviation (S_e) are known, the probabilities associated with completing the project within a certain length of time may be calculated with confidence. Turning again to statistics, we see that, although the shape of individual activity frequency distributions may be quite skewed and different from each other, the overall project frequency distribution can be closely approximated by the well-known normal distribution (the familiar bell-shaped curve also known as the Gaussian distribution). The inference of this is known as the central limit theorem and is covered in detail in texts dealing with probability and statistics. The shape of the normal distribution curve is shown in Fig. 2-15. It is symmetrical about the average and can be *completely* described by only two values: its average and its standard deviation. In other words, two normal distributions are identical if their respective averages and standard deviations are the same. Knowing these two values, the probability (area) under any portion of the curve can be calculated. For example, using the overall expected project duration and standard deviation, we see that

$T_e \pm 1.0S_T$ will account for 68.2 percent of the area under the curve

$T_e \pm 2.0S_T$ will account for 95.4 percent of the area under the curve

$T_e \pm 3.0S_T$ will account for 99.9+ percent of the area under the curve

Table 2-9 contains the areas for portions of the normal distribution

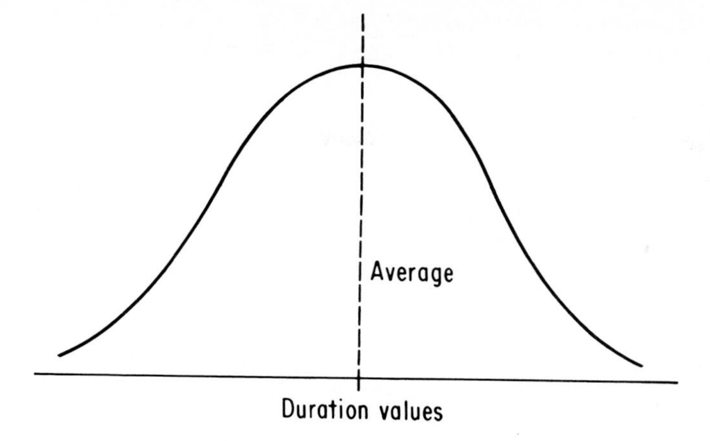

Figure 2-15 The normal probability distribution.

Table 2-9 Area under one-half the normal distribution for various durations expressed as Z, where $Z = (D - T_e)/S_T$

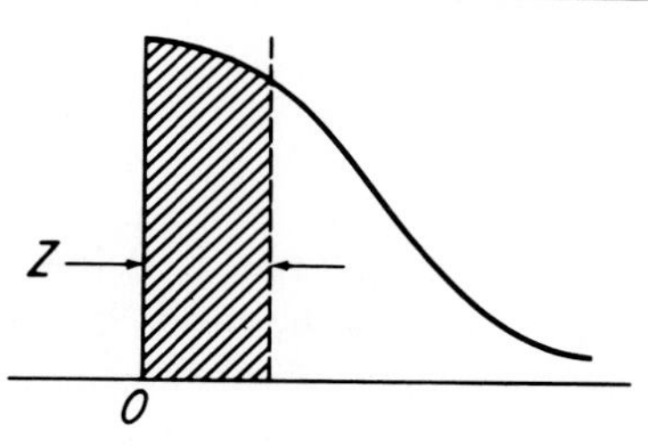

Z	Area	Z	Area	Z	Area
0.0	0.0	1.0	0.341	2.0	0.477
0.1	0.040	1.1	0.364	2.1	0.482
0.2	0.079	1.2	0.385	2.2	0.486
0.3	0.118	1.3	0.403	2.3	0.489
0.4	0.155	1.4	0.419	2.4	0.492
0.5	0.192	1.5	0.433	2.5	0.494
0.6	0.226	1.6	0.475	2.6	0.495
0.7	0.258	1.7	0.455	2.7	0.496
0.8	0.288	1.8	0.464	2.8	0.497
0.9	0.316	1.9	0.471	2.9	0.4998
				3.0	0.4999

Note: As the curve is symmetrical, the same value occurs for $-Z$ as for $+Z$. Also note that the total area under one-half the curve equals 0.5.

function, expressed in terms of the difference between a particular duration D and the expected project duration T_e divided by the project standard deviation S_T:

$$Z = \frac{D - T_e}{S_T} \tag{2-7}$$

where Z = a dimensionless parameter expressing the horizontal axis of the standardized normal distribution function.

To illustrate the use of this table, suppose a project had a T_e of 243 days with a S_T of 22.4. From this information we know that there is a 50 percent probability of finishing in 243 days *or less* and a 50 percent probability of finishing in 243 days *or more*. What is the probability of finishing in 275 days or less? To answer this question, first calculate Z, which is $(275-243)/22.4 = 1.43$. From Table 2-9, the area under the right-hand portion of the distribution is approximately 0.46. Combining that area with the left half of the curve yields a probability of $0.50 + 0.46 = 0.96$, or a 96 percent probability of finishing *within* 275 days. Remember, the calculated probability of finishing in *exactly* 275 days is zero, which is quite different from finishing "within" 275 days (which means 275 days or less).

Turning to the example PERT problem already worked, we can now calculate the probabilities associated with finishing within any given time. For example, if the planner wants to know the probability of finishing within 45 days, the value of Z is

$$Z = \frac{45 - 36}{5.0} = 1.8$$

From the table the area under the right half of the curve is 0.464. The probability of finishing within 45 days is $0.5 + 0.464$ or about 0.96 (or 96 percent). This means there is a 0.04, or 4 percent, probability of finishing in 45 days *or more*. Now, what is the probability of finishing within 34 days? The calculation of Z is

$$Z = \frac{34 - 36}{5.0} = -0.40$$

Here the use of the fact that the normal curve is symmetrical comes in very handy. The area under the *left*-hand side of the symmetrical curve in this case is approximately 0.155. Since there is a 0.50 probability of finishing within 36 days, there must be a $0.50 - 0.16 = 0.34$, or 34 percent, probability of finishing within 34 days. And there is a $100 - 34$ or 66 percent probability of finishing in more than 34 days.

Thus the use of PERT can be a powerful tool for the project planner, especially if he has activities for which a range of durations can be estimated.

Problems with PERT The foremost problem with using PERT is the increased complexity, for three durations are now used for a given activity instead of only one. For a small network with only a few activities, this does not pose much difficulty, but for a large, complex network involving hundreds or even thousands of activities, the extra calculations become enormous. Fortunately, all these techniques are easily handled by computers, and today even the microcomputer can handle several hundred activities with ease. Therefore the only remaining major problem is having to input three durations instead of only one for a given activity. As more and more experience is gained with PERT, project planners are becoming more comfortable with it, especially since they can now predict, with a known level of probability, the projected completion dates of a complex project.

Another potential problem with PERT is the influence of *near* critical activities which have large standard deviations. Returning to the example PERT problem, look at activities H and I in Fig. 2-14. They both have rather large standard deviations and have only 1 day of float. What would happen if there was a 1-day delay in starting activity H? The path B, H, I would lose its float, and there would be two critical paths. Furthermore, the standard deviation of path 1-6-7-8 (activities B, H, and I) would be

$$4.00 + 21.78 + 32.11 = 58.88 = 7.6 \text{ days}$$

The value of 7.6 is *significantly* larger than the value of 5.0, and the effect of this larger standard deviation would be to lower the probabilities associated with completing the project within a specified time. In this example the probability of finishing within 45 days would reduce from 96 percent to around 88 percent.

The important point here is that "near critical" activities with large standard deviations can influence the overall project. This can "partially" be handled with the aid of the computer. Activities with large standard deviations can be flagged and examined for their float times. If their float times are short, the overall project standard deviation can be calculated for the near critical path and compared with the critical path standard deviation. Given this information, the project planner can use good engineering judgment in calculating probabilities associated with specific completion dates.

PROBLEMS

2-1 Prepare an AON (arrow) diagram, showing ES, LS, EF, and LF times and the critical path, for a project involving the activities listed in Table 2-10.

2-2 Prepare an AON (arrow) diagram, showing ES, LS, EF, and LF times and the critical path, for a project involving the activities listed in Table 2-11.

2-3 Same as Prob. 2-1, except prepare a precedence diagram.

2-4 Same as Prob. 2-2, except prepare a precedence diagram.

2-5 Prepare a time-grid diagram for the project of Prob. 2-2, for which the durations are expressed in days. Assume a week of 5 days, with work to start on the first Monday in May of the current year. Show the calendar days for the starting and finishing of each activity, with no lost time as a result of weather or other causes.

Table 2-10
List of activities for problem 2-1

Activity	Duration	Activities which immediately precede	Activities which immediately follow
A	3	None	*B, C*
B	5	*A*	*D, E*
C	4	*A*	*F, I*
D	7	*B*	*G*
E	6	*B*	*H*
F	11	*C*	*H*
G	6	*D*	*J*
H	4	*E, F*	*K*
I	3	*C*	*K, L*
J	6	*G*	*M*
K	5	*H, I*	*N*
L	7	*I*	*O*
M	5	*J*	*P*
N	3	*K*	*P*
O	2	*L*	*P*
P	4	*M, N, O*	None

2-6 Prepare a list of activities for the construction of a parking lot for a small office building. The lot is to have a layer of crushed rock for the base course and be weatherproofed with an asphaltic concrete surface. Then assign durations to each activity and prepare an arrow diagram and a list of starts, finishes, and total floats for the project. Indicate the critical path.

2-7 Same as Prob. 2-6, except prepare a precedence diagram.

Table 2-11
List of activities for problem 2-2

Activity	Duration	Activities which immediately precede	Activities which immediately follow
A	3	None	*C*
B	5	None	*D, E*
C	4	*A*	*F, G*
D	8	*B*	*F, G*
E	9	*B*	*H, I*
F	6	*C, D*	*J, K, L, M*
G	8	*C, D*	*K, L, M*
H	6	*E*	*K, L, M*
I	5	*E*	*P*
J	4	*F*	*N*
K	7	*F, G, H*	*O*
L	6	*F, G, H*	*Q*
M	7	*F, G, H*	*P*
N	4	*J*	*R*
O	8	*K*	*R*
P	4	*I, M*	*R*
Q	5	*L*	*R*
R	3	*N, O, P, Q*	None

2-8 For the precedence diagram shown in Fig. 2-16, determine the ES, LS, EF, and LF time for each activity and indicate the critical path.

2-9 Prepare a precedence diagram for the construction of a concrete retaining wall 800 ft long. The wall is to be built in sections, each 100 ft in length, using forms eight times. Use appropriate lags and leads and estimate durations for each activity. Then calculate ES, LS, EF, and LF times and the critical path.

2-10 (*a*) Calculate the expected durations, variances, and standard deviations of each of the activities listed in the following table.

(*b*) Prepare an arrow diagram showing ES, LS, EF, and LF times and the critical path.

(*c*) Calculate the expected project duration and project standard deviation.

(*d*) Determine the probability of finishing *within* 5 days beyond the expected project duration. Determine the probability of finishing in greater than 8 days beyond the expected project duration.

Activity			Durations, days		
Code	From	To	Optimistic	Most probable	Pessimistic
A	1	2	4	7	12
B	1	3	6	10	12
C	2	4	2	2	2
D	2	5	9	15	18
E	3	5	11	14	20
F	3	6	4	7	10
G	4	7	7	9	10
H	5	8	1	1	1
I	6	8	3	4	8
J	7	10	4	12	15
K	8	10	6	11	14
L	6	9	3	4	6
M	10	11	3	8	10
N	9	11	5	8	11
O	11	12	6	8	10

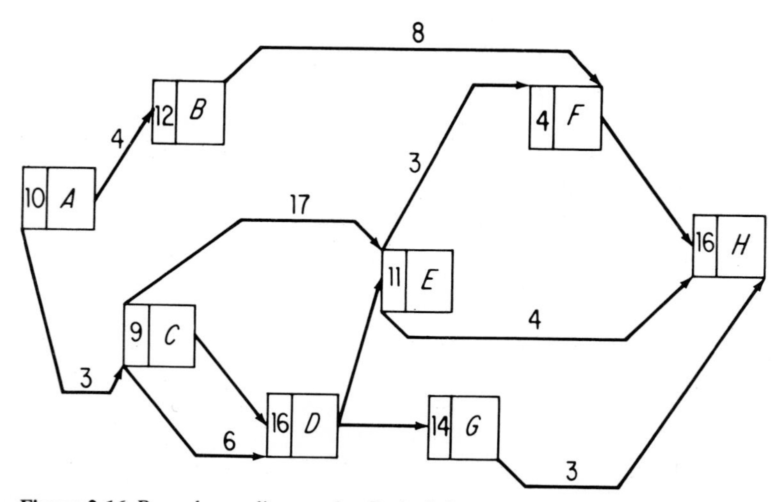

Figure 2-16 Precedence diagram for Prob. 2-8.

2-11 Using the table of activities for Prob. 2-1, assume each activity has the following normal durations/costs and crash durations/costs. Crash the project 5 days with the minimum increase in direct costs. Plot the increase in cost versus time for each day crashed.

	Normal		Crash	
Activity	Duration, days	Cost	Duration, days	Cost
A	3	$ 400	3	$ 400
B	5	870	4	1,170
C	4	1,250	3	1,375
D	7	1,400	5	1,800
E	6	1,600	4	2,200
F	11	4.500	8	5,400
G	6	2,400	5	2,450
H	4	3,200	4	3,200
I	3	2,450	1	2,500
J	6	1,970	4	2,050
K	5	3,450	4	3,475
L	7	5,670	4	5,760
M	5	2,460	5	2,460
N	3	120	3	120
O	2	340	2	340
P	4	1,260	2	2,460

REFERENCES

Project planning and management

1. Weist, Jerome D., and Ferdenand K. Levy: *A Management Guide to PERT/CPM*, 2d ed., Prentice-Hall, Englewood Cliffs, N.J., 1977, 229 p.
2. Ahuja, H. N.: *Construction Performance Control by Networks*, Wiley, New York, 1976.
3. Harris, R. B.: *Precedence and Arrow Networking Techniques for Construction*, Wiley, New York, 1978.

1st chapter

CHAPTER
THREE
FACTORS AFFECTING THE SELECTION OF CONSTRUCTION EQUIPMENT

GENERAL INFORMATION

A problem which frequently confronts a contractor as he plans to construct a project is the selection of the most suitable equipment. He should consider the money spent for equipment as an investment which he can expect to recover, with a profit, during the useful life of the equipment. A contractor does not pay for construction equipment; the equipment must pay for itself by earning for the contractor more money than it cost. Unless it can be established in advance that a unit of equipment will earn more than the cost, it should not be purchased.

A contractor can never afford to own all types or sizes of equipment that might be used for the kind of work he does. It may be possible to determine what kind and size of equipment seem most suitable for a given project, but this information alone will not necessarily justify the purchase of the equipment. Perhaps the project under consideration is not large enough to justify the purchase, because the cost cannot be recovered before the completion of the project, and it may not be possible to dispose of the equipment at the completion of the project at a reasonable price. A contractor may own a type of equipment, which is presently idle, that is less desirable than the proposed equipment, but, considering the probable heavy depreciation for the proposed equipment and the uncertainty that it can be used on future projects, the apparently ideal equipment may prove to be more expensive than equipment now owned by the contractor.

Any time a unit of equipment will pay for itself on work that is certain to be done it is good business to purchase it. For example, if a unit of equipment costing $25,000 will save $50,000 on a project, a contractor is justified in purchasing it, regardless of the prospects of using it on additional projects or the prospects of selling it at a favorable price when the project is finished.

STANDARD TYPES OF EQUIPMENT

There is no clear definition of standard equipment. Equipment that is standard for one contractor may be special equipment for another contractor. It depends on the extent to which a contractor will use it in his construction operations. Another method which is sometimes used to distinguish between standard and special equipment is the extent to which it is commonly manufactured and available to prospective purchasers. Thus, a 1-cu-yd diesel-powered crawler-mounted power shovel may be standard equipment, whereas a 30-cu-yd shovel could be classified as special equipment. The larger shovel is normally manufactured for a specific purchaser.

Contractors should confine their purchases to standard equipment unless a project definitely justifies the purchase of special equipment. Delivery of standard equipment may be obtained more quickly. Standard equipment can be used economically on more than one project. Repair parts for standard equipment may be obtained more quickly and economically than for special equipment. If a contractor no longer needs a unit of standard equipment, he can usually dispose of it more easily and at a more favorable price than a piece of special equipment.

SPECIAL EQUIPMENT

One definition of special equipment is equipment that is manufactured for use on a single project or for a special type of operation. Such equipment may not be suitable or economical for use on another project. An example of special equipment is a 90-cu-yd power shovel used to remove the overburden in strip mining for coal. Another example is the hydraulic dredge which was constructed primarily for use in building the Ft. Randall Dam.

REPLACEMENT OF PARTS

A factor which may be overlooked by a prospective purchaser of equipment is the ease and speed with which replacement parts may be obtained. All equipment parts are subject to failure, regardless of the care which they receive. A truck with a broken axle is useless until the axle is replaced. A broken part in a power shovel may delay an entire project for weeks, while the contractor waits for the part to be manufactured and shipped. Prior to purchasing equipment, the buyer should determine where spare parts are obtainable. If parts are not obtainable quickly, it may be wise to purchase other equipment, for which parts are quickly available, even though the latter seems less desirable. This is an argument for standard equipment.

THE COST OF OWNING AND OPERATING CONSTRUCTION EQUIPMENT

There are several methods of determining the probable cost of owning and operating construction equipment. No known method will give exact costs under all operating conditions. At best the estimate is only a close approximation of the cost. Carefully kept records for equipment previously used should give information which may be used as a guide for the particular equipment. But there is no assurance that similar equipment will involve similar costs, especially if the equipment is used under different conditions. Factors which affect the cost of owning and operating equipment include the cost of the equipment delivered to the owner, the severity of the conditions under which it is used, the number of hours it is used per year, the number of years it is used, the care with which the owner maintains and repairs it, and the demand for used equipment when it is sold, which will affect the salvage value.

When it is necessary to estimate the cost of owning and operating equipment prior to purchasing it, cost records, based on past performance, generally will not be available. The costs which should be considered include investment and depreciation (usually termed ownership costs) and maintenance, repairs, lubrication, and fuel (usually termed operating costs). In the following paragraphs each of these costs is discussed and methods are presented to calculate them.

INVESTMENT COSTS

It costs money to own equipment, as the discussion in Chap. 1 demonstrated. One part of ownership costs includes the interest on the money invested, taxes of all types which are assessed against the equipment, insurance, and storage. The rates for these items will vary somewhat among different owners, with location, and for other reasons. These *ownership* costs usually accrue whether or not the equipment is actually used, so it becomes very important to maximize the utilization of all such equipment.

There are several methods of determining the cost of interest paid on the money invested in equipment. Even though the owner pays cash for the equipment, he should charge interest on the investment, as the money spent for the equipment could be invested in some other asset which would return interest to the owner. The possibility of earning interest on money is lost to the equipment owner when he spends the money for equipment.

Some equipment owners charge a fixed rate of interest against the full purchase cost of the equipment each year it is owned. This method gives an annual interest cost which is higher than it should be. One realistic approach is that, for each year that equipment is used, the owner should deduct from its earnings an amount equal to the annual cost of depreciation. Since this money is retained by the owner, it reduces his net investment in the equipment. After the equipment has been used for the estimated depreciation period, expressed in years or units of production, the owner will have recovered its original cost through the reserve for depreciation. In any event, the interest charged should be based on a realistic value for the equipment, instead of its original cost.

The average annual cost of interest should be based on the average value of the equipment during its useful life. This value may be obtained by establishing a schedule of values for the beginning of each year that the equipment will be used. The calculations given below illustrate a method of determining the average value of equipment:

Example 3-1 Original cost of equipment, $25,000
Estimated useful life, 5 yr
Average annual cost of depreciation, $25,000 ÷ 5 = $5,000

(1) Beginning of year	(2) Cumulative depreciation	(3) Value of equipment
1	0	$25,000
2	$ 5,000	20,000
3	10,000	15,000
4	15,000	10,000
5	20,000	5,000
6	25,000	0

Total of values in column 3 = $75,000
Average value, $75,000 ÷ 5 = $15,000

Average value as % of original cost, $\frac{\$15{,}000 \times 100}{\$25{,}000} = 60$

Thus, the average value of equipment having an estimated life of 5 years is 60 percent of the original cost.

An alternative method of determining the average value of equipment having no salvage value, based on straight-line depreciation, is to develop an equation for this purpose. Figure 3-1 indicates the value of a unit of equipment whose initial cost is represented by P at the beginning of each year during its estimated life of N years.

$$\bar{P} = \frac{P + P/N}{2} = \frac{PN + P}{2N}$$

$$= \frac{P(N+1)}{2N} \tag{3-1}$$

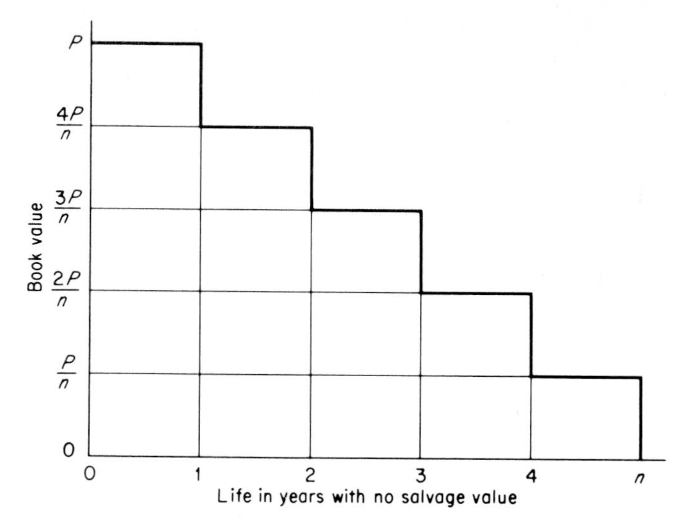

Figure 3-1 Value of equipment by year.

where P = total initial cost
$\bar{P}$ = average value
N = life in years

Table 3-1 gives the average value of equipment, with no salvage value, for various years of life, expressed as a percent of the initial cost.

If the equipment will have a salvage value at the time it is disposed of, after N years of use, its average value is determined by referring to Fig. 3-2. The

Table 3-1 Average value of equipment with no salvage value

Estimated life, yr	Average value as % of original cost
2	75.00
3	66.67
4	62.50
5	60.00
6	58.33
7	57.14
8	56.25
9	55.55
10	55.00
11	54.54
12	54.17

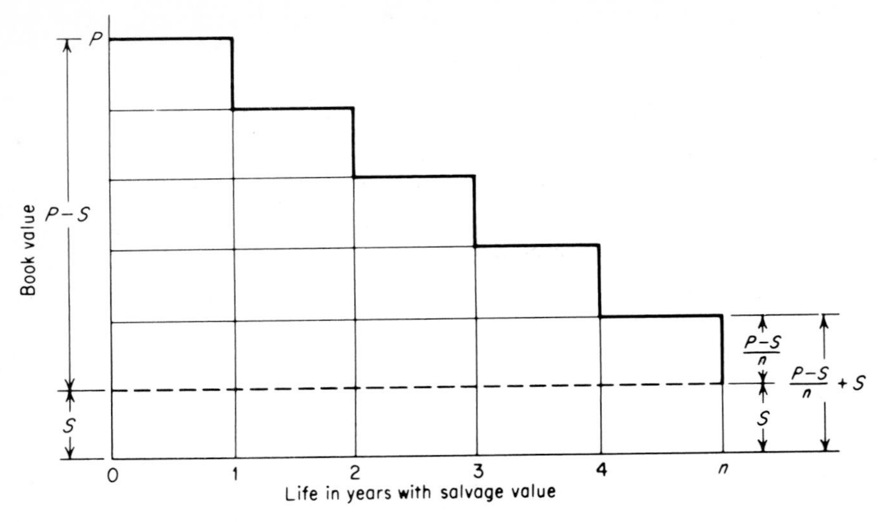

Figure 3-2 Value of equipment by year.

average value is the sum of the values at the beginning of the first year and the beginning of the last year divided by 2.

$$\bar{P} = \frac{P + (P - S)/N + S}{2} = \frac{PN + P - S + SN}{2N}$$

$$= \frac{P(N+1) + S(N-1)}{2N} \tag{3-2}$$

Example 3-2 Consider a unit of equipment costing \$25,000, with an estimated salvage value of \$5,000 after 5 years. Using Eq. (3-2), we get

$$\bar{P} = \frac{25{,}000(5+1) + 5{,}000(5-1)}{2 \times 5}$$

$$= \frac{150{,}000 + 20{,}000}{10} = \$17{,}000$$

It is common practice to combine the cost of interest, insurance, taxes, and storage and to estimate them as a fixed percent of the average value of the equipment. Until a few years ago, the national average rate was about 15 percent, which included interest of 10 percent and insurance, taxes, and storage at 5 percent per year. Today, with the vivid memories of double-digit inflation and interest rates in excess of 20 percent, it is impossible to state a national average investment cost expressed as a single percentage. The one sure thing that is known is that the investment costs will change during the life of the equipment, and the successful manager will be the one who correctly guesses these costs. For the purposes of illustration *only*, the examples in this text will employ the following rates for investment costs:

Interest on borrowed money, 12%

Taxes, insurance, and storage, 8%

The following example illustrates the calculations involved.

Example 3-3 Using the values calculated in Example 3-2, calculate the average yearly investment cost for the piece of equipment.

Average interest on borrowed money = 0.12 × $17,000 = $2,890 per year

Average taxes, insurance, etc. = 0.08 × $17,000 = $1,360 per year

DEPRECIATION

Depreciation is the loss in value of a piece of equipment over time, generally caused by wear and tear from use, deterioration, obsolescence, or reduced need. The profitable owner of equipment must recover this loss of value in equipment which he owns during its useful life. Depreciation may be tracked during the life of a piece simply by preparing a graph of the market value of the equipment with time. Caution should be exercised in selecting the appropriate "market value." For example, most equipment has a wholesale and a retail market value, which are dependent upon many factors.

The general term *depreciation* should not be confused with the specific term *depreciation accounting*. *Depreciation accounting* is the *systematic* allocation of the costs of a capital investment over some specific number of years. There are three reasons for calculating the depreciation accounting value (usually termed *book value*) of a piece of equipment. They are:

1. To provide the construction owner and project manager with an easily calculated "estimate" of the current market value of the equipment. To do this the method of depreciation accounting selected should approximate market value.
2. To provide a systematic method for allocating the depreciation portion of equipment ownership costs over a period of time and to a specific productivity rate. The method chosen should be simple and easily understood.
3. To allocate the depreciation portion of ownership costs in such a manner that the greatest tax benefits will accrue. Depreciation accounting for this purpose must follow strict legal governmental guidelines, which frequently change as new laws are enacted.

It is common to utilize at least two, and sometimes three, different depreciation accounting methods on a particular piece of equipment, reporting one value to the estimators for use on construction and another to the Internal Revenue Service to obtain the most favorable tax benefits.

To calculate the depreciation by any depreciation method, close estimates of the following items must be known.

1. The purchase price of the piece of equipment (termed P)
2. The economic life of the equipment (the optimum period of time to keep the equipment), or the "recovery" period allowed for income tax purposes (termed N)
3. The estimated resale value at the close of the economic life, known as the *salvage value* (termed F)

With these three items of information known or estimated, the depreciation can be calculated using a number of methods which have been devised. The following three methods are the most commonly used.

1. Straight-line method
2. Sum-of-the-years method
3. Declining-balance method

Straight-line method of depreciation accounting Straight-line (SL) depreciation is the easiest method to calculate and is probably the most widely used method in construction. The annual amount of depreciation D_m, for any year m, is a constant value, and thus the book value BV_m decreases at a uniform rate over the useful life of the equipment. The equations are:

$$\text{Deprecation rate, } R_m = 1/N \tag{3-3}$$

$$\text{Annual depreciation amount, } D_m = R_m(P - F) = (P - F)/N \tag{3-4}$$

$$\text{Book value at year } m,\ BV_m = P - mD_m \tag{3-5}$$

(*Note*: The value $P - F$ is often referred to as the *depreciable value* of the investment.) An example follows.

Example 3-4 (SL problem) A piece of equipment is available for purchase for \$12,000, has an estimated useful life of 5 years, and has an estimated salvage value of \$2,000. Determine the depreciation and the book value for each of the 5 years using the SL method.

$$R_m = \tfrac{1}{5} = 0.2$$

$$D_m = 0.2(12{,}000 - 2{,}000) = \$2{,}000 \text{ per year}$$

The table of values are:

m	BV_{m-1}	D_m	BV_m
0	\$ 0	\$ 0	\$12,000
1	12,000	2,000	10,000
2	10,000	2,000	8,000
3	8,000	2,000	6,000
4	6,000	2,000	4,000
5	4,000	2,000	2,000

Further let us assume the equipment is expected to be used about 1,400 hr per year. The estimated hourly depreciation portion of the ownership cost of this piece of equipment is

$$\frac{\$2{,}000}{\$1{,}400} = \$1.428 = \$1.43 \text{ per hr}$$

Sum-of-the-years method of depreciation accounting This is an *accelerated* method (fast write-off), which is a term applied to accounting methods which permit rates of depreciation faster than straight line. The rate of depreciation is a factor times the depreciable value $(P - F)$. This factor is determined as follows:

1. The denominator of the factor is the sum of the digits including 1 through the last year in the life of the piece of equipment. Thus:

$$\text{SOY} = \frac{N(N+1)}{2} \tag{3-6}$$

2. The depreciation rate R_m is

$$R_m = \frac{N - m + 1}{\text{SOY}} \tag{3-7}$$

3. The annual depreciation D_m for the mth year (at any age m) is

$$D_m = R_m(P - F) = \frac{N - m + 1}{\text{SOY}}(P - F) \tag{3-8}$$

4. The book value BV_m at the end of year m is:

$$BV_m = P - (P - F)\left[\frac{m\left(N - \frac{m}{2} + 0.5\right)}{\text{SOY}}\right] \tag{3-9}$$

An example of the use of the method follows.

Example 3-5 (SOY method) Using the same values as given in Example 3-4, calculate the allowable depreciation and the book value for each of the 5 years using the SOY method.

$$\text{SOY} = 1 + 2 + 3 + 4 + 5 = 15 \quad \text{or} \quad = \frac{5(6)}{2} = 15$$

$$R_m = \frac{5 - m + 1}{15}$$

$$D_m = R_m \times (12{,}000 - 2{,}000) = \frac{(5 - m + 1)(10{,}000)}{15}$$

Pertinent data are tabulated in the following table.

Year	R_m	D_m	BV_m
0		\$ 0	\$12,000
1	$\frac{5}{15}$	3,333	8,667
2	$\frac{4}{15}$	2,667	6,000
3	$\frac{3}{15}$	2,000	4,000
4	$\frac{2}{15}$	1,333	2,667
5	$\frac{1}{15}$	667	2,000

Notice in this case the allowable depreciation each year is different, which makes calculations of hourly costs cumbersome as they would change each year. Similarly, calculations of depreciation costs per unit of work would also be cumbersome. Consequently, most companies use the straight-line method of depreciation accounting when calculating hourly or other unit costs to recover depreciation on the equipment.

Declining-balance methods Declining balance methods also are accelerated depreciation methods that provide for even larger portions of the cost of a piece of equipment to be written off in the early years. Interestingly, this method often more nearly approximates the actual loss in market value with time. Declining methods range from 1.25 times the current book value divided by the life to 2.00 times the current book value divided by the life (the latter is termed *double declining balance*). Note that although the estimated salvage value F is not included in the calculation, the book value can *not* go below the salvage value. Following are the equations necessary to use the declining balance methods.

The symbol R is used for depreciation rate for declining balance method of depreciation.

1. For 1.25 declining balance (1.25DB) method, $R = 1.25/N$.
   For 1.50 declining balance (1.5DB) method, $R = 1.50/N$.
   For 1.75 declining balance (1.75DB) method, $R = 1.75/N$.
   For double declining balance (DDB) method, $R = 2.00/N$.
2. The allowable depreciation D_m for any year m and any depreciation rate R is

$$D_m = RP(1 - R)^{m-1} \quad \text{or} \quad D_m = (\text{BV}_{m-1})R \tag{3-10}$$

3. The book value for any year m is:

$$\text{BV}_m = P(1 - R)^m \quad \text{or} \quad \text{BV}_m = \text{BV}_{m-1} - \text{BV}_m, \text{ provided that } \text{BV}_m \geqq F \tag{3-11}$$

Since the book value can never go below the estimated salvage value, the declining balance method must be "forced" to intersect the value of F at time N. This may be accomplished by switching from the declining balance method to either the SOY method or the SL method. The following example illustrates this method of depreciation.

Example 3-6 (DDB method) For the same piece of equipment described in Example 3-4, calculate the allowable depreciation and the book value for each of the 5 years of its life.

$$R = \frac{2.0}{5} = 0.4$$

$$D_m = 0.4(\mathrm{BV}_{m-1})$$

$$\mathrm{BV}_m = \mathrm{BV}_{m-1} - D_m$$

The results of the calculation are given in the following table.

Year	D_m	BV_m
0	0	12,000
1	0.4 × 12,000 = 4,800	7,200
2	0.4 × 7,200 = 2,880	4,320
3	0.4 × 4,320 = 1,728	2,592
4	0.4 × 2,592 = 592	2,000
5	0	2,000

Note that in year 4, the calculated depreciation using the DDB method resulted in a book value lower than the estimated salvage value. Therefore, the allowable depreciation was only $592, which in effect was a straight-line depreciation taken in year 4. Then in year 5, since all the depreciation had already been taken, the allowable depreciation would equal to zero.

The allowable depreciation for the example piece of equipment using each of the three methods is shown in Fig. 3-3.

The total ownership costs would be the sum of the investment costs and the depreciation costs. The following example illustrates the calculations.

Example 3-7 A piece of equipment costing $67,000 new, with a 7-year useful life and an expected salvage value of $7,000 is being considered for purchase. Calculate the yearly ownership costs using straight-line depreciation. Also calculate the hourly ownership costs assuming the piece of equipment will be utilized an average of 1,400 hr each year.

Using Eq. 3-2, the average investment value is

$$\frac{67{,}000(8) + 7{,}000(6)}{14} = \$41{,}286$$

Using Eq. 3-4, the annual depreciation amount is

$$\frac{67{,}000 - 7{,}000}{7} = \$8{,}571$$

The total annual ownership costs would be:

Interest = 0.12 × 41,286 =	$ 4,954
Taxes, etc. = 0.08 × 41,286 =	3,302
Depreciation =	8,571
Total annual ownership costs =	$16,287

The hourly ownership costs would be:

$$\frac{\$16{,}287}{1{,}400} = \$12.02 \text{ per hour of use}$$

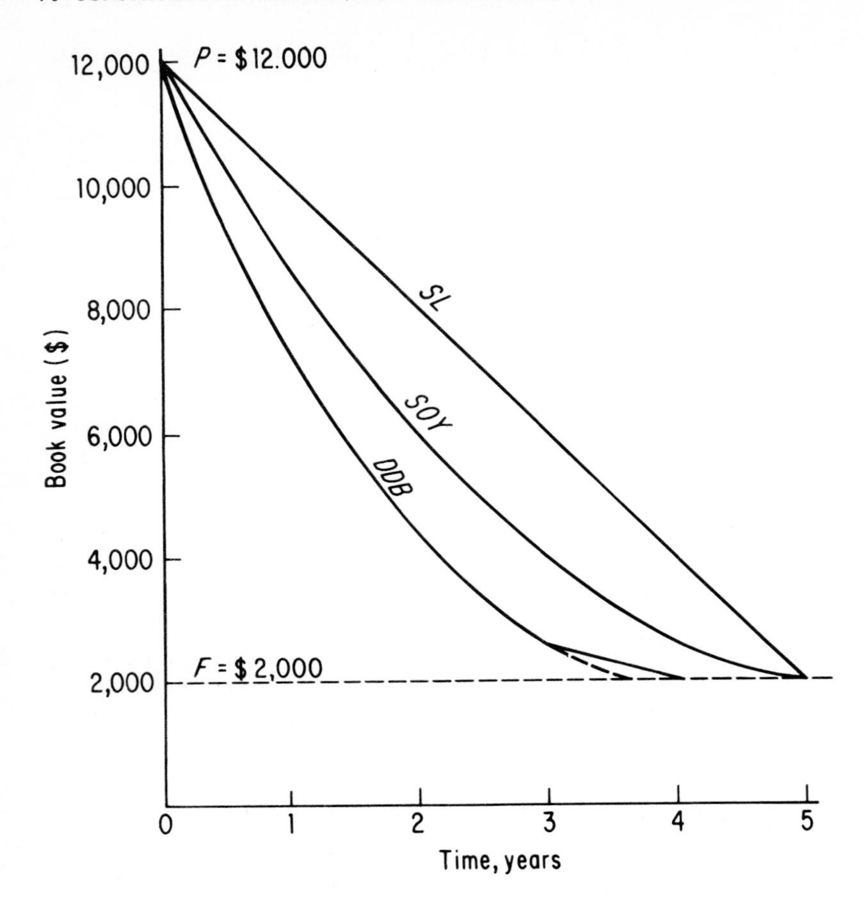

Figure 3-3 Allowable depreciation curves for three methods of depreciation.

ALTERNATIVE METHOD OF CALCULATING THE COST OF OWNERSHIP

Another way to calculate the ownership costs associated with a piece of equipment is to use the time value of money approach discussed in Chap. 1. Knowing the cost and estimated salvage value of a piece of equipment, along with the interest rate to be used in the calculation, the equivalent uniform annual ownership cost can be determined. The following example illustrates this method.

Example 3-8 Given the same piece of equipment described in Example 3-7, calculate the yearly and hourly ownership costs.

The cash flow diagram for the purchase of this equipment is given in Fig. 3-4. The equivalent uniform annual ownership cost is (for the previously discussed interest rate of 20

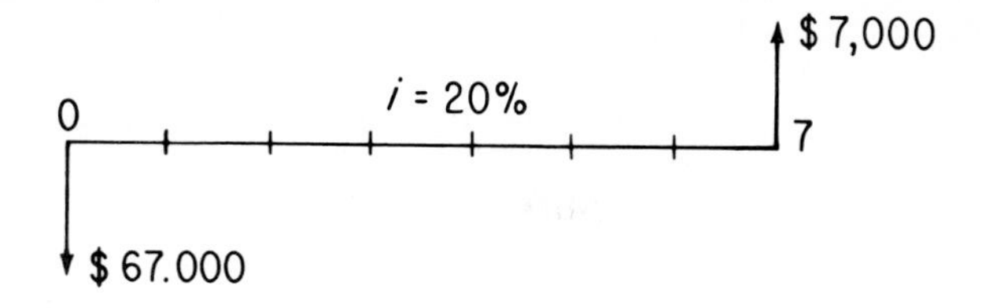

Figure 3-4 Cash flow diagram for Example 3-8.

percent):

$$A = \$67{,}000(A/P{,}20{,}7) - \$7{,}000(A/F{,}20{,}7) = \$18{,}045$$

The hourly cost would be

$$\frac{\$18{,}045}{1{,}400} = \$12.89 \text{ per hour of use}$$

There is a significant difference in the costs calculated from the two methods (compare the results from Examples 3-7 and 3-8). The alternate method will always yield a larger estimate of the ownership costs because it considers the time value of money for *all* expenditures rather than using the average value times the interest rate. While the first method has been used for many years and, when interest rates were lower, gave reasonable estimates, today the alternate method normally will yield a more realistic estimate of the ownership costs. Of course, it too is only an estimate, and the assumption that taxes, insurance, and storage costs will function in the same way as interest rates on borrowed capital may be somewhat conservative.

OPERATING COSTS

Operating costs are those costs associated with the operation of a piece of equipment. In contrast to ownership costs, which generally accrue whether or not the equipment is actually being used, operating costs usually occur *only* when the equipment is being used. Operating costs include the fuel and lubrication costs, operator costs, and minor maintenance and repair costs. The other costs involved in equipment are major maintenance, major repair, and tires. These costs are sometimes included in the cost of ownership and sometimes in the cost of operating. Either way, they represent a significant cost category and should not be overlooked. Each of the operating costs is discussed in the following paragraphs.

Maintenance and repairs The cost of maintenance and repairs will vary considerably with the type of equipment, the service to which it is assigned, and the care which it receives. If a bearing is greased and adjusted at frequent intervals, its life will be much longer than if it is neglected.

The annual cost of maintenance and repairs may be expressed as a percent of the annual cost of depreciation or it may be expressed independently of depreciation. In any event, it should be sufficient to cover the cost of keeping the equipment operating. The annual cost of maintenance and repairs for a power shovel may vary from 80 to 120 percent of the annual cost of depreciation, with 100 percent a fair average value. The annual cost for certain types of rock-crushing equipment may be much higher, while for an electric motor it will be lower. Experience records serve as a guide in estimating these costs.

Operating conditions Construction equipment which is driven by internal combustion engines requires fuel and lubricating oil, which should be considered as an operating cost. While the amounts consumed and the unit cost of each will vary with the type of equipment, the conditions under which it is used, and location, it is possible to estimate the cost reasonably accurately for a given project.

The person who is responsible for selecting the equipment should estimate the conditions under which the equipment will operate. There are at least two conditions which will apply to most projects, the extent to which the engine will operate at full power all the time, and the actual time that the unit will operate in an hour or a day.

While the power unit in a piece of equipment may be capable of developing a given horsepower when operating at maximum output, it is well known that maximum output usually will not be required at all times. For example, the full power of an engine may be required while a power shovel is loading the dipper, but during the balance of the cycle the demands on the engine are reduced considerably. The full power of a tractor will be required while it is loading a scraper with earth, and possibly while it is climbing an embankment, but for the rest of the round-trip cycle it is possible that less than the maximum power will be required. Consider the gasoline-engine-driven air compressor that is heard so frequently. For a short time the engine will operate at full power, then it will idle for a while, these conditions alternating as the air is used.

Horsepower ratings Because the horsepower ratings specified in the literature of various manufacturers are not determined under the same operating conditions, it is not possible to compare the work capability of different engines with a high degree of accuracy. The power may be specified for standard conditions, namely at a barometric pressure of 29.9 in. of mercury (in. Hg) and at a temperature of 60°F, or it may be specified for normal operating conditions, with altitudes up to 2,000 ft above sea level and at temperatures up to 85°F. The specified power may be the maximum that the bare engine can develop or it may be the flywheel power with all accessories attached to the engine. The accessories will vary with engines, but usually include a fan, generator, fuel pump, water pump, air cleaner, and lubricating oil pump. Each of these accessories requires power, for a combined demand that may equal 20 or 25 percent or more of the rated flywheel power of the engine. Thus an engine rated at 200 fwhp may develop 250 hp when rated as a bare engine.

Fuel consumed When operating under standard conditions, a gasoline engine will consume approximately 0.06 gal of fuel per flywheel horsepower hour (fwhp-hr), while a diesel engine will consume approximately 0.04 gal per fwhp-hr. A horsepower hour is a measure of the work performed by an engine.

In order to determine the work performed by an engine it is necessary to know the average power generated by the engine and the duration of this performance. Engines used in the construction industry seldom operate at a constant output or at the rated output, except for short periods of time. A tractor engine may operate at maximum power when it is loading a scraper or negotiating an adverse slope. During the balance of its cycle the demand on the engine will be reduced substantially, resulting in a decreased consumption of fuel. Also, construction equipment is seldom operated the entire 60 min in an hour.

Consider a power shovel with a diesel engine rated at 160 fwhp. When used to load trucks, the engine may operate at maximum power while filling the dipper, requiring 5 sec out of a cycle time of 20 sec. During the other 15 sec the engine may operate at not more than one-half of its rated power. Also the shovel may be idle for 10 to 15 min, or more, during an hour, with the engine providing only that power required for internal operation.

Assume that this shovel operates 50 min per hr, to give an operating factor $= \frac{50}{60} \times 100 = 83.3$ percent. The approximate amount of fuel consumed in an hour can be determined as follows:

Rated output at flywheel = 160 hp
Engine factor:

Filling the dipper, $\frac{5}{20} \times 1$	= 0.250
Rest of cycle, $\frac{15}{20} \times 0.5 \times 1$	= 0.375
Total factor	= 0.625

Time factor $= \frac{50}{60} = 0.833$
Operating factor $= 0.625 \times 0.833 = 0.520$
Fuel consumed per hr $= 0.520 \times 160 \times 0.04 = 3.33$ gal

For other operating factors the quantity of fuel consumed should be estimated in a similar manner.

Lubricating oil The quantity of lubricating oil used by an engine will vary with the size of the engine, the capacity of the crankcase, the condition of the piston rings, and the number of hours between oil changes. For extremely dusty operations it may be desirable to change oil every 50 hr, but this is an unusual condition. It is common practice to change oil every 100 to 200 hr. The quantity of oil consumed by an engine per change will include the amount added during the change plus the make-up oil between changes.

A formula which may be used to estimate the quantity of oil required

$$q = \frac{\text{hp} \times f \times 0.006 \text{ lb per hp-hr}}{7.4 \text{ lb per gal}} + \frac{c}{t} \qquad \text{(3-12)}$$

where q = quantity consumed, gph
hp = rated horsepower of engine
c = capacity of crankcase, gal
f = operating factor
t = number of hours between changes

It assumes that the quantity of oil consumed per rated horsepower hour, between changes, will be 0.006 lb. Using the formula, for a 100-hp engine with a crankcase capacity of 4 gal, requiring a change every 100 hr, the quantity consumed per hour will be

$$q = \frac{100 \times 0.6 \times 0.006}{7.4} + \frac{4}{100} = 0.049 + 0.04 = 0.089 \text{ gal per hr}$$

EXAMPLES ILLUSTRATING THE COST OF OWNING AND OPERATING CONSTRUCTION EQUIPMENT

Example 3-9 Determine the probable cost per hour for owning and operating a 3/4-cu-yd diesel-engine-powered crawler-type power shovel. The following information will apply:

Engine, 160 hp
Crankcase capacity, 6 gal
Hours between oil changes, 100
Operating factor, 0.60
Fuel consumed per hr, $160 \times 0.6 \times 0.04 = 3.9$ gal
Lubricating oil consumed per hr,

$$\frac{160 \times 0.6 \times 0.006}{7.4} + \frac{6}{100} = 0.138 \text{ gal}$$

Useful life, 5 yr, with no salvage value
Hours operated per yr, 1,400
Shipping weight, 56,000 lb

Cost to owner:		
List price, f.o.b factory, including dipper	=	$119,350
Freight cost, 56,000 lb @ $2.40 per cwt	=	1,344
Taxes, $119,350 @ 5%	=	5,967
Unloading and assembling at destination	=	359
Total cost to owner	=	$127,010
Average investment, 0.60 × $127,010		76,207
Annual cost:		
Depreciation, $127,010 ÷ 5 yr	=	$25,402
Maintenance and repairs, 100% of depreciation	=	25,402
Investment, 0.20 × $76,207	=	15,241
Total annual fixed cost	=	$66,045
Hourly cost:		
Ownership and maintenance, $66,045/1,400 hr = 47.175	=	$47.18
Fuel, 3.9 gal @ $1.25 = $4.875	=	4.88
Lubricating oil, 0.138 gal @ $2.00	=	0.28
Grease, 0.5 lb @ $0.50	=	0.25
Total cost per hr, excluding labor	=	$52.59

Using the alternative method of calculating ownership costs (see page 71) results in the following:

Annual ownership costs:

$$\$127{,}010(A/P{,}20{,}5) = 127{,}010 \times 0.33438 = \$42{,}470$$

Hourly ownership costs:

$$\frac{\$42{,}470}{1{,}400} = \$30.336 = \$30.34$$

Hourly operating costs:

$$\text{Maintenance} = \frac{\$25{,}402}{1{,}400} = \$18.14$$

Other =	$ 5.41
Total hourly costs =	$53.89

Example 3-10 Determine the probable cost per hour for owning and operating a 25-cu-yd heaped capacity bottom-dump wagon with six rubber tires. Because the tires will have a different life than the wagon, they should be treated separately. The following information will apply:

Engine, 250-hp diesel
Crankcase capacity, 14 gal
Time between oil changes, 80 hr
Operating factor, 0.60
Fuel consumed per hr, $250 \times 0.6 \times 0.04 = 6.0$ gal
Lubricating oil consumed per hr,

$$\frac{250 \times 0.6 \times 0.006}{7.4} + \frac{14}{80} = 0.30 \text{ gal}$$

Other lubricants used per hr, 0.50 lb
Useful life, 5 yr, with no salvage value
Life of tires, 5,000 hr
Repairs to tires, 15% of cost of depreciation of tires

Cost to owner:	
Cost delivered, including freight and taxes	= $92,623
Less cost of tires	= 12,113
Net cost less tires	= $80,510
Average investment, 0.6 × $92,623	= $55,744
Annual cost:	
Depreciation, $80,510 ÷ 5 yr	= $16,102
Maintenance and repairs, 50% of depreciation	= 8,051
Investment, 0.20 × $55,744	= 11,149
Total annual fixed costs	= $35,302
Hourly cost:	
Cost, $35,302 ÷ 1,400	= $25.22
Tire depreciation, $12,113 ÷ 5,000 hr	= 2.42
Tire repairs, 0.15 × $2.42	= 0.36
Fuel, 6.0 gal @ $1.25	= 7.50
Lubricating oil, 0.30 gal @ $2.00	= 0.60
Grease, 0.50 lb @ $0.50	= 0.25
Total cost per hr, excluding labor	= $36.35

Again, the alternate method of calculating ownership costs can be employed in this example.

Annual ownership costs (less tires):

$$\$80{,}510(A/P{,}20{,}5) = 80{,}510 \times 0.33438 = \$26{,}291$$

Annual cost of tires $\left(\text{life} = \dfrac{5{,}000}{1{,}400} = 3.57 \text{ yr}\right)$:

$$\$12{,}133(A/P{,}20{,}3.57) = 12{,}133 \times 0.39432 = \$4{,}784$$

Hourly ownership costs:

$$\frac{26{,}291 + 4{,}784}{1{,}400} = \$22.20$$

Hourly operating costs:

$$\text{Maintenance} = \frac{8{,}051}{1{,}400} = 5.75$$

$$\text{Other} = \underline{8.71}$$

$$\text{Total cost per hour, excluding labor} = \$36.66$$

Note that in each of these two examples the alternate method of calculating ownership costs are higher than when they are calculated in the traditional manner. The alternate method is believed to be the more realistic of the two and is recommended so long as we face high interest rates.

As noted, the costs determined in the previous examples do not include any allowances for the salvage value of the equipment, if any, at the end of its useful life. It is anticipated that there will be a realizable salvage value at the end of the indicated useful life, the costs should be adjusted accordingly.

The hourly cost of owning and operating construction equipment, as illustrated in the previous examples, will vary with the conditions under which the equipment is operated, and the job planner should analyze each job to determine the probable conditions.

If a power shovel is used to excavate a soft material, the life of the dipper teeth, wire rope, and other parts which are affected by the wear and strain will be relatively long. Repair costs will be relatively low. However, if the shovel is used to excavate rock or other hard materials, the dipper teeth, wire rope, clutch linings, and certain gear parts will be subjected to greater strains and the life of each will be reduced. Repair costs will be correspondingly increased. Likewise, the consumption of fuel will be affected by digging conditions.

If trucks are operated over straight, reasonably level, smooth roads, the cost of repairs will be lower than when the same trucks are operated over poorly maintained roads, with steep hills, ruts, or deep sand. A study of statistical information showing the cost per mile for operating automotive equipment over roads having different types of surfaces will reveal surprisingly large variations in the costs. These variations will apply to trucks used for hauling materials.

ECOMOMIC LIFE OF CONSTRUCTION EQUIPMENT

The owner of construction equipment should be interested in obtaining the lowest possible cost per unit of production. In order to accomplish this objective he must follow an informed program of equipment replacement. When should equipment be replaced? If the owner replaces it too soon, he will experience an unnecessary capital loss, whereas, if he waits too long, the equipment will have passed its period of economic operation.

In order to determine the most economical time to replace equipment, accurate records of maintenance and repair costs and downtime must be kept for each machine. The owner must consider all costs related to the ownership and operation of the equipment, and the effect which continued use will have on these costs.

The costs to be considered are:

1. Depreciation and replacement
2. Investment
3. Maintenance and repairs
4. Downtime
5. Obsolescence

An analysis of the effect which hours of usage will have on each of these costs will establish the time at which a machine should be replaced. The following example will illustrate a method of conducting an analysis to determine each of these costs.

Example 3-9

Initial cost of equipment, $20,000
No. of hours used per yr, 2,000

Costs of depreciation and replacement When one considers the replacement of equipment, it is necessary to know the salvage value of the machine and the replacement cost of an equal machine. Because the average cost of construction equipment has been increasing at a rate of approximately 5 percent per year during the past 10 years, and it appears that this rate of increase will continue, it is necessary to reflect this increase in a depreciation and replacement analysis. The salvage value should be the actual amount that can be realized on a trade-in for a replacement machine.

Table 3-2 includes the necessary information for this analysis. Note that the replacement cost is increased $1,000 per year to provide for increases in the cost of equipment.

Costs of investment Table 3-3 gives the cumulative costs per hour for this equipment. The investment is assumed to be 15 percent per year on the value of the equipment at the beginning of the year (see Fig. 3-5). The owner of equipment should use an interest rate which is appropriate to his operations.

Maintenance and repair costs Table 3-4 lists the costs of maintenance and repairs for the equipment and the calculated cost per cumulative hour. Because of the large variation in the cost of maintenance and repairs, depending on the conditions under which equipment is used, it is important to keep accurate records of these costs.

Table 3-2 Depreciation and replacement costs

End of year	Replacement cost	Salvage value	Loss on replacement	Cumulative hours of use	Cumulative cost per hr
0	$20,000	$20,000	0	0	0
1	21,000	15,000	$ 6,000	2,000	$3.00
2	22,000	12,000	10,000	4,000	2.50
3	23,000	10,000	13,000	6,000	2.17
4	24,000	8,500	15,500	8,000	1.94
5	25,000	7,000	18,000	10,000	1.80
6	26,000	6,000	20,000	12,000	1.67
7	27,000	5,200	21,800	14,000	1.56
8	28,000	4,500	23,500	16,000	1.47

Table 3-3 Investment costs

Year	Investment, start of year	Depreciation	Investment, end of year	Investment cost	Cumulative investment cost	Cumulative use, hr	Cumulative cost per hr
1	$20,000	$5,000	$15,000	$3,000	$3,000	2,000	$1.50
2	15,000	3,000	12,000	2,250	5,250	4,000	1.32
3	12,000	2,000	10,000	1,800	7,050	6,000	1.18
4	10,000	1,500	8,500	1,500	8,550	8,000	1.07
5	8,500	1,500	7,000	1,275	9,825	10,000	0.98
6	7,000	1,000	6,000	1,050	10,875	12,000	0.91
7	6,000	800	5,200	900	11,775	14,000	0.84
8	5,200	700	4,500	780	12,555	16,000	0.79

Table 3-4 Maintenance and repair costs

Year	Annual cost	Cumulative cost	Cumulative use, hr	Cumulative cost per hr
1	$ 880	$ 800	2,000	$0.44
2	1,620	2,500	4,000	0.63
3	2,250	4,750	6,000	0.79
4	2,740	7,490	8,000	0.94
5	3,360	10,850	10,000	1.09
6	3,870	14,720	12,000	1.23
7	4,740	19,460	14,000	1.39
8	5,480	24,940	16,000	1.56

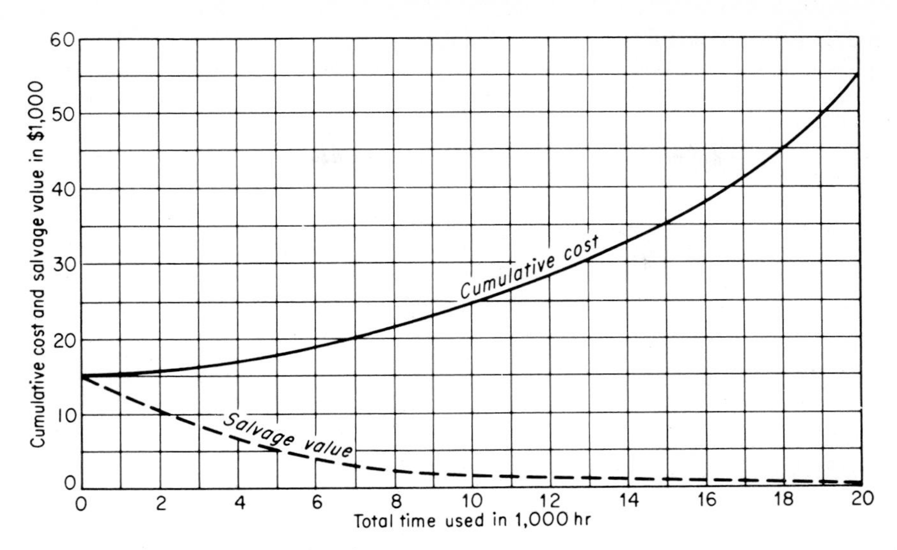

Figure 3-5 Investment cost of equipment.

Downtime costs Downtime is the time that a machine is not working because it is undergoing repairs or adjustments. Downtime tends to increase with usage. Availability is a term that indicates the portion of the time that a machine is in actual production or is available for production, expressed as a percent. For example, if a machine is down 5 percent of the time, its availability is 95 percent.

If a machine whose operating cost is $6.00 per hour has an average downtime of 5 percent, the cost per hour for this downtime will be $0.05 \times \$6.00 = \$0.30$. If the machine is used 2,000 hr per year, the annual cost will be $2{,}000 \times \$0.30 = \$600$. Table 3-5 shows the downtime and the cumulative cost per hour for this example.

Productivity is a measure of the ability of equipment to produce at its original rate. If the productivity of a machine decreases with usage, the effect of this decrease is to increase the cost of production, which is equivalent to an increase in the cost per hour for continuing to use the equipment. For example, if the cumulative cost per hour in the seventh column of Table 3-5 is $0.68 and the productivity factor is 0.90, this decrease in productivity has the effect of increasing the cumulative cost per hour to $\$0.68 \div 0.90 = \$0.76$ per hour. The costs shown in the last column should be used for equipment whose productivity decreases with usage.

Obsolescence costs It has been the history of construction equipment that continuing improvements in the productive capacities have resulted in lower production costs. These improvements, whose advantages can be gained only by the replacement of older equipment with newer equipment, decrease the desirability of continuing to use the older equipment. For example, if a new machine will reduce production costs by 10 percent, when compared with production costs for an existing machine, the existing machine will suffer a loss in value equal to 10 percent. This is defined as an obsolescence loss. Failure to take advantage of this potential reduction in production cost through the acquisition of new equipment will result in a production cost that is higher than necessary for the owner of old equipment.

During the past years productive improvements of construction equipment have averaged about 5 percent per year, and it appears that future improvements may continue at the same rate.

Table 3-6 illustrates the effect of obsolescence costs resulting from the continued use of

Table 3-5 Downtime costs

Year	Downtime, %	Cost per hr	Downtime cost, yr	Cumulative downtime cost	Cumulative hours	Cumulative cost per hr	Productivity factor	Cumulative cost per hr
1	3	$0.18	$ 360	$ 360	2,000	$0.18	1.00	$0.18
2	6	0.36	720	1,080	4,000	0.27	0.99	0.27
3	8	0.48	960	2,040	6,000	0.34	0.98	0.35
4	10	0.60	1,200	3,240	8,000	0.41	0.96	0.42
5	12	0.72	1,440	4,680	10,000	0.47	0.96	0.49
6	14	0.84	1,680	6,360	12,000	0.53	0.94	0.56
7	17	1.02	2,040	8,400	14,000	0.60	0.92	0.65
8	20	1.20	2,400	10,800	16,000	0.68	0.90	0.76

equipment that might be replaced with newer equipment which is capable of producing at a lower cost.

Summary of costs Table 3-7 lists the five separate costs that should be considered in determining the desirability of replacing the machine described in Example 3-9.

Because some, but not all, types of equipment experience obsolescence losses resulting from improvements in production, the table gives the summarized results in two sets of values. If obsolescence is not a factor in the decision to replace a machine, the most economic life is 5 years. However, if the replacement is delayed to the end of the sixth year, the additional cost is not significant.

If the machine is subject to a 5 percent obsolescence loss per year, it should be replaced at the end of the third year.

Table 3-6 Obsolescence costs per hour for the life of the equipment

Year	Obsolescence factor	Equipment cost per hr	Obsolescence		Cumulative cost	Cumulative	
			Cost per hr	Cost per yr		Use, hr	Cost per hr
1	0	$6.00	$ 0	$ 0	$ 0	2,000	$0.00
2	0.05	6.00	0.30	600	600	4,000	0.15
3	0.10	6.00	0.60	1,200	1,800	6,000	0.30
4	0.15	6.00	0.90	1,800	3,600	8,000	0.45
5	0.20	6.00	1.20	2,400	6,000	10,000	0.60
6	0.25	6.00	1.50	3,000	9,000	12,000	0.75
7	0.30	6.00	1.80	3,600	12,600	14,000	0.90
8	0.35	6.00	2.10	4,200	16,800	16,000	1.15

Table 3-7 Summary of cumulative costs per hour

Item	Year 1	2	3	4	5	6	7	8
Depreciation and replacement	$3.00	$2.50	$2.17	$1.94	$1.80	$1.67	$1.56	$1.47
Investment	1.50	1.32	1.18	1.07	0.98	0.91	0.84	0.79
Maintenance and repairs	0.44	0.63	0.79	0.94	1.09	1.23	1.39	1.56
Downtime	0.18	0.27	0.35	0.42	0.49	0.56	0.65	0.76
Subtotal	5.12	4.72	4.49	4.37	4.36	4.37	4.44	4.58
Obsolescence	0.00	0.15	0.30	0.45	0.60	0.75	0.90	1.15
Total	5.12	4.87	4.79	4.82	4.96	5.12	5.34	5.73

Table 3-8 lists the annual and cumulative losses resulting from replacing the equipment too early or too late with obsolescence losses excluded. Table 3-9 shows the same results with obsolescence losses included.

The extra cost per hour appearing in column 5 of Tables 3-8 and 3-9 is obtained by subtracting the cost in column 4 from the cost in column 3 for the given year, which latter costs are obtained from Table 3-7.

Economic life of equipment that serves other equipment If a machine works alone, the method of determining its economic life illustrated in Tables 3-2 through 3-9 will apply. An example of this condition is a dragline excavating a drainage ditch, wasting the excavated earth in a spoil bank adjacent to the ditch. However, if this dragline is used to load trucks, a delay in production will

Table 3-8 Losses resulting from improper replacement
Excluding obsolescence losses

Replaced at end of year	Cumulative hours	Cumulative cost per hr	Minimum cost per hr	Extra cost per hr	Total loss
1	2,000	$5.12	$4.36	$0.76	$1,520
2	4,000	4.72	4.36	0.36	1,440
3	6,000	4.49	4.36	0.13	780
4	8,000	4.37	4.36	0.01	80
5	10,000	4.36	4.36	0.00	0
6	12,000	4.37	4.36	0.01	120
7	14,000	4.44	4.36	0.08	1,120
8	16,000	4.58	4.36	0.22	3,520

Table 3-9 Losses resulting from improper equipment replacement

Including obsolescence losses

Replaced at end of year	Cumulative hours	Cumulative cost per hr	Minimum cost per hr	Extra cost per hr	Total loss
1	2,000	$5.12	$4.79	$0.33	$ 660
2	4,000	4.87	4.79	0.08	320
3	6,000	4.79	4.79	0	0
4	8,000	4.82	4.79	0.03	240
5	10,000	4.96	4.79	0.17	1,700
6	12,000	5.12	4.79	0.33	3,960
7	14,000	5.34	4.79	0.55	7,700
8	16,000	5.73	4.79	0.94	15,040

idle the trucks, perhaps with the cost of the trucks and drivers continuing during the delay. For this condition the total cost of the delay is chargeable to the dragline. The effect of considering this cost is to increase the unit cost of production, and to reduce the economic life of the dragline when it is used as a service unit.

One method of alleviating this increasing cost per unit of production is to replace the older machine with one having a higher availability factor, and to assign the older machine to operations where it can work alone.

Another method of alleviating the increased costs is to maintain a high availability factor for the machine by adopting a good maintenance program and replacing worn parts prior to failure at the end of a shift or over a weekend, when the machine is not in use.

If a tractor-mounted front-end loader, whose cost is $16.00 per hr, including the operator, is used to load four trucks, whose costs are $10.00 per hr each, including the drivers, the information appearing in Table 3-5 should be

Table 3-10 Downtime costs considering the cost of time lost by serviced equipment

Based on working 2,000 hr per yr

				Cumulative		
Year	Downtime, percent	Cost per hr	Downtime cost per yr	Downtime cost	Hours	Cost per hr
1	3	$ 1.68	$ 3,360	$ 3,360	2,000	$1.68
2	6	3.36	6,720	10,080	4,000	2.52
3	8	4.48	8,960	19,040	6,000	3.17
4	10	5.60	11,200	30,240	8,000	3.78
5	12	6.72	13,440	43,680	10,000	4.37
6	14	7.84	15,680	59,360	12,000	4.93
7	17	9.52	19,040	78,400	14,000	5.60
8	20	11.20	22,400	100,800	16,000	6.30

modified to reflect the higher total cost, namely $16.00 + (4 × $10.00) = $56.00 per hour. Table 3-10 shows the effect of combining the cost of the loader and the cost of the nonproductive time of the trucks resulting from the downtime of the loader. This is the proper method of determining the cost to a project when a primary unit of equipment, which serves other equipment, experiences a breakdown.

SOURCES OF CONSTRUCTION EQUIPMENT

Contractors and other users of construction equipment frequently are concerned with a decision as to whether to purchase or rent (lease) equipment. Under certain conditions it is financially advantageous to purchase, whereas under other conditions it is more economical and satisfactory to rent it. It is the purpose of this section to assist a user in determining which method is the more economical.

There are at least three methods under which a contractor may secure the use of construction equipment. He may:

1. Purchase it
2. Rent it (usually using a lease agreement)
3. Rent it with an option to purchase it at a later date

The method selected should be the one that will provide the use of the equipment at the lowest total cost, consistent with the use that the contractor will make of the equipment. Each method has both advantages and disadvantages which should be considered prior to making a decision. If cost is the only factor to be considered, an analysis of the cost under each method should give the answer. If other factors should be considered, they should be evaluated and applied to the cost as a basis on which to reach a decision. The correct decision for one contractor will not necessarily apply for another contractor. For example, a given contractor may engage in work that requires the use of wellpoint systems for most of his projects, while another may require the use of such a system only once every 2 or 3 years. It is probable that the former should purchase the equipment, while the latter should rent it. Thus, a contractor probably should purchase equipment that he will use frequently, and he should rent equipment that he will use only rarely.

The purchase of equipment, as compared with renting it, has several advantages, including the following:

1. It is more economical if the equipment is used sufficiently.
2. It is more likely to be available for use when needed.
3. Because ownership should assure better maintenance and care, purchased equipment should be kept in better mechanical condition.

Among the disadvantages of owning equipment are the following:

1. It may be more expensive than renting.
2. The purchase of equipment may require a substantial investment of money or credit that may be needed for other purposes.
3. The ownership of equipment may influence a contractor to continue using obsolete equipment after superior equipment has been introduced.
4. The ownership of equipment designed primarily for a given type of work may induce a contractor to continue doing that type of work, whereas other work requiring different types of equipment might be available at a higher profit.
5. The ownership of equipment might influence a contractor to continue using the equipment beyond its economical life, thereby increasing the cost of production unnecessarily.

After all is said and done, by far the most influential factor in deciding whether to purchase or rent a piece of equipment is its expected long-term utilization. Obviously, renters of equipment must charge enough to generate a profit, and thus their hourly charge would be higher than the comparable cost to an owner—*if the owner used the equipment extensively.* If the expected use is short-term or sporadic, then renting is usually the less costly alternative.

Once the decision is made whether to purchase or rent, the next decision to be made is whether to simply rent or rent with an option to purchase. The latter alternative will result in a higher rental cost as some of the periodic rental charges will be applicable toward the purchase price of the equipment. This is an attractive alternative if the renter of the equipment believes he may have enough use for the equipment to purchase it, but is not too sure that the utilization will be as high as predicted. Such rental agreements result in higher hourly charges than straight rental agreements.

While there are no uniformly established rental rates for construction equipment, the Associated Equipment Distributors (P.O. Box 97724, Chicago, IL 60690) publishes annually a booklet entitled *Nationally Averaged Rental Rates for Construction Equipment,* which gives representative rates found to be in effect in the United States.

PROBLEMS

Note: Assume an investment cost equal to 20% per year for the average value of the equipment.

3-1 Using the straight-line method of depreciating equipment, determine the annual cost of depreciation for a tractor whose total initial cost is $56,580 if the assumed life is 5 years, with an estimated salvage value of $6,000.

3-2 Using the double-declining-balance method, determine the cost of depreciation each year for 5 years for the tractor of Prob. 3-1.

3-3 Using the sum-of-the-years-digits method, determine the cost of depreciation each year for 5 years for the tractor of Prob. 3-1. What is the book value of the tractor at the end of 3 years?

3-4 A power shovel whose total initial cost was $86,340 was assumed to have a useful life of 5 years, with a salvage value of $8,000. It has been depreciated by the double-declining-balance method for 4 years. What is its book value?

3-5 A power shovel whose total initial cost is $72,390 has an estimated useful life of 6 years, with an estimated salvage value of $7,000. Prepare a table listing the annual costs of depreciation and the book value at the end of each of the 6 years based on the straight-line, double-declining-balance, and sum-of-the-years-digits methods of computing the annual cost of depreciation.

3-6 Prepare a graph with curves showing the book values of the power shovel of Prob. 3-5 during the 6 years of its life.

3-7 Determine the average value of a unit of equipment whose total initial cost is $36,000, with an estimated salvage value of $4,000 after 6 years.

3-8 Determine the probable cost per hour for owning and operating a power shovel for the following conditions using straight-line depreciation. (Use both methods.)

Engine, 180-hp diesel
Crankcase capacity, 5 gal
Time between oil changes, 100 hr
Operating factor, 0.50
Useful life, 5 yr
Hours used per year, 1,400
Total initial cost, $92,480
Estimated salvage value after 5 yr, $8,000
Annual cost of maintenance and repairs equals 80% of annual depreciation
Cost of fuel, $1.25 per gal
Cost of lubricating oil, $3.00 per gal
Cost of other oils and grease, $0.50 per hr

3-9 Determine the probable cost per hour for owning and operating a wheel-type tractor-pulled scraper for the following conditions using straight-line depreciation. (Use both methods.)

Engine, 240-hp diesel
Crankcase capacity, 6 gal
Time between oil changes, 80 hr
Operating factor, 0.60
Useful life, 5 yr
Hours used per year, 1,400
Annual cost of maintenance and repairs, 75% of annual depreciation
Life of tires, 5,000 hr
Repairs to tires, 15% of the depreciation of tires
Total initial cost, $76,620
Cost of tires, $18,240
Estimated salvage value, $8,000
Cost of fuel, $1.25 per gal
Cost of lubricating oil, $3.00 per gal
Cost of other oils and grease, $0.50 per hr

3-10 Determine the probable cost per hour for owning and operating the tractor-pulled scraper of Prob. 3-9 if it will be used only 1,000 hr per year, with all other conditions remaining the same.

CHAPTER

FOUR

ENGINEERING FUNDAMENTALS OF MOVING EARTH

GENERAL INFORMATION

In this chapter many problems related to excavating, hauling, and placing earth will be discussed. With the constantly growing volume of earthwork for dams, levees, highways, airports, and other projects the need for selecting the most suitable construction equipment is becoming increasingly important. Persons in the construction industry, including contractors and engineers, should understand the effects which the selection of equipment and methods has on the cost of handling earth. It is hoped that the analyses of problems related to earthwork will assist in demonstrating how effective the application of engineering can be in determining the cost of earthwork.

ROLLING RESISTANCE

Rolling resistance is a resistance which is encountered by a vehicle in moving over a road or surface. This resistance varies considerably with the type and condition of the surface over which a vehicle moves. Soft earth offers a higher resistance than hard-surfaced roads such as concrete pavement. For vehicles which move on rubber tires the rolling resistance varies with the size, pressure, and the tread design of the tires. For equipment which moves on crawler tracks, such as tractors, the resistance varies primarily with the type and condition of the road surface. If a truck is driven off a hard-surfaced highway into a field of soft earth, the resistance to moving is increased materially, as all drivers know. If a loaded wheelbarrow has a well-inflated pneumatic tire, it is much more easily pushed along a concrete sidewalk than when the tire is semideflated or soft. The difference is due to changes in the rolling resistance. A narrow-tread high-pressure tire gives lower rolling resistance than a broad-tread low-pressure tire on a hard-surfaced road. This is the result of the smaller area of contact between the tire and the road surface. However, if the road surface is soft and the tire tends to sink into the earth, a broad-tread low-pressure tire will offer a lower rolling resistance than a narrow-tread high-pressure tire. The reason for this condition is that the narrow tire sinks into the earth more deeply than the broad tire and thus is always having to climb out of a deeper hole, which is equivalent to climbing a steeper grade. As explained later in this book, the type and size tires selected for earth-hauling equipment should be determined after the condition of the haul road is known.

The rolling resistance of an earth-haul road probably will not remain constant under varying climatic conditions or for varying types of soil which exist along the road. If the earth is stable, highly compacted, and well maintained with a grader, and if the moisture content is kept near the optimum, it is possible to provide a surface with a rolling resistance about as low as for concrete and asphalt. It is possible to add moisture, but following an extended period of rain it may be difficult to remove the excess moisture, and the haul road will become muddy, with an increase in rolling resistance. Providing good surface drainage will speed the removal of the water and should permit the road to be reconditioned more quickly. For a major earth project it is good economy to provide a patrol grader, sprinkler trucks, and probably rollers to keep the haul road in good condition. As illustrated under the subject of trucks, the maintenance of low rolling resistance is one of the best financial investments an earth-moving contractor can make.

Although it is impossible to give completely accurate values for the rolling resistances for all types of haul roads and wheels, the values given in Table 4-1 are reasonably accurate and may be used for estimating purposes. Rolling resistance is expressed in pounds of tractive pull required to move each gross ton over a level surface of the specified type or condition. For example, if a loaded truck having a gross weight equal to 20 tons is moving over a level road whose rolling resistance is 100 lb per ton, the tractive effort required to keep the truck moving at a uniform speed will be

$$20 \text{ tons} \times 100 \text{ lb per ton} = 2{,}000 \text{ lb}$$

If it is desirable to determine the rolling resistance of a haul road, this can be done by towing a truck or other vehicle whose gross weight is known along a level section of the haul road at a uniform speed. The tow cable should be equipped with a dynamometer or some other device which will permit the average tension in the cable to be determined. This tension is the total rolling resistance of the gross weight of the truck. The rolling resistance in pounds per

Table 4-1 Representative rolling resistances for various types of wheels and surfaces

In pounds per 2,00 lb-ton or kilograms per metric ton of gross load

Type of surface	Steel tires, plain bearings	Crawler-type track and wheel	Rubber tires, anti-friction bearings High pressure	Low pressure
Smooth concrete	40	55	35	45
	(20)	(27)	(18)	(23)
Good asphalt	50–70	60–70	40–65	50–60
	(25–35)	(30–35)	(20–33)	(25–30)
Earth, compacted	60–100	60–80	40–70	50–70
and maintained	(30–50)	(30–40)	(20–35)	(25–35)
Earth, poorly	100–150	80–110	100–140	70–100
maintained	(50–75)	(40–55)	(50–70)	(35–50)
Earth, rutted,	200–250	140–180	180–220	150–200
muddy, no	(100–125)	(70–90)	(90–110)	(75–100)
maintenance	280–320	160–200	260–290	220–260
Loose sand and gravel	(140–160)	(80–100)	(130–145)	(110–130)
Earth, very muddy,	350–400	200–240	300–400	280–340
rutted, soft	(175–200)	(100–120)	(150–200)	(140–170)

gross ton will be

$$R = \frac{P}{W} \tag{4-1}$$

where R = rolling resistance, lb per ton
P = total tension in tow cable, lb
W = gross weight of truck, tons

If it is necessary to tow the loaded truck up or down a sloping haul road, an appropriate correction for the effect of the slope may be applied to the tension in the tow cable, as explained in the following section. In order to apply a correction it is necessary to know the grade of the haul road over which the test is being conducted.

THE EFFECT OF GRADE ON REQUIRED TRACTIVE EFFORT

When a vehicle moves up a sloping road, the total tractive effort required to keep the vehicle moving is increased approximately in proportion to the slope of the road. If a vehicle moves down a sloping road, the total tractive effort required to keep the vehicle moving is reduced in proportion to the slope of the road. The most common method of expressing a slope is by percent. A 1 percent slope is one where the surface rises or drops 1 ft vertically in a horizontal distance of 100 ft. If the slope is 5 percent, the surface rises or drops

5 ft per 100 ft of horizontal distance. If the surface rises, the slope is defined as plus, while if it drops, the slope is defined as minus. All automobile drivers know that a plus slope retards, while a minus slope aids, an automobile traveling along a highway. The same forces apply to construction equipment moving over a road. This is a physical property which is not affected by the type of equipment or the condition or type of the road.

The effect of grade is to increase, for a plus slope, or decrease, for a minus slope, the required tractive effort by 20 lb per gross ton of weight for each 1 percent of grade. While this amount is not strictly correct for all slopes, it is sufficiently accurate for most construction projects.

Figure 4-1 illustrates the method of determining the effect of grade on tractive effort. The line *AB* is horizontal. The slope of *AC* is 1 percent. *DE* is perpendicular to *AB*. *DF* is perpendicular to *AC*. *EF* is parallel to *AC*. Triangle *DEF* is similar to triangle *ABC*. For practical purposes the length of *AC* is 100 ft. *W* is a 1-ton weight, represented by the vector *DE*. *P* is the component of *W* parallel to *AC*.

From the similarity of triangles,

$$\frac{EF}{ED} = \frac{P}{W} = \frac{BC}{AC} \quad \text{or} \quad P = W\frac{BC}{AC} = 2{,}000 \text{ lb} \times \frac{1}{100} = 20 \text{ lb}$$

If *BC* is increased to 2 ft,

$$P = 2{,}000 \text{ lb} \times \frac{2}{100} = 40 \text{ lb}$$

For any given slope the approximate value of *P* in pounds per ton is

$$P = 2{,}000 \text{ lb} \times \frac{\% \text{ slope}}{100} = \frac{20 \text{ lb} \times \% \text{ slope}}{\text{ton}} \tag{4-2}$$

Example 4-1 Consider the effect of grade on the total tractive effort of a truck whose gross weight is 20 tons. The truck will be driven up a road whose slope is 5 percent. The additional tractive effort resulting from the slope is

$$P = 20 \text{ tons} \times 20 \text{ lb per ton} \times 5\% = 2{,}000 \text{ lb}$$

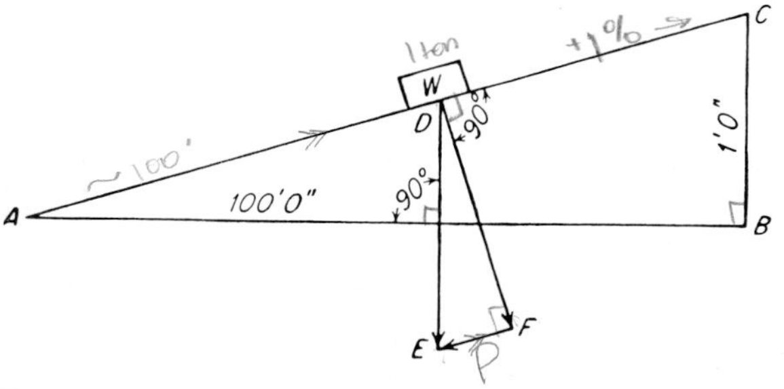

Figure 4-1 The effect of grade on the performance of a tractor or truck.

Table 4-2 The effect of grade on the tractive effort of vehicles
In pounds per ton or kilograms per metric ton of gross weight

Slope, %	Lb per ton	Kg per m ton	Slope, %	Lb per ton	Kg per m ton
1	20.0	10.0	12	238.4	119.2
2	40.0	20.0	13	257.8	128.9
3	60.0	30.0	14	277.4	138.7
4	80.0	40.0	15	296.6	148.3
5	100.0	50.0	20	392.3	196.1
6	119.8	59.9	25	485.2	242.6
7	139.8	69.9	30	574.7	287.3
8	159.2	79.6	35	660.6	330.3
9	179.2	89.6	40	742.8	371.4
10	199.0	99.5	45	820.8	410.4
11	218.0	109.0	50	894.4	447.2

Thus, the truck engine must continually deliver to the driving wheels 2,000 lb of rimpull to overcome the effect of the slope. If the truck is moving down the same slope, the effect of the grade will be to help the engine and truck, which is equivalent to adding 2,000 lb to the rimpull of the track.

If a tractor is towing a load, the combined gross weights of the tractor and its towed load should be used in determining the effect of the grade.

Table 4-2 gives values for the effect of slope, expressed in pounds per gross ton or kilograms per metric ton (m ton) of weight of the vehicle.

THE EFFECT OF GRADE IN LOCATING A BORROW PIT

Sometimes engineers and contractors do not give sufficient consideration to the grade or slope of the haul road in locating borrow pits. It is desirable, when possible, to locate a borrow pit at a higher elevation than the fill, in order that the slope down the haul road may help the loaded trucks or other hauling equipment by permitting them to carry larger loads or to travel at higher speeds. Since the vehicles will be empty when returning up the haul road from the fill to the borrow pit, the effect of the grade will be considerably less. This item is discussed in detail in Chap. 9.

COEFFICIENT OF TRACTION

The total energy of an engine in any unit of equipment designed primarily for pulling a load can be converted into tractive effort only if sufficient traction can be developed between the driving wheels or tracks and the haul surface. If there is not sufficient traction, the full power of the engine cannot be used. The wheels or tracks will slip on the surface. Thus, the coefficient of traction

Table 4-3 Coefficients of traction for various road surfaces

Surface	Rubber tires	Crawler tracks
Dry, rough concrete	0.80–1.00	0.45
Dry clay loam	0.50–0.70	0.90
Wet clay loam	0.40–0.50	0.70
Wet sand and gravel	0.30–0.40	0.35
Loose, dry sand	0.20–0.30	0.30
Dry snow	0.20	0.15–0.35
Ice	0.10	0.10–0.25

between rubber tires or crawler tracks and different haul surfaces is important to the operators of hauling units.

The coefficient of traction may be defined as the factor by which the total load on a driving tire or track should be multiplied in order to determine the maximum possible tractive force between the tire or track and the surface just before slipping will occur. For example, the driving tires of a truck rest on a level haul road of dry clay. The total pressure between the tires and the road surface is 8,000 lb. In testing the tires for slippage by applying a driving force to the wheels, it is found that slippage will occur when the tractive force between the tires and the surface is 4,800 lb. The coefficient of traction is $4{,}800 \div 8{,}000 = 0.60$.

The coefficient of traction between rubber tires and road surfaces will vary with the type of tread on the tires and with the road surface. For crawler tracks it will vary with the design of the grouser and the road surface. These variations are such that exact values cannot be given. Table 4-3 gives approximate values for the coefficient of traction between rubber tires or crawler tracks and road surfaces which are sufficiently accurate for most estimating purposes.

Example 4-2 Assume that a rubber-tired tractor has a total weight of 18,000 lb on the two driving tires. The maximum rimpull in low gear is 9,000 lb. If the tractor is operating in wet sand, with a coefficient of traction of 0.30, the maximum possible rimpull prior to slippage of the tires will be $0.30 \times 18{,}000$ lb $= 5{,}400$ lb. Regardless of the power of the engine, not more than 5,400 lb of tractive effort may be used because of the slippage of the wheels. If the same tractor is operating on dry clay, with a coefficient of traction of 0.60, the maximum possible rimpull prior to slippage of the tires will be $0.60 \times 18{,}000$ lb $= 10{,}800$ lb. For this surface the engine will not be able to cause the tires to slip. Thus, the full power of the engine may be used.

THE EFFECT OF ALTITUDE ON THE PERFORMANCE OF INTERNAL-COMBUSTION ENGINES

An internal-combustion engine operates by combining oxygen from the air with the fuel and then burning the mixture to convert latent energy into mechanical energy. The power of an engine is a measure of the rate at which it can produce energy from fuel. For each charge of fuel and air into a cylinder there must be a correct ratio between the quantity of fuel and air if the maximum efficiency and power are to be obtained from the engine. The ratio between the quantities should be that which will provide just enough oxygen to supply the requirements of the fuel for complete combustion. If the density of the air is reduced because of altitude, the quantity of oxygen in a given volume of air will be less than for the same volume of air at sea level. As each cylinder of an engine draws in a given volume of air prior to the firing stroke, there will be less oxygen in the cylinder if the density of the air is reduced. Since the ratio of the oxygen and fuel should remain constant, it will be necessary to reduce the quantity of fuel supplied to an engine at high altitudes. This is usually done by adjusting the carburetor. The effect on the engine is to reduce the power. A human being experiences the same effect when he engages in physical work at a high altitude. Although he breathes the same volume of air, he may not get enough oxygen to supply his requirements.

If the density of the air decreased uniformly with the altitude, it would be possible to express the loss in power of an engine due to altitude by means of a simple formula with a high degree of accuracy. Actually this is not true.

For most practical purposes it is sufficiently accurate to assume that for four-cycle gasoline and diesel engines the loss in power due to altitude will be equal to approximately 3 percent of the sea-level horsepower for each 1,000 ft above the first 1,000 ft. Thus, for a four-cycle engine with 100 belt hp at sea level, the power at 10,000 ft above sea level would be determined as follows:

$$\begin{array}{lr}
\text{Sea-level power} & = 100 \text{ hp} \\
\text{Loss due to altitude, } \dfrac{0.03 \times 100 \times (10{,}000 - 1{,}000)}{1{,}000} & = \quad 27 \text{ hp} \\
\text{Effective power} & = \overline{\quad 73 \text{ hp}}
\end{array}$$

For the two-cycle engine, which is becoming increasingly more popular in the diesel field, the loss in power due to altitude is approximately 1 percent of the sea-level horsepower for each 1,000 ft above the first 1,000 ft. This type engine has its air supplied, under a slight pressure, by a blower, whereas the four-cycle engine depends on the suction of the pistons for the supply of air. If the engine described in the previous paragraph is a two-cycle, the power at 10,000 ft above sea level would be determined as follows:

$$\begin{array}{lr}
\text{Sea-level power} & = 100 \text{ hp} \\
\text{Loss due to altitude, } \dfrac{0.01 \times 100 \times (10{,}000 - 1{,}000)}{1{,}000} & = \quad 9 \text{ hp} \\
\text{Effective power} & = \overline{\quad 91 \text{ hp}}
\end{array}$$

The two previous problems indicate that, other factors being equal, at high altitude a two-cycle engine will give better performance than a four-cycle engine.

The effect of the loss in power due to altitude may be eliminated by the installation of a supercharger. This is a mechanical unit which will increase the pressure of the air supplied to the engine, thus permitting sea-level performance at any altitude. If equipment is to be used at high altitudes for long periods of time, the increased performance probably will more than pay for the installed cost of a supercharger.

A contractor who has established production rates for his equipment at or near sea level will make a serious mistake if he uses those production rates in bidding a job to be constructed at a high altitude. He must install superchargers or apply a correction factor, which is more fully explained under the subjects of trucks and tractors.

THE EFFECT OF TEMPERATURE OF THE PERFORMANCE OF INTERNAL-COMBUSTION ENGINES

Many persons who have driven an automobile through a desert during a hot afternoon have noticed that the performance of the automobile seems sluggish. If driving was continued into the night after the temperature had decreased appreciably, the performance of the engine seemed to improve noticeably. This experience was not imaginary. An internal-combustion engine will develop a higher horsepower at a low air temperature than at a high temperature. The effect of temperature on the performance of an internal-combustion engine has been determined from laboratory tests. The next section discusses the combined effect of pressure and temperature on the performance of internal-combustion engines.

THE COMBINED EFFECT OF PRESSURE AND TEMPERATURE ON THE PERFORMANCE OF INTERNAL-COMBUSTION ENGINES

When an internal-combustion engine is tested to determine its power, it is necessary to conduct the tests under standard conditions in order that the results may be significant. Standard conditions mean a temperature of 60°F and average sea-level barometric pressure, equivalent to 29.92 in. of mercury (in. Hg). As the power of the engine usually is determined with a brake or a dynamometer, the result is expressed as the brake horsepower (bhp) of the engine or as flywheel horsepower (fwhp).

If a test must be conducted under other than standard conditions, the horsepower for a four-cycle engine may be determined by using the equation

$$H_c = H_o \frac{P_s}{P_o} \sqrt{\frac{T_o}{T_s}} \tag{4-3}$$

where H_c = corrected bhp for standard conditions
H_o = observed bhp, as determined from tests

P_s = standard barometric pressure, 29.92 in. Hg
P_o = observed barometric pressure, in. Hg, at time of test
T_o = absolute temperature, °F, equal to 460 + observed temperature
T_s = absolute temperature for standard conditions, equal to 460 + 60 = 520°F

Example 4-3 A gasoline engine was tested under the given conditions and was found to develop the indicated horsepower. It is desired to convert the results to bhp for standard conditions.

Observed hp, 86.43
Observed pressure, 29.52 in. Hg
Observed temperature, 42°F

Substituting these values in Eq. (4-3), we get

$$H_c = 86.43 \times \frac{29.92}{29.52}\sqrt{\frac{460+42}{520}} = 86.07 \text{ hp}$$

Thus, this engine should develop 86.07 bhp if tested under standard conditions.

Equation (4-3) may be used to determine the probable effective horsepower of a four-cycle engine at any temperature and altitude. From Table 4-4 determine the probable barometric pressure for the given altitude. Estimate the probable temperature. Apply this information to Eq. (4-3), and solve for the effective horsepower.

Example 4-4 A tractor is operated by a four-cycle diesel engine. When tested under standard conditions, the engine developed 130 fwhp. What is the probable horsepower at an altitude of 3,660 ft, where the average daily temperature is 72°F?

The information for use in Eq. (4-2) will be as follows:

Table 4-4 Average barometric pressures for various altitudes above sea level
In inches of mercury

Altitude above sea level, ft	Barometric pressure in. Hg
0	29.92
1,000	28.86
2,000	27.82
3,000	26.80
4,000	25.82
5,000	24.87
6,000	23.95
7,000	23.07
8,000	22.21
9,000	21.36
10,000	20.55

$H_c = 130$
$P_s = 29.92$ in.
$P_o = 26.14$ in. (from Table 4-4)
$T_s = 520°F$
$T_o = 460 + 72 = 532°F$

Find H_o.

Rewriting Eq. (4-3) and substituting the given information, we get

$$H_o = H_c \frac{P_o}{P_s} \sqrt{\frac{T_s}{T_o}}$$

$$= 130 \times \frac{26.14}{29.92} \sqrt{\frac{520}{532}} = 112.7 \text{ hp}$$

Thus the probable horsepower of the engine will be reduced to 112.7 as a result of the increased altitude and temperature.

Table 4-5 gives factors by which the horsepower of a four-cycle engine, as determined under standard conditions, may be multiplied to obtain the probable horsepower for various altitudes and temperatures. Owing to variations in the barometric pressure at any altitude as the result of changes in climatic conditions, the factors may vary slightly with climatic conditions.

The two-cycle diesel engine operates under different conditions from those which apply to a four-cycle engine. Therefore, the correction factors given in Table 4-5 will not apply to two-cycle engines. If similar information is desired, it should be requested from the manufacturer.

Table 4-5 Correction factors for determining the effective horsepower of four-cycle engines

For various altitudes and temperatures

Altitude above sea level, ft	Temperatures, °F								
	110	90	70	60	50	40	20	0	−20
0	0.954	0.971	0.991	1.000	1.008	1.018	1.039	1.062	1.085
1,000	0.920	0.937	0.955	0.964	0.974	0.984	1.003	1.025	1.048
2,000	0.887	0.904	0.921	0.930	0.938	0.948	0.968	0.988	1.010
3,000	0.855	0.872	0.888	0.896	0.905	0.914	0.933	0.952	0.974
4,000	0.825	0.840	0.856	0.865	0.873	0.882	0.899	0.918	0.938
5,000	0.795	0.809	0.825	0.833	0.842	0.849	0.867	0.885	0.904
6,000	0.767	0.781	0.795	0.803	0.811	0.820	0.836	0.853	0.872
7,000	0.738	0.752	0.767	0.775	0.782	0.790	0.806	0.823	0.840
8,000	0.712	0.725	0.739	0.746	0.754	0.762	0.776	0.793	0.811
9,000	0.686	0.699	0.713	0.720	0.727	0.734	0.748	0.764	0.782
10,000	0.682	0.675	0.687	0.693	0.707	0.707	0.722	0.737	0.753

DRAWBAR PULL

The available pull which a crawler tractor can exert on a load that is being towed is referred to as the drawbar pull of the tractor. The pull is expressed in pounds. From the total pulling effort of an engine there must be deducted the pull required to move the tractor over a level haul road before the drawbar pull can be determined. If a crawler tractor tows a load up a slope, its drawbar pull will be reduced by 20 lb for each ton of weight of the tractor for each 1 percent slope.

The performance of crawler tractors, as reported in the specifications supplied by the manufacturer, is usually based on the Nebraska tests. In testing a tractor to determine its maximum drawbar pull at each of the available speeds, the haul road is calculated to have a rolling resistance of 110 lb per ton. If a tractor is used on a haul road whose rolling resistance is higher or lower than 110 lb per ton, the drawbar pull will be reduced or increased, respectively, by an amount equal to the weight of the tractor in tons multiplied by the variation of the haul road from 110 lb per ton.

Example 4-5 A tractor whose weight is 15 tons has a drawbar pull of 5,684 lb in sixth gear when operated on a level road having a rolling resistance of 110 lb per ton. If the tractor is operated on a level road having a rolling resistance of 180 lb per ton, the drawbar pull will be reduced by 15 tons × (180 − 110) = 1,050 lb. Thus, the effective drawbar pull will be 5,684 − 1,050 = 4,634 lb.

The drawbar pull of a crawler tractor will vary indirectly with the speed of each gear. It is highest in the first gear and lowest in the top gear. The specifications supplied by the manufacturer should give the maximum speed and drawbar pull for each of the several gears. The following is an example:

Gear	Speed, mph	Drawbar pull, lb
1st	1.72	28,019
2d	2.18	22,699
3d	2.76	17,265
4th	3.50	13,769
5th	4.36	10,074
6th	7.00	5,579

RIMPULL

Rimpull is a term which is used to designate the tractive force between the rubber tires of driving wheels and the surface on which they travel. If the coefficient of traction is high enough to eliminate tire slippage, the maximum rimpull is a function of the power of the engine and the gear ratios between the engine and the driving wheels. If the driving wheels slip on the haul surface, the maximum effective rimpull will be equal to the total pressure between the tires and the surface multiplied by the coefficient of traction. Rimpull is expressed in pounds.

If the rimpull of a vehicle is not known, it may be determined from the equation

$$\text{Rimpull} = \frac{375 \times \text{hp} \times \text{efficiency}}{\text{speed, mph}} \text{ lb} \tag{4-4}$$

The efficiency of most tractors and trucks will range from 80 to 85 percent. For a rubber-tired tractor with a 140-hp engine and a maximum speed of 3.25 mph in first gear, the rimpull will be

$$\text{Rimpull} = \frac{375 \times 140 \times 0.85}{3.25} = 13{,}730 \text{ lb}$$

The maximum rimpull in all gear ranges for this tractor will be as follows:

Gear	Speed, mph	Rimpull, lb
1st	3.25	13,730
2d	7.10	6,285
3d	12.48	3,576
4th	21.54	2,072
5th	33.86	1,319

In computing the pull which a tractor can exert on a towed load, it is necessary to deduct from the rimpull of the tractor the tractive force required to overcome the rolling resistance plus any grade resistance for the tractor. It will be noted that the rubber-tired tractor differs from the crawler tractor in this respect. For example, if a tractor whose maximum rimpull in the first gear is 13,730 lb weighs 12.4 tons and is operated up a haul road with a slope of 2 percent and a rolling resistance of 100 lb per ton, the pull available for towing a load will be determined as follows:

Max rimpull	= 13,730 lb
Pull required to overcome grade, $12.4 \times 20 \times 2 = 496$ lb	
Pull required to overcome rolling resistance, $12.4 \times 100 = 1{,}240$ lb	
Total pull to be deducted	= 1,736 lb
Pull available for towing a load	= 11,994 lb

PROBLEMS

4-1 A four-wheel tractor whose operating weight is 46,284 lb is pulled up a road whose slope is +4 percent at a uniform speed. If the average tension in the towing cable is 4,680 lb, what is the rolling resistance of the road?

4-2 Consider a wheel-type tractor-pulled scraper whose gross weight is 94,170 lb, including the tractor, scraper, and its load. What is the equivalent gain in horsepower resulting from operating this vehicle down a 4 percent slope instead of up the same slope at a speed of 12 mph?

(*Note*: 1 hp equals 33,000 ft-lb of work per minute.)

4-3 A wheel-type tractor-pulled scraper having a combined weight of 138,000 lb is push-loaded down a 6 percent slope by a crawler tractor whose weight is 56,240 lb. What is the equivalent gain in loading force for the tractor and scraper resulting from loading the scraper downslope instead of upslope?

4-4 A tractor has a 300-hp engine under standard conditions. What is the power of the engine when it is operating at an altitude 8,000 ft above sea level and at a temperature of 85°F?

4-5 A four-cycle gasoline engine was tested under the given conditions and was found to develop the indicated horsepower. Determine the horsepower for standard conditions.

Observed hp, 112.56
Observed temperature, 70°F
Observed atmospheric pressure, 22.14 in. Hg

4-6 A wheel-type tractor with a 210-hp engine has a maximum speed of 4.65 mph in first gear. Determine the maximum rimpull of this tractor in each of the indicated gears if the efficiency is 80 percent.

Gear	Speed, mph
1st	4.65
2d	7.60
3d	11.50
4th	17.40
5th	26.80

4-7 If the tractor of Prob. 4-6 weighs 21.4 tons and is operated over a haul road whose slope is +3 percent with rolling resistance of 80 lb per ton, determine the maximum external pull by the tractor in each of the five gears.

4-8 If the tractor of Probs. 4-6 and 4-7 is operated down a 4 percent slope whose rolling resistance is 80 lb per ton, determine the maximum external pull by the tractor in each of the five gears.

CHAPTER

FIVE

SOIL STABILIZATION AND COMPACTION

INTRODUCTION

Soils are used extensively in and with many types of construction; they are used to support structures, to support pavements for highways and airports, and as dams and levees to resist the passage of water. Some soils may be suitable for use in their natural state, while others must be excavated, processed, and compacted in order to serve their purposes.

A knowledge of the properties, characteristics, and behavior of soils is highly important to those persons who are associated with the design or construction of projects involving the use of soils. A great deal of useful knowledge related to the properties and characteristics of soils has been developed by Mr. R. R. Proctor since he initiated a scientific study to determine the density-moisture relationship of soils in 1933. His original methods or modifications thereof are now used in specifying the methods of processing soils used for construction purposes.

GLOSSARY OF TERMS

The following glossary is used to define the terms that are used with soils.

AASHTO American Association of State Highway and Transportation Officials.

Aggregate, coarse Crushed rock or gravel.

Aggregate, fine Sand or fine crushed stone used for filling voids in coarse aggregate.

Backfill Material used in refilling a cut or other excavation.

Bank A mass of soil rising above an average level, or any soil which is dug from its original position.

Bank measure A measure of the volume of earth in its natural position before it is excavated.

Base The layer of material in a roadway or airport runway section on which the pavement is placed.

Binder Fine aggregate or other materials which fill voids and hold coarse aggregate together.

Borrow pit An excavation from which fill material is excavated.

Clay A soil composed of particles less than about 5 μm in size.

Cohesion The quality of some soil particles to be attracted to like particles, manifested in a tendency to stick together, as in clay.

Cohesive materials A soil having properties of cohesion.

Compacted volume A measurement of the volume of a soil after it has been subjected to compaction.

Core The impervious portion of an embankment, such as a dam.

Density The ratio of the weight of a material to its volume.

Embankment A fill whose top is higher than the adjoining natural ground.

Fines Soil or crushed stone whose particles are small in size.

Grain-size curve A graph of the analysis of a soil showing the percentage of sizes by weight.

Granular material A soil, such as sand, whose particle sizes are such that they do not stick together.

Impervious A material that resists the flow of water through it.

In situ Soil in its original or undisturbed position.

Lift A layer of soil placed on top of soil previously placed in an embankment.

Liquid limit The water content, expressed as a percent of the weight of water to the dry weight of the soil, at which the soil passes from a plastic to a liquid state.

Optimum moisture content The percent of moisture, by weight, at which the greatest density of a soil can be obtained by compaction.

Pass A working trip or passage of an excavating, grading, or compaction machine.

Plasticity index The numerical difference between a soil's liquid limit and its plastic limit.

Plastic limit The lowest water content, expressed as a percent of the ratio of the weight of moisture to the dry weight of soil, at which the soil remains in a plastic state.

Proctor, or Proctor test A method developed by R. R. Proctor for determining the moisture-density relationship in soils subjected to compaction.

Proctor, modified A moisture-density test involving a higher compactive effort than the Proctor test.

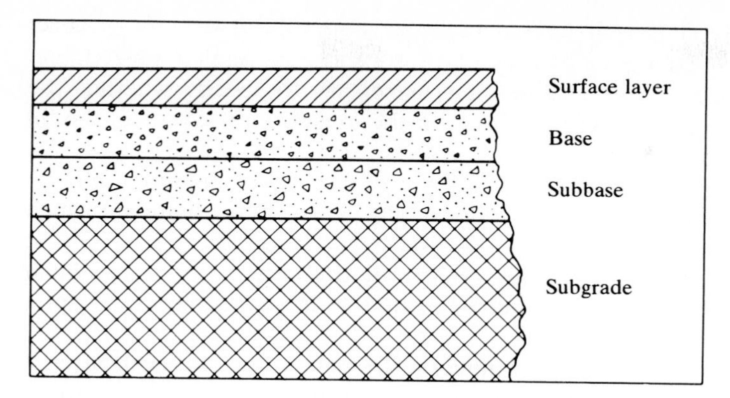

Figure 5-1 Layers for constructing a high-quality pavement.

Rock The hard, mineral matter of the earth's crust, occurring in masses and broken up in sizes generally requiring blasting before excavating in the field.

Silt A fine-grained soil composed of particles between about 0.08 and 0.005 mm.

Soil The loose surface material of the earth's crust, created naturally from the disintegration of rocks or decay of vegetation, that can be excavated easily using power equipment in the field.

Stabilize To make the soil more firm, increase its strength and stiffness, and decrease its sensitivity to volume changes with changes in moisture content.

Subbase The layer of selected material placed to furnish strength to the base of a road. In areas where the construction goes through marshy, swampy, unstable land it is often necessary to excavate the natural materials in the area of the roadway and replace them with more stable materials. The material used to replace the unstable natural soils is generally called subbase material, and when compacted it is known as the subbase.

Subgrade The surface produced by grading native earth, or cheap imported materials which serve as a base for a more expensive paving.

Surface layer The top layer of a road, street, parking lot, runway, etc., that covers and protects the sublayers from the action of traffic and the weather. If the layer has structural strength properties, it is often referred to as a pavement layer.

PROPERTIES OF SOILS

Prior to discussing earth handling or analyzing problems involving earthwork it is desirable to become more familiar with some of the physical properties of earth. These properties have a direct effect on the ease or difficulty of handling earth, the selection of equipment, and the production rates of the equipment.

Swell and shrinkage It is well known that the volume and density of earth undergo considerable changes when the earth is excavated, hauled, placed, and compacted. Because of these changes it is necessary to specify whether the volume is measured in its original position, in the loose condition, or in the fill after compaction.

The bank-measure volume is the volume of the earth measured in the borrow pit, trench, canal, or cut prior to loosening. This is the volume on which payment usually is based.

The loose-measure volume is the volume of the earth after it has been removed from its natural position and deposited in trucks, scrapers, or spoil piles.

The compacted volume, or fill volume, is the volume of the earth after it has been placed in a fill, such as a dam or road, and compacted. For projects requiring compacted earth fill the volume in the fill may be used as the basis of payment.

The volume of earth should be expressed in cubic yards, regardless of whether it is bank-measure, loose-measure, or compacted.

When the volume of earth increases because of loosening, this increase is defined as swell. It is expressed as a percent of the original undisturbed volume. Thus, if the earth removed from a hole having a volume of 1 cu yd is found to have a loose volume of 1.25 cu yd, the gain in volume is 0.25 cu yd, or 25 percent. This particular earth is then said to have a swell of 25 percent. The values of swell vary considerably for different classes of earth, as indicated in Table 5-1.

When earth is placed in a fill and compacted under modern construction methods, it will often have a smaller volume than in its original condition. This reduction in volume is the result of an increase in the density and is illustrated by the difficulty frequently encountered in driving wood stakes into a fill after the earth has been thoroughly compacted by sheep's-foot tamping rollers, pneumatic tires, or other compacting equipment. This reduction in volume from the bank-measure volume is defined as shrinkage. It is expressed as a percent of the original undisturbed volume. Thus if the earth removed from a hole having a volume of 1 cu yd is found to have a compacted volume of 0.9 cu yd, the loss is 0.1 cu yd, or 10 percent. For this condition the earth is said to have a shrinkage of 10 percent. For any given class of earth the percent of shrinkage will vary with the extent and degree of compaction and the amount of moisture present during compaction.

Table 5-1 gives representative values for swell for different classes of earth. These values will vary with the extent of loosening and compaction. If more accurate values are desired for a specific project, tests should be made on several samples of the earth taken from different depths or from different locations within the proposed cut. The test may be made by weighing a given volume of undisturbed, loose, and compacted earth. A container having the same volume should be used in determining the weight for each of the three conditions.

Table 5-1 Representative properties of earth and rock*

Material	Weight, lb per cu yd (kg per m³)		Percent swell	Swell factor*
	Bank	Loose		
Clay, dry	2,700	2,000	35	0.74
	(1,600)	(1,185)	35	0.74
Clay, wet	3,000	2,200	35	0.74
	(1,780)	(1,305)	35	0.74
Earth, dry	2,800	2,240	25	0.80
	(1,660)	(1,325)	25	0.80
Earth, wet	3,200	2,580	25	0.80
	(1,895)	(1,528)	25	0.80
Earth and gravel	3,200	2,600	20	0.83
	(1,895)	(1,575)	20	0.83
Gravel, dry	2,800	2,490	12	0.89
	(1,660)	(1,475)	12	0.89
Gravel, wet	3,400	2,980	14	0.88
	(2,020)	(1,765)	14	0.88
Limestone	4,400	2,750	60	0.63
	(2,610)	(1,630)	60	0.63
Rock, well blasted	4,200	2,640	60	0.63
	(2,490)	(1,565)	60	0.63
Sand, dry	2,600	2,260	15	0.87
	(1,542)	(1,340)	15	0.87
Sand, wet	2,700	2,360	15	0.87
	(1,600)	(1,400)	15	0.87
Shale	3,500	2,480	40	0.71
	(2,075)	(1,470)	40	0.71

* The swell factor is equal to the loose weight divided by the bank weight per unit of volume.

The percent swell and shrinkage may be determined from Eqs. (5-1) and (5-2), respectively.

$$S_w = \left(\frac{B}{L} - 1\right) \times 100 \tag{5-1}$$

$$S_h = \left(1 - \frac{B}{C}\right) \times 100 \tag{5-2}$$

where S_w = % swell
S_h = % shrinkage
B = density of undisturbed earth
L = density of loose earth
C = density of compacted earth

The densities of earth usually are expressed in pounds per cubic foot.

Example 5-1 Determine the percent swell and shrinkage for earth whose densities are as follows:

Undisturbed, 92 lb per cu ft
Loose, 76 lb per cu ft
Compacted, 108 lb per cu ft

The percent swell will be

$$S_w = (\tfrac{92}{76} - 1) \times 100$$

$$= (1.21 - 1) \times 100 = 21\%$$

The percent shrinkage will be

$$S_h = (1 - \tfrac{92}{108}) \times 100$$

$$= (1 - 0.85) \times 100 = 15\%$$

Similar results may be obtained by using a calibrated container to measure the volume of a given quantity of earth in the undisturbed, loose, and compacted states.

To illustrate the use of Table 5-1, 10 cu yd of earth in a borrow pit may occupy 12.5 cu yd in a truck and 9 cu yd in a compacted fill.

Types of soils. Soils may be classified according to the sizes of the particles of which they are composed, by their physical properties, or by their behavior when the moisture content varies.

A contractor is concerned primarily with five types of soils: gravel, sand, silt, clay, organic matter, and combinations of these types. Different agencies and specification groups denote the sizes of these types of soil differently, causing some confusion. The following size limits represent those most widely used today.

Gravel is a rocklike material whose particles are larger than $\frac{1}{4}$ in. (0.6 mm). Sizes larger than around 10 in. are usually called boulders.

Sand is a disintegrated rock whose particles vary in sizes from those of gravel down to 0.002 in. (0.05 mm). It may be classified as coarse or fine sand, depending on the sizes of the grains. Sand is a granular, or noncohesive, material whose strength is not affected by its moisture content.

Silt is a very fine sand, and is thus a granular material whose particles are smaller than 0.002 in. (0.05 mm) and larger than about 0.005 mm. It is a noncohesive material, and it has little or no strength. It compacts very poorly.

Clay is a cohesive material whose particles are microscopic in size, less than about 0.005 mm. The cohesion between the particles gives clays a high strength when dry. Clays are subject to considerable changes in volume with variations in moisture content. When clays are combined with granular soils, the strengths of such soils are increased greatly.

Organic matter is a partly decomposed vegetable matter. If it is present in soil that is used for construction purposes, it should be removed and replaced with a more suitable soil.

Soils existing under natural conditions may not contain the several types in the desired ratios to produce the necessary properties for construction purposes. For this reason it may be necessary to obtain soils from several sources and then blend them as they are placed in a fill.

If the material in a borrow pit consists of layers of different types of soils, the specifications for the project may require the use of excavating equipment that will dig through the several layers in order to mix the soil.

Soil tests Prior to preparing the specifications for a project representative samples of soil are collected and tested in laboratories to determine their properties, including the dry unit weight and the percent of moisture required for maximum compacted density. The optimum moisture content is the ratio of the weight of water to the dry weight of soil, expressed as a percent, that will permit the soil to be compacted to the maximum density with the least effort.

Figure 5-2 shows moisture-density curves which illustrate the effect of varying amounts of moisture on the density of a soil subjected to equal compactive efforts, for the standard and modified Proctor tests. It will be noted that the modified Proctor gives a higher density at a lower moisture content than the standard Proctor. The optimum moisture for the standard Proctor is 17 percent, versus 14 percent for the modified Proctor. The modified Proctor test may be more nearly correct because it is more reflective of the heavier equipment used today in compacting soils.

Laboratory tests The laboratory test that is accepted by highway departments and other agencies is the Proctor test. For this test a sample of soil consisting of $\frac{1}{4}$ in. and finer material is used. The sample is divided into three equal parts, and moisture is added. Each part is placed separately in a cylindrical steel mold whose inside diameter is 4.0 in. and whose height is 4.59 in., and then compacted by dropping a 5.5-lb rammer, with a circular base, 25 times from a height of 12 in. above the specimen. The specimen is removed

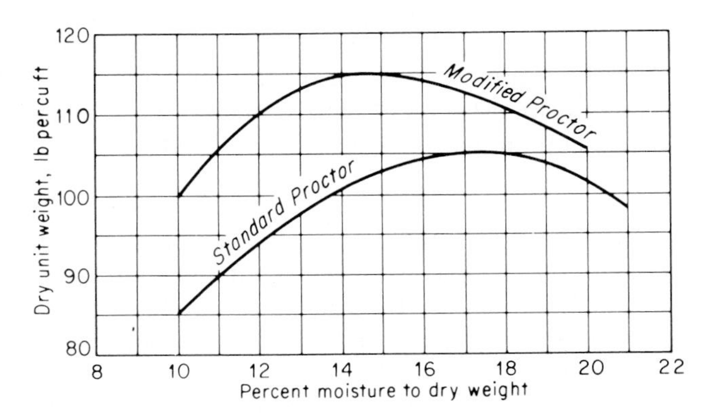

Figure 5-2 Moisture-density curves for fixed compaction.

from the mold and weighed immediately; then it is dried to a constant weight to remove all moisture and weighed again to determine the moisture content. The test is repeated, using varying amounts of water in the samples, until the moisture content that produces the maximum density is determined. This test is designated as AASHTO T 99-70.

The modified Proctor test, designated as AASHTO T 180-70, is performed in a similar manner, except that a rammer whose weight is 10 lb is dropped 18 in.

Field test The specifications for a project may require a contractor to compact the soil to a 100 percent relative density, based on the standard Proctor test. If the laboratory density of the soil is determined to be 120 lb per cu ft, the contractor must compact the soil to a density of 120 lb per cu ft.

Field tests can be conducted by removing samples of compacted soil from the fill at random locations, and then determining the damp and dry weights of each sample. The volume of the hole can be determined by several methods. The most common method is to fill the hole with dry sand from a container of known weight. The difference in the weight of the container, before and after the hole is filled, will determine the weight of sand used, from which the volume of the hole can be determined. If the wet weight and the dry weight of the sample and the volume of the hole from which it was removed are known, the dry-weight density can be determined. Also, the moisture content of the sample can be determined.

Nuclear determination of moisture density of soils Nuclear methods are used extensively to determine the moisture-density of soils. The instrument required for this test can be transported readily to the fill, placed at a location where a test is to be conducted, and within a few minutes the results can be read directly from the indicators (see Fig. 5-3).

The device utilizes the Compton effect of gamma-ray scattering for density determinations and hydrogenous thermalization of fast neutrons for moisture determinations. The emitted rays enter the ground, where they are partially absorbed and partially reflected. Reflected rays pass through Geiger-Müller tubes in the surface gauge. Counts per minute are read directly on a reflected-ray counter gauge and are related to moisture and density calibration curves.

Advantages of the nuclear method when compared with the Proctor method include the following:

1. Decreases the time required for a test from as much as a day to a few minutes, thereby eliminating potentially excessive delays for the contractor.
2. Does not require the removal of soil samples from the site of the tests
3. Provides a means of performing density tests on soils containing large-sized aggregates and on frozen materials
4. Reduces or eliminates the effect of the personal element, and possible errors, that may occur in performing Proctor tests

Figure 5-3 Density-moisture gauge and scaler. *(Troxler Electronic Laboratories, Inc.)*

Instruments are available that will measure moisture and density of soils at depths up to 200 ft or more below the surface of the ground. The measurements are performed by drilling holes in the soil to the desired depths, installing aluminum tubing in the holes temporarily, then lowering nuclear sensing probes down the tubing and making the tests at the desired depths.

Because nuclear tests are conducted with instruments that present a potential source of radiation, an operator should exercise reasonable care to assure that no harm can result from the use of the instruments. However, by following the instructions furnished with the instruments and exercising proper care, exposure can be maintained well below the limits set by the Nuclear Regulating Commission (NRC), formerly the Atomic Energy Commission. In the United States a license is required to own, possess, or use nuclear-type density- and moisture-measuring instruments. A license may be obtained from the Nuclear Regulating Commission, and where required from state and/or local government agencies.

SOIL STABILIZATION

Many soils are subject to differential expansion and shrinkage when they undergo changes in moisture content. Many soils also move and rut when subjected to moving wheel loads. If pavements are to be constructed on such soils, it is usually necessary to stabilize them to reduce the volume changes and strengthen them to the point where then can carry the imposed load, even when they are saturated. In the broadest sense, stabilization refers to any treatment of the soil which renders it more stable, and thus there are two kinds of stabilization—mechanical and chemical. In engineering construction, however, stabilization is most often referred to when compaction is preceded by the addition and mixing of an inexpensive admixture, termed a stabilization agent, which alters the chemical makeup of the soil, resulting in a more stable material.

Stabilization may be applied in place to a soil in its natural position or as it is placed in a fill. Also, stabilization may be applied in a plant and then transported to the job site for placement and compaction.

Methods of stabilizing soils include, but are not limited to, the following operations:

1. Blending and mixing heterogeneous soils to produce more homogeneous soils
2. Incorporating lime or lime–fly ash into soils that are high in clay content
3. Blending asphalt with the soil
4. Incorporating portland cement (with or without fly ash) into soils that are largely granular in nature
5. Incorporating various salts into the soil
6. Incorporating certain chemicals into the soil
7. Compacting the soils after they are processed

Blending and mixing soils If the soils that are to be used in a fill are heterogeneous in their original states, such as in a borrow pit, they may be mixed during excavation by using equipment such as a power shovel or a deep-cutting belt loader to excavate through several layers in one operation. When such material is placed on a fill, it may be subjected to further blending by several passes with a disk harrow, as illustrated in Fig. 5-4.

Stabilizing soils with lime In combination with compaction, soil stabilization with lime involves a chemical process where the soil is improved with the addition of lime. In this context, the most troublesome soils are the clays and silty clays with plasticity indexes (PI) greater than about 10. Unless stabilized, these soils usually become very soft when water is introduced. Lime, in its hydrated form [$Ca(OH)_2$], will rapidly cause cation exchange and flocculation/agglomeration, provided it is intimately mixed with the soil (see Fig. 5-5). The clay-type soil will then behave much more like a silt-type soil. This reaction begins to occur within an hour after mixing, and significant changes are realized within a very few days, depending upon the PI of the soil and the amount of lime used. The observed effect in the field is one of a drying action. According to Krebs and Walker [1], "Clay otherwise in a plastic condition may

Figure 5-4 Disk harrow used to blend soil. *(Rome Industries.)*

Figure 5-5 Self-propelled soil pulverized-stabilizer. *(RayGo, Inc.)*

become semisolid or friable in consistency. Moreover, normal clay-water interactions are inhibited. Plasticity index decreases.... The reactions that cause amelioration effects in lime-clay mixtures are not clearly known, but it is thought that sufficient lime dissolves in the soil water to create a highly alkaline environment and to crowd calcium ions onto the clay exchange complex, causing severe flocculation." Following this rapid soil improvement is a longer, slower, soil improvement termed pozzolanic reaction. In this reaction, the lime chemically combines with siliceous and aluminous constituents in the soil to cement the soil together. Here some confusion exists. Some people refer to this as a cementious reaction, which is a term normally associated with the hydraulic action occurring between portland cement and water, in which the two constituents chemically combine to form a hard, strong product. The confusion is increased by the fact that almost two-thirds of portland cement is lime (CaO). But the lime in portland cement starts out already chemically combined during manufacture with silicates and aluminates and thus is not in an available or "free" state to combine with the clay.

The cementing reaction of the lime, as $Ca(OH)_2$, with the clay is a very slow process, quite different from the reaction of portland cement and water, and the final form of the products are thought to be somewhat different. The slow strength with time experienced with lime stabilization of clay provides flexibility in manipulation of the soil. Lime can be added and the soil mixed and compacted, initially drying the soil and flocculating it. Several days to several weeks later the soil can be remixed and compacted to form a dense stabilized layer that will continue to gain strength for many years. The resulting stabilized soils have been shown to be extremely durable [2]. For example, runways and taxiways constructed in Texas in 1943 using 2 percent lime to reduce the plasticity of caliche gravel have been giving excellent service ever since. Texas pioneered in the use of lime to stabilize clays, and their roads and streets have been in service since the late 1940s.

Lime–fly ash stabilization This type of stabilization, although not new, is just now beginning to be used widely, the reason is that fly ash, which is the residue that would "fly" out the stack in a coal-fired power plant if it were not captured, is becoming extremely plentiful throughout the United States and, in fact, throughout many parts of the world. Fly ash is a by-product in the production of electricity using coal. As such, it can be a highly variable product, and its engineering usefulness can range from superior to extremely poor. Today, in the United States alone, there is in excess of 90 million tons of fly ash being collected each year [3]. The newer, more modern power plants, literally pulverize the coal until almost all of it passes the No. 200-mesh sieve before using it as a fuel. The resulting fly ash is extremely fine in size (often finer than portland cement) and contains the silicates and aluminates necessary to combine with the lime in soil stabilization. Laboratory and field results indicate that fly ash, of suitable quality, can replace a *portion* of the lime necessary to stabilize a clay-type soil [4, 5]. Because lime is relatively expensive and fly ash often quite inexpensive, lime–fly ash stabilization of soils is being increasingly utilized in many parts of the world with a high degree of success. The major drawback to the use of fly ash is that two stabilizing agents are being used instead of only one, which means more manipulation of the soil and more chance for error. There are a number of excellent references on the use of lime–fly ash stabilization, a few of which are included at the close of this chapter (see [4] and [5]).

Asphalt-soil stabilization When asphalts, such as an emulsion or a cutback, are mixed with granular soils, usually in amounts of 5 to 7 percent of the volume of the soil, this treatment will produce a much more durable and stable soil. Some soils have been stabilized by adding 10 to 15 percent of minus No. 200-mesh fines to fill the interstices in the soils and then mixing this blend with asphalt.

The moisture content of the soil must be low at the time the asphalt is added. Also, it is necessary to allow the volatile oils to evaporate from the bitumen before finishing and rolling the material.

Soils treated in this manner may be used as finished surfaces for low-traffic-density secondary roads, or they may serve as base courses for high-type pavements.

Cement-soil stabilization Stabilizing soils with portland cement is an effective method of strengthening certain soils. As long as they are predominately granular with only minor amounts of clay particles, the use of portland cement has been found to be effective. A good rule of thumb is that soils with PI less than about 10 are likely candidates for this type of stabilization. Soils with higher amounts of clay-sized particles are very difficult to manipulate and mix thoroughly with the cement before the cement sets. The terms "soil cement" and "cement-treated base" are often used interchangeably, and generally describe this type of stabilization. The amount of cement in the soil is usually 5 to 7 percent by dry weight of the soil.

As discussed in connection with lime stabilization, fly ash is becoming plentiful in many areas of the world, and it can be effectively utilized to replace a portion of the portland cement in a soil cement (see Chap. 20 for a more complete discussion of fly ash and portland cement). Replacement percentages on an equal weight basis or on a 1.25 : 1.0 fly ash/portland cement replacement basis have been used. There are a number of excellent references on this subject, three of which are listed at the close of this chapter [4, 5, 6].

The construction methods involve spreading the portland cement uniformly over the surface of the soil, then mixing it into the soil, preferably with a pulverizer-type machine, to the specified depth, followed by fine grading and compaction. If the moisture content of the soil is low, it will be necessary to sprinkle the surface with water during the processing operation. The material should be compacted within 30 min after it is mixed, using tamping or pneumatic-tired rollers, followed by final rolling with smooth-wheel rollers. It may be necessary to apply a seal of asphalt or other acceptable material to the surface to retain the moisture in the mix.

SOIL COMPACTION

By far the most widely used method of soil strengthening for use as a subgrade under a pavement structure or other foundation is compaction of the soil at optimum moisture. The benefits of proper compaction are enormous, far outweighing their costs. Typically, a uniform layer, or lift, of from 4 to 12 in. of soil is compacted by means of several passes of heavy mechanized compaction equipment. An often overlooked fact is that the denser the soil is compacted, the better it will perform in service. Of course compaction costs money, and the owner is interested in achieving the most economical construction that will perform as intended. The contractor is concerned with meeting the specifications. What is altogether too often overlooked is the increased benefits to be derived by obtaining greater densification, whether chemically stabilized or not.

Specifications governing compaction may be one of the following types:

1. Method only (often termed "recipe")
2. Method and end result
3. Suggested method and end result
4. End result only (often termed "performance")

Method only specifications If the specifications for a project direct the contractor to place the soil in lifts of a specified depth, with the soil having a specified moisture content, with the provision that a specified type of roller having a specified weight is to be used to compact the soil by making a specified number of passes over each lift, the contractor will have no choice except to comply with the requirements of the specifications. If the owner prescribes this type of specifications, he will be obligated to accept the responsibility for the results.

Method and end results specifications For most projects this is not a satisfactory specification. Unless extensive predesign tests have been performed on soil samples, which eliminate the possibility of the soil behaving differently than is expected, it is probable that a specified method of compacting the soil will result in excessive costs because compacting operations will be continued after adequate compaction is attained, or compaction operations may be discontinued before adequate density is attained. In any event, the contractor should not be held responsible for the end results.

A further objection to the use of this type of specification is that it may not permit a contractor to make use of methods which he has found to be economical and effective. Thus, the use of this type of specification may result in an unnecessarily high cost for the project.

Suggested method and end result specifications This type of specification seems to be more desirable than the two types previously described. It leaves a contractor free to select any reasonable method and equipment which he may have learned from vast experience will provide the required density. Experienced contractors are quite ingenious, and if given an opportunity to use this experience, they frequently can attain the specified results at a significant reduction in costs.

At the same time this type of specification can serve as a guide to a less-experienced contractor.

End result only specifications Several states and agencies are moving toward a policy of using this type of specification. The argument for the use of this policy is that the owner is interested primarily or solely in the end result. For example, the specifications dictate that the soil shall be compacted to 95 percent relative density, based on the modified Proctor test. Unless there are justified reasons for prescribing the methods to be used, the contractor should be permitted to select his own methods, which may be substantially less expensive than other prescribed methods.

TYPES OF COMPACTING EQUIPMENT

Compaction is attained by applying energy to a soil by one or more of the following methods:

1. Kneading action
2. Static weight
3. Vibration
4. Impact
5. Explosives

Many types of compacting equipment are available, including the following:

1. Tamping rollers
2. Smooth-wheel rollers
3. Pneumatic-tired rollers
4. Vibrating rollers, including tamping, smooth-wheel, and pneumatic
5. Self-propelled vibrating plates and/or shoes
6. Manually propelled vibrating plates
7. Manually propelled compactors
8. Vibratory compactors for deep sand

Table 5-2 summarizes the types of equipment suited for compacting soils. On some projects it may be desirable to use more than one type of equipment to attain the desired results and to effect the greatest economy.

Tamping rollers Tamping rollers are of the sheep's-foot type or modifications thereof. This roller, which may be towed by a tractor or self-propelled, consists of a hollow steel drum on whose outer surface there are welded a number of projecting steel feet, which on different pieces may be of varying lengths and cross sections. A unit may consist of one or several drums mounted on one or more horizontal axles. The weight of a drum may be varied by adding water or sand to produce unit pressures under the feet up to 750 psi or more.

As a tamping roller moves over the surface, the feet penetrate the soil to produce a kneading action and a pressure to mix and compact the soil from the bottom to the top of the layer. With repeated passages of the roller over the surface, the penetration of the feet decreases until the roller is said to walk out of the fill.

The specifications may prescribe one of the following as a means of

Table 5-2 Types of equipment suited for compacting soils

Type compactor	Soil best suited for	Maximum effect in loose lift, in.	Density gained in lift	Maximum weight, tons
Sheep's foot	Clay, silty clay, gravel with clay binder	7 to 12	Nearly uniform	20
Steel tandem two-axle	Sandy silts, most granular material with some clay binder	4 to 8	Average*	16
Steel tandem three-axle	Same as above	4 to 8	Average*	20
Steel three-wheel	Granular or granular-plastic material	4 to 8	Average* to uniform	20
Pneumatic, small-tire	Sandy silts, sandy clays, gravelly sand and clays with few fines	4 to 8	Average* to uniform	12
Pneumatic large-tire	All types	? to 24	Uniform	50
Vibratory	Sand, silty sands, silty gravels	3 to 6	Uniform	30
Combinations	All	3 to 6	Uniform	20

*The density may decrease with depth.

attaining the desired compaction:

1. The number of passes of a roller, producing a specified unit pressure under the feet, over each layer of soil
2. Repeated passes of a roller, producing a specified unit pressure under the feet, over each layer of soil until the penetration of the feet does not exceed a stated depth
3. Repeated passes of a roller over each layer until the soil is compacted to a specified density

Sheep's-foot rollers are quite effective in compacting clays and mixtures of sand and clay. However, they cannot compact granular soils such as sand and gravel. Also, the depth of a layer of soil to be compacted is limited to approximately the length of the feet.

Figure 5-6 illustrates tractor-pulled sheep's-foot rollers on a project, and Fig. 5-7 illustrates a self-propelled tamping roller and dozer.

Modified tamping rollers A modification of the tamping roller, designated as a grid roller, is illustrated in Fig. 5-8. When this roller is ballasted with concrete blocks, it is capable of producing very high soil pressures, and when it is used to compact soil containing rocks, the high concentration of pressure on rocks projecting above the surface of the soil is effective in shattering the rocks and forcing the broken pieces into the soil to produce a relatively smooth surface.

Figure 5-6 Tractor-pulled sheep's-foot rollers.

Figure 5-7 Self-propelled tamping roller and dozer. *(RayGo, Inc.)*

Figure 5-8 Tractor-pulled ballasted grid roller. *(Hyster Company.)*

Smooth-wheel rollers These rollers may be classified by type or by weight. A diesel-powered dual-drum self-propelled roller is shown in Fig. 5-9. Note that the front wheel is used for steering while the rear wheel is powered for driving.

Smooth-wheel rollers may be classified by weight, which is usually stated in tons. The rolls are steel drums, which may be ballasted with water or sand to increase the weights. If a machine is designated as 8–14 tons, it means that the minimum weight of the machine only is 8 tons and that it can be ballasted to give a maximum weight of 14 tons.

Specifications governing these rollers may be of two types, one type simply designating the weight, and the other type designating the weight per linear inch of roll, such as 300 lb per inch of roller width. Specifying the weight only does not necessarily indicate the compressive pressure under the wheels. Specifying the minimum weight per linear inch of width is a more definitive method, and appears to be superior.

When compacting cohesive soils, these rollers tend to form a crust over the surface, which may prevent adequate compaction in the lower portions of a lift. However, these rollers are effective in compacting granular soils, such as sand, gravel, and crushed stone, and they are also effective in smoothing surfaces of soils that have been compacted by tamping rollers.

Pneumatic-tired rollers These are surface rollers which apply the principle of kneading action to effect compaction below the surface. They may be self-propelled or towed. They may be small- or large-tired units (see Fig. 5-10).

Figure 5-9 Smooth-wheel roller, diesel powered. *(Ferguson Manufacturing and Equipment Co., Inc.)*

Figure 5-10 Self-propelled pneumatic roller. *(Ferguson Manufacturing and Equipment Co., Inc.)*

The small-tired units usually have two tandem axles with four to nine tires on each axle. The rear wheels are spaced to travel over the surfaces between the front wheels, which produces a complete coverage of the surface. The wheels may be mounted in a manner that will give them a wobbly-wheel effect to increase the kneading action of the soil. Usually the weight of a unit may be varied by adding ballast to suit the material being compacted.

Large-tired rollers are available in sizes varying from 15 to 200 tons gross weight. They utilize two or more big earth-moving tires on a single axle. The air pressure in the tires may vary from 80 to 150 psi. Because of the heavy loads and high tire pressures, they are capable of compacting all types of soils to greater depths. These units are frequently used to proof roll subgrades and bases on airfields and earth-fill dams.

There are at least four methods of indicating the compacting ability of pneumatic rollers; these are

1. The gross weight of the unit
2. The gross weight per wheel
3. The weight per inch of tire width
4. The air pressure in the tires

Because the area of contact between a tire and the ground surface over which it passes varies with the air pressure in the tire, specifying the total weight or the weight per wheel is not necessarily a satisfactory method of indicating the compacting ability of a roller. A more definitive method of designating the compacting ability is to specify the gross weight, the number and sizes of tires, and the tire inflation pressure.

Table 5-3 illustrates the effect of gross vehicle weight and tire inflation pressure on the ground contact pressure and the load per inch of tire width.

Figure 5-11 illustrates a graphical method of determining the ground contact pressure for a 13.00 × 24 18-ply smooth compactor tire subjected to varying loads and inflated to varying air pressures. Similar information is available from tire manufacturers for other tire sizes and loads (see Fig. 5-12).

As indicated by the dashed lines in Fig. 5-11, a wheel load of 8,000 lb and

Table 5-3 Effect of variations in gross weight and tire inflation pressure on ground contact pressure

Gross weight, lb		7,650		15,300		22,500		25,000	
	Inflation	Ground contact pressure							
Tire size	pressure, psi	Psi*	Pli†	Psi	Pli	Psi	Pli	Psi	Pli
7.50 × 15 4 or 6 ply	35	33	125	39	237	44	333	46	369
	45	38	127	45	241	50	338	51	374
	55	44	129	50	243	55	341	57	377
	60	46	131	53	245	58	344	61	380
7.50 × 15 10 ply	50	43	145	50	250	56	342	58	378
	60	47	152	54	254	60	347	62	382
	70	50	162	58	258	64	350	66	385
	80	54	175	62	264	68	354	70	389
	90	58	183	65	272	71	359	74	392
7.50 × 15 12 ply	50	43	153	50	250	57	343	59	378
	60	47	164	55	256	61	347	64	383
	70	51	170	59	264	66	351	68	386
	80	55	184	62	270	70	357	72	392
	90	58	202	66	276	73	364	76	397
	100	62	218	69	289	76	369	79	402
	110	65	224	72	293	79	375	82	406
7.50 × 15 14 ply	50	47	158	57	253	63	348	65	385
	60	50	170	59	260	67	353	68	389
	70	52	181	62	268	69	358	72	394
	80	55	192	65	276	73	365	75	399
	90	57	210	68	281	76	370	78	405
	100	61	225	71	290	79	377	82	408
	110	65	230	75	293	83	385	85	417
	120	68	239	79	301	87	391	89	423
	130	71	243	82	318	90	400	93	431

Source: Firestone Tire and Rubber Company.

* Ground contact pressure in psi.

† Ground contact pressure in lbs per in. of tire width.

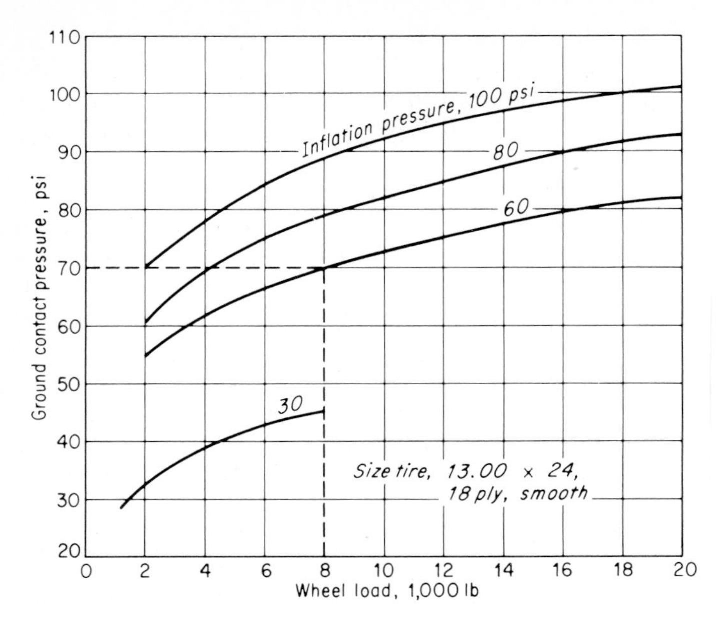

Figure 5-11 Ground pressure at varying wheel loads and air pressures.

an inflation pressure of 60 psi gives a ground contact pressure of 70 psi. The contact area for the tire will be 8,000 lb ÷ 70 psi = 114.3 sq in.

Pressure bulb theory of load distribution This theory is related to the distribution of a load, and thus to the unit soil pressure, when the load is applied

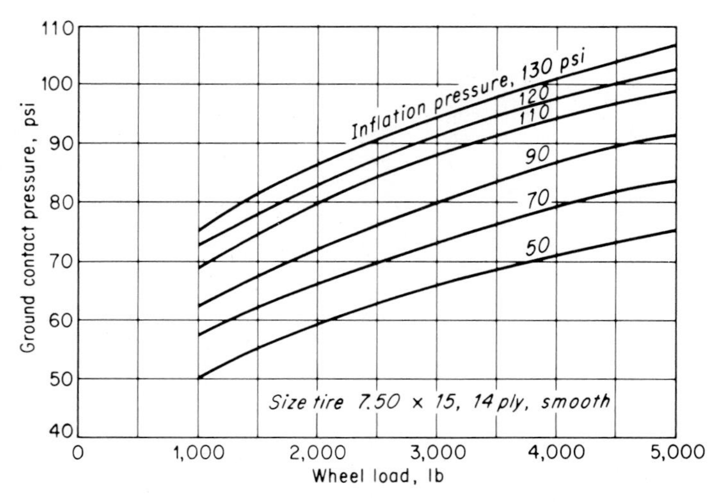

Figure 5-12 Ground pressure at varying wheel loads and air pressures.

to the soil through a circular object. Because the contact area between a tire and the ground approximates a circle, the theory can be applied to pressures in the soil under tires with slight modifications. Figure 5-13 illustrates the ratios of unit pressures to ground contact pressure at varying depths below the surface of the ground.

Pneumatic-tire rollers with variable inflation pressures When a pneumatic-tired roller is used to compact soil through all stages, the first passes over a lift should be made with relatively low tire pressures to increase flotation and ground coverage. However, as the soil is compacted, the air pressure in the tires should be increased up to the maximum specified value for the final pass. Prior to the development of a method of varying the air pressure while a roller is in operation it was necessary to (1) vary the pressure in the tires, (2) vary the weight of the ballast on the roller, or (3) keep rollers of different weights and tire pressures on a project in order to provide units to fit the particular needs of a given compaction condition.

Several manufacturers produce rollers that are equipped to permit the operator to vary the tire pressure without stopping the machine. The first passes are made with relatively low tire pressures. As the soil is compacted, the tire pressure is increased to suit the particular conditions of the soil. The use of this type of roller usually permits adequate compaction with fewer passes than are required by the constant pressure rollers.

Vibrating compactors Certain types of soils such as sand, gravel, and relatively large stones respond quite well to compaction produced by a combination of pressure and vibration. When these materials are vibrated, the particles shift

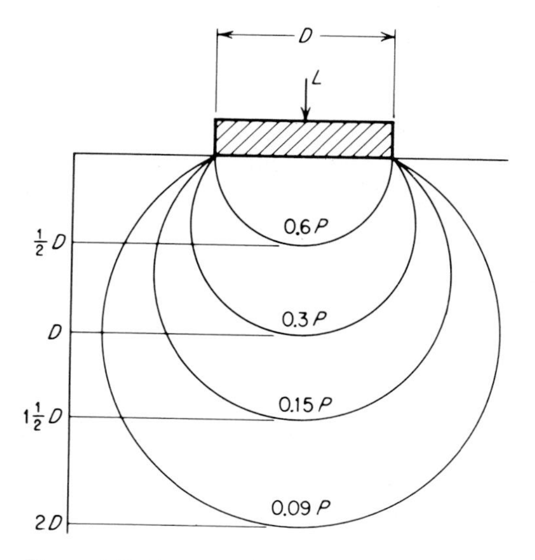

D = diameter of circle
L = load
A = area of circle
P = unit pressure under area

Example: $D = 10$ in.
$A = 78.5$ sq in.
$L = 4.710$ lb

$$P = \frac{4.710}{78.5} = 60 \text{ psi}$$

Distance under surface, in.	Factor	Soil pressure, psi
0	1.00	60.0
5	0.60	36.0
10	0.30	30.0
15	0.15	9.0
20	0.09	5.4

Figure 5-13 Variations in pressure with depth under a load.

Figure 5-14 Towed-type vibrating sheep's-foot roller.

their positions and nestle more closely with adjacent particles to increase the density of the mass.

Several types of compactors have demonstrated their abilities to produce excellent densification of these soils. They include

1. Vibrating sheep's-foot rollers (Fig. 5-14)
2. Vibrating steel-drum rollers (Fig. 5-15)
3. Vibrating pneumatic-tired rollers
4. Vibrating plates or shoes (Fig. 5-16)

Vibrating sheep's-foot, steel-drum, pad-type, and pneumatic-tired rollers are actuated by separate engines mounted on the rollers, or in some cases by hydraulic drives which rotate horizontal shafts on which one or more eccentric weights are mounted. Vibrations may vary from 1,000 to 5,000 per min, with the actual number corresponding to the natural resonant frequency of vibrations for the given soil. In addition to compacting the soil, the drum- and pad-type rollers tend to shatter the rock particles near the surface and thus leave a relatively smooth surface.

During the construction of the Cougar Dam on the McKenzie River in Oregon the U.S. Army Corps of Engineers conducted tests using 5- and 10-ton vibratory steel-drum rollers to determine the effectiveness of these machines in compacting the rockfill portion of the dam [7]. The larger roller, having a static weight of 10 tons and a dynamic force of 20 tons, demonstrated its ability to consolidate lifts of rock 36 in. thick, with individual rocks varying in sizes up to

Figure 5-15 Vibrating steel-drum roller. *(RayGo, Inc.)*

24 in. The rollers were towed over the fills at speeds between 1.5 and 2 mph, while vibrating at frequencies of approximately 1,400 vibrations per minute (vpm). Settlement was measured after two, four, and six passes, with the following results:

Number of passes	Relative % of compaction
2	60
4	83
6	100

In general, better compaction efficiencies and economy are attained by moving vibrating compactors at relatively slow speeds, 1.5 to 2.5 mph. Slow speeds permit a greater flow of vibratory energy into the soil.

Manually operated vibratory plate compactors Figure 5-16 illustrates a self-propelled vibratory-plate compactor used for consolidating soils and asphalt (hot or cold mix) in locations where large units are not practical. These gasoline- or diesel-powered units are rated by centrifugal force, exciter revolutions per minute, depth of vibration penetration (lift), feet per minute travel, and area of coverage per hour.

Figure 5-16 Self-propelled Vibro-plate. *(Wacker Corporation.)*

Manually operated vibratory tamping compactors Figure 5-17 illustrates a compacting unit of this type which may be used in locations where larger units are not practical.

Manually operated rammer compactors Figure 5-18 illustrates a gasoline-engine-driven rammer used for compacting cohesive or mixed soils in confined areas. These units range in impact from about 300 to 900 or more ft-lb per sec at an impact rate up to 850 per min, depending on the specific model. Performance criteria include pounds per blow, area covered per hour, and depth of compaction (lift) in inches. Rammers are self-propelled in that each blow moves them ahead slightly to contact new soil.

Densification of soils by explosive vibrations [8] In a loose, saturated, granular soil, a sudden shock or vibration causes localized spontaneous liquefaction and displacement of the soil grains. The weight of the soil is temporarily transferred to the liquid, and the soil particles fall into a much denser pattern, aided by the weight of the soil above. In a dry soil, the shock and vibration cause the

Figure 5-17 Self-propelled vibrating compactor. *(RayGo, Inc.)*

particles to move to a new denser and more compact pattern. The moment soil particles are freed or loosened from their initial orientation, even small pressures are effective in realigning them in a more compact mass. This process is not reversible, and the new density is permanent.

The maximum result is obtained in a soil that is either dry or completely saturated. As the water content of the soil decreases from 100 percent saturation to the retained or absorbed water content around each grain of soil, less densification is obtained because of the increase in capillary tension between the grains. This effect becomes more pronounced as the soil becomes finer. The effect of capillary tension can be overcome by ponding or flooding the area, then allowing adequate time for the water to seep downward to meet the existing water level. Although the quantity of water required to saturate the soil is not great, an increased head will reduce the time required to attain full saturation.

A slight amount of superimposed weight is advantageous in reorienting the soil particles. The upper 2 to 3 ft (0.6 to 0.9 m) of soil usually undergoes only a slight increase in density unless completely saturated to the surface. Even when the soil is saturated, the increase in density is not as pronounced as it is at deeper elevations. If it is necessary to compact this upper layer of soil, vibratory compaction may be attained by using conventional mechanical equipment.

After the initial shock, other charges exploded in the same area will cause

Figure 5-18 Manually operated rammer. *(Wacker Corporation.)*

further settling, but each successive charge produces a smaller effect until no appreciable or useful settling can be obtained. The ratio for any series of charges, empirically obtained, shows that the first quarter of a charge causes approximately 60 percent densification, the second quarter causes 25 percent more, the third quarter 10 percent more, and last quarter 5 percent more. This condition is one of the factors affecting the spacing of the holes to assure that a sufficient number of overlapping effects will be applied to each location in any area.

Spacings, depths, and sizes of explosive charges The densification of soils by means of explosives is different from regular blasting in that no craters can be blown in the soil and there is no debris from the explosion of the charges. The energy from an explosion must be contained entirely within the ground.

The approximate sizes and depths of charges can be determined from existing formulas, tempered with experience and checked by trial tests conducted at the site of the operation. The amount of each charge will vary according to the type of soil, depth of strata, desired amount of densification, spacing of holes, present ground-water level, nearness of structures, overlapping effect of charges, and type of explosive used. Experience has indicated that the center of an explosive charge should be below the center of the mass of soil to be densified at approximately the two-thirds point down. It may be necessary to have separate charges in the same hole when cohesionless strata are separated by layers of cohesive soil or when operating near existing buildings or structures. For such conditions, delay electric blasting caps may be used to permit firing at delayed intervals, thereby reducing the energy per firing.

Horizontal spacings of holes may vary from 10 to 25 ft (3.0 to 7.5 m) and are governed by the depth of the strata, the size of the charge, and the overlapping effect of adjacent charges. Spacings closer than 10 ft (3.0 m) in saturated soils should be avoided, unless carefully investigated for safety, because of possible propagation of sensitive explosions of adjacent charges.

The firing pattern should allow a number of charges to act on one particular area. However, the pattern should leave an area on one or two sides to permit excess pore water to escape from a square. Reduction of the voids in a volume of soil results in release or displacement of a large volume of water, the quantity depending on the depth, the original void ratio, and the amount of densification of the soil.

Installation of the explosives After the spacings, depths, and sizes of charges have been determined and a sequence of shots or firing pattern has been established, the locations of the holes are marked out. One method of producing the holes for the charges is to use one or more nonsparking plastic or aluminum pipes of the desired sizes, which can be self-jettied into the soil to the correct depth. An explosive charge with an electric blasting cap and lead wires is then placed in each hole through the pipe. After the pipe is withdrawn, the hole is filled and tamped with a wooden rod.

After the charges for a given pattern have been installed, the blasting caps are tested individually using a galvanometer. Then the charges are fired according to the schedule of delays previously determined.

Publications are available which furnish dependable information on the transportation, storage, handling, and use of explosives [9, 10]. Most manufacturers of explosives can furnish useful information on this subject. Local regulations, licenses, permits, and prejudices must also be respected when using explosives.

Deep densification of soils using a terra-probe vibrator [11] This method of vibrator compaction is accomplished by the use of a vibratory pile-driving apparatus together with an open-end tubular probe; the probe is driven into and extracted from the soil to be compacted on certain modular spacings [12]. When the driving and extracting phases are accomplished, densification of the soil occurs both inside and outside the probe, with the concentration of vibratory energy creating extreme densification inside the probe and with densification outside the probe diminishing with distance (see Fig. 5-19).

To date, the vibratory pile-driving apparatuses used have been in the frequency ranges of 720 to 1,100 cpm (11 to 19 Hz), the normal operating

Figure 5-19 Vibratory probe in position to densify the soil. *(L. B. Foster Company.)*

frequency being 900 cycles per min (15 Hz). The vibrator has been able to create amplitudes of 3/8 to 1 in. (9.5 to 25 mm). The vibrator creates vertical energy by counter-rotating eccentric weights, which cancel out the horizontal effects and give vertical vibrations only.

The best probe material used has been an open-end 30-in. (760 mm) pipe of 3/8-in. (9.5-mm) wall thickness with 4- to 6-in. (100- to 150-mm) wide and 1/2-in. (13-mm) thick steel bands spaced 5 to 10 ft (1.5 to 3 m) apart on the outside of the pipe, together with wider driving and clamping bands at the bottom and top of the probe. Tests using other diameters have revealed that smaller diameters give less densification inside the pipe, while larger diameters require more vibratory energy and thicker and heavier probe material. The probe is usually 10 to 15 ft (3 to 4.6 m) longer than the maximum penetration depth, to allow for any flexing of the probe, particularly when probes more than 50 to 55 ft long (15 to 17 m) are used. This also allows for any cut-off requirements during application.

The probe is attached to the vibrator by means of a hydraulic clamp; this permits the vertical vibratory energy produced to travel to the probe material undiminished, as the probe, hydraulic clamping head, and vibrating transmission case act as a unit.

A mobile crane of sufficient size and capacity is required to handle the vibrating unit and the probe length during driving and extracting operations. An overburden of sand is required before beginning the operation to compensate for the settling that will result from the compaction. About a 12 percent shrinkage allowance has been satisfactory for most applications, but a hydraulic fill with a relatively low density may require a 15 percent allowance.

Spacing of the probes The dimensions of the modular spacings of the probes are dependent on the required relative density of the soil. Test patterns of several different spacings should be run initially to determine the required spacing to give the desired density. Square patterns have been used with spacings varying from 3 to 8 ft (0.9 to 2.4 m). Square spacings seem to offer better results and faster operations than other patterns. Also, if additional probes are needed, they can be placed in the centers of the square patterns. Small lathing stakes are generally used to indicate the location of each probe setting.

Production rates obtained with the terra-probe Compaction by this method is very expeditious, resulting in an average rate of about 15 probes per hour. For projects requiring shallow probes in loose soils the rate could be higher, while for projects requiring deeper probes in denser soils the rate could be lower.

During earlier applications of this method, tests were conducted by allowing the probes to sink to the lowest desired depth and then continuing the vibrations for several minutes to determine if additional densification was produced. The tests reveal that little, if any, additional densification resulted.

Limitations on the classifications of soils that may be compacted by the terra-probe. The use of this method should be limited to saturated cohesionless soils. Because densification of soils by this method causes the grains to assume new positions that reduce the volume of voids in the mass, it is necessary for the grains to be free at least temporarily and to be submerged in water; only saturated cohesionless soils will respond to the vibrations effectively. If unsaturated sands are to be densified by this method, they should be submerged or saturated by adding sufficient water to bring the upper level to near the surface of the ground.

In some test applications, this method was unsuccessful in densifying the soils. These soil conditions were:

1. Saturated sand with sieve analysis indicating 60 percent passing the No. 100 sieve, 52 percent passing the No. 120 sieve, and 5 percent passing the No. 200 sieve
2. Dry soil with 80 percent passing the No. 100 sieve and 50 percent passing the No. 200 sieve
3. Dry material for an 11-ft (3.1-m) shallow depth, with the sieve analysis indicating 85 percent passing the No. 40 sieve, 25 percent passing the No. 100 sieve, and 12 percent passing the No. 200 sieve

In some projects, in which the upper layers of soil consist of mud, muck, and silt, it will be necessary to remove all the undesirable material, because it will not be densified by the vibratory method. The replacement material should be a granular cohesionless soil that will respond to vibratory densification. In some instances it may be necessary to screen the replacement material in order to remove any silt or clay balls.

Cost considerations In all applications of the numerous techniques and methods available, consideration of costs becomes important. The terra-probe method, because vibratory energy accelerates its applications, has some interesting cost considerations. An overall cost consideration should include the following items:

1. Cost of soil removal, if required
2. Cost of replacement soil, if required
3. Cost of necessary soil overburden, if required
4. Cost of a testing program
5. Cost of the vibratory compaction method

The first four costs will be directly related to each project site, as each may or may not be applicable.

The cost of the vibratory compaction method can vary depending on the size of the area, the initial soil density, the required density, and the depth to be compacted. Because the mobilization and demobilization costs would be the same for a small or a large project, the unit cost for a small project would be higher. With other conditions remaining the same, the cost per unit volume should be less for projects requiring deep probes. A higher specified density, requiring closer spacing of probes, will result in a higher cost per unit of volume than the wider spacing permitted with lower-density requirements.

The depth of compaction is a factor of consideration, but it is less significant than other factors because of the speed provided by the vibratory device used. Overall probing time for a hole 25 ft (7.6 m) deep would not be doubled if the depth were increased to 50 ft (15 m) under the same soil conditions because the time required to move between the probes would be the same for both depths and the time required to penetrate the additional depth would be a matter of a few seconds for most projects.

The costs of the vibratory compaction method have varied for the reasons given heretofore, but a bracketing of the cost range has been established for several projects. The costs have ranged, in 1973 dollars, from about \$0.70 per cu yd (\$0.91 per m^3) to about \$1.10 per cu yd (\$1.40 per m^3) of compacted soil. This cost range was inclusive of the varying density requirements and included a marine-type operation involving more equipment than a land operation. None of these applications was larger than 100,000 cu yd (76,000 m^3) in volume. Therefore, it is realistic to assume that larger projects would have a lower unit cost. None of these costs included the contractor's profit.

Conclusions regarding the method of deep-sand vibratory compaction This method of vibratory compaction, though simple, represents an advance in the engineering field and in the construction industry. The availability of vibratory pile-driving devices, the employment of their best features learned through job experiences, and their effectiveness in cohesionless soils have led to this method of compacting deep sand.

The employment of this method provides the following advantages:

1. An effective means for compacting a range of saturated sands
2. An expedient method of compaction because of the speed of driving and extracting made available by the vibratory pile-driving device
3. Less expensive compaction than that obtained by other means of compaction because of the high production rates
4. Adjustable modular spacings to adapt to final density requirements and job site conditions
5. An effective method of compacting soil to substantial depths
6. A compaction method where lower initial density can make the method more expedient than higher initial density
7. A means of densification of some soils to reduce the soil liquefaction hazards of earthquakes
8. A method of compaction for use in land reclamation.

In order to increase the effectiveness of this method the following recommendations are made:

1. That an *in situ* test pattern procedure of two or three modular spacings be used to establish a grid pattern for probe holes for each application
2. That a quality control of the highest order be used in dredging out undesirable material, in cleaning the bottom of the excavation, and in the

dredge fill operation to ensure that only good compactable material is used for replacement

3. That this method not be used for compacting soils less than 12 to 13 ft (3.7 to 4.0 m) deep, because other methods of compacting generally are more economical
4. That when this method is used to compact soils 12 to 20 ft (3.7 to 6.1 m) deep, closer spacings of probes be used than for holes more than 20 ft (6.1 m) deep.

PROBLEMS

5-1 A given soil weighs 110 lb per cu ft *in situ*, 96 lb per cu ft loose, and 118 lb per cu ft when compacted in a fill. Determine the percent swell and shrinkage for the soil.

5-2 The capacity of a truck is 16 cu yd of earth. If the earth weighs 110 lb per cu ft *in situ* and 94 lb per cu ft loose, what is the capacity of the truck expressed in cubic yards bank measure?

5-3 If earth is placed in fill at the rate of 190 cu yd per hr, compacted measure at 10 percent moisture content, and the dry weight of the compacted earth is 2,890 lb per cu yd, how many gallons of water must be supplied each hour to increase the moisture content of the earth from 4 to 10 percent by weight?

5-4 Earth whose *in situ* weight is 112 lb per cu ft, loose weight is 95 lb per cu ft, and compacted weight is 120 lb per cu ft is placed in a fill at the rate of 240 cu yd per hr, measured as compacted earth, in layers whose compacted thickness is 6 in. Sheep's-foot roller drums, each 5 ft wide, are pulled by a tractor at a speed of 2 mph, with an operating factor of 75 percent. Determine the number of drums required to provide the necessary compaction if eight drum passes are specified for each layer of earth.

5-5 If a multiwheel pneumatic roller whose 7.50 × 15 14-ply tires are inflated to 90 psi, and whose wheel loads are 2,800 lb each, is used to compact a soil, what is the maximum compacted depth of a layer of earth that can be compacted to a unit pressure of not less than 50 psi at the bottom of the layer? Assume that the ground contact area for a tire is a circle whose area in square inches equals the wheel load divided by the ground contact pressure.

REFERENCES

1. Krebs, Robert D., and Richard D. Walker: *Highway Materials*, McGraw-Hill, New York, 1971, 428 p.
2. Kelley, Conrad M.: "A Long Range Durability Study of Lime Stabilized Bases," *Bulletin No. 328,* National Lime Association, Washington, D.C., 1977.
3. Covey, James N.: "An Overview of Ash Utilization in the United States," *Proceedings of the Fly Ash Applications in 1980 Conference*, Texas A&M University, May 1980.
4. Meyers, J. F., R. Pichumami, and B. S. Kapples: "Fly Ash as a Highway Construction Material," U.S. Department of Transportation, Federal Highway Administration, June 1976.
5. Terrel, R. L.: "A Guide Users Manual for Soil Stabilization," U.S. Department of Transportation, Federal Highway Administration, April 1979.
6. McKerall, W. C., and W. B. Ledbetter: Variability and Control of Class C Fly Ash, *Cement, Concrete, and Aggregates*, CCAGDP, vol. 4, no. 2, Winter 1982, ASTM, Philadelphia, Pa.
7. Bertram, George E.: *Proceedings of the Second Panamanian Conference on Soil Mechanics and Foundations*, Sao Paulo, Brazil, vol. I, pp. 444–452, 1963.

8. Prugh, Byron J.: Densification of Soils by Explosive Vibrations, *Journal of the Construction Division, Proceedings of the American Society of Civil Engineers,* vol. 89, pp. 79–100, March 1963.
9. "Blasters' Handbook," 12th ed., E. I. Du Pont de Nemours & Company, Inc., Wilmington, DE 19898, 1949.
10. "Safety in the Handling and Use of Explosives," Pamphlet No. 17, Institute of Makers of Explosives, 420 Lexington Avenue, New York, NY 10017, 1966.
11. Jones, H. W.: Densification of Sand for Drydock by Terra-Probe, *Journal of the Soils Mechanics and Foundation Division, Proceedings of the American Society of Civil Engineers,* vol. 99, pp. 451–470, June 1973.
12. Anderson, Robert D.: New Method for Deep Sand Vibratory Compaction, *Journal of the Construction Division, Proceedings of the American Society of Civil Engineers,* vol. 100, pp. 79–95, March 1974.
13. Ferguson Manufacturing and Equipment Co., Inc., 4900 Harry Hines Blvd., Dallas TX 75221.
14. RayGo, Inc., 9401 85th Ave., Minneapolis, MN 55440.
15. Wacker Corporation, 3808 Elm Street, Milwaukee, WI 53209.

CHAPTER
SIX

TRACTORS AND RELATED EQUIPMENT

TRACTORS

TRACTOR USES

Tractors have many uses as construction equipment. While their primary purpose may be to pull or push loads (Fig. 6-1), they are also used as mounts for many types of accessories, such as front-end shovels, rippers, bulldozer blades, sidebooms, hoes, trenchers, and others. There are sizes and types to fit almost any job for which they are usable.

TYPES OF TRACTORS

Tractors may be divided into two major types:

1. Crawler
2. Wheel

Wheel tractors are either two-wheel or four-wheel.

In selecting a tractor several factors should be considered; they include, but are not limited to, the following:

1. The size required for a given job
2. The kind of job for which it will be used—bulldozing, pulling a scraper, ripping, clearing land, etc.
3. The type of footing over which it will operate, i.e., high-tractive or low-tractive efficiency

Figure 6-1 Crawler tractor pushing a self-loading scraper. *(Caterpillar Tractor Co.)*

4. The firmness of the haul road
5. The smoothness of the haul road
6. The slope of the haul road
7. The length of haul
8. The type of work it will do after this job is completed

Crawler tractors Crawler tractors are usually rated by size or weight and power. The weight is important on many projects because the maximum tractive effort that a unit can provide is limited to the product of the weight times the coefficient of traction for the unit and the particular road surface, regardless of the power supplied by the engine. Table 4-3 gives the coefficients of traction for various surfaces.

Most manufacturers make crawler tractors with some or all models equipped with a choice of direct drive or torque converter and power-shift drives.

Crawler tractors with direct drive Table 6-1 gives pertinent information and performance data for tractors equipped with direct drives. Some manufacturers' specifications list two sets of drawbar pulls—rated and maximum. The rated is the drawbar pull that can be sustained for continuous operation, while the maximum is the drawbar pull that the tractor can exert for a short period while lugging the engine, such as when passing over a soft spot in the ground, which requires a temporary higher tractive effort. Thus, the rated pull should be used for continuous operation. Also, the available drawbar pull is subject to the limitations on traction developed between the tracks and the ground.

Some manufacturers rate their engines under standard conditions, namely at 60°F and sea-level elevation, while others rate their engines under more representative operating conditions, such as 85°F and up to 2,500 ft elevation above sea level.

Table 6-1 Representative specifications and performance data for crawler tractor equipped with direct drive

Approximate operating weight, lb	18,300	32,000	47,000
Flywheel hp	93	160	235
Drawbar hp	75	128	187
Ratio, lb per hp	197	200	200

Performance data

	Speed, mph	Speed, fpm	Drawbar pull, lb	Speed, mph	Speed, fpm	Drawbar, pull, lb	Speed, mph	Speed, fpm	Drawbar, pull, lb
Gear, forward									
1st	1.7	150	17,240	1.5	132	32,500	1.5	132	44,400
2d	2.7	238	10,470	2.2	193	22,700	1.9	167	34,500
3d	3.7	326	7,090	3.1	272	15,000	2.7	238	24,100
4th	5.2	458	4,670	4.6	405	9,390	3.5	307	17,750
5th	6.8	598	3,190	5.9	518	6,770	4.6	405	13,000
6th							6.3	555	8,450
Gear, reverse									
1st	2.1	185	13,670	1.8	158	28,470	1.5	132	43,700
2d	3.3	290	8,180	2.5	220	18,935	2.0	176	33,900
3d	4.6	405	5,440	3.7	325	12,390	2.7	238	23,700
4th	6.4	563	3,480	5.4	475	7,620	3.6	317	17,400
5th							4.6	405	12,700
6th							6.4	563	8,250

Crawler tractors with torque converter and power-shift transmissions Many crawler tractors are available with torque converter drives and power-shift transmissions, which eliminate shifting gears. These drives provide an efficient flow of power from the engine to the tracks by automatically selecting the speed which is most suitable for the load pulled by the tractor.

Figure 6-2 illustrates the performance curves for a tractor equipped with a torque converter and a three-speed transmission. Assume that the tractor must provide a drawbar pull of 50,000 lb. The figure indicates that the tractor should be operated in first gear and that it will have a maximum speed of 1.33 mph. If the required drawbar pull is 10,000 lb, this pull can be provided by each of the three gears. However, because the tractor can provide the required pull in third gear at a speed of 5 mph, it should be operated in this gear. The gear selected should be the highest one that will provide the required pull in order to operate at the highest possible speed.

The equivalent drawbar pull which a tractor must provide, regardless of whether it is a direct-drive or power-shift type, is the algebraic sum of the pull required by the towed load, the effect of grade on the tractor, and the effect of increased or decreased rolling resistance on the tractor.

Figure 6-3 illustrates a two-wheel tractor-pulled scraper.

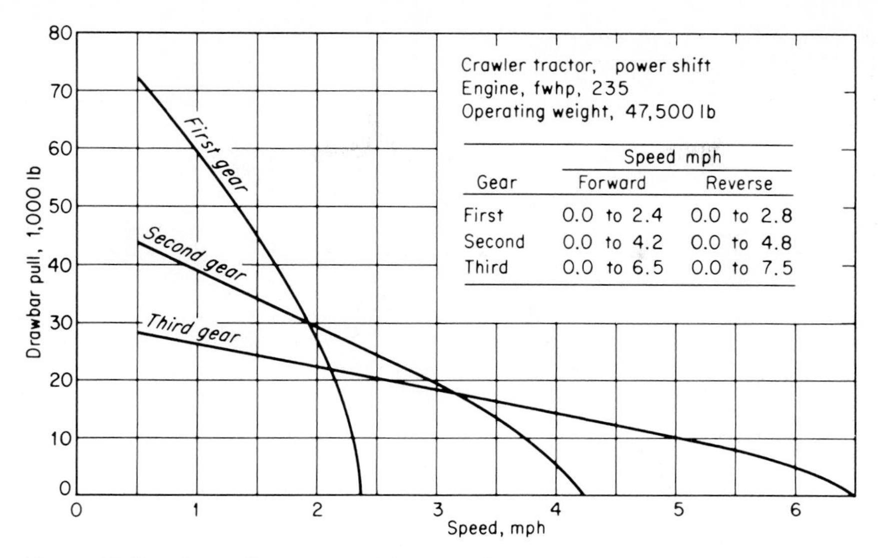

Figure 6-2 Drawbar pull versus speed for a crawler tractor.

Example 6-1 Consider the tractor whose performance data is illustrated in Fig. 6-2. The tractor must provide a drawbar pull of 12,000 lb to tow an attached load up a 6 percent grade over a haul road having a rolling resistance of 80 lb per ton. Determine the equivalent required drawbar pull and the maximum speed for the tractor.

The weight of the tractor is 23.75 tons.

The equivalent drawbar pull is as follows:

1. Required by towed load	=	12,000 lb
2. Required by grade, 23.75 × 6 × 20	=	2,850 lb
Subtotal	=	14,850 lb
3. Reduction in rolling resistance, 23.75(80 − 110)	=	− 712 lb
Total	=	14,138 lb

An examination of Fig. 6-2 indicates that the tractor can operate in third gear at a speed of 4 mph.

Wheel tractors One of the primary advantages of a wheel tractor compared with a crawler tractor is the higher speed possible with the former, in excess of 30 mph for some models. However, in order to attain a higher speed a wheel tractor must sacrifice pulling effort. Also, because of the lower coefficient of traction between rubber tires and some soil surfaces, the wheel tractor may slip its wheels before developing its rated pulling effort.

The traction developed by a wheel tractor is expressed in pounds of rimpull. This is a measure of the tractive effort which the engine is capable of delivering to the surface supporting the driving wheels. The net drawbar pull of a wheel tractor is obtained by deducting from the available rimpull and pull required to overcome the rolling resistance of the unit when it is traveling on a level haul surface or by deducting from the product of the coefficient of traction and the gross weight on the pulling wheels the pull required to overcome the rolling

Figure 6-3 Two-wheel tractor-pulled scraper.

Table 6-2 Representative specifications for two-wheel tractors

Approximate weight, lb (kg):	32,200 (14,560)	17,740 (8,050)
Engine hp (kW)	275 (205)	180 (134)
Ratio, lb per hp (kg/kW)	117 (71)	98 (60)
Tire sizes, in. (mm)	24.00 × 29 (610 × 738)	21.00 × 25 (534 × 635)

Performance data

Speed Gear	Speed, mph (km/h)*	Rimpull, lb (kg)†	Speed, mph (km/h)*	Rimpull, lb (kg)†
1st	2.16 (3.48)	25,000‡ (11,380)	3.41 (5.50)	15,850 (7,175)
2nd	4.18 (6.73)	17,100 (7,785)	7.25 (11.70)	7,450 (3,380)
3rd	7.15 (11.50)	10,050 (4,560)	12.63 (20.35)	4,280 (1,945)
4th	12.18 (19.60)	5,880 (2,670)	22.28 (35.90)	2,420 (1,100)
5th	20.00 (32.20)	3,580 (1,620)	35.03 (56.35)	1,540 (700)
Reverse	2.79 (4.49)	25,000‡ (11,380)	4.35 (7.00)	12,440 (5,650)

* To convert mph to km/h, multiply mph by 1.609.

† To convert lb to kg, multiply lb by 0.454.

‡ These rimpulls are limited by the maximum traction resulting from the weights on the tires, when pulling loaded scrapers.

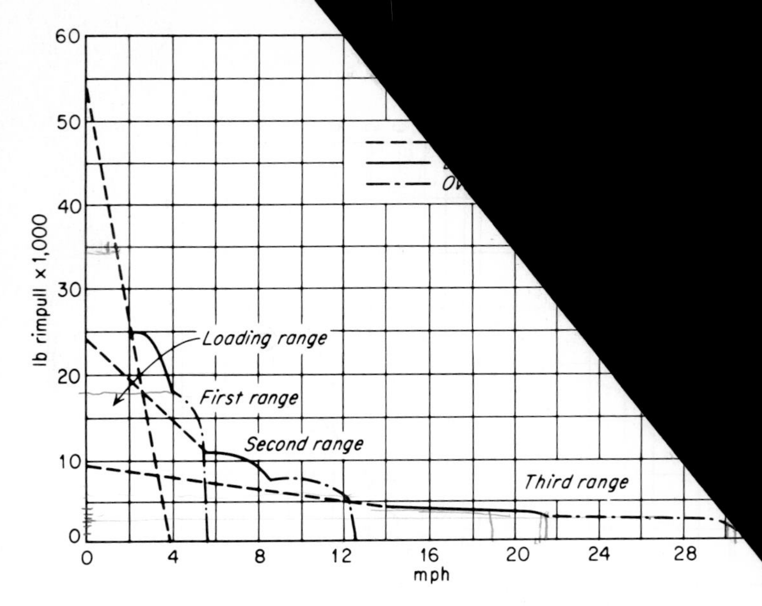

Figure 6-4 Rimpull-speed chart for wheel tractor and scraper with power-shift tran
rimpull will depend on traction available and total weight of tractor drive wheels.

resistance, using whichever value is the smaller. As the speed is in
through the selection of higher gears, the rimpull will be decreased i
proximately the same proportion. Thus, for a given unit whose engine
operated at a rated power, the product of the speed times the rimpull w
remain approximately constant.

Performance data for wheel tractors While most wheel tractors are equipped with torque converters and power-shift transmissions, some are equipped for direct drive. For this reason performance data will be presented for both types.

Table 6-2 illustrates the type of specifications provided for wheel tractors equipped for direct drive.

Figure 6-4 illustrates the type of information that may be provided for a wheel tractor by the manufactuer. The unit consists of a two-wheel tractor and a two-wheel scraper having the following specifications:

Engine fwhp, 250
Scraper capacity
  Struck, 14 cu yd
  Heaped, 18 cu yd

Weight of unit, lb:	Tractor	Scraper	Total
Empty	33,570	14,730	48,300
Loaded	49,670	40,630	90,300
Distribution loaded	55%	45%	100%

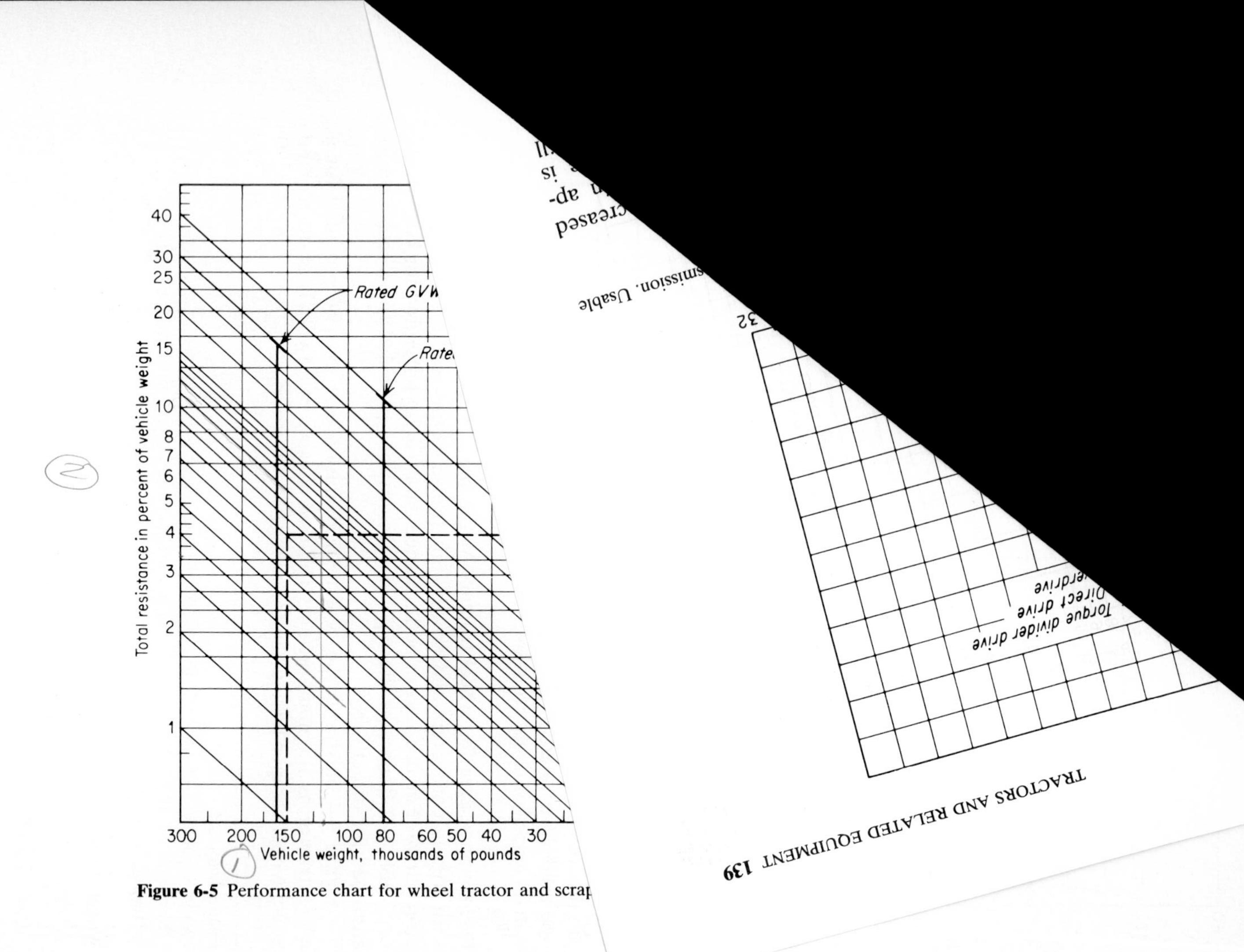

Figure 6-5 Performance chart for wheel tractor and scrap

The tractor is equipped to permit torque divider drive, direct drive, or overdrive in each gear range. When an engine is capable of providing the required rimpull, the overdrive should be used in order to make use of the higher speed.

The figure is a performance chart whose purpose is to enable the user of the unit to determine the maximum possible speed for a given load and haul condition, as illustrated by the following example.

Example 6-2 Determine the maximum speed for the tractor of Fig. 6-4 for the stated conditions.

Gross vehicle weight loaded, 88,000 lb	
Grade	= 6%
Rolling resistance, 80 lb per ton	= 4%
Total resistance	= 10%
Required rimpull, 88,000 × 0.10	= 8,800 lb

Figure 6-4 reveals that this rimpull can be provided by any one of the three gear ranges. However, the maximum speed, 8 mph, is obtained in the second range using direct drive.

Another type of performance chart is illustrated in Fig. 6-5. This chart is applicable for a two-wheel tractor and a two-wheel scraper having the following specifications:

Engine fwhp at 2,100 rpm, 398			
Scraper capacity			
Struck, 24 cu yd			
Heaped, 32 cu yd			
Weight of unit, lb:	Tractor	Scraper	Total
Empty	51,700	27,650	79,350
Loaded	81,270	78,080	159,350
Distribution loaded	51%	49%	100%

The tractor is equipped with a six-speed power-shift transmission with an overdrive in the top five speed ranges. The total resistance in percent of vehicle weight is the algebraic sum of the resistances resulting from grade and rolling resistance. The maximum speed is determined by applying the following steps:

1. Start with the appropriate vehicle weight on the lower left horizontal scale.
2. Read up this weight line to the intersection with the sloping total resistance line.
3. From this intersection read horizontally to the right to the intersection with the speed range performance curve.
4. From this intersection read down to the lower right scale to determine the vehicle speed.

If the application of step 3 results in the intersection of two speed range curves, use the curve which gives the higher speed.

The following example illustrates the method of using Fig. 6-5.

Example 6-3 Determine the maximum speed for the tractor for the stated conditions.

Gross vehicle weight loaded, 150,000 lb
Grade = 5%
Rolling resistance, 60 lb per ton = 3%
Total resistance = 8%

The two dashed lines, representing steps 1, 2, and 3, indicate that the tractor can operate in the third speed range, using normal or overdrive position. Because the use of overdrive will permit a speed of 9.5 mph, whereas normal will permit a speed of only 8 mph, the former should be used.

GRADABILITY

Gradability is defined as the maximum slope, expressed as a percent, up which a crawler or wheel-type prime mover may move at a uniform speed. The gradability may be determined for an empty or a loaded vehicle. Thus, the gradability of a tractor only will be greater than for a tractor that is pulling a loaded vehicle. Gradability may be specified for any desired gear.

The forward motion of a prime mover is limited by the following factors:

1. The power developed by the engine and available as drawbar pull or rimpull.
2. The rolling resistance of the haul road.
3. The gross weight of the prime mover and its load.
4. The grade to be negotiated. Adverse grade adds to the resistance, while favorable grade subtracts from the resistance.

The gradability of a crawler tractor is determined by subtracting from the available drawbar pull the total pull required to overcome the rolling resistance on the unit and any load that it will pull. The surplus drawbar pull, if it is less than the coefficient of traction multiplied by the weight of the tractor, is then available to negotiate a grade. As the drawbar pull of a crawler tractor, taken from the manufacturer's specifications, is usually based on a rolling resistance of 110 lb per ton, any rolling resistance in excess of this amount should be applied to the weight of the tractor. The entire rolling resistance on the towed load should be used. In order to provide a reasonable factor of safety, not more than 85 percent of the rated drawbar pull of a tractor should be used in determining the gradability of the unit.

Example 6-4 Determine the gradability of a crawler tractor pulling a high-pressure rubber-tired self-loading scraper and its load. The following information is available:

Tractor horsepower, 180
Weight of tractor, 40,500 lb or 20.25 tons
Drawbar pull in 1st gear, 33,714 lb
Available drawbar pull, $0.85 \times 33{,}714 = 28{,}600$ lb
Weight of loaded scraper, 78,960 lb or 39.48 tons
Haul road, rutted, uneven earth
Rolling resistance for tractor, 160 lb per ton
Excess rolling resistance for tractor, 50 lb per ton
Rolling resistance for scraper, 210 lb per ton

The gradability is determined as follows:

Rolling resistance of tractor, $20.25 \times 50 = 1{,}012$ lb
Rolling resistance of scraper, $39.48 \times 210 = 8{,}291$ lb
Combined rolling resistance = 9,303 lb
Drawbar pull available to overcome grade:
Maximum available drawbar pull = 28,600 lb
Required for rolling resistance = −9,303 lb
Pull available for grade = 19,297 lb

Combined weight of tractor and loaded scraper:
Tractor, 20.25 tons
Scraper, 39.48 tons
Total 59.73 tons
Pull required per ton per 1% grade, 20 lb
Pull required per 1% grade for the total load, $20 \times 59.73 = 1{,}195$ lb
Maximum possible grade, $\dfrac{19{,}297}{1{,}195} = 16\%$

For the tractor alone the maximum possible grade will be:

Maximum available drawbar pull = 28,600 lb
Pull required for rolling resistance = −1,012 lb
Pull available for grade = 27,588 lb

Pull required per 1% grade, $20 \times 20.25 = 405$ lb
Maximum possible grade, $\dfrac{27{,}588}{405} = 68\%$, provided the tracks do not slip

The gradability of a wheel-type tractor or a truck can be determined in the same manner as for a crawler tractor. However, if the manufacturer's specifications furnish sufficient information, the gradability for any gear can be determined from the equation

$$K = \frac{972 \times T \times G}{R \times W} - \frac{N}{20} \tag{6-1}$$

where K = gradability, %
T = rated engine torque, lb-ft
G = total gear reduction for particular gear selected
R = rolling radius, the radius of the loaded driving wheels, in., measured from center of axle to surface of ground
W = gross weight of complete unit, lb
N = rolling resistance, lb per ton

Example 6-5 Determine the gradability of a wheel-tractor-pulled wagon and its load, when operating in third gear at sea level. The following information will apply:

Rated torque at 2,100 rpm, 750 lb-ft
Total gear reduction, 41.0
Rolling radius, loaded, 29.38 in.
Gross weight, 138,500 lb
Rolling resistance, 50 lb per ton

$$K = \frac{972 \times 750 \times 41.0}{29.38 \times 138{,}500} - \frac{50}{20} = 7.3 - 2.5 = 4.8\%$$

The value, 4.8 percent, is obtained by using the full torque of the engine. If only 85 percent of the torque is considered, as a safety precaution, the gradability will be

$$K = 7.3 \times 0.85 - 2.5 = 6.2 - 2.5 = 3.7\%$$

If the rolling resistance of the haul road is permitted to increase to 80 lb per ton, and 85 percent of the torque is considered, the gradability will be

$$K = 6.2 - \tfrac{80}{20} = 6.2 - 4.0 = 2.2\%$$

Table 6-3 gives representative gradability for a wheel-tractor-pulled wagon, both empty and loaded, operating at sea level, based on using the full engine torque. In addition, the following information is applicable:

Engine, 300 belt hp
Rated torque at 2,100 rpm, 750 lb-ft
Rolling radius, 29.38 in.
Empty weight of tractor and wagon, 58,500 lb
Gross weight with load, 138,500 lb
Coefficient of traction, 0.6
Rolling resistance, 40 lb per ton

An appropriate performance chart, such as Fig. 6-5, can also be used to determine the gradability of a tractor, as illustrated by the following example.

Table 6-3 Representative gradability of a wheel tractor and wagon

	Speed			Gradability, %	
Gear	mph	(km/h)	Gear reduction	Empty	Loaded
1st	3.1	(5.0)	116.9:1	22.0*	18.8
2d	5.2	(8.3)	70.3:1	22.0*	10.5
3d	9.0	(14.5)	41.0:1	15.3	5.3
4th	15.7	(25.2)	23.5:1	7.8	2.2
5th	24.7	(39.7)	14.9:1	4.3	0.6
Reverse	4.1	(6.6)	90.0:1	22.0*	14.0

* These values are limited by the maximum traction between the tires and the haul road.

Example 6-6 Determine the gradability of the tractor represented in Fig. 6-5 when it is operating with a gross vehicle weight of 150,000 lb at a uniform speed of 10 mph. The steps are as follows:

1. Read up the 10-mph line to the intersection with the normal drive curve for the fourth speed range.
2. Read horizontally to the left to the intersection with the 150,000-lb vehicle weight line.
3. Read upward to the left, parallel with the sloping total resistance lines, to the value 6.7 percent, which is the combined resistance of the grade and the rolling resistance.
4. Deduct the rolling resistance of the haul road, expressed as a percent, from the 6.7 percent. The remainder is the maximum grade up which this vehicle can operate at a uniform speed. For example, if the rolling resistance is 60 lb per ton, equal to 3 percent, the maximum grade will be $6.7 - 3.0 = 3.7$ percent.

BULLDOZERS

GENERAL INFORMATION

The term bulldozer may be used in a broad sense to include both a bulldozer and an angledozer. These machines may be further divided, on the basis of their mountings, into crawler-tractor- or wheel-tractor-mounted. Based on the method of raising and lowering the blade, a bulldozer may be classified as cable-controlled or as hydraulically controlled (see Fig. 6-6). Each type of

Figure 6-6 Hydraulic-controlled bulldozer. *(Caterpillar Tractor Co.)*

equipment has a place in the construction industry. For some projects either type will be satisfactory, while for other projects one type will be superior.

Bulldozers are versatile machines on many construction projects, where they may be used from the start to the finish for such operations as:

1. Clearing land of timber and stumps
2. Opening up pilot roads through mountains and rocky terrain
3. Moving earth for haul distances up to approximately 300 ft
4. Helping load tractor-pulled scrapers
5. Spreading earth fill
6. Backfilling trenches
7. Clearing construction sites of debris
8. Maintaining haul roads
9. Clearing the floors of borrow and quarry pits

Bulldozers are mounted with blades perpendicular to the direction of travel, while angledozers are mounted with the blades set at an angle with the direction of travel. The former push the earth forward, while the latter push it forward and to one side. Some blades may be adjusted to permit their use as bulldozers or angledozers. The size of a bulldozer is indicated by the length and height of the blade. Plates may be installed at the ends of a blade to reduce the spillage when a machine is used for moving earth.

CRAWLER-MOUNTED VERSUS WHEEL-MOUNTED BULLDOZERS

At one time bulldozers were mounted on crawler tractors only. However, with the development of wheel tractors, bulldozers have been mounted on them also. Each type of mounting has advantages under certain conditions. For some jobs the conditions are such that either type may be used satisfactorily.

Among the advantages claimed for the crawler-mounted bulldozer are the following:

1. Ability to deliver greater tractive effort, especially in operating on soft footing, such as loose or muddy soil
2. Ability to travel over muddy surfaces
3. Ability to operate in rocky formations, where rubber tires might be seriously damaged
4. Ability to travel over rough surfaces, which may reduce the cost of maintaining haul roads
5. Greater flotation because of the lower pressures under the tracks
6. Greater use versatility on jobs

Among the advantages claimed for wheel-mounted bulldozers are the following:

1. Higher travel speeds on the job or from one job to another
2. Elimination of hauling equipment to transport the bulldozer to a job
3. Greater output, especially when considerable traveling is necessary
4. Less operator fatigue
5. Ability to travel on paved highways without damaging the surface

If the equipment user has a job which is large enough to justify the purchase of special equipment, he should select the equipment that is most suitable for the particular job. However, since small jobs will seldom justify the purchase of special equipment, it is desirable to select equipment which can be used on other jobs. Under the latter conditions the selection of versatile equipment will usually be a wise choice.

Figure 6-7 illustrates a crawler-mounted bulldozer opening up construction on a precarious mountain road-building job where sharp rocks predominate. For work of this type the crawler-mounted bulldozer is usually superior to the wheel-mounted. Figure 6-8 illustrates a wheel-mounted bulldozer.

MOVING EARTH WITH BULLDOZERS

Under certain conditions bulldozers are satisfactory machines for moving earth for such jobs as excavating ponds for stock water, trench silos, and highway cuts, stripping the topsoil from land or ore deposits, constructing low levees, backfilling trenches, spreading material on fills, etc. In general, haul distances should be less than 300 ft. Either a crawler-mounted or a wheel-mounted tractor may be used, a crawler-mounted machine having an advantage on short hauls with soft or muddy ground, and a wheel-mounted machine possibly having an advantage on longer hauls and firm ground.

The output of a bulldozer will vary with the conditions under which it operates. During the first passes over a given lane most of the initial earth will spill off the ends of the blade to form a windrow on each side of the lane. After these windrows have been built up to form a trench, further end spillage will be reduced or eliminated, with a substantial increase in output. Steel plates on the ends of a blade will reduce end spillage. On some jobs two bulldozers, working side by side, with adjacent ends of the blades in contact, have been used to increase the output as much as 50 percent over the combined output of two machines working separately. If earth can be pushed downhill, the output of a machine will be increased substantially because of the advantage of the favorable grade and the ability to float larger quantities of earth ahead of the machine. Figure 6-9 illustrates two bulldozers operating under favorable conditions, namely, downhill dozing, slot excavation, side-by-side operation, and floating extra earth ahead of the machine.

Figure 6-7 Crawler-tractor-mounted bulldozer. *(Caterpillar Tractor Co.)*

Figure 6-8 Wheel-tractor-mounted bulldozer. *(Caterpillar Tractor Co.)*

Figure 6-9 Operating two bulldozers side by side.

THE OUTPUT OF BULLDOZERS

The blade of a bulldozer has a theoretical capacity which varies with the class of earth and the size of the blade. If the capacity of a blade is known, one can determine the approximate output of a machine by estimating the number of passes it will make in an hour.

Example 6-7 Estimate the approximate output of a bulldozer for the following conditions:

Material, sandy loam topsoil, weight 2,700 lb per cu yd bm
Swell, 25%
Haul distance, 100 ft, over level ground, with bulldozer operating in a slot
Crawler tractor, 72 drawbar hp
Moldboard size, 9 ft 6 in. long, 3 ft 0 in. high
Rated moldboard capacity, 3.6 cu yd loose volume
Net moldboard capacity, 3.6 ÷ 1.25 = 2.9 cu yd bm
Operating factor, 50-min hr
Probable round-trip time

Pushing, 100 ft @ 1.5 mph	= 0.758 min
Returning, 100 ft @ 3.5 mph	= 0.324 min
Fixed time, loading and shifting gears	= 0.320 min
Total time	= 1.402 min

Trip per hr, 50 ÷ 1.402 = 35.7
Output per hr, 35.7 trips @ 2.9 cu yd = 103.4 cu yd bm

The output given in the example is based on favorable operating conditions which permit a load equal to the maximum capacity of the dozer. For most projects the load will be less than the maximum possible capacity. For example, if the earth is ordinary soil, the load might be reduced to 2.0 cu yd bank measure. With other conditions remaining the same, the output per hour will be

35.7 trips @ 2.0 cu yd = 71.4 cu yd bm

The approximate capacity of a bulldozer blade may be determined from the size of the load pushed by the blade. Actual measurements of representative loads will give better results than estimates. For example, if a blade 9 ft 6 in. long by 3 ft 0 in. high is used to push earth in a slot or trench whose height is about equal to that of the blade, it is possible to fill the blade to full length and height. Although the shape of the front slope of the earth will be irregular, assume that it is equivalent to a 2:1 slope, whose cross section is a triangle having a base width of 3 ft. The loose volume of the load is given by the equation

$$V = \frac{L \times H \times W}{2 \times 27}$$

where L = length of load, ft
H = height of load, ft
W = width of load, ft

For the stated conditions the volume will be

$$V = \frac{9.5 \times 3 \times 6}{2 \times 27} = 3.2 \text{ cu yd}$$

Table 6-4 Representative blade capacities and bulldozer output
In cubic yards bank measure

Blade length	Blade height, in.	Tractor, drawbar hp	Forward speed, fpm	Reverse speed, fpm	Blade capacity, cu yd	Output, cu yd per hr — Haul distance, ft: 100	200	300	400
11 ft 3 in.	$45\frac{1}{2}$	130	150	326	4.8	184	105	74	57
10 ft 3 in.	$45\frac{1}{2}$	80	123	334	4.4	152	86	60	46
9 ft 6 in.	38	65	123	343	2.8	98	55	38	29
8 ft 2 in.	38	65	123	343	2.4	84	47	33	25
7 ft 2 in.	$32\frac{1}{2}$	43	150	167	1.5	47	26	18	14
5 ft 8 in.	$27\frac{1}{2}$	32	150	185	0.9	29	16	11	9
11 ft 2 in.	43	210*	141	712	4.2	178	103	73	56
11 ft 3 in.	36	122*	141	712	3.0	127	74	52	40

* These values are flywheel horsepower for wheel-mounted dozers.

For a swell of 25 percent the net volume will be

$$3.2 \div 1.25 = 2.56 \text{ cu yd bm}$$

If the dozing is done without slots, the capacity of the blade will be reduced by approximately 25 percent. Also, if the earth is so hard that a full load cannot be moved, the capacity must be reduced accordingly.

Table 6-4 gives approximate blade capacities and outputs in cubic yards bank measure for various sizes of blades and tractors. The information given in the table is based on pushing full loads in slots. It is assumed that the tractors will push the loads forward in first gear, then return for another load in reverse gear. It is assumed that the tractors will operate 50 min per hr. For other job conditions the outputs given in the table must be modified.

CLEARING LAND

LAND CLEARING OPERATORS

Clearing land may be divided into several operations, depending on the type of vegetation, the condition of the soil and topography, the amount of clearing required, and the purpose for which the clearing is done, as listed below.

1. Removing all trees and stumps, including roots
2. Removing all vegetation above the surface of the ground only, leaving stumps and roots in the ground
3. Disposing of vegetation by stacking and burning it
4. Knocking all vegetation down, then chopping or crushing it to or into the surface of the ground, or burning it later
5. Killing or retarding the growth of brush by cutting the roots below the surface of the ground

TYPES OF EQUIPMENT USED

Several types of equipment are used for clearing land, with varying degrees of success. Included are the following:

1. Tractor-mounted bulldozers
2. Tractor-mounted special blades
3. Tractor-mounted rakes
4. Tractor-pulled chains and steel cables

Tractor-mounted bulldozers. Whereas bulldozers were used extensively during the past to clear land, they are now being replaced by special blades mounted on tractors. There are at least two valid objections to the use of bulldozers. Prior to felling large trees they must excavate earth from around the tree and cut the main roots, which leaves objectionable holes in the ground, and requires much time. Also, when stacking the felled trees and other vegetation they transport considerable earth to the piles, which makes burning more difficult.

Tractor-mounted special blades Two types of special blades are used to fell trees; both are mounted on the front ends of tractors.

One is a single-angle blade with a projecting stinger on the lead side, extending ahead of the blade, so that it may be forced into and through a tree to split and weaken it. Thus, if a tree is too large to be felled in one pass, the trunk may be split and removed in parts. Also, the tractor may make a pass around a tree with the stinger penetrating the ground to cut the main horizontal roots of the tree. It may be used to remove stumps and to stack material for burning.

Another type of special blade is a V blade, with a protruding stinger at its lead point, as illustrated in Fig. 6-10. The sole effect of the blade permits it to

Figure 6-10 Tractor-mounted V blade for clearing land. *(Fleco Corporation.)*

Figure 6-11 Tractor-mounted V blade splitting a large tree. *(Fleco Corporation.)*

Figure 6-12 Tractor-mounted clearing blade. *(Rome Industries.)*

slide along the surface of the ground, thereby cutting vegetation flush with the surface. However, it can be lowered below the surface to remove stumps. Also, the blade may be raised to permit the stinger to pierce a tree above the surface of the ground. Other special blades are shown in Figs. 6-11 and 6-12.

Tractor-mounted rakes Figure 6-13 illustrates a tractor-mounted rake which can be used to grub and pile trees, boulders, and similar materials without transporting excess quantities of soil. Granular material, such as sand and gravel, flows between the teeth readily. Optional tooth spacing is available for use under varying soil conditions. However, some plastic material may tend to combine with the vegetation and clog the spaces between the teeth.

This rake can be an effective tool when stacking the cleared material in piles for burning.

Tractor-mounted clamp rakes Figure 6-14 illustrates a tractor-mounted clamp rake which can be used to pick up felled trees and brush and transport them to sites for burning or to other sites for disposal. For some projects this method of handling the material is better than using a tractor-mounted rake to push it over the surface of the ground. Using this type of rake reduces or eliminates the soil transported to the stack. Also, because of its higher reach, the clamp rake can be more effective in reshaping a pile of material to increase the rate of burning.

Tractor-pulled chains Figure 6-15 illustrates a heavy chain pulled by two crawler tractors, which is effective in felling trees and partially eliminating

Figure 6-13 Tractor-mounted land-clearing rake. *(Fleco Corporation.)*

Figure 6-14 Tractor-mounted clamp rake. *(Fleco Corporation.)*

Figure 6-15 Tractor-pulled chain used to clear land. *(Caterpillar Tractor Co.)*

Figure 6-16 Tractor-pulled root plow.

brush when used to clear semiarid land. The effectiveness of the chain may be increased by welding steel sections, such as short lengths of rail, to the links perpendicular to the chain. The extra weight holds the chain closer to the surface of the ground and removes more of the smaller brush and other vegetation.

Figure 6-17 Tractor-mounted grapple shear cutting a tree. *(Rome Industries.)*

Figure 6-18 Burning brush with forced draft and fuel oil. *(Fleco Corporation.)*

A second trip over the previously chained area several months after the first pass, with the chain pulled in the opposite direction, will further reduce the quantity of surviving vegetation.

Tractor-pulled root plows Figure 6-16 illustrates a tractor-mounted root plow which has been very effective in killing small trees and brush, especially when it is used in clearing semiarid land. The plow floats under the surface of the ground at a predetermined depth to slice the roots of the plants below the bud zones, thereby preventing many of them from resprouting.

This method of clearing land is economical and effective when used on a project in which it is not necessary to disturb the surface of the ground, such as clearing the right-of-way for a transmission line, or to increase the grazing capacity of range land.

Cutting trees with a shear Figure 6-17 illustrates a tractor-mounted shear which may be used to cut trees above the surface of the ground. This machine is especially useful for felling only selected trees without disturbing others.

DISPOSAL OF BRUSH

When brush is to be disposed of by burning, it should be piled in stacks and windrows, with a minimum amount of soil. Shaking a rake while it is moving the brush will reduce the amount of soil present.

If the brush and trees are burned while the moisture content is high, it may be necessary to provide a continuous external source of fuel, such as oil, to maintain satisfactory combustion. The burner illustrated in Fig. 6-18, which consists of a gasoline-engine-driven pump and a propeller, is capable of maintaining a fire even under adverse conditions. The liquid fuel is blown into the pile of material as a stream while the propeller furnishes a supply of air to assure vigorous burning.

PRODUCTION RATES

As previously stated, the rate of clearing land will depend on several variables, including, but not limited to, the following: (1) density of vegetation, (2) sizes and kinds of trees, (3) kind of soil, (4) topography, (5) rainfall, (6) types of equipment used, (7) skill of equipment operators, and (8) requirements of the specifications governing the project.

Equation (6-2) may be used as a guide in estimating the required time to fell trees only, using a shear-type cutting blade illustrated in Fig. 6-10, mounted on a crawler tractor of the size indicated in Table 6-5 [1]. Prior to preparing an estimate, the estimator should visit the project to be cleared in order to obtain information needed to evaluate the variable factors in the equation. With this information reasonably applicable values can be assigned to the factors listed in Table 6-5. Thus we have

$$T = B + M_1N_1 + M_2N_2 + M_3N_3 + M_4N_4 + DF \tag{6-2}$$

where T = time per acre, min

B = base time required for a tractor to cover an acre with no trees requiring splitting or individual treatment, min

M = time required per tree in each diameter range, min

Table 6-5 Representative times in minutes for cutting trees with tractor-mounted blades

Size tractor, fwhp	Base time B	Time to cut a tree* 1–2 ft dia. M_1	2–3 ft dia. M_2	3–4 ft dia. M_3	4–6 ft dia. M_4	Time per foot for diameters above 6 ft F
93	40	0.8	4.0	8.0	25	. . .
130	28	0.5	2.0	4.0	12	4.0
190	21	0.3	1.5	2.5	7	2.0
320	18	0.3	0.5	1.5	4	1.2

* The times listed are for cutting trees flush with the surface of the ground. If it is necessary to remove the stumps, the times should be increased by 50 percent.

Table 6-6 Representative times in minutes for stacking trees with tractor-mounted blades

Size tractor, fwhp	Base time B	Time to stack a tree 1–2 ft dia. M_1	2–3 ft dia. M_2	3–4 ft dia. M_3	4–6 ft dia. M_4	Time per foot for diameters above 6 ft F
93	35	0.3	0.6	2.5	...	...
130	28	0.2	0.4	1.5	3.0	...
190	24	0.1	0.3	1.0	2.0	0.4
320	20	0.0	0.1	0.7	1.2	0.2

N = number of trees per acre in each diameter range obtained from a field survey
D = sum of diameter in feet of all trees per acre, if any, larger than 6 ft in diameter at ground level
F = time required per foot of diameter to fell trees larger than 6 ft in diameter, min

Equation (6-2) may also be used to estimate the time required to stack felled trees into windrows spaced approximately 200 ft apart, by letting M_1, M_2, etc., represent the time required to move a tree into a windrow. Table 6-6 gives representative values for the time required to pile trees.

COST OF CLEARING LAND

The cost of clearing land varies considerably with the factors previously listed. Very little information on the subject has been released. However, in 1958 the Agricultural Experiment Station of Auburn University, Auburn, Alabama, conducted tests to determine the cost of clearing land using three sizes of crawler tractors, equipped with bulldozer blades and with shearing blades, such as the one illustrated in Fig. 6-10. The results of the tests have been published in a booklet [2].

For test purposes an area of 24 acres was divided into 12 plots of 2 acres each, with dimensions 198 ft wide by 440 ft long. Each size tractor cleared two plots using a bulldozer blade and two plots using a shearing blade. The net time required to fell, stack, and burn the material from each plot was determined. The trees consisted of pine, oak, hickory, and gum, distributed by species, size, and density as listed in Table 6-7. The diameters of the trees were measured at breast height.

The trees were felled, and then pushed along the surface of the ground and stacked in windrows not more than 198 ft apart, after which they were burned. During the burning operation the timber was pushed into tighter stacks to increase the burning effectiveness, using a tractor-mounted blade.

Table 6-7 Types of equipment used, species, sizes, and densities of trees

Plot no.	Blade used*	% by species		% by size trees, in.		No. trees per acre
		Hardwood	Pine	To 6	Above 6	
1	*B*	79	21	87	13	375
2	*B*	98	2	74	26	285
3	*B*	97	3	76	24	385
4	*B*	56	44	87	13	585
5	*B*	53	47	93	7	680
6	*B*	78	22	87	13	755
7	*S*	80	20	86	14	690
8	*S*	29	71	98	2	1,545
9	*S*	72	28	82	18	445
10	*S*	60	40	98	2	710
11	*S*	89	11	72	28	410
12	*S*	75	25	76	24	400

* *B* denotes a bulldozer blade and *S* denotes a shearing blade.

Table 6-8 Average machine time required, in hours, to clear an acre of land based on size tractor and blade used

	Time per acre, hr					
	93 fwhp		130 fwhp		190 fwhp	
Operation	*B**	*S*	*B*	*S*	*B*	*S*
Felling	2.19	1.58	1.71	1.14	0.92	0.71
Stacking	0.52	0.55	0.56	0.60	0.48	0.46
Disposal	1.75	0.84	1.80	0.78	1.93	0.70
Total	4.46	2.97	4.07	2.52	3.33	1.87

* *B* denotes a bulldozer blade and *S* denotes a shearing blade.

Table 6-8 shows the average time required by each size crawler tractor and type of blade to fell, stack, and dispose of an acre of timber. The smaller amounts of time required to dispose of trees felled with the shearing blades were the result of the smaller amounts of soil in the roots of trees felled with this type of blade.

RIPPING ROCK

GENERAL INFORMATION

Although rock has been ripped with varying degrees of success for many years, recent developments in methods, equipment, and knowledge have greatly increased the extent of ripping today. Rock that was considered to be unrippable a few years ago is now ripped with relative ease, and at cost reductions, including ripping and hauling with scrapers, amounting to as much as 50 percent when compared with the costs of drilling, blasting, loading with shovels, and hauling with trucks.

The major developments that are responsible for the increase in ripping rock include:

1. More powerful tractors
2. Improvements in the sizes and performances of rippers
3. Better instruments for determining the rippability of rocks
4. Improved techniques in using instruments and equipment

DETERMINING THE RIPPABILITY OF ROCK

Prior to selecting the method of excavating and hauling rock it is desirable to determine if the rock can be ripped or if it will be necessary to drill and blast it. Because the rippability of most types of rocks is related to the speed at which sound waves travel through rock it is possible to use seismographic methods to determine with reasonable accuracy if a rock can be ripped.

Rocks which propagate sound waves at low velocities are rippable, whereas rocks which propagate waves at high velocities are not rippable. Rocks having intermediate velocities are classified as marginal. Figure 6-19 indicates the

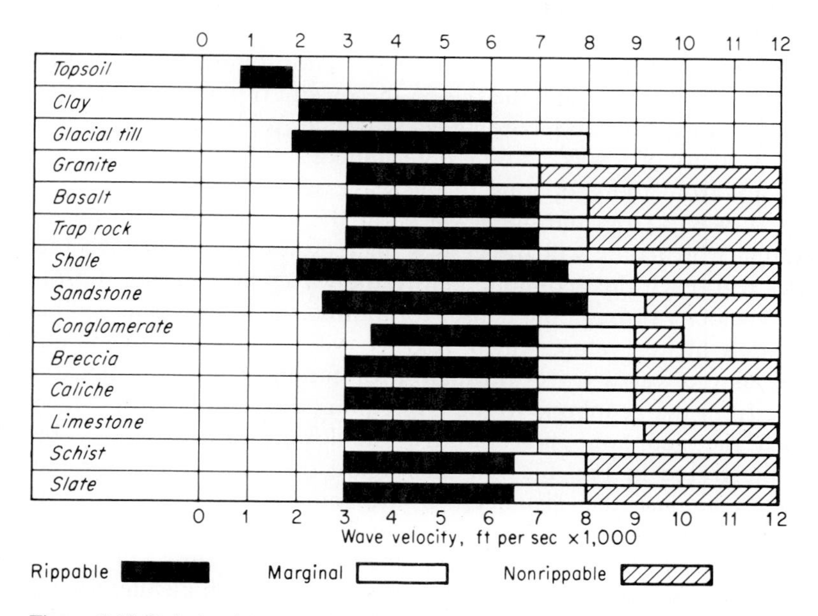

Figure 6-19 Relation between rippability of rock and velocity of sound waves.

velocity ranges for various types of soils and rocks encountered on construction projects. The indication that a rock may be rippable, marginal, or nonrippable is based on using a single-shank hydraulic ripper mounted on a crawler tractor whose engine develops about 385 fwhp. If two such tractors are used in tandem to pull one ripper, the rippability range can be increased somewhat, up to velocities as high as 8,000 to 10,000 fps in some instances. The information appearing in the figure should be used as a guide only. The decision to rip or not to rip rock should be based on the relative costs of excavating, using the methods under consideration, and the equipment available. If smaller tractors are to be used, the upper limits on the velocities of rock to be ripped will be less than those appearing in Fig. 6-19. Field tests may be necessary to determine if a given rock can be ripped economically.

DETERMINING THE SPEED OF SOUND WAVES IN ROCK

Figure 6-20 shows a type of geophysical equipment that is used to determine the velocity of sound waves in soil. Figure 6-21 illustrates the paths followed by sound waves from the wave-generating source through a formation to the detecting instruments. A geophone, which is a sound sensor, is driven into the ground at station 0. Equally spaced points 1, 2, 3, etc., are located along a line, as indicated. A wire is connected from the geophone to the seismic timer, and another wire is connected from the timer to a sledge hammer, or another impact-producing tool. A steel plate is placed on the ground at stations 1 through 8, in successive order. When the hammer strikes the steel plate, a switch closes instantly to send an electric signal to the timer, which starts the timer. At the same instant the blow from the hammer sends sound waves into the formation, which travel to the geophone. Upon the receipt of the first wave the geophone signals the timer to stop recording elapsed time. With the distance and time known, the velocity of a wave can be determined.

As the distance from the geophone to the wave source, namely the steel plate, is increased, waves from the plate will enter the lower and more dense formation, through which they will travel at a higher speed than through the topsoil, and thus will reach the geophone before the waves through the topsoil arrive. When the velocity through the denser formation is determined, it will be noted that it has a higher value, which velocity will remain approximately constant as long as the waves travel through a formation of uniform density.

The distance from the geophone to the point along the stations from which the waves reach the geophone first through the lower formation may be determined as illustrated in Fig. 6-22. In this figure the travel times for the waves are plotted against the distances to the impact stations. For each formation the velocity is essentially a straight line. The point of intersection of the velocity lines indicates the critical distance from the geophone.

The depth to the surface separating the two strata depends on the critical distance and the velocities in the two materials. It can be computed from the

Figure 6-20 Geophysical equipment used to determine the velocity of sound waves in soil. *(Soiltest, Inc.)*

equation

$$D_1 = \frac{L_1}{2}\sqrt{\frac{V_2 - V_1}{V_2 + V_1}} \tag{6-3}$$

where D = depth, ft
L_1 = critical distance, ft
V_1 = velocity of wave in top stratum, fps
V_2 = velocity of wave in lower stratum, fps

Solving formula (6-3) for D_1 gives

$$D_1 = \frac{36}{2}\sqrt{\frac{3{,}000 - 1{,}000}{3{,}000 + 1{,}000}} = 13 \text{ ft}$$

Thus the topsoil has an apparent depth of 13 ft.

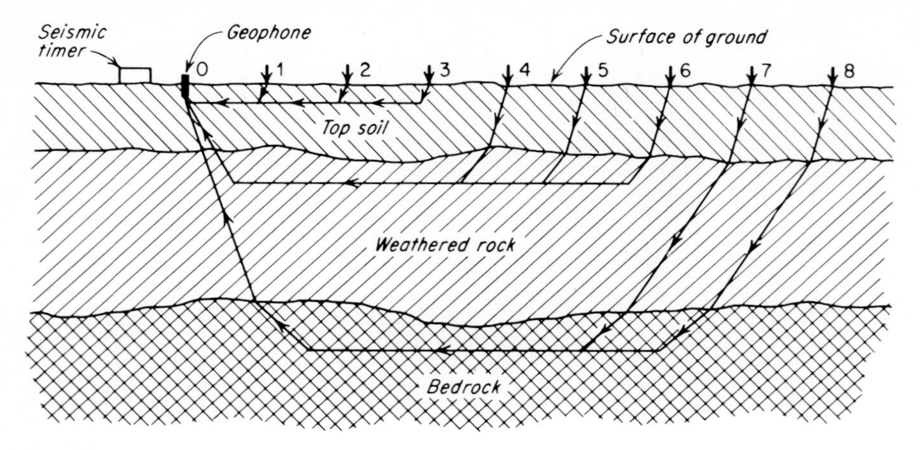

Figure 6-21 Paths of sound waves through formations.

Equation (6-4) may be used to determine the apparent depth of the two top strata, namely the topsoil and the weathered rock.

$$D_2 = \frac{C_2}{2}\sqrt{\frac{V_3 - V_2}{V_3 + V_2}} + D_1\left[1 - \frac{V_2\sqrt{V_3^2 - V_1^2} - V_3\sqrt{V_2^2 - V_1^2}}{V_1\sqrt{V_3^2 - V_2^2}}\right] \tag{6-4}$$

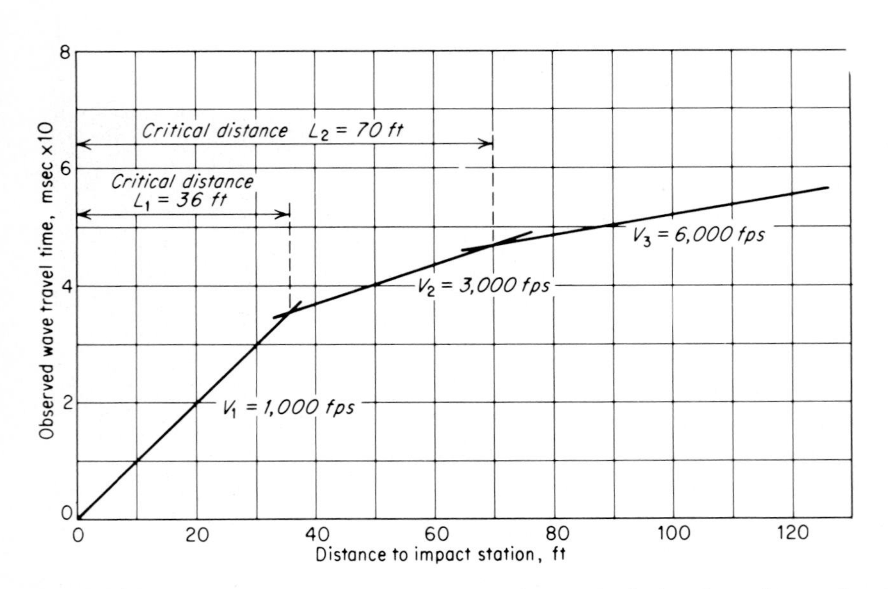

Figure 6-22 Relationship between distance, travel time, and velocity of sound waves in soil.

where $C_2 = 70$ ft
$D_1 = 13$ ft
$V_1 = 1{,}000$ fps
$V_2 = 3{,}000$ fps
$V_3 = 6{,}000$ fps
Solving,

$$D_2 = 31 \text{ ft}$$

TYPES OF RIPPERS

Figure 6-23 illustrates a type of tractor-mounted hydraulically operated ripper, which is more commonly used than the towed type. The ripper in Fig. 6-23 is using three shanks. The number of shanks used depends on the size of the tractor, the depth of penetration desired, the resistance of the material being ripped, and the degree of breakage of the material desired. If the material is to be excavated by self-loading scrapers, it should be broken into particles that can be loaded into scrapers, usually not more than 24 to 30 in. maximum sizes. Only a field test conducted at the project will demonstrate which method, depth, and degree of breakage is most satisfactory and economical.

Another method of classifying rippers is illustrated in Fig. 6-24. The shank

Figure 6-23 Tractor-mounted hydraulically operated triple-shank ripper. *(Fiat-Allis Construction Machinery, Inc.)*

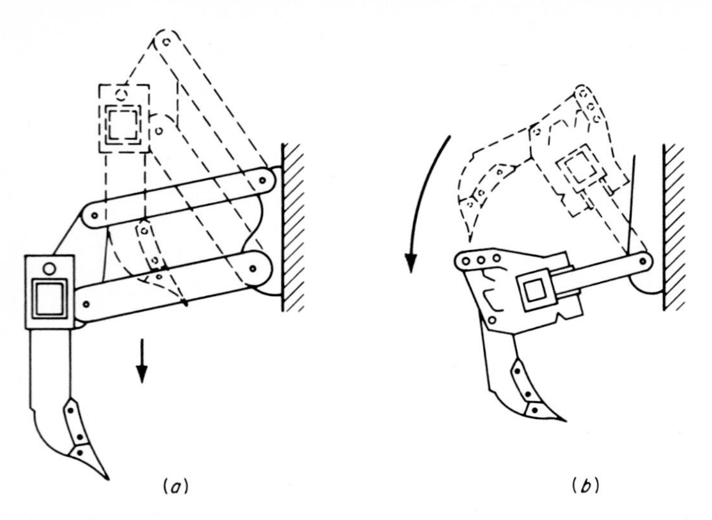

Figure 6-24 Two types of linkage used to mount rippers on tractors. (*a*) Parallel: By raising and lowering shank through an almost vertical plane, parallel-type linkage keeps points at constant angle at any penetration depth. This holds the tip at its best cutting angle, reduces wear, and smooths output. (*b*) Hinge: The traditional linkage arrangement swings the clevis and shank down through an arc, changing the point angle at various depths. Hinged rippers offer some advantages in boulder-strewn soil.

in (*a*) is attached to the tractor with a parallel-type linkage, while the shank in (*b*) is attached with a hinge- or radial-type linkage.

As the depth of penetration of the parallel-type linkage is varied, the point is kept at a constant angle, which reduces the wear and stabilizes the production.

The angle of the point of the hinge-type linkage will vary as the depth of penetration is varied, which may be a disadvantage with some types of rock. However, hinge-type rippers may offer some advantages when ripping soil containing boulders.

ECONOMY OF RIPPING ROCK

Although the cost of excavating rock by ripping and scraper loading is considerably higher than for earth that requires no ripping, it may be much less expensive than using an alternate method, such as drilling, blasting, shovel loading, and truck hauling. An example of a reduction in cost effected by the use of ripping and scraper hauling is illustrated by the experience of a contractor who constructed a section of Interstate 5 highway in southern Oregon [4]. The rock was sandstone and volcanic agglomerate with some decomposed granite and basalt.

In preparing his estimate the contractor planned to handle most of the material by drilling, blasting, shovel loading, and truck hauling, as indicated in

Table 6-9 Ripping rock costs versus alternate methods*

	Estimated methods, quantities, and costs			Actual methods and costs	
Method	Cost per cu yd	Volume, cu yd	Total cost	Volume, cu yd	Total cost
Blast, shovel, truck	$0.86	3,100,000	$2,666,000	100,000	$ 86,000
Blast, scraper	0.68	none	none	900,000	612,000
Ripper, scraper	0.46	700,000	322,000	2,800,000	1,288,000
Total		3,800,000	$2,988,000	3,800,000	$1,986,000

* The costs shown were correct at the time the project was constructed.

the accompanying table. However, job experience demonstrated that he could handle most of the rock by ripping and scraper loading, with a substantial reduction in the cost, as shown in Table 6-9, which is based on revised estimates made prior to completing the project.

Although the scrapers were strengthened considerably for use on this project, the cost of repairs was approximately double the cost for scrapers used on earth projects. The life of scraper tires was reduced from about 4,000 hr to 1,000 and 1,500 hr, depending on where the scrapers were working. It was necessary to limit the scraper loads to approximately 10 percent below their normal struck capacities. Even under these conditions scrapers maintained an average availability factor of 91.5 percent.

The bibliography at the end of this chapter lists sources containing cost information on ripping rock.

FRONT-END LOADERS

USES

Front-end loaders are used extensively in construction work to handle and transport bulk material, such as earth and rock, to load trucks, to excavate earth, as bulldozers, etc. They are both satisfactory and economical when used for such purposes.

TYPES AND SIZES

There are basically two types of front-end loaders, the crawler-tractor-mounted type and the wheel-tractor-mounted type, as illustrated in Figs. 6-25 and 6-26,

Figure 6-25 Track-type loader. *(Caterpillar Tractor Co.)*

Figure 6-26 Wheel-tractor-mounted front-end loader. *(Caterpillar Tractor Co.)*

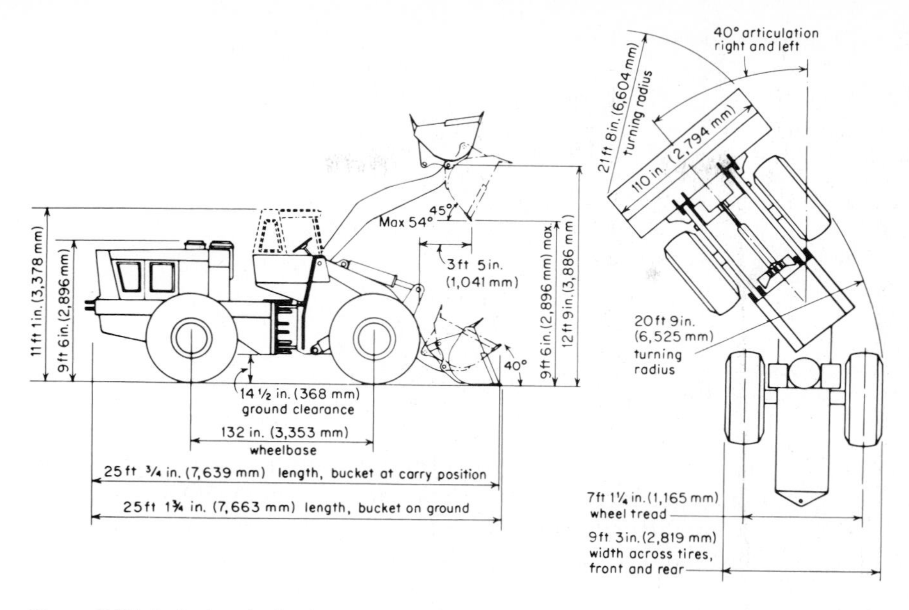

Figure 6-27 Articulated wheel-tractor-mounted loader. *(International Harvester Company.)*

respectively. They may be further classified by the capacities of the buckets or the weights that the buckets can lift. Wheel-mounted units may be steered by the rear wheels, or they may be articulated, to permit steering as indicated in Fig. 6-27, which also gives important dimensional specifications.

OPERATING SPECIFICATIONS

Representative operating specifications for a wheel-tractor loader furnish information such as that listed below.

Engine flywheel hp @ 2,300 rpm, 119
Speeds, forward and reverse:
Low, 0 to 3.9 mph
Intermediate, 0 to 11.1 mph
High, 0 to 29.5 mph
Operating load (SAE)*, 6,800 lb
Tipping load, straight ahead, 17,400 lb
Tipping load, full turn, 16,800 lb
Lifting capacity, 18,600 lb
Breakout force, maximum, 30,000 lb

*The operating load is rated at 50 percent of the tipping load, considering the combined weight of the bucket and the load, measured from the center of gravity of the extended bucket at its maximum reach with standard counterweights and nonballasted tires.

Table 6-10 Bucket selection chart

SAE capacity, cu yd Struck	Heaped	Weight of material,* lb per cu yd	Weight of material, lb
4	$4\frac{1}{2}$	1,500	6,750
$2\frac{1}{2}$	3	2,200	6,600
$2\frac{1}{4}$	$2\frac{1}{2}$	2,700	6,750
2	$2\frac{1}{4}$	3,000	6,750
$1\frac{3}{4}$	2	3,300	6,600

* These are loose-measure weights.

The maximum capacity bucket selected for use with this tractor will depend on the weight of the material to be handled, as indicated in the bucket selection chart (Table 6-10). The heaped capacity is based on a slope of 2:1, as approved by the Society of Automotive Engineers (SAE).

PRODUCTION RATES FOR CRAWLER-TRACTOR LOADERS

The production rate for a tractor loader will depend on the: (1) fixed time required to load the bucket, shift gears, turn, and dump the load, (2) time required to travel from the loading to the dumping position, (3) time required to return to the loading position, and (4) the actual volume of material hauled each trip. Figure 6-28 illustrates a typical loading situation, using a crawler-tractor-mounted loader, having the following specifications:

Bucket capacity, heaped, $2\frac{1}{4}$ cu yd

Travel speed by gear:

Forward	mph	fpm
1st	1.9	167
2d	2.9	255
3d	4.0	352
Reverse		
1st	2.3	202
2d	3.6	317
3d	5.0	440

Assume that the tractor will travel at an average of 80 percent of the specified speeds in 2d gear, forward and reverse. The fixed time should be based on time studies for the particular equipment and job.

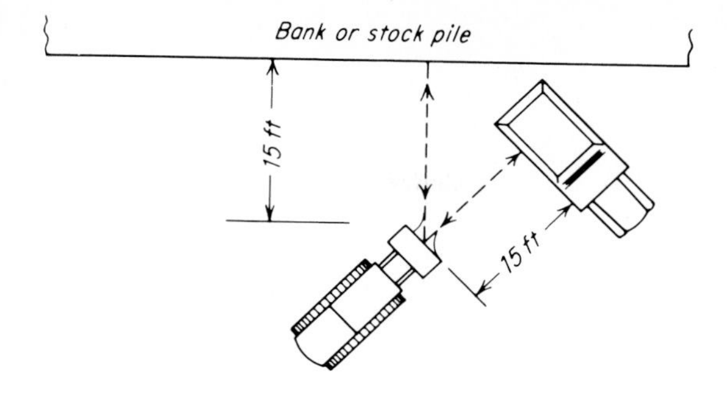

Figure 6-28 Tractor loader loading a truck.

The cycle time per load will be, in minutes

Fixed time to load, shift, turn, and dump	= 0.40
Haul time, 30 ft ÷ 204 fpm	= 0.15
Return, 30 ÷ 252 fpm	= 0.12
Cycle time	= 0.67 min

Although the rated heaped capacity of the bucket is $2\frac{1}{4}$ cu yd, it is probable that the average volume will be about 90 percent of this capacity for sustained loads. Assume an average capacity of $0.9 \times 2.25 = 2.03$ cu yd, loose volume.

The production in a 60-min hour will be as follows:

$$\text{No. cycles, } \frac{60}{0.67} = 90$$

$$\text{Volume, } 90 \times 2.03 = 182.7 \text{ cu yd}$$

If the material has a swell of 25 percent, and the tractor has an operating factor of 45 min per hr, the volume per hour in bank measure will be

$$V = \frac{182.7}{1.25} \times \frac{45}{60} = 110 \text{ cu yd}$$

The chart in Fig. 6-29 gives the production rates for crawler-tractor-mounted loaders based on handling earth having a swell of 25 percent and an operating factor of 45 min per hr. The loose weight of the earth is 2,700 lb per cu yd. It is assumed that the actual average volume of earth in a bucket is 90 percent of its rated heaped capacity. The production rates are determined as follows:

Fixed time, 0.40 min
Haul speed, in 2d gear, $0.8 \times 255 = 204$ fpm
Return speed, in 3d gear, $0.8 \times 440 = 352$ fpm

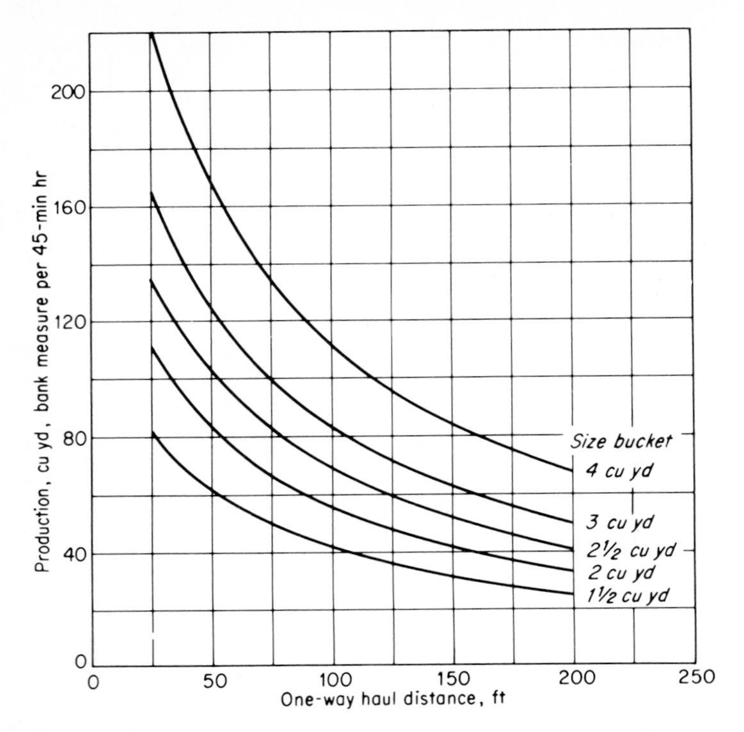

Figure 6-29 Production rates for crawler-tractor-mounted loaders.

PRODUCTION RATES FOR WHEEL-TRACTOR LOADERS

The production rates for wheel-tractor loaders are determined in the same manner as for crawler-tractor loaders. However, because they are more maneuverable and can travel faster on smooth haul surfaces, the production rates for wheel units should be higher than for crawler units under favorable conditions.

Consider a wheel unit with a $2\frac{1}{2}$-cu-yd heaped capacity bucket, handling material weighing 2,700 lb per cu yd loose volume, for which the swell is 25 percent. This unit, equipped with a torque converter and a power-shift transmission, has the following speed ranges, forward and reverse:

Low range, 0 to 3.9 mph
Intermediate range, 0 to 11.1 mph
High range, 0 to 29.5 mph

When hauling a loaded bucket, the unit should travel at an average speed of about 80 percent of its maximum speed in the low range. When returning empty, the unit should travel at an average speed of about 60 percent of its maximum speed in the intermediate range for distances less than 100 ft, and at

Cycle time for one-way distance, ft

Haul distance, ft	25	50	100	150	200
Fixed time	0.40	0.40	0.40	0.40	0.40
Haul time	0.12	0.24	0.49	0.73	0.98
Return time	0.07	0.14	0.28	0.42	0.56
Cycle time, min	0.59	0.78	1.17	1.55	1.94
Trips per hr	76.3	57.8	38.5	29.1	23.2

Volume hauled per hour, in cubic yards bank measure, by sizes of buckets

Size bucket, cu yd		One-way haul distance, ft				
Loose	Bank*	25	50	100	150	200
1½	1.08	82.3	62.5	41.6	31.5	25.1
2	1.44	110.0	83.5	55.5	42.0	33.5
2½	1.80	133.0	104.0	69.5	52.5	40.8
3	2.16	164.6	125.0	83.2	63.0	50.2
4	2.88	220.0	167.0	111.0	84.0	67.0

* Based on a swell of 25 percent and an average load equal to 90 percent of the rated capacity.

about 80 percent ot its maximum speed in the same range for distances of 100 ft and over. The average speeds should be about as follows:

$$\begin{aligned}
&\text{Hauling, all distances, } 0.8 \times 3.9 \times 88 &= 274 \text{ fpm}\\
&\text{Returning, 0 to 100 ft, } 0.6 \times 11.1 \times 88 &= 585 \text{ fpm}\\
&\text{Returning, 100 ft and over, } 0.8 \times 11.1 \times 88 &= 780 \text{ fpm}
\end{aligned}$$

If the haul surface is not well maintained, or is rough, these speeds should be reduced to realistic values.

Cycle time

Haul distance, ft	25	50	100	150	200
Fixed time	0.35	0.35	0.35	0.35	0.35
Haul time	0.09	0.18	0.36	0.55	0.73
Return time	0.05	0.09	0.13	0.19	0.26
Cycle time, min	0.49	0.62	0.84	1.09	1.34
Trips per 45-min hr	92.0	72.6	53.7	41.2	33.6

Volume hauled per hour, in cubic yards bank measure, by size of buckets

Size bucket, cu yd		One-way haul distance, ft				
Loose	Bank*	25	50	100	150	200
2	1.44	132.5	104.5	77.2	59.5	40.5
3	2.16	198.0	157.0	116.0	89.0	72.5
4	2.88	264.0	204.5	154.0	118.5	96.6
5	3.60	331.0	261.0	193.0	148.0	122.0
6	4.32	397.0	313.0	231.0	177.0	144.0

* Based on a swell of 25 percent and an average load equal to 90 percent of the rated capacity.

Because of the greater maneuverability of the wheel-loader, its fixed time should be slightly less than for a crawler-loader. Assume a fixed time of 0.35 min.

Figure 6-30 illustrates the variations in production rates by sizes of buckets and one-way haul distances.

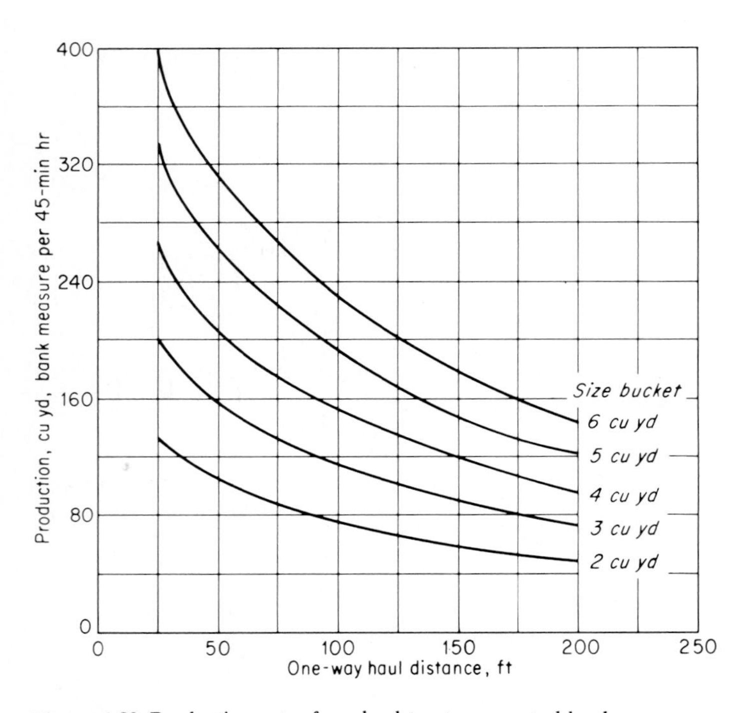

Figure 6-30 Production rates for wheel-tractor-mounted loaders.

PROBLEMS

6-1 If the tractor-pulled scraper for which the rimpull and speed chart of Fig. 6-4 applies hauls a gross vehicle weight of 75,000 lb up a 4 percent slope over a haul road whose rolling resistance is 70 lb per ton, determine the maximum speed range and speed for the vehicle.

6-2 The tractor-pulled scraper for which Fig. 6-4 applies will operate under the following conditions:

Gross vehicle weight, 75,000 lb
Weight on driving wheels, 60 percent of gross weight
Coefficient of traction, 0.5
Rolling resistance of hauling road, 80 lb per ton

Determine the maximum grade up which the vehicle can operate in the first speed range in either direct drive or overdrive, considering the available rimpull as limited by the tractor or the coefficient of traction.

6-3 If the tractor-pulled scraper for which Fig. 6-5 applies hauls a gross vehicle weight of 120,000 lb up a 4 percent grade on a haul road whose rolling resistance is 70 lb per ton, determine the maximum speed range and speed for the vehicle.

6-4 Determine the maximum speed of the tractor-pulled scraper for which Fig. 6-5 applies for each of the following conditions:

Gross vehicle weight, 120,000 lb
Weight on driving wheels, 70,000 lb
Coefficient of traction, 0.6

(*a*) Haul road is level, with a rolling resistance of 70 lb per ton.
(*b*) Haul road is up a 7 percent grade, with a rolling resistance of 80 lb per ton.
(*c*) Haul road is down a 4 percent grade, with a rolling resistance of 80 lb per ton.

REFERENCES

1. "Latin-American Land Development Seminar, Program and Proceedings," Rome Industries, Cedartown, GA 30125.
2. "Rome Training Presentations No. 11-A, 11-C, 11-E, 11-F," Rome Industries, Cedartown, GA 30125.
3. Fleco Corporation, P.O. Box 3270, Jacksonville, FL 32203.
4. "From Brush to Grass," *Form AEO-30060,* Caterpillar Tractor Co., Peoria, IL 61629.
5. "Land Improvement Contractors," *Form AEO-30049-01*, Caterpillar Tractor Co., Peoria, IL 61629.
6. "Land Clearing," Caterpillar Tractor Co., Peoria, IL 61629.
7. "Cost of Clearing Land," *Circular 133*, Agricultural Experiment Station of Auburn University, Auburn, Ala., June 1959.
8. New Developments in Earthmoving-Rippers, *Construction Methods and Equipment*, vol. 47, pp. 149–168, April 1965.
9. How to Rip Economically, *Roads and Streets*, vol. 107, pp. 43–52, December 1964.
10. When to Move Rock with Scrapers, *Roads and Streets*, vol. 108, pp. 44–95, July 1965.
11. Rip Basalt with Big Tractor, *Roads and Streets*, vol. 116, pp. 114–115, October 1973.
12. Alloy Points Rip Hard Sandstone Fast, *Roads and Streets*, vol. 116, p. 114, September 1973.
13. Ripping Increases Production, Saves Trouble, Money, *Roads and Streets*, vol. 115, pp. 78–79, August 1972.

14. Church, Horace K.: 433 Seismic Excavation Studies: What They Tell About Rippability, *Roads and Streets*, vol. 115, pp. 86–92, January 1972.
15. Contractor's Own Seismic Study Helps to Set Job Stategy, *Roads and Streets*, vol. 115, pp. 26–28, January 1972.
16. Church, Horace K.: Seismic Exploration Yields Data, *Engineering News-Record*, vol. 175, pp. 62–66, August 12, 1965.
17. Crice, Douglas B.: The Geophysical Half of Geotechnical Engineering, *Civil Engineering*, vol. 45, pp. 62–65, October 1965.
18. "Handbook of Ripping, A Guide to Greater Profits," 5th ed., Caterpillar Tractor Co., Peoria, IL 61629, 1975.

CHAPTER

SEVEN

SCRAPERS

GENERAL INFORMATION

Tractor-pulled scrapers have established an important position in the earth-moving field. As they are self-operating to the extent that they can load, haul, and discharge material, they are not dependent on other equipment. If one of them experiences a temporary breakdown, it is not necessary to stop the job, as would be the case for a machine which is used exclusively for loading earth into hauling units, for if the loader breaks down, the entire job must stop until repairs can be made. The self-loading scrapers are available with capacities up to 50 cu yd or more.

These machines are the result of a compromise between the best loading and the best hauling machines, and, as must be expected of any composite machine, they are not superior to other equipment in both loading and hauling. Power shovels, draglines, and belt loaders usually will surpass them in loading only, while trucks may surpass them in hauling only, especially when long, well-maintained haul roads are used. However, their ability to load and haul earth gives them a definite advantage on many projects. The development of high-speed wheel-type tractors has increased the economic haul distance for this type of equipment up to a mile or more on many projects.

The ability of these machines to deposit their loads in uniformly thick layers will facilitate the succeeding spreading operations. On the return trips to borrow pits the cutting blades of scrapers may be lowered enough to remove high spots, thereby assisting in maintaining the haul roads.

Earth frequently is found in stratified layers, which must be blended by mixing the materials from several layers. The limited depth of cut will not permit scrapers to mix the layers satisfactorily. For this reason shovels and trucks sometimes are used, even though scrapers will handle the earth more economically.

The advantages and disadvantages previously given for crawler compared with wheel tractors will, in most instances, apply to tractor-pulled scrapers.

TYPES AND SIZES OF SCRAPERS

There are two types of scrapers, based on the type of tractor used to pull them—the crawler-tractor-pulled and the wheel-tractor-pulled types. The latter type may be further subdivided, as indicated and described.

1. Crawler-tractor-pulled
2. Wheel-tractor-pulled
   a. Single-engine
   b. Twin-engine
   c. Two-bowl tandem
   d. Elevating scraper

These various types are illustrated in Figs. 7-1 through 7-5.

Figure 7-1 Crawler tractor and self-loading scraper.

Figure 7-2 Single-engine wheel-type tractor-pulled scraper.

Figure 7-3 Twin-engine tractor and scraper.

Figure 7-4 Two-bowl tandem scraper unit.

Figure 7-5 Wheel-type tractor-pulled elevating scraper.

Crawler-tractor scraper For relatively short haul distances the crawler-type tractor, pulling a rubber-tired self-loading scraper, can move earth economically. The high drawbar pull in loading a scraper, combined with good traction, even on poor haul roads, gives the crawler tractor an advantage for short hauls. However, as the haul distance is increased, the low speed of a crawler tractor is a disadvantage compared with a wheel tractor.

Unless the loading operation is difficult, a crawler tractor can load a scraper without the aid of a bulldozer. However, if there are several scraper units on a job, the increased output resulting from using a bulldozer to help load the scrapers usually will justify the use of a bulldozer.

Wheel-tractor scrapers For longer haul distances the higher speed of a wheel-type tractor-pulled self-loading scraper will permit it to move earth more economically than a crawler-type tractor. Although the wheel-type tractor cannot deliver as great a tractive effort in loading a scraper, the higher travel speed, which may exceed 30 mph for some models, will offset the disadvantage in loading when the haul distance is sufficiently long.

The break-even distance, the haul distance at which the cost of hauling with a crawler or a wheel tractor will be the same, may be determined by making an analysis of a given job. The analysis should consider the class of soil, the condition of the borrow pit, the condition, length, and slope of the haul road, the nature of the fill, and weather.

The size of a scraper. The size of a scraper may be specified as the struck, or heaped, capacity of the bowl, expressed in cubic yards. The struck capacity is the volume of the material that a scraper will hold when the top of the material is struck off even with the top of the bowl. In specifying the heaped capacity of a scraper some manufacturers specify the slope of the material above the sides of the bowl with the designation SAE. The SAE (Society of Automotive Engineers) specifies a slope of 2:1, measured horizontally and vertically, respectively. Because the slope will vary with the class of material being hauled, the heaped capacity is only an approximate value.

The capacity of a scraper, expressed in cubic yards bank measure, is obtained by multiplying the loose volume in the scraper by an appropriate swell factor, as given in Table 5-1, or more accurately one that is known to apply to the particular soil being handled. Owing to the compacting effect on the earth in a scraper, resulting from the pressure required to force additional earth into the bowl, the swell usually is less than for earth deposited into a truck by a power shovel. Tests indicate that the swell factors given in Table 5-1 should be increased by 10 percent for earth loaded into a conventional scraper. When computing the bank measure volume for an elevating scraper, the swell factors given in Table 5-1 should be used.

If a conventional scraper hauls an average heaped load of 22.5 cu yd of wet earth, for which the adjusted swell factor is $0.80 + 0.08 = 0.88$, the bank-measure volume will be $22.5 \times 0.88 = 19.8$ cu yd.

OPERATING A SCRAPER

A scraper is loaded by lowering the front end of the bowl until the cutting edge, which is attached to and extends across the width of the bowl, enters the ground and, at the same time, raising the front apron to provide an open slot through which the earth may flow into the bowl. As the scraper is pulled forward, a strip of earth is forced into the bowl. This operation is continued until the bowl is filled or until no more earth may be forced in. The cutting edge is raised and the apron is lowered to prevent spillage during the haul trip.

The dumping operation consists of lowering the cutting edge to the desired height above the fill, raising the apron, and forcing the earth out between the blade and the apron by means of a movable ejector mounted at the rear of the bowl.

The elevating scraper illustrated in Fig. 7-5 is equipped with horizontal slats which are operated by two endless chains, to which the ends of the slats are connected. As the scraper moves forward with its cutting edges digging into and loosening the earth, the slats rake the earth upward and into the bowl of the scraper. This action requires less energy than pushing earth upward through material already in the bowl. As a result, this scraper is capable of loading the bowl without assistance from a pusher tractor for some types of soils. Also, the pulverizing action of the slats permits a more complete filling of the bowl, and it permits a more uniform spreading action on the fill.

PERFORMANCE CHARTS

Manufacturers of wheel-type scrapers provide a performance chart for each of their units. This chart contains information that may be used to analyze the performance of a unit under various operating conditions. Figure 7-6 is a performance chart for a scraper whose specifications are as follows:

Engine fwhp at 2,100 rpm, 441
Transmission, torque converter and power-shift, with overdrive and six speeds forward
Capacity of scraper:
  Struck, 28 cu yd
  Heaped, 3:1 slope, 32 cu yd
  Heaped, 1:1 slope, 38 cu yd

Weights (net weight distribution, empty):	
Drive axle, 60.4%	= 55,800 lb
Scraper axle, 39.6%	= 36,600 lb
Total	= 92,400 lb
Payload	= 94,000 lb
Total load	= 186,400 lb
Gross weight distribution	
Drive axle, 50%	= 93,200 lb
Scraper axle, 50%	= 93,200 lb

Assume that the scraper is hauling a gross load of 186,400 lb. Use the chart to determine the highest gear and the maximum speed at which it may operate when traveling up a 4 percent grade on a haul road having a rolling resistance of 40 lb per ton, equal to a 2 percent grade. Because the chart is based on zero rolling resistance it will be necessary to add the rolling resistance to the grade resistance to determine the total resistance, which is 6 percent.

The chart is used by applying the following four steps:

1. Find the gross vehicle weight on the lower left horizontal scale, using the weight of 186,400 lb.
2. Read up to slanted total resistance line, namely 6 percent.
3. From the intersection of the two lines read horizontally to the right to the interception with the performance curve. This line intercepts the third speed range overdrive curve.
4. Read down from this interception to the lower scale to determine the maximum vehicle speed, which is 10.6 mph.

The scraper can also operate in the direct drive of the third speed range, but its maximum speed will be slightly less than 9 mph.

The vertical row of figures at the right of the chart indicates the rimpull in 1,000 lb required at 6 percent total resistance to move the unit at a uniform speed slightly larger than 11,000 lb.

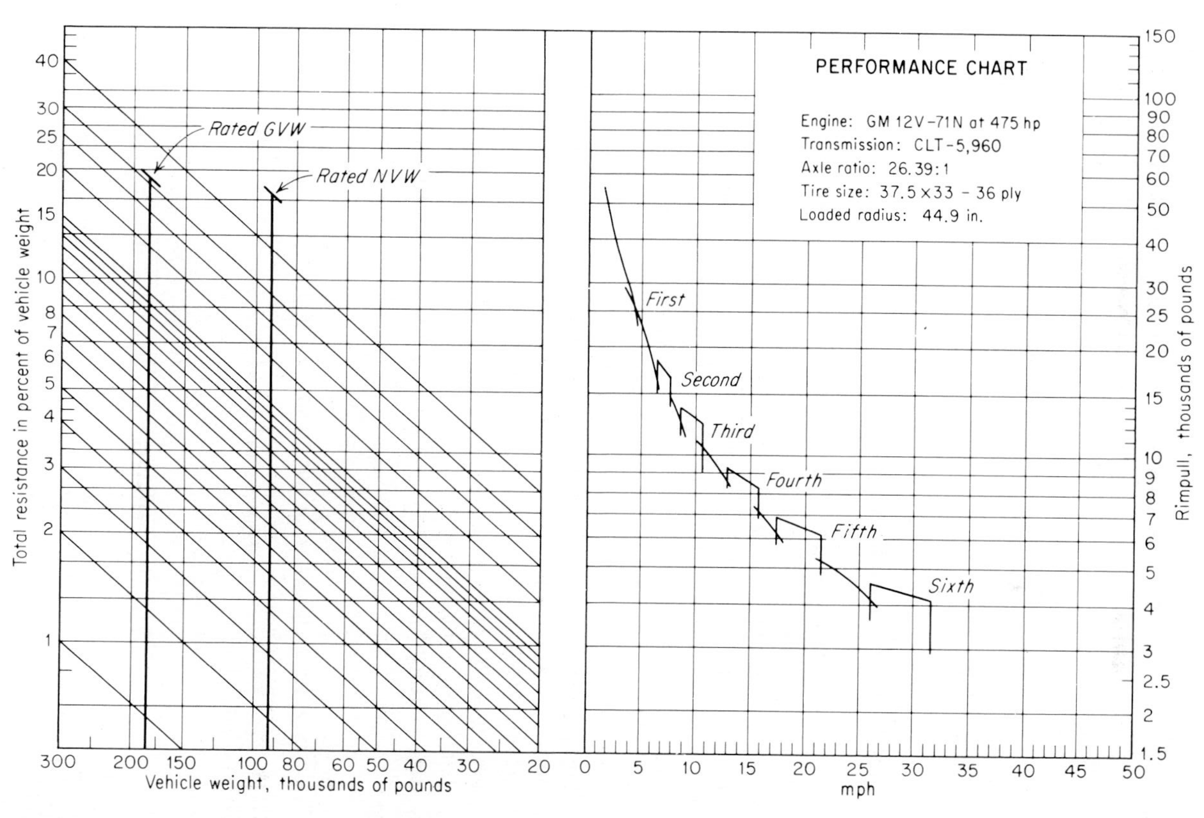

Figure 7-6 Performance chart for wheel-type tractor and scraper.

If the gradability of this scraper in direct drive third speed range is desired, it may be determined from the chart by using the following steps:

1. Read horizontally to the left from the top of the curve for the gear and speed range to the intersection with the vertical line indicating the gross weight of the vehicle.
2. Use this point of intersection to determine the total resistance value from the appropriate slanted resistance line or lines. For the stated conditions the value is about 7.5 percent. Deducting the 2 percent for rolling resistance, the maximum grade is about 5.5 percent.

CYCLE TIME FOR A SCRAPER

The cycle time for a scraper is the time required to load, haul to the fill, dump, and return to the loading position again. The cycle time may be divided into two elements, fixed time and variable time.

The fixed time is the time devoted to other than hauling or returning empty. It includes the time for loading, dumping, turning, accelerating, and decelerating, all of which are reasonably constant under uniform operating conditions. Table 7-1 gives representative time values for each of these fixed-time elements based on favorable, average, and unfavorable conditions. It will be noted that the time indicated for accelerating and decelerating varies with the maximum speed at which a vehicle will travel on the haul road. It takes longer to accelerate to a speed of 20 mph than to a speed of 10 mph, and also it takes longer to decelerate from the higher speed. The times are based on using a pusher tractor of adequate size.

The time required to haul and return depends on the distance traveled and

Table 7-1 Fixed time elements, in minutes, for wheel scrapers

	Hauling speed ranges, mph (km/h)								
	5 to 8 (8 to 13)			8 to 15 (13 to 24)			15 to 30 (24 to 48)		
Element	(1)*	(2)	(3)	(1)	(2)	(3)	(1)	(2)	(3)
Loading	0.8	1.0	1.4	0.8	1.0	1.4	0.8	1.0	1.4
Dumping-turning	0.4	0.5	0.6	0.4	0.5	0.6	0.4	0.5	0.6
Accelerating-decelerating	0.3	0.4	0.6	0.6	0.8	1.0	1.0	1.5	2.0
Total time	1.5	1.9	2.6	1.8	2.3	3.0	2.2	3.0	4.0

* Columns 1, 2, and 3 indicate the times for favorable, average, and unfavorable conditions, respectively, which will vary with the size and condition of the loading pit and the dumping area.

the average speed of the vehicle. Because hauling and returning are usually at different speed ranges, it is necessary to determine the time for each separately.

Example 7-1 Determine the cycle time for a scraper to haul earth from a pit to a fill 2,000 ft distant under average fixed-time conditions, with an average haul speed of 12 mph and an average return speed of 24 mph.

Because the hauling speed governs the fixed time elements in Table 7-1, the fixed time will be 2.3 min. The individual times are determined as follows:

$$\begin{aligned}
&\text{Fixed time} && = 2.3 \text{ min} \\
&\text{Haul time,} \quad \frac{2{,}000}{12 \times 88} && = 1.9 \text{ min} \\
&\text{Return time,} \quad \frac{2{,}000}{24 \times 88} && = 1.0 \text{ min} \\
&\quad\text{Total time} && = 5.2 \text{ min}
\end{aligned}$$

OPERATING EFFICIENCY AND PRODUCTION

If the scraper cycle time of 5.2 min developed in the previous article could be maintained for a period of 60 min, the unit would make 60/5.2 = 11.5 trips, and the volume of material hauled would equal the product of the number of trips times the average volume per load. However, scrapers and other types of construction equipment do not operate 60 min per hr. This introduces terms defined as operating efficiency and operating factor. If the scraper operates on the average for a sustained period of time 50 min per hr, its operating factor is 50/60 = 0.83. The actual operating time of a machine can be determined from a stop-watch study of the machine, conducted for an interval, or intervals, of time. Studies conducted at several different times during a day, and repeated over a period of several days, will give more accurate results than a single study. Many time studies have been made for all or most types of construction equipment, and the results have been very helpful in enabling contractors to increase their operating efficiencies. The studies are more useful if the duration and cause of each delay are recorded for future analysis.

Based on these studies the operating efficiencies of equipment such as scrapers may be classified as follows:

Classification	Operating efficiency, min per hr	Operating factor
Excellent	55	0.92
Average	50	0.83
Fair	45	0.75
Unfavorable	40	0.67

The actual rate of production for any unit of construction equipment should be estimated or determined by applying an appropriate operating factor to its maximum rate of production.

NUMBER OF SCRAPERS SERVED BY A PUSHDOZER

If wheel-type tractor-pulled scrapers are to attain their maximum hauling capacities, they need the assistance of one or more push tractors during the loading operation to reduce the loading and cycle time. Although crawler-type tractor-pulled scrapers are frequently referred to as self-loading units, it may be economically desirable to provide push tractors for them. If using a push tractor will increase the job production enough to more than pay the cost of the tractor, it is good business to use one, regardless of the type of scraper units used.

When using push tractors, it is desirable to match the number of pushers with the number of scrapers. If a pusher or a scraper must wait for the other, it reduces the operating efficiency of the waiting unit and the project and results in an increased production cost.

The pusher cycle time includes the time required to load a scraper plus the time required to move into position to load another scraper. With the cycle time for the scraper and the push tractor determined, Eq. (7-1) may be used to determine the number of scrapers that a tractor may serve.

$$N = \frac{T_s}{T_p} \tag{7-1}$$

where N = number of scrapers served
T_s = cycle time for scraper
T_p = cycle time for pusher tractor

Table 7-2 Representative pusher tractor cycle times

Method of loading	Loading conditions	Cycle time, min
Back track	Favorable	1.7
	Average	2.5
	Unfavorable	3.0
Chain	Favorable	1.2
	Average	1.6
	Unfavorable	2.0
Shuttle	Favorable	1.2
	Average	1.6
	Unfavorable	2.0

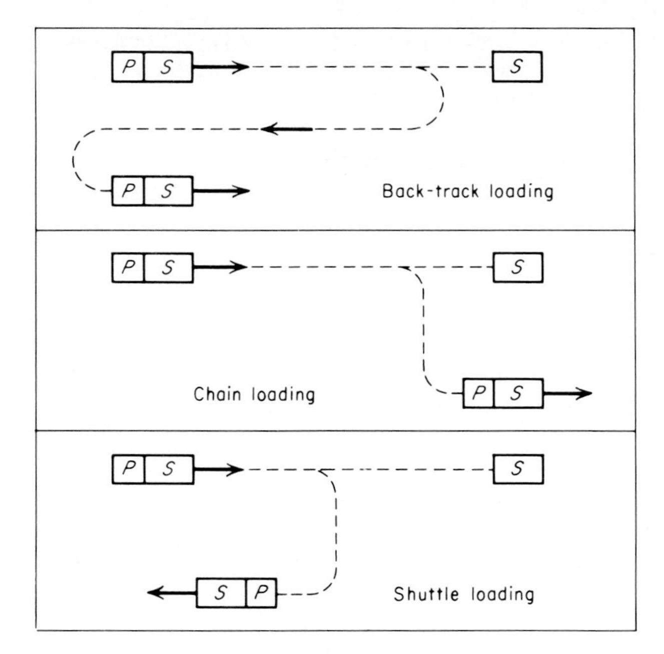

Figure 7-7 Methods of push loading scrapers.

The cycle time for a pusher tractor will vary with the conditions in the loading pit, the relative size of the tractor and the scraper unit, and the loading method used by the tractor. Figure 7-7 shows three loading methods that are used, and Table 7-2 lists representative cycle times for each method for favorable, average, and unfavorable conditions.

Favorable loading conditions include loading in a large pit or cut, ripping hard soil prior to loading, loading down grade, maintaining a smooth loading surface, and using a pusher tractor whose power is matched with the size of the scraper. Unfavorable loading conditions are the opposite of those for favorable conditions.

INCREASING THE PRODUCTION RATES OF SCRAPERS

There are at least two methods that a contractor may use to obtain a higher profit on a project involving earthwork. One method is to increase the bid prices on the work. However, competition usually limits the price which he may bid. The alternate method is to organize and operate his equipment in a manner that will assure the maximum production at the lowest cost. The latter method usually offers the best opportunity for attainment. Thus, a contractor should strive to increase the production without increasing his costs. There are several methods whereby he may attain this objective.

Ripping Most types of tight soils will load more easily if they are ripped ahead of the scraper. If the value of the increased production resulting from ripping exceeds the cost of ripping, the soil should be ripped.

When rock is ripped for scraper loading, the depth ripped should always exceed the depth excavated to leave a loose layer of material under the tracks and tires to provide good traction and to reduce the wear on the tracks and tires.

Prewetting the soil Some soils will load more easily if they are reasonably moist. Prewetting can be done in conjunction with ripping, ahead of loading, to permit a uniform penetration of the moisture into the soil.

Also, prewetting soil in the cut can reduce or eliminate the use of water trucks on the fill, thereby reducing the possible congestion of equipment on the fill, and the elimination of excess moisture on the surface of the fill may facilitate the movement of the scrapers on the fill.

Loading down grade [4] When it is practicable to do so, scrapers should be loaded down grade. Each 1 percent of favorable grade is the equivalent of increasing the loading force by 20 lb per ton of gross weight of the push tractor and scraper unit.

Consider a wheel-type scraper unit whose capacity is 30 cu yd, with a net empty weight of 90,000 lb and a gross loaded weight of 170,000 lb. This unit is push loaded by a tractor whose weight is 70,000 lb. The combined empty weight will be 160,000 lb and the gross weight will be 240,000 lb. Assume that loading is done down a 10 percent grade. The increased force available to assist in loading the scraper will be determined by using Eq. 4-2.

$$\text{Initial force with the scraper empty} = \frac{160{,}000}{2{,}000} \times 10 \times 20 = 16{,}000 \text{ lb}$$

$$\text{Final force with the scraper loaded} = \frac{240{,}000}{2{,}000} \times 10 \times 20 = 24{,}000 \text{ lb}$$

If the combined drawbar pull and rimpull of the two tractors is 140,000 lb during the loading operation, the effect of downgrade loading is to increase the available power by about 11 percent up to 17 percent.

APPLYING THE LOAD-GROWTH CURVE TO SCRAPER LOADING

Without a critical evaluation of available information it might appear that the lowest cost of moving earth with scrapers is to load every scraper to its maximum capacity before it leaves the cut. However, numerous studies of loading practices have revealed that loading scrapers to their maximum capacities usually will reduce, rather than increase, the rate of production.

When a scraper starts loading, the earth flows into it rapidly and easily, but as the quantity of earth in the bowl increases, the incoming earth encounters greater resistance, and the rate of loading decreases quite rapidly, as illustrated in Fig. 7-8. This figure is a load-growth curve, which shows the relation between the load in a scraper and the loading time. An examination of the curve reveals that during the first 0.5 min the scraper loads about 17 cu yd of earth. During the next 0.5 min it loads an additional 2 cu yd, and, if loading is continued to 1.4 min, the gain in volume during the last 0.4 min is less than 1 cu yd.

The information shown in Fig. 7-8 and Table 7-3 was obtained from a field study of scraper production. The basic information and calculations are given in the following example.

Example 7-2 The equipment used and job conditions were as follows:

Equipment:
- Pusher tractor, 335 fwhp, with power shift
- Scraper tractor, 345 fwhp, two-wheel
- Scraper capacity, struck, 19.5 cu yd, heaped, 27 cu yd
- Net weight of scraper unit, 60,000 lb

Job conditions:
- Earth, sandy clay, swell 33 percent, weight, 3,050 lb per cu yd bm
- Haul distances varied from 500 to 10,000 ft
- Haul-road rolling resistance 65 lb per ton; grade, 2 percent adverse for loaded unit
- Combined rolling and grade resistance
  - Loaded, 65 + 40 = 105 lb per ton, or 5.25 percent
  - Empty, 65 − 40 = 25 lb per ton, or 1.25 percent

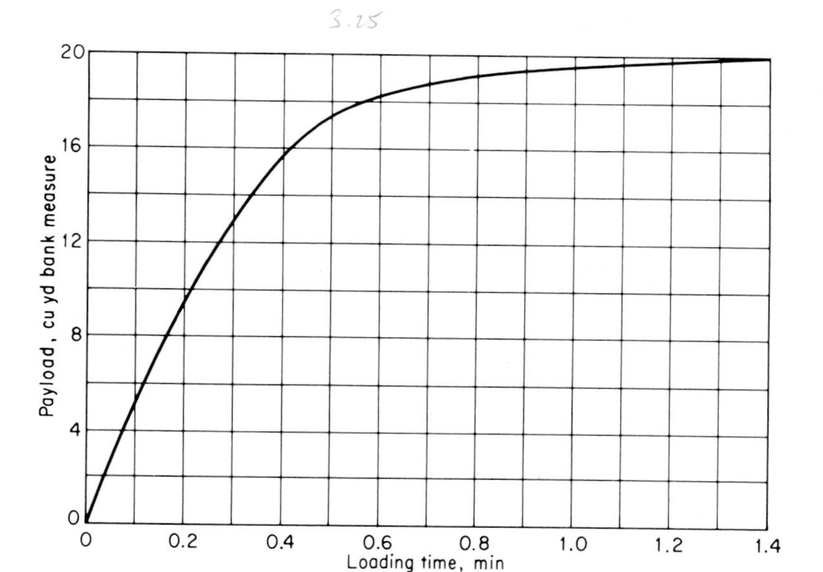

Figure 7-8 Load-growth curve for scraper loading.

Table 7-3 Variations in the rates of production of scrapers with loading times*,†

Loading time, min	Other time, min	Cycle time, min	Number trips per hr	Payload,‡ cu yd	(cu m)	Production per hr, cu yd	(cu m)
0.5	5.7	6.2	8.07	17.4	(13.3)	140	(107)
0.6	5.7	6.3	7.93	18.3	(14.0)	145	(111)
0.7	5.7	6.4	7.81	18.9	(14.5)	147	(112)
0.8	5.7	6.5	7.70	19.2	(14.7)	148*	(113)
0.9	5.7	6.6	7.57	19.5	(14.9)	147	(112)
1.0	5.7	6.7	7.46	19.6	(15.0)	146	(112)
1.1	5.7	6.8	7.35	19.7	(15.1)	145	(111)
1.2	5.7	6.9	7.25	19.8	(15.2)	143	(109)
1.3	5.7	7.0	7.15	19.9	(15.2)	142	(109)
1.4	5.7	7.1	7.05	20.0	(15.3)	141	(108)

* For a 50-min hour.
† Determined from measured performance.
‡ The economical loading time is 0.8 min.

The calculations and results for a haul distance of 2,500 ft are as follows, based on a 19-cu-yd load, bank measure.

Weight of hauling unit, empty = 30 ton

Weight of load, $\dfrac{19 \times 3,050}{2,000} = 29$ ton

Gross vehicle weight = 59 ton

? 105 lb/ton

Rimpull required, loaded, 59 × 105 = 6,190 lb
Maximum speed, 13.8 mph (from specifications)
Assumed actual speed, 0.9 × 13.8 = 12.4 mph
Rimpull required, empty, 30 × 25 = 750 lb
Actual speed, 22.6 mph
Cycle time, less loading time:

Accelerating and decelerating = 1.0 min
Turning, dumping, spotting, and boosting = 1.1 min
Hauling, $\dfrac{2,500}{12.4 \times 88}$ = 2.3 min
Returning, $\dfrac{2,500}{22.6 \times 88}$ = 1.3 min
Total time, excluding loading = 5.7 min

The information appearing in Table 7-3 is based on a 50-min hour.

Figure 7-9 shows that for this equipment and project the economical loading time increases with an increase in the haul distance.

THE EFFECT OF ROLLING RESISTANCE ON THE PRODUCTION OF SCRAPERS

A job condition that is sometimes neglected is the effect of the rolling

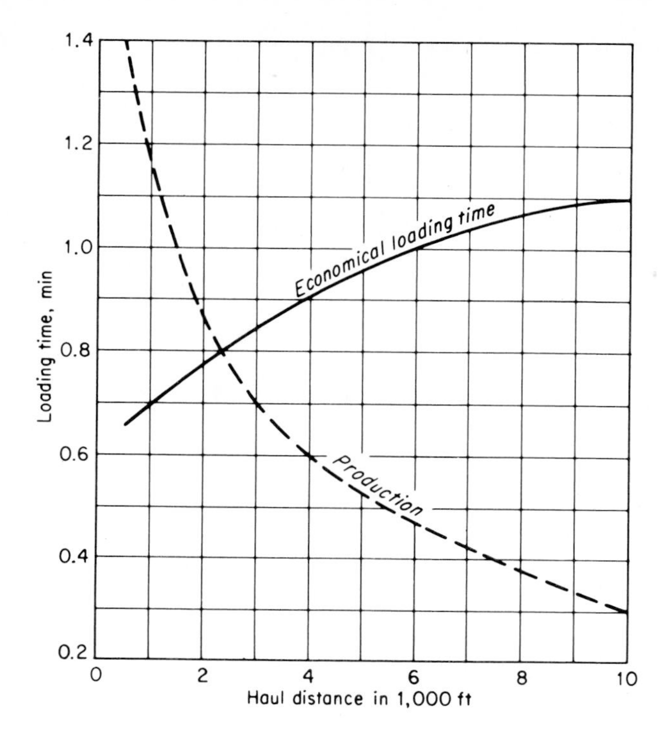

Figure 7-9 The effect of haul distance on the economical loading time of a scraper.

resistance of the haul road on the production of scrapers, and the cost of hauling earth. A well-maintained road permits faster travel speeds and reduces the costs of maintenance and repairs for the scrapers.

Consider an analysis of the performance of four wheel-type scrapers, hauling earth 1 mile over two level roads whose rolling resistances are 60 lb per ton and 100 lb per ton. The former road receives good maintenance while the latter receives no maintenance. The results of the analysis are given in Table 7-4. The table reflects an increase in the cost of maintenance and repairs for scrapers operating on the poorly maintained road.

ANALYZING THE PERFORMANCE OF A WHEEL–TYPE SCRAPER

The following example anayzes the performance of a wheel-type scraper for the specified equipment and job conditions.

Example 7-3 See Fig. 7-6 for performance chart.

Table 7-4. Effect of haul-road maintenance on the production and cost of hauling earth

	Haul-road maintenance	
Item	Good	Poor
Fixed scraper time, min	2.4	2.4
Travel time, min	4.2	6.2
Total cycle time, min	6.6	8.6
Trips per 50-min hr	7.58	5.82
Payload, cu yd bm	16	16
Production, cu yd per hr	121	93
Production, cu yd per hr for 4 scrapers	484	372
Cost per hr for scrapers:		
4 × $47.30	$189.20	
4 × $51.10		$204.40
Cost per hr for pusher loader	49.40	49.40
Cost per hr for motor grader	24.30	none
Total cost per hr	$262.90	$253.80
Cost per cu yd	$0.544	$0.682

Tractor engine, 441 fwhp
Scraper capacity
Struck, 28 cu yd
Heaped 3:1 slope 32 cu yd
Net weight of empty unit = 92,400 lb
Maximum payload = 94,000 lb
Maximum gross weight = 186,400 lb
Weight on drive axle = 93,200 lb
Weight on scraper axle = 93,200 lb
Total length of haul, 4,000 ft as follows:
1,200 ft of +4% grade loaded
1,400 ft of +2% grade loaded
1,400 ft of −2% grade loaded
Type soil, sandy clay, weight 3,100 lb per cu yd bm, swell 33%
Loose weight of soil, 3,100 × 0.75* = 2,320 lb per cu yd
Rolling resistance, 80 lb per ton = 4%

Figure 7-9, which probably applies to this scraper with reasonable accuracy, indicates an economical loading time of 0.9 min, and Fig. 7-8 indicates a load equal to 96 percent of the rated capacity, namely 0.96 × 32 = 30.7 cu yd, loose measure. The weight of the load will be 30.7 × 2,320 = 71,200 lb, which is below the specified maximum weight.

The actual weight will be

Empty = 92,400 lb
Load = 71,200 lb
Gross weight = 163,600 lb

Use Fig. 7-6 to determine the speed for each section of the road, hauling and returning.

* See Chap. 5 for a discussion of the swell factor.

Distance, ft	Grade, %	Total resistance, %	Speed, mph	Travel time, min
Hauling				
1,200	4	8	9.5	1.44
1,400	2	6	11.0	1.45
1,400	−2	2	31.5	0.60
Returning				
1,400	2	6	21.5	0.75
1,400	−2	2	31.5	0.60
1,200	−4	0	31.5	0.60
Total travel time				5.44

Table 7-1 indicates a fixed time of 3.0 min for a speed range up to 30 mph and average conditions. The cycle time will be

Fixed time, min = 3.00
Travel time, min = 5.44
Cycle time, min = 8.44

The speeds shown in the above table can be attained provided there are no obstacles or conditions to cause delays along the haul road. Thus, the cycle time of 8.44 min is the minimum for the specified conditions. If delays are anticipated, the travel speeds and the cycle time should be adjusted to reflect the anticipated conditions. The production per 50-min hour is determined as follows:

Volume per load, 30.7 cu yd loose measure
Swell factor, 0.75 (see Chap. 5)
Payload, $30.7 \times 0.75 = 23.0$ cu yd bm

Number of trips per hr, $\dfrac{50}{8.44} = 5.93$

Volume hauled per hr, $5.93 \times 23 = 136$ cu yd

PROBLEMS

7-1 The tractor-pulled scraper for which Fig. 6-4 applies weighs 44,620 lb empty. Its average load of earth will be 15 cu yd, loose measure, whose weight will average 2,780 lb per cu yd. Under average conditions the scraper will be loaded with the assistance of a bulldozer. After the scraper is loaded, it will travel up a 4 percent haul road, whose rolling resistance is 70 lb per ton, to a fill 3,000 ft distant, where it will deposit its load. Determine the probable cycle time required for this unit to make a round trip.

(*a*) If the operating factor for the unit is 0.75, how many trips should it make in 1 hr?

(*b*) If the earth swells 23 percent, what volume of earth should the scraper haul per hour, expressed as cubic yards, bank measure?

7-2 If the scraper of Prob. 7-1 is push loaded by a bulldozer of adequate size, how many scrapers should a bulldozer serve under favorable, average, and unfavorable conditions when using the back-track and chain methods of loading?

7-3 If the scraper of Prob. 7-1 hauls earth, as previously specified, under the following job conditions, determine the probable cycle time per trip based on an operating factor of 1.0. The total one-way haul distance will be 2,800 ft and will consist of three sections whose conditions are as follows:

Section	Distance, ft	Slope, %	Rolling resistance, lb per ton
1	1,000	0	80
2	1,200	−3	75
3	600	−2	90

The scraper will return along the same haul road. Assume average conditions for determining the fixed time. If the operating factor is 0.83, what is the probable volume of earth hauled per hour by each scraper, expressed in cubic yards bank measure?

7-4 If the scraper for which Table 7-3 applies spends 1.2 min obtaining its load, what is the lost income per hour based on a contract price of $0.86 per cu yd for the earth when compared with the production possible under the most economical loading time?

7-5 Two contractors use the same types of scrapers, for which Table 7-3 applies, and the same types of bulldozers to help load the scrapers. The loading pits and the haul roads are the same.

Contractor A uses a chain method of push loading his scrapers for 0.8 min. He has a fleet of six scrapers and as many bulldozers as required to help load the scrapers.

Contractor B uses a back-track method of loading the scrapers for 1.2 min. He has a fleet of six scrapers and as many bulldozers as required to help load the scrapers.

The cost per hour for a scraper is $72.25 and that for a bulldozer is $61.75, including the operators. Using the production and cycle times given in Table 7-3, determine the cost of loading and hauling a cubic yard of earth for each contractor.

REFERENCES

1. International Harvester Company, 401 North Michigan Avenue, Chicago, IL 60611.
2. Euclid, Inc., 22221 St. Clair Avenue, Cleveland, OH 44117.
3. WABCO Construction and Mining Equipment Group, 2300 N. E. Adams Street, Peoria, IL 61639.
4. Three Scrapers in Tandem Make Fast Work Moving Earth at Building Site, *Construction Methods and Equipment*, vol. 51, pp. 66–69, September 1969.

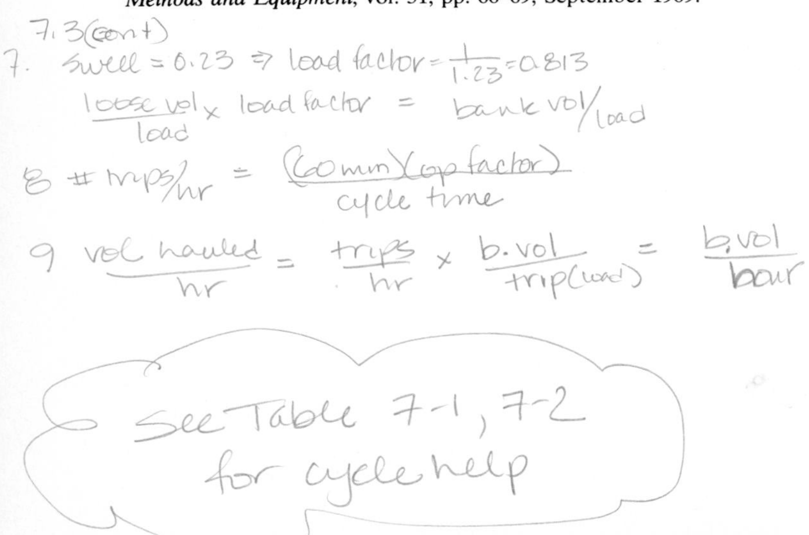

CHAPTER

EIGHT

EXCAVATING EQUIPMENT

INTRODUCTION

This chapter will deal with the listed types of equipment which are used to excavate earth and related materials and to lift items frequently used in construction operations. The equipment includes the following machines:

1. Power shovels
2. Backhoes
3. Draglines
4. Clamshells and cranes
5. Trenching machines
6. Wheel-mounted belt loaders

The first four machines belong to a group which is frequently identified as the Power Crane and Shovel Association family [1]. This association has conducted and supervised studies and tests which have provided considerable information related to the performance, operating conditions, production rates, economic life, and cost of owning and operating equipment in this group. Also, the association has participated in establishing and adopting certain standards that are applicable to this equipment. The results of the studies, conclusions and actions, and the standards have been published in technical bulletins and booklets.

Some of the information published by the Power Crane and Shovel Association is reproduced in this book, with the permission of the association.

Production rates appearing in this chapter are based on a 50-min hour, unless stated otherwise.

USEFUL LIVES OF POWER SHOVELS AND BACKHOES, DRAGLINES AND CLAMSHELLS, AND CRANES

Table 8-1 lists illustrative useful lives of these machines in both years and working hours based on the assumption that they will be used 1,800 hr per year. These values are presented primarily for illustrative purposes, which should not imply that all machines of these types will experience the same usages and useful lives. Therefore it is recommended that each owner of such equipment adopt a realistic expected useful life for the equipment that will be used under conditions that will apply to his operations.

In two-shift operation, a machine might be expected to last about half as many years as indicated in the tables, but it should last about as many working hours as indicated.

POWER SHOVELS

GENERAL INFORMATION

Power shovels are used primarily to excavate earth and load it into trucks or tractor-pulled wagons or onto conveyor belts. They are capable of excavating all classes of earth, except solid rock, without prior loosening. They may be mounted on crawler tracks, in which case they are referred to as crawler-mounted shovels. Such shovels have very low travel speeds, but the wide treads give low soil pressures, which permit them to operate on soft ground. They may be mounted on rubber-tired wheels. Single-engine self-propelled units are powered and operated from the excavator cab. The non-self-propelled units mounted on the rear of trucks, which are referred to as truck-mounted, have separate engines for operating them. Rubber-tire-mounted shovels, which have higher travel speeds than the crawler-mounted units, are useful for small jobs where considerable traveling is necessary and where the road surfaces and ground are firm. Figure 8-1 illustrates a crawler-mounted shovel.

THE SIZE OF A POWER SHOVEL

The size of a power shovel is indicated by the size of the dipper, expressed in cubic yards. In measuring the size of the dipper the earth is struck even with the contour of the dipper. This is referred to as the struck volume, as

Table 8-1 Illustrative useful life figures for power shovels and backhoes, draglines and clamshells, and cranes*

Shovels and backhoes

Machine sizes, cu yd		Useful life			
		Crawler-mounted		Rubber-tire-mounted	
Over	Through	Years	Working hours*	Years	Working hours*
0	$\frac{5}{8}$	8	14,400	10	18,000
$\frac{5}{8}$	1	10	18,000	11	19,800
1	$1\frac{3}{4}$	11	19,800	13	23,400
$1\frac{3}{4}$	$2\frac{1}{2}$	13	23,400		
$2\frac{1}{2}$	$3\frac{1}{2}$	15	27,000		
$3\frac{1}{2}$	5	16	28,800		

Draglines and clamshells

Machine sizes, cu yd		Useful life			
		Crawler-mounted		Rubber-tire-mounted	
Over	Through	Years	Working hours*	Years	Working hours*
0	$\frac{5}{8}$	10	18,000	10	18,000
$\frac{5}{8}$	1	11	19,800	13	23,400
1	$1\frac{3}{4}$	13	23,400	15	27,000
$1\frac{3}{4}$	$2\frac{1}{2}$	14	25,200	17	30,600
$2\frac{1}{2}$	$3\frac{1}{2}$	16	28,800	18	32,400
$3\frac{1}{2}$	5	17	30,600	19	34,200

Cranes

Machine sizes, tons		Useful life			
		Crawler-mounted		Rubber-tire-mounted	
Over	Through	Years	Working hours*	Years	Working hours*
0	18	12	21,600	13	23,400
18	35	14	25,200	15	27,000
35	60	16	28,800	17	30,600
60	90	18	32,400	18	32,400
90	120	19	34,200	19	34,200
Over 120		20	36,000	20	36,000

* The total working hours are based on 1,800 working hours per year. A working hour is considered to be 50 min.

Source: Power Crane and Shovel Association [1b].

Figure 8-1 Cable-operated crawler-mounted power shovel. *(Northwest Engineering Company.)*

distinguished from the heaped volume which a dipper may pick up in loose soil. Owing to the swelling of a soil when it is loosened, the bank-measure volume of a dipper will be less than the loose volume. It is possible that a dipper may be heaped sufficiently to give a bank-measure volume equal to the rated size of the dipper. However, this condition will not occur except for easy digging soils, under favorable conditions, and the assumption should not be made unless field tests indicate it to be correct. If a 2-cu-yd dipper, excavating a soil whose swell is 25 percent, is able to fill the dipper to its struck volume only, the bank-measure volume will be $2 \div 1.25 = 1.6$ cu yd.

Power shovels are commonly available in the following sizes: $\frac{3}{8}$, $\frac{1}{2}$, $\frac{3}{4}$, 1, $1\frac{1}{4}$, $1\frac{1}{2}$, 2, and $2\frac{1}{2}$ cu yd, which are classified by the Power Crane and Shovel Association as commercial sizes. Larger sizes may be available, or they can be manufactured on special order.

THE BASIC PARTS AND OPERATION OF A SHOVEL

The basic parts of a power shovel include the mounting, cab, boom, dipper stick, and dipper. These parts for a cable-controlled shovel are illustrated in Fig. 8-2.

With a shovel in the correct position, near the face of the earth to be excavated, the dipper is lowered to the floor of the pit, with the teeth pointing into the face. A crowding force is applied through the shipper shaft, and at the same time tension is applied to the hoisting line to pull the dipper up the face of the pit. If the depth of the face is just right, considering the type of soil and the size of the dipper, the dipper will be filled as it reaches the top of the face. If the depth of the face, referred to as the depth of cut, is too shallow, it will

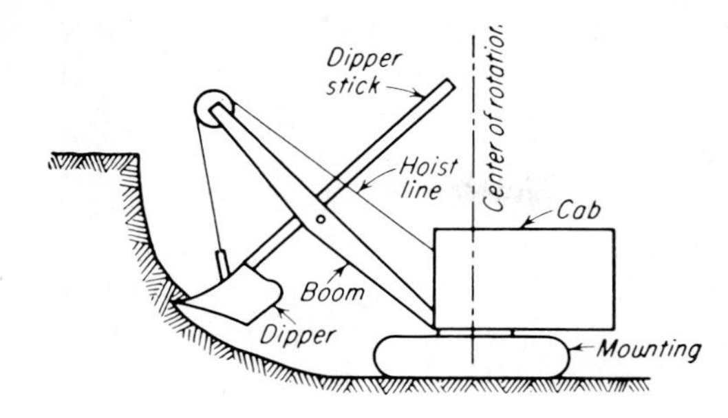

Figure 8-2 Basic parts and operation of a cable-operated power shovel.

not be possible to fill the dipper completely without excessive crowding and hoisting tension, and possibly not at all. This subjects the equipment to excessive strains and reduces the output of the unit. If the depth of the face is greater than is required to fill the dipper, when operating under favorable crowd and hoist, it will be necessary to reduce the depth of penetration of the dipper into the face if the full face is to be excavated or to start the excavation above the floor of the pit. The material left near the floor of the pit will be excavated after the upper portion of the face is removed.

The general operations of a hydraulically controlled shovel differ from those of a cable-controlled shovel primarily in that the operating forces of the former are produced by pistons instead of by cables.

SELECTING THE TYPE AND SIZE POWER SHOVEL

One of the problems which confronts the purchaser of a power shovel is the selection of the type and size. Several factors will affect the selection.

In selecting the type of shovel the prospective purchaser should consider the probable concentration of work to be performed. If there will be numerous small jobs in different locations, the mobility of the rubber-tired-mounted shovel will be a distinct advantage. If the work will be concentrated in large jobs, mobility will be of less importance and the crawler-mounted shovel will be more desirable. A crawler-mounted shovel usually is less expensive than the rubber-tired-mounted unit and can operate on ground surfaces which are not firm enough to support the latter type unit.

In selecting the size of a shovel, the two primary factors which should be considered are the cost per cubic yard of material excavated and the job conditions under which the shovel will operate.

In estimating the cost per cubic yard the following factors should be considered:

1. The size of the job, as a larger job may justify the higher cost of a large shovel.
2. The cost of transporting a large shovel will be higher than for a small one.

3. The depreciation rate for a large shovel may be higher than for a small one, especially if it is to be sold at the end of a job, owing to the probable greater difficulty of selling a large shovel.
4. The cost of downtime for repairs for a large shovel may be considerably greater than for a small one, owing to increased delays in obtaining parts for a large shovel, especially if the parts must be manufactured to order.
5. The combined cost of drilling, blasting, and excavating rock for a large shovel may be less than for a small shovel, as a large machine will handle bigger rocks than a small one. This may permit a saving in the cost of drilling and blasting.
6. The cost of wages per cubic yard will be less for a large shovel than for a small one.

The following job conditions should be considered in selecting the size of a shovel:

1. High lifts to deposit earth from a basement or trench into trucks at natural ground level will require the long reach of a large shovel.
2. If blasted rock is to be excavated, the large-size dipper will handle bigger rocks.
3. If the material to be excavated is hard and tough, the dipper of the large shovel, which exerts higher digging pressures, will handle the material more easily.
4. If the time allotted for the completion of a project requires a high hourly output, a large shovel must be used.
5. The size of available hauling units should be considered in selecting the size of a shovel. If small hauling units must be used, the size of the shovel should be small, whereas if large hauling units are available, a large shovel should be used.
6. The weight limitations imposed by most states for hauling on highways may restrict the size of a shovel if it is to be hauled over state highways. Also, the clearance of bridges and underpasses may restrict the size.

SHOVEL DIMENSIONS AND CLEARANCES

In considering the size of a power shovel for a project it may be desirable to know the dimensions of the boom and the dipper stick and the maximum cutting height, digging radius, dumping radius, and dumping height. Figure 8-3A is a clearance diagram for a cable-operated power shovel. Table 8-2 gives representative dimensions and clearances for power shovels. The clearances are for a boom angle of 45°. For boom angles other than 45° the clearances will be more or less than the values given in Table 8-2. Manufacturers' specifications should be consulted for exact values of the clearances. The maximum dumping

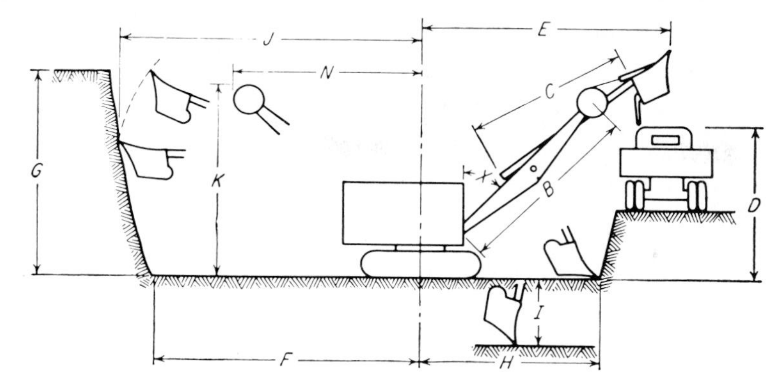

Figure 8-3A Clearance diagram for a cable-operated power shovel.

height and radius are especially important when a shovel in a pit is loading trucks at natural ground level.

OPTIMUM DEPTH OF CUT

The optimum depth of cut is that depth which produces the greatest output and at which the dipper comes up with a full load without undue crowding. The depth varies with the class of soil and the size of the dipper. Values of optimum depths for various classes of soils and sizes of dippers are given in Table 8-4.

Figure 8-3B A hydraulic-operated front shovel loading material into a truck. *(Caterpillar Tactor Co.)*

Table 8-2 Dimensions and clearances for power shovels for 45° boom angle

Size dipper, cu yd	Length std boom, ft	Length std handle, ft	Max cutting height, ft	Max digging radius, ft	Max dumping height, ft	Max dumping radius, ft
$\frac{3}{8}$	13–15	11–13	17–19	22	13–15	18–20
$\frac{1}{2}$	15–16	12–13	19–24	21–24	14–16	19–20
$\frac{3}{4}$	17–18	13–15	21–27	25–28	15–17	22–24
1	20	16	23–27	31–32	15–18	23–25
$1\frac{1}{4}$	21	16	23–27	31–32	16–19	24–27
$1\frac{1}{2}$	21–23	16–18	24–29	32–33	18–20	28–30
$1\frac{3}{4}$	22–24	16–18	26–30	32–33	18–20	28–30
2	22–25	17–19	26–30	33–36	19–20	30–33
$2\frac{1}{2}$	25–26	18–19	28–35	35–38	19–21	32–34

FRONT SHOVELS

Figure 8-3A illustrates a hydraulic-operated crawler-mounted front shovel loading material into a truck. Two models of this shovel are available, namely, model 235 and model 245. Table 8-4A gives limited information for each of the

Table 8-3 Approximate shovel digging and loading cycles for various angles of swing*

Size of shovel dipper, cu yd	Easy digging Moist loam, light sandy clay Angle of swing, deg				Medium digging Good common earth Angle of swing, deg				Hard, tough digging Hard tough clay Angle of swing, deg			
	45	90	135	180	45	90	135	180	45	90	135	180
$\frac{3}{8}$	12	16	19	22	15	19	23	26	19	24	29	33
$\frac{1}{2}$	12	16	19	22	15	19	23	26	19	24	29	33
$\frac{3}{4}$	13	17	20	23	16	20	24	27	20	25	30	34
1	14	18	21	25	17	21	25	29	21	26	31	36
$1\frac{1}{4}$	14	18	21	25	17	21	25	29	21	26	31	36
$1\frac{1}{2}$	15	19	23	27	18	23	27	31	22	28	33	38
$1\frac{3}{4}$	16	20	24	28	19	24	28	32	23	29	34	39
2	17	21	25	30	20	25	29	34	24	30	35	41
$2\frac{1}{2}$	18	22	27	32	21	26	31	36	25	31	37	43

* The time is in seconds with no delays when digging in optimum depths of cut and loading trucks on the same grade as the shovel.

Source: Power Crane and Shovel Association [1c].

Table 8-4 Ideal outputs of cable-operated power shovels in cubic yards (cubic meters) per 60-min hour, bank measure

Class of material	Size of shovel, cu yd (cu m)								
	$\frac{3}{8}$ (0.29)*	$\frac{1}{2}$ (0.38)*	$\frac{3}{4}$ (0.57)*	1 (0.76)*	$1\frac{1}{4}$ (0.95)*	$1\frac{1}{2}$ (1.14)*	$1\frac{3}{4}$ (1.33)*	2 (1.53)*	$2\frac{1}{2}$ (1.91)*
Moist loam, or light sandy clay	3.8	4.6	5.3	6.0	6.5	7.0	7.4	7.8	8.4
	(1.1)†	(1.4)†	(1.6)†	(1.8)†	(2.0)†	(2.1)†	(2.2)†	(2.4)†	(2.6)†
	85	115	165	205	250	285	320	355	405
	(65)‡	(88)‡	(126)‡	(157)‡	(190)‡	(218)‡	(244)‡	(272)‡	(309)‡
Sand and gravel	3.8	4.6	5.3	6.0	6.5	7.0	7.4	7.8	8.4
	(1.1)	(1.4)	(1.6)	(1.8)	(2.0)	(2.1)	(2.2)	(2.4)	(2.6)
	80	110	155	200	230	270	300	330	390
	(61)	(84)	(118)	(153)	(176)	(206)	(229)	(252)	(298)
Good common earth	4.5	5.7	6.8	7.8	8.5	9.2	9.7	10.2	11.2
	(1.4)	(1.7)	(2.1)	(2.4)	(2.6)	(2.8)	(2.9)	(3.1)	(3.4)
	70	95	135	175	210	240	270	300	350
	(54)	(73)	(103)	(134)	(160)	(183)	(206)	(229)	(268)
Hard, tough clay	6.0	7.0	8.0	9.0	9.8	10.7	11.5	12.2	13.3
	(1.8)	(2.1)	(2.4)	(2.7)	(3.0)	(3.3)	(3.5)	(3.7)	(4.0)
	50	75	110	145	180	210	235	265	310
	(38)	(57)	(84)	(111)	(137)	(156)	(180)	(202)	(236)
Wet, sticky clay	6.0	7.0	8.0	9.0	9.8	10.7	11.5	12.2	13.3
	(1.8)	(2.1)	(2.4)	(2.7)	(3.0)	(3.3)	(3.5)	(3.7)	(4.0)
	25	40	70	95	120	145	165	185	230
	(19)	(30)	(53)	(72)	(91)	(110)	(125)	(141)	(175)
Well-blasted rock	40	60	95	125	155	180	205	230	275
	(30)	(46)	(72)	(95)	(118)	(137)	(156)	(175)	(210)
Poorly blasted rock	15	25	50	75	95	115	140	160	195
	(11)	(19)	(38)	(57)	(73)	(88)	(107)	(122)	(149)

* These values are the sizes of shovels in cubic meters.
† These values are the depths of cut in meters.
‡ These values are the ideal outputs in cubic meters.

two shovels. For more complete information on the shovels, the reader is referred to the publication *Caterpillar Performance Handbook Edition* 14.

THE OUTPUT OF POWER SHOVELS

The actual output of a power shovel is affected by numerous factors, including the following:

1. Class of material

2. Depth of cut
3. Angle of swing
4. Job conditions
5. Management conditions
6. Size of hauling units
7. Skill of the operator
8. Physical condition of the shovel

Table 8-4A The rated capacities of the buckets of hydraulic-operated front shovels*

	Model 235	
Capacity	Front dump	Bottom dump
Heaped	3.0 cu yd (2.3 m^3)	2.38 cu yd (1.8 m^3)
Struck	2.5 cu yd (1.9 m^3)	1.94 cu yd (1.5 m^3)
Flywheel power	195 hp (145 KWH)	
	Model 245	
Heaped	5.0 cu yd (3.8 m^3)	4.0 cu yd (3.1 m^3)
Struck	4.4 cu yd (3.34 m^3)	3.48 cu yd (2.64 m^3)
Flywheel power	325 hp (242 KWH)	

* Caterpillar Tractor Co.

The output of a shovel should be expressed in cubic yards per hour based on bank-measure volume. The capacity of a dipper is based on its struck volume. In excavating some classes of materials, it is possible for a dipper to pick up a heaping volume which may exceed the struck volume. In order to obtain the bank-measure volume of a dipper of earth, the average loose volume should be divided by 1 plus the swell, expressed as a fraction. For example, if a 2-cu-yd dipper, excavating material whose swell is 25 percent, will handle an average loose volume of 2.25 cu yd, the bank-measure volume will be $2.25 \div 1.25 = 1.8$ cu yd. If this shovel can make 2.5 cycles per min, which includes no allowance for lost time, the output will be $2.5 \times 1.8 = 4.5$ cu yd per min, or 270 cu yd per hr. This is an ideal output, which will seldom, if ever, be experienced on a project. Table 8-4 gives the ideal outputs of power shovels, expressed in cubic yards and cubic meters bank measure, for various classes of materials, based on digging at optimum depth with a 90° swing and no delays. In Table 8-4 the upper figure is the optimum depth in feet and the lower figure is the ideal output in cubic yards.

The actual outputs of power shovels will be significantly lower than the values in Table 8-4 because operators do not generally operate continuously at peak efficiency, nor do they operate a full 60 min in an hour. This is discussed in more detail in the later sections of this book.

THE EFFECT OF THE DEPTH OF CUT ON THE OUTPUT OF A POWER SHOVEL

If the depth of the face from which a shovel is excavating material is too shallow, it will be difficult or impossible to fill the dipper in one pass up the face. The operator will have a choice of making more than one pass to fill the dipper, which will increase the time per cycle, or he may carry a partly filled dipper to the hauling unit each cycle. In either case the effect will be to reduce the output of the shovel.

If the depth of the face is greater than the minimum required to fill the dipper, with favorable crowding and hoisting forces, the operator may do one of three things. He may reduce the depth of penetration of the dipper into the face in order to fill the dipper in one full stroke. This will increase the time for a cycle. He may start digging above the base of the face, and then remove the lower portion of the face later. Or he may run the dipper up the full height of the face and let the excess earth spill down to the bottom of the face, to be picked up later. The choice of any one of the procedures will result in some lost time, based on the time required to fill the dipper when it is digging at optimum depth. As indicated in Table 8-4, the optimum depth varies with the class of material and the size of the dipper.

The effect of the depth of cut on the output of a shovel is illustrated in Table 8-5. In the table the percent of optimum depth of cut is obtained by dividing the actual depth of cut by the optimum depth for the given material and dipper, then multiplying the result by 100. Thus, if the actual depth of cut is 6 ft and the optimum depth is 10 ft, the percent of optimum depth of cut is $\frac{6}{10} \times 100 = 60$.

Table 8-5 Conversion factors for depth of cut and angle of swing for a power shovel

Percent of optimum depth	Angle of swing, deg						
	45	60	75	90	120	150	180
40	0.93	0.89	0.85	0.80	0.72	0.65	0.59
60	1.10	1.03	0.96	0.91	0.81	0.73	0.66
80	1.22	1.12	1.04	0.98	0.86	0.77	0.69
100	1.26	1.16	1.07	1.00	0.88	0.79	0.71
120	1.20	1.11	1.03	0.97	0.86	0.77	0.70
140	1.12	1.04	0.97	0.91	0.81	0.73	0.66
160	1.03	0.96	0.90	0.85	0.75	0.67	0.62

THE EFFECT OF THE ANGLE OF SWING ON THE OUTPUT OF A POWER SHOVEL

The angle of swing of a power shovel is the horizontal angle, expressed in degrees, between the position of the dipper when it is excavating and the position when it is discharging the load. The total time in a cycle includes digging, swinging to the dumping position, dumping, and returning to the digging position. If the angle of swing is increased, the time for a cycle will be increased, while if the angle of swing is decreased, the time for a cycle will be decreased. The effect of the angle of swing on the output of a shovel is illustrated in Table 8-5. For example, if a shovel which is digging at optimum depth has the angle of swing reduced from 90 to 60°, the output will be increased by 16 percent.

The output of a shovel operating at 90° swing and optimum depth, which is obtained from Table 8-4, should be multiplied by the proper conversion factor from Table 8-5 in order to obtain the probable output for any given depth and angle of swing.

Example 8-1 The use of the tables is illustrated by considering a 2-cu-yd shovel excavating common earth, with a depth of cut of 12 ft and an angle of swing of 60°. The percent of optimum depth is $12/10.2 \times 100 = 118$.

By interpolating in Table 8-5 the factor is found to be 1.115. However, it is doubtful that values beyond two decimal places are significant. Therefore, 1.11 is sufficiently accurate for practical purposes. The probable output of the shovel will be $300 \times 1.11 = 333$ cu yd per 60-min hour.

Although the information given in Tables 8-4 and 8-5 is based on extensive field studies, the reader is cautioned against using it too literally without adjusting it for conditions which will probably exist on a project. As explained later, additional factors must be applied to whatever extent they are necessary in the judgment of the project planner.

THE EFFECT OF JOB CONDITIONS ON THE OUTPUT OF A POWER SHOVEL

As every owner of a power shovel knows, no two excavating jobs are alike. There are certain conditions at every job over which the owner of the shovel has no control. These conditions must be considered in estimating the probable output of a shovel.

A shovel may operate in a large, open pit, with a firm, well-drained floor, where trucks can be spotted on either side of the shovel to eliminate lost time waiting for hauling units. The terrain of the natural ground may be uniformly level, so that the depth of cut will always be optimum. The haul road is not affected by climatic conditions, such as rains. A job of this type is large enough

Table 8-6 Factors for job and management conditions

	Management conditions			
Job conditions	Excellent	Good	Fair	Poor
Excellent	0.84	0.81	0.76	0.70
Good	0.78	0.75	0.71	0.65
Fair	0.72	0.69	0.65	0.60
Poor	0.63	0.61	0.57	0.52

* Values are based on a 60-min hour.

to justify the selection of balanced hauling units. Such a project might be classified as having excellent job conditions.

Another shovel may be used to excavate material for a highway cut through a hill. The depth of cut may vary from zero to considerably more than the optimum depth. The sides of the cut must be carefully sloped. The cut may be so narrow that a loaded truck must move out before an empty truck can back into loading position. As the truck must be spotted behind the shovel, the angle of swing will approximate 180°. The floor of the cut may be muddy, which will delay the movement of the trucks. Light rains may delay operations for several days. A project of this type might be classified as having poor job conditions.

In excavating a basement, which requires the trucks to travel up an earth ramp, a power shovel may be delayed considerably by ground water or rain, by the difficulty of getting hauling units in and out, and by the difficulty of excavating the corners.

Job conditions may be classified as excellent, good, fair, and poor. There is no uniform standard which may be used as a guide in classifying a job. Each job planner must used his own judgment and experience in deciding which condition best represents his job. Table 8-6 illustrates the effect of job conditions on the output of a power shovel.

THE EFFECT OF MANAGEMENT CONDITIONS ON THE OUTPUT OF A POWER SHOVEL

The attitude of the owner of a shovel in establishing the conditions under which a shovel is operated will affect the output of the shovel. While the owner may not be able to improve job conditions, he may take several steps to improve management conditions, including the following:

1. Greasing and lubricating the shovel frequently

2. Checking the shovel parts that are subject to the greatest wear, and replacing worn parts while the shovel is not being operated, as at the end of a shift
3. Replacing badly worn wire rope between shifts
4. Replacing dull dipper teeth with sharp ones, as required
5. Giving the shovel a major overhaul between jobs, if necessary
6. Keeping at the job extra parts that are subject to the greatest wear
7. Keeping the pit floor clean and smooth to permit better truck spotting and to reduce the angle of swing
8. Providing adequate trucks of the correct size to eliminate lost time in loading and waiting for trucks
9. Paying a bonus to the crew for production in excess of an agreed amount to encourage high production
10. Providing a competent supervisor to keep the job running smoothly

Management conditions may be classified as excellent, good, fair, and poor. Table 8-6 illustrates the effect of management conditions on the output of a power shovel.

EXAMPLES ILLUSTRATING THE EFFECT OF THE VARIOUS FACTORS ON THE OUTPUT OF A POWER SHOVEL

It is doubtful that a job planner will be able to select exactly the correct factors to be used in estimating the output of a shovel. As a result, the actual output of a shovel may vary from the estimated output. Experience and good judgment are essential to the selection of the correct factors. If the output is found to fall below that estimated, it may be possible to increase it by modifying the operating conditions.

Example 8-2 To illustrate the use of the information in Tables 8-4 to 8-6, consider a 1-cu-yd power shovel for excavating hard clay with a depth of cut of 7.5 ft. An analysis of the project indicates an average angle of swing of 75°, job conditions will be fair, and management will be good. Determine the probable output in cubic yards per hour bank measure.

From Table 8-4 the ideal output will be 145 cu yd per hr. The optimum depth is 9 ft.

Percent of optimum depth is $7.5/9 \times 100 = 83.3$.
From Table 8-5 the depth-swing factor is 1.04.
From Table 8-6 the job-management factor is 0.69.
The probable output per hr, $145 \times 1.04 \times 0.69 = 104$ cu yd.
For a 50-min hr the probable output will be $\frac{50}{60} \times 104 = 86$ cu yd.

Example 8-3 This example illustrates the effect of the various conditions on the output of a shovel. Determine the probable output, in cubic yards per hour bank measure, for a 1-cu-yd shovel for each of the conditions given in the table.

Factors involved	Moist loam	Common earth	Hard clay	Wet clay	Poorly blasted rock
	Class of material				
Depth, ft	6.0	10.0	8.0	12.0	Varies
Angle of swing, deg	60	90	120	180	120
Job conditions	Good	Fair	Fair	Poor	Fair
Management conditions	Good	Good	Fair	Poor	Good
Ideal output, cu yd per hr*	200	175	145	95	75
Optimum depth, ft	6.0	7.8	9.0	9.0	
Percent optimum depth	100	128	89	133	
Depth-swing factor	1.16	0.94	0.87	0.67	
Job-management factor	0.75	0.69	0.65	0.52	0.69
Probable output, cu yd per hr†	174	114	82	33	52

* Based on a 60-min hour.
† Based on a 50-min hour.

METHODS OF INCREASING THE OUTPUT OF A POWER SHOVEL

A problem which frequently confronts a consultant on the selection and operation of excavating equipment is to analyze a project which is not being operated satisfactorily in order to recommend corrective steps to increase the output and reduce the cost of handling the material.

Example 8-4 On one such project, where the cost was exceeding the estimate, an analysis was made to determine methods of reducing the cost of excavating and hauling the earth. The material was common earth. The analysis of the operations revealed the following information:

Size of power shovel, $1\frac{1}{2}$ cu yd
Depth of cut, 12 ft
Angle of swing, 120°
Size of trucks, 6 cu yd bm
Round-trip time for a truck, 19 min
No. trucks, 8

The time spent by the shovel in cleaning up the floor of the pit, moving, and undergoing repairs reduced the actual excavating time to about 30 min per hr.

The floor of the pit was rough, muddy, and heavily rutted because of inadequate drainage, which reduced the efficiency of the hauling units.

The output averaged 108 cu yd per hr.

The direct cost of excavating and hauling the earth was determined as follows:

Shovel, operator and oiler	= $ 87.50 per hr
Trucks and drivers, 8 @ $28.80	= 232.40 per hr
Direct overhead and supervision	= 31.20 per hr
Total cost	= $349.10 per hr
Cost per cu yd, $349.10 ÷ 108	= $ 3.23

The analysis indicates that the output could be increased by taking the following steps:

1. Use a small bulldozer to keep the floor of the pit clean and well drained.
2. Reduce the depth of cut to the optimum.
3. Reduce the angle of swing to 75° by improving the floor of the pit.
4. Improve the job conditions to good by proper maintenance of the pit and haul roads and by excavating at optimum depth.
5. Improve the management conditions to good by properly servicing the equipment at the end of the shifts and by paying a bonus of $0.04 per cu yd, to be divided among the workers, for all production in excess of 120 cu yd per hr.
6. Reduce the round-trip time of the trucks to 15 min by improving the haul road and the pit floor.
7. Provide additional trucks to haul the increased output of the shovel.

If the recommended steps are taken, the probable output of the shovel will be as follows:

Estimated actual excavating time, 50 min per hr
Ideal output, 240 cu yd per hr
Depth-swing factor, 1.07
Job-management factor, 0.75
Probable output, $240 \times 1.07 \times 0.75 = 193$ cu yd per hr

The number of trucks required to haul the earth will be calculated as follows:

Assume trucks operate 50 min per hr
No. trips per hr per truck, $\frac{50}{15} = 3.33$
Volume hauled per hr per truck, $3.33 \times 6 = 20$ cu yd
No. trucks needed, $193 \div 20 = 9.6$
Ten trucks are needed.

The revised direct cost of excavating and hauling the earth will be as follows:

Shovel, operator and oiler	= $ 87.50 per hr
Trucks and drivers, 10 @ $28.80	= 288.00 per hr
Direct overhead and supervision	= 31.20 per hr
Cost of bulldozer and operator	= 27.75 per hr
Cost of bonus, 73 cu yd @ $0.04	= 2.92 per hr
Total cost	= $437.37 per hr
Cost per cu yd, $437.37 ÷ 193	= $ 2.24
Net reduction in the cost per cu yd, $3.23 − $2.24	= $ 0.99

This saving is sufficiently large to demonstrate the financial effect of applying intelligent engineering in the selection of equipment and in analyzing an operation. The failure to apply engineering analysis to the operation of a project is one reason why a contractor may complete a project with a loss, while another will complete a similar project with a profit.

DRAGLINES

GENERAL INFORMATION

Draglines are used to excavate earth and load it into hauling units, such as trucks or tractor-pulled wagons, or to deposit it in levees, dams, and spoil banks near the pits from which it is excavated. In general, a power shovel up to a capacity of $2\frac{1}{2}$ cu yd can be converted into a dragline by replacing the boom of the shovel with a crane boom and substituting a dragline bucket for the shovel dipper.

For some projects either a power shovel or a dragline may be used to excavate materials, but for others the dragline will have a distinct advantage compared with a shovel. A dragline usually does not have to go into a pit or hole in order to excavate. It may operate on natural ground while excavating material from a pit with its bucket. This will be very advantageous when earth is removed from a ditch, canal, or pit containing water. If the earth is hauled with trucks, they do not have to go into the pit and contend with mud. If the earth can be deposited along a canal or ditch or near a pit, it frequently is possible to use a dragline with a boom long enough to dispose of the earth in one operation, eliminating the need for hauling units, which will reduce the cost of handling the earth. Draglines are excellent units for excavating trenches when the sides are permitted to establish their angles of repose, without shoring.

One disadvantage in using a dragline compared with a power shovel is the reduced output of the dragline. A comparison of the ideal output of various sizes of draglines with the output of power shovels shows that a dragline will excavate approximately 75 to 80 percent as much earth as a shovel of the same size.

TYPES OF DRAGLINES

Draglines may be divided into three types, as follows:

1. Crawler-mounted (see Fig. 8-4)
2. Wheel-mounted, self-propelled (see Fig. 8-5)
3. Truck-mounted (see Fig. 8-6)

Crawler-mounted draglines can operate on surfaces which are too soft for wheel- or truck-mounted equipment, but their speeds are so slow, frequently less than 1 mph, that it may be necessary to use auxiliary hauling equipment to transport them from one job to another, especially if the distance is great. Wheel- and truck-mounted units may have travel speeds in excess of 30 mph.

THE SIZE OF A DRAGLINE

The size of a dragline is indicated by the size of the bucket, expressed in cubic yards, which, in general, is the same size as the dipper of the power shovel into which it may be converted. However, most draglines may handle more than one size bucket, depending on the length of the boom and the class of material

Figure 8-4 Crawler-mounted dragline. *(Northwest Engineering Company.)*

excavated. Because the maximum lifting capacity of a dragline is limited by the force which will tilt the machine over, it is necessary to reduce the size of the bucket when a long boom is used or when the material has a high specific gravity. In practice the combined weight of the bucket and its load should produce a tilting force not greater than 75 percent of the force required to tilt the machine over. A longer boom, with a smaller bucket, should be used when it is necessary to increase the digging reach or the dumping radius.

If the material is difficult to excavate, the use of a smaller bucket, which will reduce the digging resistance, may permit an increase in the output of a dragline.

Typical working ranges for a dragline that will handle buckets varying in sizes from $1\frac{1}{4}$ to $2\frac{1}{2}$ cu yd are given in Table 8-7 (see Fig. 8-7 for the dimensions given in the table).

Figure 8-5 Wheel-mounted self-propelled dragline.

Figure 8-6 Truck-mounted dragline.

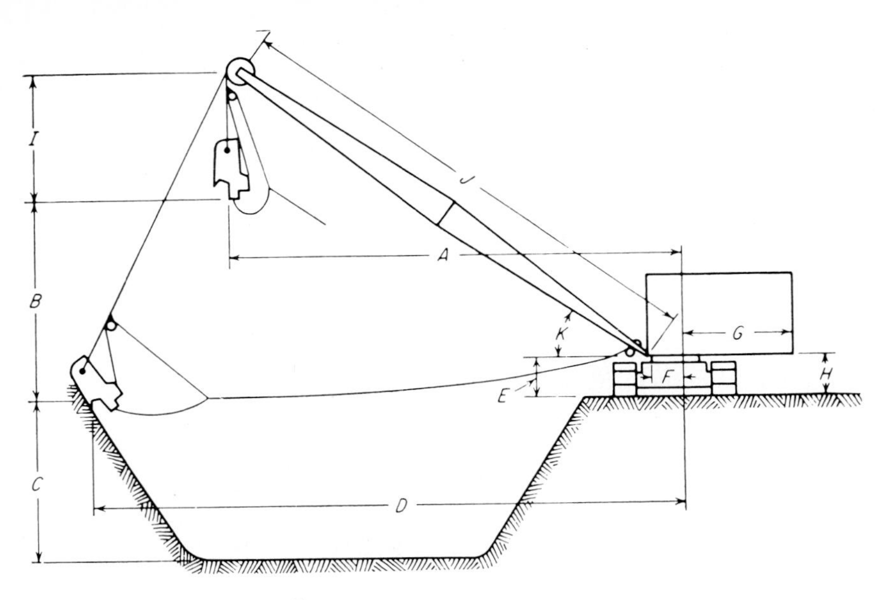

Figure 8-7 Dragline range diagram.

THE BASIC PARTS AND OPERATION OF A CABLE-CONTROLLED DRAGLINE

The basic parts of a dragline are illustrated in Fig. 8-8.

Excavating is started by swinging the empty bucket to the digging position, at the same time slacking off the drag and the hoist cables. Separate drums on the basic unit are available for each of these cables so that they may be coordinated into a smooth operation. Excavating is accomplished by pulling the bucket toward the machine while regulating the digging depth by means of the tension maintained in the hoist cable. When the bucket is filled, the operator takes in on the hoist line while playing out the drag cable. The bucket is so constructed that it will not dump its contents until it is desired. Hoisting, swinging, and dumping of the loaded bucket follow in that order; then the cycle is repeated. Dumping is accomplished by releasing the drag cable. An experienced operator can cast the excavated material beyond the end of the boom.

Since it is more difficult to control the accuracy in dumping from a dragline as compared with a power shovel, it is desirable to use larger hauling units for dragline loading in order to reduce the spillage. A size ratio equal to at least five to six times the capacity of the bucket is recommended.

Figure 8-9 shows the dragline digging zones. The work should be planned to permit most of the digging to be done in the zones which permit the best digging, with the poor digging zone used as little as possible.

Table 8-7 Typical working ranges for a cable-controlled dragline with maximum counterweights

J, boom length, 50 ft:						
Capacity, lb*	12,000	12,000	12,000	12,000	12,000	12,000
K, boom angle, deg	20	25	30	35	40	45
A, dumping radius, ft	55	50	50	45	45	40
B, dumping height, ft	10	14	18	22	24	27
C, max digging depth, ft	40	36	32	28	24	20
J, boom length, 60 ft:						
Capacity, lb*	10,500	11,000	11,800	12,000	12,000	12,000
K, boom angle, deg	20	25	30	35	40	45
A, dumping radius, ft	65	60	55	55	52	50
B, dumping height, ft	13	18	22	26	31	35
C, max digging depth, ft	40	36	32	28	24	20
J, boom length, 70 ft:						
Capacity, lb*	8,000	8,500	9,200	10,000	11,000	11,800
K, boom angle, deg	20	25	30	35	40	45
A, dumping radius, ft	75	73	70	65	60	55
B, dumping height, ft	18	23	28	32	37	42
C, max digging depth, ft	40	36	32	28	24	20
J, boom length, 80 ft:						
Capacity, lb*	6,000	6,700	7,200	7,900	8,600	9,800
K, boom angle, deg	20	25	30	35	40	45
A, dumping radius, ft	86	81	79	75	70	65
B, dumping height, ft	22	27	33	39	42	47
C, max digging depth, ft	40	36	32	28	24	20
D, digging reach	Depends on working conditions and operator's skill with bucket					

* Combined weight of bucket and material must not exceed capacity.

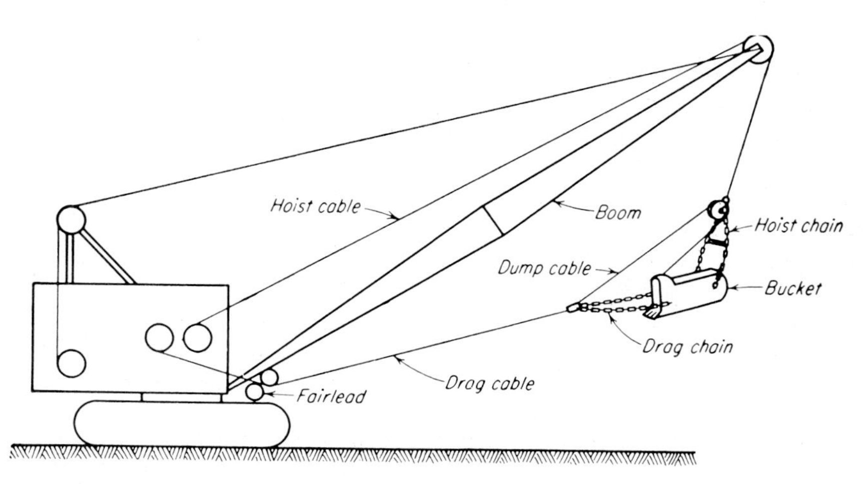

Figure 8-8 Basic parts of a dragline.

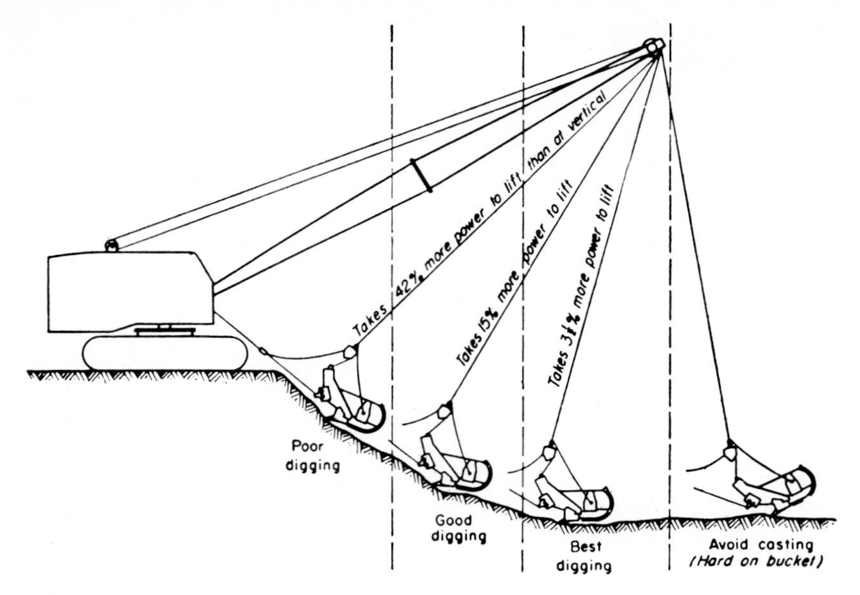

Figure 8-9 Dragline digging zones.

OPTIMUM DEPTH OF CUT

A dragline will produce its greatest output if the job is planned to permit the earth to be excavated at the optimum depth where possible. Table 8-9 gives the optimum depth of cut for various sizes of buckets and classes of materials, using short-boom draglines.

THE OUTPUT OF THE DRAGLINES

The output of a dragline will vary with the following factors:

1. Class of material
2. Depth of cut
3. Angle of swing
4. Size and type of bucket
5. Length of boom
6. Job conditions
7. Management conditions
8. Method of disposal, casting or loading trucks
9. Size of hauling units, if used
10. Skill of operator
11. Physical condition of the machine

Table 8-8 Approximate dragline digging and loading cycles for various angles of swing*

Size of drag-line bucket, cu yd	Easy digging Light moist clay or loam Angle of swing, deg				Sand or gravel Angle of swing, deg				Good common earth Angle of swing, deg			
	45	90	135	180	45	90	135	180	45	90	135	180
$\frac{3}{8}$	16	19	22	25	17	20	24	27	20	24	28	31
$\frac{1}{2}$	16	19	22	25	17	20	24	27	20	24	28	31
$\frac{3}{4}$	17	20	24	27	18	22	26	29	21	26	30	33
1	19	22	26	29	20	24	28	31	23	28	33	36
$1\frac{1}{4}$	19	23	27	30	20	25	29	32	23	28	33	36
$1\frac{1}{2}$	21	25	29	32	22	27	31	34	25	30	35	38
$1\frac{3}{4}$	22	26	30	33	23	28	32	35	26	31	36	39
2	23	27	31	35	24	29	33	37	27	32	37	41
$2\frac{1}{2}$	25	29	34	38	26	31	36	40	29	34	40	44

* The time is in seconds with no delays when digging at optimum depths of cut and loading trucks on the same grade as the shovel.

Source: Power Crane and Shovel Association.

Table 8-8 gives approximate dragline digging and loading cycles for various angles of swing.

The output of a dragline should be expressed in cubic yards per hour bank measure. This quantity may be obtained from field observations, or it may be estimated by multiplying the average loose volume per bucket by the number of cycles per hour and dividing by 1 plus the swell factor for the earth, expressed as a fraction. For example, if a 2-cu-yd bucket, excavating material whose swell is 25 percent, will handle an average loose volume of 2.4 cu yd, the bank-measure volume will be $2.4 \div 1.25 = 1.92$ cu yd. If the dragline can make 2 cycles per min, the output will be $2 \times 1.92 = 3.84$ cu yd per min or 230 cu yd per hr bank measure. This is an ideal output, which will seldom, if ever, be experienced on a project. Table 8-9 gives the ideal outputs of short-boom draglines, expressed in cubic yards bank measure, for various classes of materials, based on digging at optimum depth, with a 90° swing, and no delays. In Table 8-9 the upper figure is the optimum depth in feet and the lower figure is the ideal output in cubic yards.

THE EFFECT OF THE DEPTH OF CUT AND THE ANGLE OF SWING ON THE OUTPUT OF A DRAGLINE

The outputs of draglines given in Table 8-9 are based on digging at optimum depths with an angle of swing of 90°. For any other depth or angle of swing the ideal output of the particular unit should be multiplied by an appropriate

Table 8-9 Ideal outputs of short-boom draglines, in cubic yards (cubic meters) per 60-min hour, bank measure

Class of material	Size of bucket, cu yd (cu m)*								
	3.8 (0.29)*	1/2 (0.38)*	3/4 (0.57)*	1 (0.76)*	1 1/4 (0.95)*	1 1/2 (1.14)*	1 3/4 (1.33)*	2 (1.53)*	2 1/2 (1.91)*
Moist loam or light sandy clay	5.0	5.5	6.0	6.6	7.0	7.4	7.7	8.0	8.5
	(1.5)†	(1.7)†	(1.8)†	(2.0)†	(2.1)†	(2.2)†	(2.4)†	(2.5)†	(2.6)†
	70	95	130	160	195	220	245	265	305
	(53)‡	(72)‡	(99)‡	(122)‡	(149)‡	(168)‡	(187)‡	(202)‡	(233)‡
Sand and gravel	5.0	5.5	6.0	6.6	7.0	7.4	7.7	8.0	8.5
	(1.5)	(1.7)	(1.8)	(2.0)	(2.1)	(2.2)	(2.4)	(2.5)	(2.6)
	65	90	125	155	185	210	235	255	295
	(49)	(69)	(95)	(118)	(141)	(160)	(180)	(195)	(225)
Good common earth	6.0	6.7	7.4	8.0	8.5	9.0	9.5	9.9	10.5
	(1.8)	(2.0)	(2.4)	(2.5)	(2.6)	(2.7)	(2.8)	(3.0)	(3.2)
	55	75	105	135	165	190	210	230	265
	(42)	(57)	(81)	(104)	(127)	(147)	(162)	(177)	(204)
Hard, tough clay	7.3	8.0	8.7	9.3	10.0	10.7	11.3	11.8	12.3
	(2.2)	(2.5)	(2.7)	(2.8)	(3.1)	(3.3)	(3.5)	(3.6)	(3.8)
	35	55	90	110	135	160	180	195	230
	(27)	(42)	(69)	(85)	(104)	(123)	(139)	(150)	(177)
Wet, sticky clay	7.3	8.0	8.7	9.3	10.0	10.7	11.3	11.8	12.3
	(2.2)	(2.5)	(2.7)	(2.8)	(3.1)	(3.3)	(3.5)	(3.6)	(3.8)
	20	30	55	75	95	110	130	145	175
	(15)	(23)	(42)	(58)	(73)	(85)	(100)	(112)	(135)

* These values are the sizes of the buckets in cubic meters.
† These values are the depths of cut in meters.
‡ These values are the ideal outputs in cubic meters.

Table 8-10 The effect of the depth of cut and angle of swing on the output of draglines

Percent of optimum depth	Angle of swing, deg							
	30	45	60	75	90	120	150	180
20	1.06	0.99	0.94	0.90	0.87	0.81	0.75	0.70
40	1.17	1.08	1.02	0.97	0.93	0.85	0.78	0.72
60	1.24	1.13	1.06	1.01	0.97	0.88	0.80	0.74
80	1.29	1.17	1.09	1.04	0.99	0.90	0.82	0.76
100	1.32	1.19	1.11	1.05	1.00	0.91	0.83	0.77
120	1.29	1.17	1.09	1.03	0.98	0.90	0.82	0.76
140	1.25	1.14	1.06	1.00	0.96	0.88	0.81	0.75
160	1.20	1.10	1.02	0.97	0.93	0.85	0.79	0.73
180	1.15	1.05	0.98	0.94	0.90	0.82	0.76	0.71
200	1.10	1.00	0.94	0.90	0.87	0.79	0.73	0.69

depth-swing factor. The effect of the depth of cut and the angle of swing on the output of a dragline is given in Table 8-10.

Example 8-5 A 2-cu-yd short-boom dragline is to be used to excavate hard, tough clay. The depth of cut will be 15.4 ft, and the angle of swing will be 120°. Determine the probable output of the dragline if there are no other factors to affect the output.

The percent of optimum depth, $15.4/11.8 \times 100 = 130$. From Table 8-10 the correction factor is 0.89. The probable output will be $195 \times 0.89 = 173$ cu yd per hr. For a 50-min hour the probable output will be

$$\text{Output} = 0.83 \times 173 = 143 \text{ cu yd}$$

THE EFFECT OF JOB AND MANAGEMENT CONDITIONS ON THE OUTPUT OF A DRAGLINE

The effect of job and management conditions on the output of a dragline will be about the same as for a power shovel. The information is given in Table 8-6.

THE EFFECT OF THE SIZE OF THE BUCKET AND THE LENGTH OF THE BOOM ON THE OUTPUT OF A DRAGLINE

In selecting the size and type bucket, the dragline and bucket should be matched properly in order to obtain the best action and the greatest operating efficiency, which will produce the greatest output of material. Buckets are generally available in three types: light-duty, medium-duty, and heavy-duty. Light-duty buckets are used for excavating materials which are dug easily, such as sandy loam, sandy clay, or sand. Medium-duty buckets are used for general excavating service such as digging clay, soft shale, or loose gravel. Heavy-duty buckets are used for mine stripping, handling blasted rock, and excavating hardpan and highly abrasive materials. Buckets are sometimes perforated to permit excess water to drain from the loads. Figure 8-10 shows a medium-duty dragline bucket. Figure 8-11 shows a 1-cu-yd dragline bucket dumping its load on a spoil bank.

Table 8-11 gives representative capacities, weight, and dimensions for dragline buckets.

The normal size of a dragline bucket is based on its struck capacity, which is expressed more accurately in cubic feet. In selecting the most suitable size bucket for use with a given dragline, it is desirable to know the weight of the loosened material to be handled, expressed in pounds per cubic foot. While it is desirable to use the largest size bucket possible in the interest of increasing the output, care should be exercised to see that the combined weight of the load and the bucket does not exceed the safe load recommended for the dragline.

Example 8-6 The importance of this analysis is illustrated by referring to the information given in Table 8-7. Assume that the material to be handled has a loose weight of

Figure 8-10 Medium-duty dragline bucket.

Figure 8-11 Dragline bucket dumping its load.

Table 8-11 Representative capacities, weights, and dimensions of dragline buckets

Size, cu yd	Struck capacity, cu ft	Weight of bucket, lb			Dimension, in.		
		Light-duty	Medium-duty	Heavy-duty	Length	Width	Height
$\frac{3}{8}$	11	760	880		35	28	20
$\frac{1}{2}$	17	1,275	1,460	2,100	40	36	23
$\frac{3}{4}$	24	1,640	1,850	2,875	45	41	25
1	32	2,220	2,945	3,700	48	45	27
$1\frac{1}{4}$	39	2,410	3,300	4,260	49	45	31
$1\frac{1}{2}$	47	3,010	3,750	4,525	53	48	32
$1\frac{3}{4}$	53	3,375	4,030	4,800	54	48	36
2	60	3,925	4,825	5,400	54	51	38
$2\frac{1}{4}$	67	4,100	5,350	6,250	56	53	39
$2\frac{1}{2}$	74	4,310	5,675	6,540	61	53	40
$2\frac{3}{4}$	82	4,950	6,225	7,390	63	55	41
3	90	5,560	6,660	7,920	65	55	43

90 lb per cu ft. The use of a 2-cu-yd medium-duty bucket will be considered. If the dragline is to be operated with an 80-ft boom at a 40° angle, the maximum safe load will be 8,600 lb. The approximate weight of the bucket and its load will be

Bucket, from Table 8-11	= 4,825 lb
Earth, 60 cu ft @ 90 lb per cu ft	= 5,400 lb
Combined weight	= 10,225 lb
Maximum safe load	= 8,600 lb

As this weight will exceed the safe load on the dragline, it will be necessary to use a smaller bucket. Try a $1\frac{1}{2}$-cu-yd bucket, whose combined weight will be

Bucket	= 3,750 lb
Earth, 47 cu ft @ 90 lb per cu ft	= 4,230 lb
Combined weight	= 7,980 lb

If a $1\frac{1}{2}$-cu-yd bucket is used, it may be filled to heaping capacity, without exceeding the safe load of the dragline.

If a 70-ft boom, whose maximum safe load is 11,000 lb, will provide sufficient working range for excavating and disposing of the earth, a 2-cu-yd bucket may be used and filled to heaping capacity. The reduced cycle time, in using the 70-ft boom, will probably offset the increased time required to fill the 2-cu-yd bucket. The ratio of the output resulting from the use of a 70-ft boom and a 2-cu-yd bucket, compared with a $1\frac{1}{2}$-cu-yd bucket, should be approximately as follows:

$$\text{Output ratio, } \frac{60 \text{ cu ft}}{47 \text{ cu ft}} \times 100 = 127\%$$

$$\text{Increase in output} = 27\%$$

The previous example illustrates the importance of analyzing a job prior to selecting the size excavator to be used. The haphazard selection of equipment can result in a substantial increase in the cost of handling of earth.

THE EFFECT OF THE CLASS OF MATERIAL ON THE COST OF EXCAVATING EARTH

Figure 8-12 illustrates the effect which the class of material has on the cost per cubic yard bank measure in excavating with draglines. The hourly cost of a machine includes fixed-machine, variable-machine, and labor costs. Each machine is assumed to operate 2,000 hr per year at 75 percent efficiency. Thus, the probable hourly output of any given size machine is obtained by multiplying the ideal output, as given in Table 8-9, by 75 percent. For example, the cost of excavating good common earth using a 1-cu-yd machine is determined as follows:

Operating cost per hr = \$27.60
Ideal output per hr = 135 cu yd
Probable output, 0.75×135 = 101 cu yd
Cost per cu yd, $\$27.60 \div 101$ = \$0.272

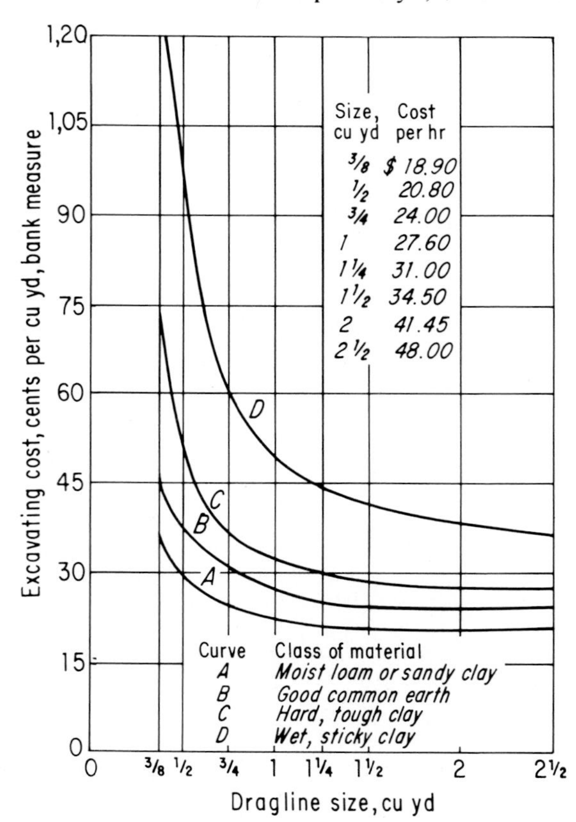

Figure 8-12 The effect of the class of material and the size of the bucket on the cost of excavating earth with a dragline.

Example 8-7 This example illustrates a method of analyzing a project to determine the size dragline required. Select a crawler-mounted dragline to excavate 234,000 cu yd bank measure of common earth in digging a canal. The dimensions of the canal will be:

Bottom width, 20 ft
Top width, 44 ft
Depth, 12 ft
Side slopes, 1:1

The excavated earth will be cast into a levee along one side of the canal, with a berm of at least 20 ft between the toe of the levee and the nearest edge of the canal. The cross-sectional area of the canal will be $(20 + 44)/2 \times 12 = 384$ sq ft. If the earth swells 25 percent when it is loosened, the cross-section area of the levee will be $384 \times 1.25 = 480$ sq ft. The dimensions will be

Height, 12 ft
Base width, 64 ft
Crest width, 16 ft
Side slope, 2:1

The total width from the outside of the levee to the outside of the canal will be:

Width of levee	=	64 ft
Width of berm	=	20 ft
Width of canal	=	44 ft
Total	=	128 ft

It will require a dragline with a boom length of 70 ft to furnish the necessary digging and dumping reaches, which will permit adequate dumping height and digging depth, with a boom angle of 30°.

The project must be completed in 1 year. Assume that weather conditions, holidays, and other major losses in time will reduce the operating time to 44 weeks of 40 hr each, or a total of 1,760 working hours. The required output per working hour will be 133 cu yd. It should be possible to operate with a 150° maximum angle of swing. The management factor should be approximately 0.80.

The required output divided by job-management factor is $133 \div 0.80 = 167$ cu yd per hr. Assume a depth swing factor of 0.81. The required ideal output is $167 \div 0.81 = 206$ cu yd per hr.

Reference to Table 8-9 indicates a $1\frac{3}{4}$-cu-yd medium-duty bucket. The combined weight of the bucket and load will be:

Weight of load, 53 cu ft @ 80 lb per cu ft	=	4,240 lb
Weight of bucket	=	4,030 lb
Total weight	=	8,270 lb
Maximum safe load, from Table 8-5	=	9,200 lb

The equipment selected should be checked to verify whether it will produce the required output.

Ideal output, 210 cu yd per hr

Percent of optimum depth, $\frac{12.0}{9.5} \times 100 = 126$

Depth-swing factor, 0.82

Job-management factor, 0.80

Probable output, $210 \times 0.82 \times 0.08 = 138$ cu yd per hr

Thus the equipment should produce the required output of 133 cu yd, with a slight surplus capacity.

CRANES

SAFE LIFTING CAPACITIES OF CRANES

Because cranes are used to hoist and move loads from one location to another, it is necessary to know the lifting capacity and working range of a crane selected to perform a given service. Manufacturers and suppliers furnish this information in literature describing their products.

When a crane lifts a load attached to the hoist line that passes over a sheave located at the boom point of the machine, there is a tendency to tip the machine over. This introduces what is defined as the tipping condition. A machine is considered to be at the point of tipping when a balance is reached between the overturning moment of the load and the stabilizing moment of the machine when the crane is on a firm, level supporting surface [2].

During a test to determine the tipping load for a crane, the outriggers, if used, should be lowered to relieve the wheels or crawler tracks of all weight on the supporting surface or ground.

The radius of the load is the horizontal distance from the axis of rotation of the crane to the center of the vertical hoist line or tackle with the load applied.

The tipping load is the load that produces a tipping condition at a specified radius. The load includes the weight of the item being lifted plus the weights of the hooks, hook blocks, slings, and any other items used in hoisting the load but excludes the weight of the hoist rope.

RATED LOADS

The lifting crane rated loads should not exceed the following percentages of tipping loads at specified radii [1i].

1. Crawler-mounted machines, 75 percent
2. Rubber-tire-mounted machines, 85 percent
3. Machines on outriggers, 85 percent

The rated loads should be based on the direction of minimum stability from the mounting, unless otherwise specified. No load should be lifted over the front area of the machine except as approved by the crane manufacturer.

CLASSIFICATIONS OF CRANES

Cranes used for lifting shall be classified by a symbol consisting of two numbers based on the rated loads of the crane in the direction of least stability, with the outriggers set if the crane is so equipped.

1. The first number of the group shall be the crane rating radius, expressed in feet, for the maximum rated load, when the crane is equipped with the base boom length.
2. The second number of the group shall be the rated load (expressed in pounds divided by 100 and rounded off to the nearest whole number) when the load is applied at a 40-ft radius, with the use of a 50-ft boom length.

Example 8-8 To illustrate this method of classifying a crane, assume that a truck-mounted crane is rated at 40 tons at a 12-ft radius when equipped with its base boom length and is rated at 19,600 lb at a 40-ft radius when equipped with a 50-ft-long boom. The classification of this crane would be 40-ton truck-mounted crane, class 12-196.

The number 12 represents the radius in feet for the 40-ton rated load, and the number 196 represents the rated load in pounds divided by 100 at a 40-ft radius.

The literature issued by the manufacturers of cranes, specifying the class of a given unit will generally describe it as a 30-ton truck-mounted crane PCSA Class 12-105, for example. The meanings of the two numerals 12 and 105 are as stated in the foregoing paragraph. The initials represent the Power Crane and Shovel Association.

Manufacturers may furnish literature describing a unit as a 15-ton crawler-mounted crane PCSA Class 10-56. This crane has a rated load of 15 tons at a radius of 10 ft instead of at 12 ft, and a rated load of 5,600 lb at a radius of 40 ft.

SPECIFICATIONS FOR CRANES

Table 8-12 illustrates the kind of information appearing in the specifications issued by the manufacturers of cranes. The crane for which the information appearing in Table 8-12 applies is described as a 30-ton crawler-mounted cable-controlled crane PCSA Class 12-105.

WORKING RANGES OF CRANES

Figure 8-13 shows graphically the height of the boom point above the surface supporting the crane and the distance from the center of rotation for the crane for various boom angles for the crane whose lifting capacities are given in Table 8-12.

Table 8-12 Lifting capacities in pounds for a 30-ton crane PCSA class 12-105

Specified crane capacities based on 75 percent of tipping load

Length of boom, ft*	Radius, ft	Rated lifting capacity, lb		
		Crane	Clamshell	Dragline
40	12	60,000		
	15	40,000		
	20	27,000		
	25	19,900	10,300	
	30	15,600	10,300	8,800
	35	12,800	10,300	8,800
	40	10,700	9,630	8,800
50	15	40,610		
	20	26,810		
	25	19,710		
	30	15,410	10,300	
	35	12,610	10,300	
	40	10,510	9,460	8,800
	45	8,930	8,035	8,800
	50	7,730	6,955	7,730
60	15	40,420		
	20	26,620		
	25	19,520		
	30	15,220		
	35	12,420	10,300	
	40	10,320	9,290	8,800
60	45	8,740	7,865	8,740
	50	7,540	6,785	7,540
	55	6,600	5,940	5,800
	60	5,800	5,220	5,800

* When used in crane service, this machine can handle a boom whose maximum length is 100 ft.

The maximum boom length for this crane for service with a clamshell, magnet, or dragline is 60 ft. However, if the machine is used for crane service only, the maximum length of the boom may be increased to 100 ft, as illustrated in Fig. 8-13. The length of the boom is increased by adding sections at or near the midlength of the boom, usually in 5-ft or 10-ft increments.

Example 8-9 As an example in using the information in Fig. 8-13, determine the minimum boom length that will permit the crane to lift a load 4 ft high to a position 36 ft above the surface on which the crane is operating. The length of the block, hook, and slings that are required to attach the hoist rope to the load is 6 ft. The location of the project will require the

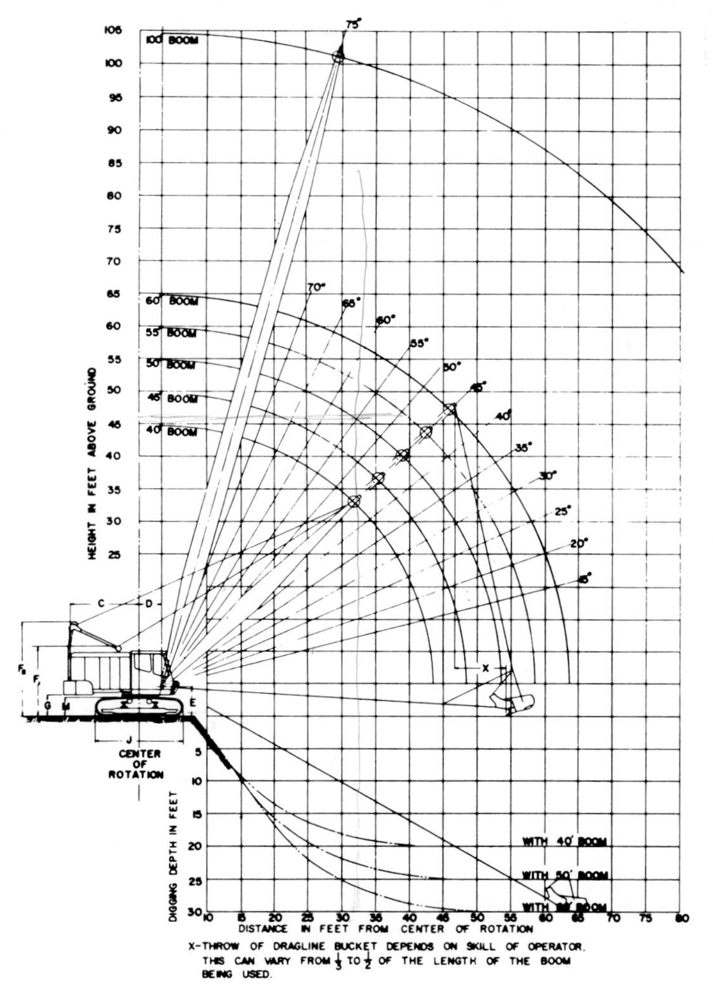

Figure 8–13 Working ranges for a 30-ton crane, PCSA Class 12–105, 1- to $1\frac{1}{2}$-cu-yd excavator, nominal rating.

crane to pick up the load from a truck at a distance of 25 ft from the center of rotation of the crane. Thus the operating radius will be 25 ft.

In order to lift the load to the specified location, the minimum height of the boom point of the crane must be at least $36 + 4 + 6 = 46$ ft above the ground supporting the crane. An examination of the diagram in Fig. 8-13 reveals that for a radius of 25 ft the height of the boom point for a boom 50 ft long is high enough.

If the block, hook, and slings weigh 1,000 lb, determine the maximum net weight of the load that can be hoisted. It will be necessary to interpolate between the values appearing in Table 8-12 for the answer. For a boom length of 50 ft and a radius of 25 ft the maximum total load is 19,710 lb. For a boom length of 60 ft and a radius of 25 ft the maximum total load is 19,520 lb. By interpolation the load should be (19,710 + 19,520)/2 = 19,635 lb. If the weight of the block, hook, and slings is deducted from the total load, the net weight of the lifted object will be 18,635 lb, which is the maximum safe weight of the object.

RATED LOADS FOR HYDRAULIC CRANES

The rated loads for hydraulic cranes are determined and indicated as for

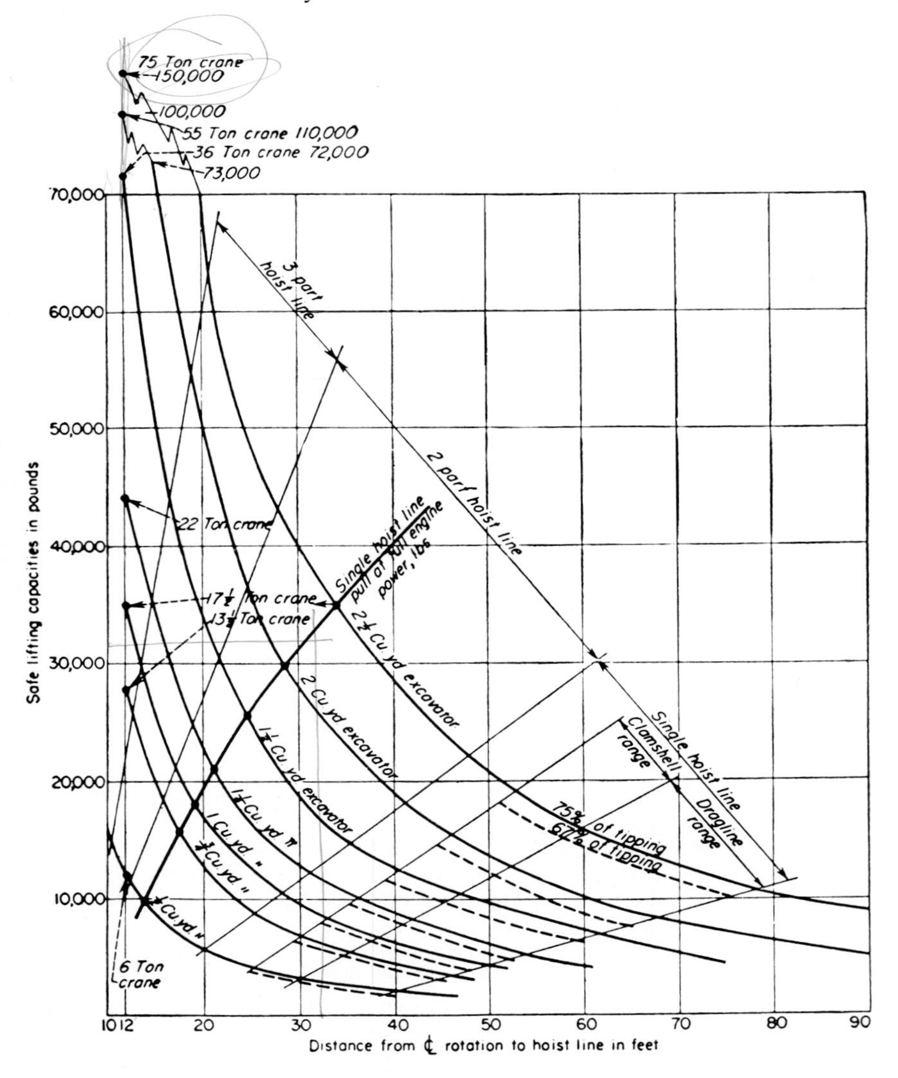

Figure 8-14 Safe lifting capacities of cranes.

Table 8-13 Lifting capacities in pounds for a 25-ton truck-mounted hydraulic crane PCSA class 12-88
Specified crane capacities based on 85 percent of tipping loads

Load radius, ft	Lifting capacity, lb* Boom length, ft						
	31.5	40	48	56	64	72	80
12	50,000	45,000	38,700				
15	41,500	39,000	34,400	30,000			
20	29,500	29,500	27,000	24,800	22,700	21,100	
25	19,600	19,900	20,100	20,100	19,100	17,700	17,100
30		14,500	14,700	14,700	14,800	14,800	14,200
35			11,200	11,300	11,400	11,400	11,400
40			8,800	8,900	9,000	9,000	9,000
45				7,200	7,300	7,300	7,300
50				5,800	5,900	6,000	6,000
55					4,800	4,900	4,900
60					4,000	4,000	4,000
65						3,100	3,300
70							2,700
75							2,200

* The loads appearing above the solid line are limited by the machine stability. The values appearing below the solid line are limited by factors other than machine stability.

cable-controlled cranes. Table 8-13 lists the rated loads, or maximum safe loads, for a 25-ton truck-mounted crane PCSA Class 12-88. The specified loads are limited to 85 percent of the tipping loads over the sides or rear, with the machine supported and leveled on fully extended outriggers, and standing on a firm, uniform supporting surface.

CLAMSHELLS

GENERAL INFORMATION

Clamshells are used primarily for handling loose materials such as sand, gravel, crushed stone, coal, etc., and for removing materials from cofferdams, pier foundations, sewer manholes, sheet-lined trenches, etc. They are especially suited to vertically lifting materials from one location to another, as in charging hoppers and overhead bins. The limits of vertical movement may be relatively large when they are used with long crane booms.

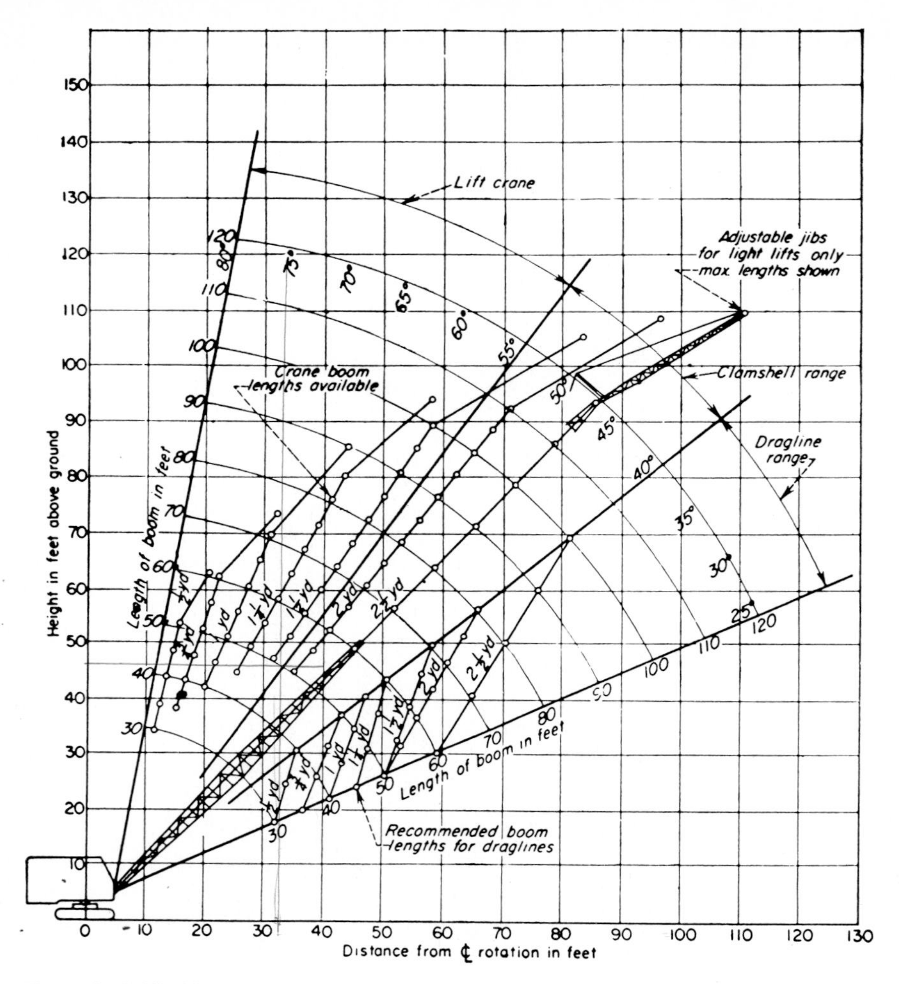

Figure 8-15 Working ranges of cranes.

CLAMSHELL BUCKETS

Clamshell buckets are available in various sizes, and in heavy-duty types for digging, medium-weight types for general-purpose uses, and lightweight types for rehandling light materials. Manufacturers supply buckets either with teeth that can be removed easily or without teeth. Teeth are used in digging the harder types of materials but are not required when a bucket is used for rehandling purposes. Figure 8-18 illustrates a rehandling and a heavy-duty digging bucket.

The capacity of a clamshell bucket is usually given in cubic yards. A more accurate capacity is given as water-level, plate-line, or heaped-measure,

Figure 8-16 Wheel-mounted hydraulic crane with telescoping boom. *(Bucyrus-Erie Company.)*

generally expressed in cubic feet. The water-level capacity is the capacity of the bucket if it were hung level and filled with water. The plate-line capacity indicates the capacity of the bucket following a line along the tops of the clams. The heaped capacity is the capacity of the bucket when it is filled to the maximum angle of repose for the given material. In specifying the heaped capacity the angle of repose usually is assumed to be 45°. The deck area indicates the number of square feet covered by the bucket when it is fully open. Table 8-14 gives representative specifications for medium-weight general-purpose-type buckets furnished by one manufacturer. Figure 8-19 illustrates the use of a clamshell to charge a batching plant.

Figure 8-17 Truck-mounted crane used to drive piles. *(Northwest Engineering Company.)*

PRODUCTION RATES FOR CLAMSHELLS

Because of the variable factors which affect the operations of a clamshell, it is difficult to give production rates that are dependable. These factors include the difficulty of loading the bucket, the size load obtainable, the height of lift, the angle of swing, the method of disposing of the load, and the experience of the operator. For example, if the material must be discharged into a hopper, the

Figure 8-18 (*a*) Wide rehandling clamshell bucket. (*b*) Heavy-duty clamshell bucket.

Figure 8-19 Feeding a batching plant with a clamshell.

Table 8-14 Representative specifications for medium-weight general-purpose-type clamshell buckets

	Size, cu yd								
	$\frac{3}{8}$	$\frac{1}{2}$	$\frac{3}{4}$	1	$1\frac{1}{4}$	$1\frac{1}{2}$	$1\frac{3}{4}$	2	$2\frac{1}{2}$
Capacity, cu ft:									
Water-level	8.0	11.8	15.6	23.2	27.6	33.0	38.0	47.0	52.0
Plate-line	11.0	15.6	21.9	32.2	37.6	43.7	51.5	60.0	75.4
Heaped	13.0	18.8	27.7	37.4	45.8	55.0	64.8	74.0	90.2
Weights, lb:									
Bucket only	1,662	2,120	2,920	3,870	4,400	5,310	5,440	6,000	7,775
Counterweights	230	300	400	400	400	500	500	600	600
Teeth	180	180	180	180	180	190	266	300	390
Complete	2,072	2,600	3,500	4,450	4,980	6,000	6,206	6,900	8,765
Dimensions:									
Deck area, sq ft	13.7	16.0	21.8	24.0	29.0	33.4	36.6	40.0	44.6
Width	2′6″	2′6″	3′0″	3′0″	3′5″	3′9″	4′0″	4′3″	4′6″
Length, open	5′5″	6′5″	7′3″	7′10″	8′5″	9′0″	9′2″	9′4″	9′11″
Length, closed	4′9″	5′7″	6′3″	6′9″	7′1″	7′6″	7′11″	8′0″	9′3″
Height, open	7′1″	7′10″	9′1″	9′9″	10′3″	10′9″	10′3″	11′6″	13′0″
Height, closed	5′9″	6′4″	7′4″	7′10″	8′3″	8′9″	8′9″	9′3″	10′4″

time required to spot the bucket over the hopper and discharge the load will be greater than when the material is discharged onto a large spoil bank. The following example will illustrate a method of estimating the probable output of a clamshell.

Example 8-10 A $1\frac{1}{2}$-cu-yd rehandling-type bucket, whose empty weight is 4,300 lb, will be used to transfer sand from a stock pile into a hopper, 25 ft above the ground. The angle of swing will average 90°. The average loose capacity of the bucket will be 48 cu ft.

The specifications for the crane unit give the following information:

Speed of hoist line, 153 fpm
Swing speed, 4 rpm
Time per cycle (approx.):

Loading bucket	= 6 sec
Lifting and swinging load, 25 ft @ 153 fpm	= 10 sec*
Dumping load	= 6 sec
Swinging back to stock pile	= 4 sec
Lost time, accelerating, etc.	= 4 sec
Total time	= 30 sec

$$\text{Maximum number of cycles per hr, } \frac{60 \times 60}{30} = 120$$

$$\text{Maximum volume per hr } \frac{120 \times 48}{27} = 213 \text{ cu yd}$$

* A skilled operator should lift and swing simultaneously. If this is not possible, additional time should be allowed for swinging the load.

If the unit operates 45 min per hr, the probable output will be (213 × 45)/60 = 159 cu yd per hr loose volume.

If the same equipment is used with a general-purpose bucket to dredge muck and sand from a sheet-piling cofferdam partly filled with water, requiring a total vertical lift of 40 ft, and to discharge it into a barge, the production rate previously determined will not apply. It will be necessary to lift the bucket above the top of the dam prior to starting the swing, which will increase the time cycle. Because of the nature of the material the load will probably be limited to the water-filled capacity of the bucket, which is 33 cu ft. The time per cycle should be about as follows:

Loading bucket	= 8 sec
Lifting load, 40 ft @ 153 fpm	= 16 sec
Swinging, 90° @ 4 rpm	= 4 sec
Dumping load	= 4 sec
Swinging back	= 4 sec
Lowering bucket, 40 ft @ 350 fpm	= 7 sec
Lost time, accelerating, etc	= 10 sec
Total time	= 53 sec

Maximum number of cycles per hr, $\frac{60 \times 60}{53} = 68$

Maximum volume per hr, $\frac{68 \times 33}{27} = 83$ cu yd

If the unit operates 45 min per hr, the probable output will be (83 × 45)/60 = 62 cu yd per hr loose volume.

HOES

GENERAL INFORMATION

The term hoe applies to an excavating machine of the power-shovel group. It is referred to by several names, such as hoe, backhoe, back shovel, and pull shovel. Figures 8-20 through 8-22 illustrate types of hoes available. As illustrated in Fig. 8-23, a power shovel is converted into a hoe by installing a dipper stick and a dipper at the end of the shovel boom. A hoe frequently is equipped with a gooseneck boom to increase the digging depth of the machine.

Hoes are used primarily to excavate below the natural surface of the ground on which the machine rests. They are adapted to excavating trenches, pits for basements, and general grading work, which requires precise control of depths. Because of their rigidity they are superior to draglines in operating on close-range work and dumping into trucks. Because of the direct pull on the dipper, hoes may exert greater tooth pressure than power shovels.

In some respects hoes are superior to wheel- or ladder-type trenching machines, especially in digging utility trenches whose banks are permitted to establish natural slopes and for which trench shoring will not be used. Hoes can remove the earth as it caves in to establish natural slopes, whereas trenching

Figure 8-20 Crawler-mounted hydraulically operated hoe. *(Drott Manufacturing Company.)*

Figure 8-21 Wheel-mounted hydraulically operated hoe. *(Drott Manufacturing Company.)*

Figure 8-22 Crawler-mounted hydraulically operated hoe loading a truck. *(Bucyrus-Erie Company.)*

machines cannot do this easily. The reduction in construction costs resulting from the elimination of shoring may be a significant item.

THE BASIC PARTS OF A CABLE-OPERATED HOE

The basic parts of a cable-operated hoe are illustrated in Fig. 8-23. The machine is placed in operation by setting the boom at the desired angle and pulling in on the hoist cable, while releasing the drag cable, to move the dipper out to the desired position. The free end of the boom is lowered by releasing the tension in the hoist cable until the dipper teeth engage the material to be dug. As the cable is pulled in, the dipper is filled. The dipper is lifted by raising the boom, and then swinging to the dumping position, which may be over a spoil bank or a truck.

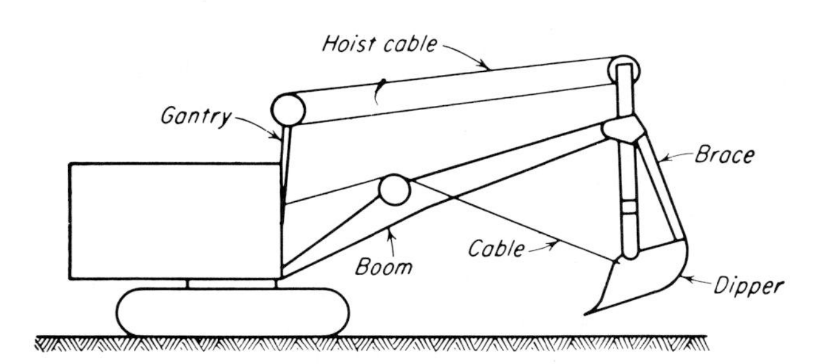

Figure 8-23 Basic parts of a cable-operated hoe.

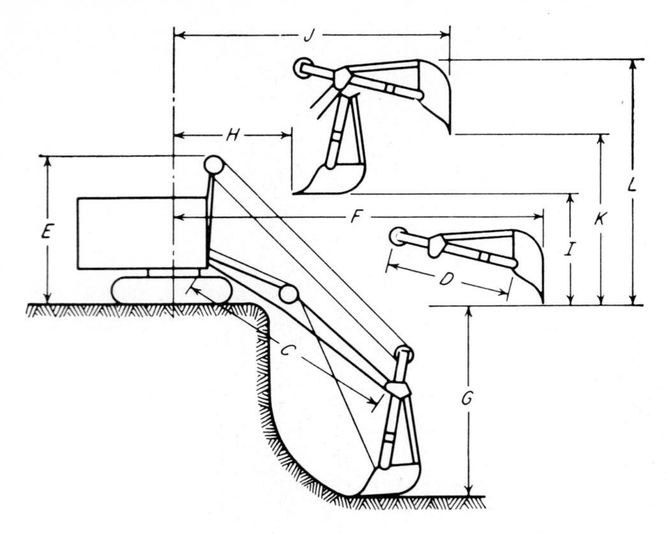

Figure 8-24 Clearance diagram for a cable-operated hoe.

WORKING RANGES OF HOES

Figure 8-24 illustrates the terms which are commonly used to identify the dimensions and working ranges of hoes. Table 8-15 gives representative dimensions and clearances for hoes. Dippers are available in various widths to suit the needs of the owner.

Figures 8-25 and 8-26 illustrate a machine which is used to shape ditches and slopes, and for other types of earthwork. Its dipper can rotate (that is its arm can twist) 90° or more, allowing the hoe to be effective in limited areas and where special shaping of the side slopes is required (note its use in Fig. 8-25).

Table 8-15 Representative dimensions and clearances for cable-operated hoes

Size dipper, cu yd	Length of boom, ft	Length of stick, ft	Max digging radius, ft	Max digging depth, ft	Radius, ft, at beginning of dump	Radius, ft, at end of dump	Clearance, ft, under dipper at beginning of dump	Clearance, ft, under dipper at end of dump
$\frac{3}{8}$	14–15	6–8	23–25	11–12	8–10	17–18	9–10	15–17
$\frac{1}{2}$	16–17	6–8	26–27	15–18	8–10	19–22	9–11	15–18
$\frac{3}{4}$	16–20	7–9	28–33	17–22	8–13	20–27	10–12	16–22
1	18–21	8–10	30–34	20–23	9–11	22–26	12–14	18–21
$1\frac{1}{2}$	22–26	9–11	36–42	25–28	13–15	28–32	14–16	27–30
$1\frac{3}{4}$	24–27	9–12	38–43	26–29	14–16	29–33	15–17	28–31

Figure 8-25 Gradall shaping the side slopes of a ditch.

Figure 8-26 Gradall clearing a drainage ditch.

Figure 8-27 Hoe used to dig a trench.

OUTPUT OF HOES

When a hoe is used to dig at moderate depths, the output may approach the output of a power shovel of comparable size digging in the same class of material. However, as the depth is increased, the output of a hoe will decrease considerably. The most effective digging action occurs when the dipper stick is at right angles to the boom. The greatest output will be obtained if digging is done near the machine, because of the reduced cycle time, and because the material rolls back into the dipper better when the dipper is pulled upward near the machine. Figure 8-27 illustrates the use of a hoe to excavate a trench.

TRENCHING MACHINES

GENERAL INFORMATION

The term trenching machine, as used in this book, applies to the wheel-and-ladder-type machines shown in Figs. 8-28 and 8-29, respectively. These

Figure 8-28 Wheel-type trenching machine.

Figure 8-29 A wheel-type trenching machine casting the excavated earth onto a spoil pile. *(Barber-Greene Company.)*

machines are satisfactory for digging utility trenches for water, gas, and oil pipelines, telephone cables, drainage ditches, and sewers where the job and soil conditions are such that they may be used. They provide relatively fast digging, with positive controls of depths and widths of trenches, which reduce expensive hand finishing to a minimum. They are capable of digging any type soil except rock. They are available in various sizes for digging trenches of varying depths and widths. They are usually crawler-mounted to increase their stability and to distribute the weight over a greater area.

WHEEL-TYPE TRENCHING MACHINES

Figure 8-28 illustrates a wheel-type trenching machine. These machines are available with maximum cutting depths exceeding 8 ft, with trench widths varying from 12 in. or less to approximately 60 in. Many of them are available with 25 or more digging speeds to permit the selection of the most suitable speed for any job condition.

The excavating part of the machine consists of a power-driven wheel, on which are mounted a number of removable buckets, equipped with cutter teeth. Buckets are available in varying widths, to which there may be attached

Table 8-16 Representative specifications for wheel-type trenching machines

Max trench depth, ft (m)	Trench width, in. (mm)	Engine power, hp (kW)	Wheel speed, fpm (m/sec)	Travel speed, mph (km/hr)	Digging speed, fpm (m/min)
5.5 (1.67)	15–18–21 (380–450–532) 20–23–26 (507–583–660)	55 (41)	36–266 (0.18–1.35)	0.5–2.7 (0.8–4.3)	0.2–10 (0.06–0.30)
6.0 (1.82)	16–18–20 (405–457–517) 20–22–24 (507–559–610) 24–26–28 (610–660–710) 28–30 (710–760)	67 (50)	153–410 (0.78–2.08)	0.16–4.6 (0.26–7.4)	2.8–57.5 (0.08–17.4)
8.5 (2.58)	38–40 (965–1,015) 40–51 (1,015–1,290)	110 (82)	243 (1.23)	1.9 (3.1)	1.3–35.0 (0.42–10.8)

Figure 8-30 A wheel-type excavator depositing the earth into a truck. *(Barber-Greene Company.)*

side cutters when it is necessary to increase the width of a trench. The machine is operated by lowering the rotating wheel to the desired depth, while the unit moves forward slowly. The earth is picked up by the buckets and deposited onto an endless belt conveyor, which can be adjusted to discharge the earth on either side of the trench.

Table 8-16 gives representative specifications for wheel-type trenching machines. As these specifications do not necessarily include all machines that are available, a prospective purchaser should consult the manufacturer's specifications for the particular machine under consideration. The various trench widths for a given machine are obtained by using different bucket widths and installing side cutters.

Wheel-type machines are especially suited to excavating trenches for water, gas, and oil pipelines, buried telephone cables, and pipe drains which are placed in relatively shallow trenches. They also may be used to excavate trenches for sewer pipes up to the maximum digging depths.

Figures 8-29 and 8-30 illustrate two wheel-type trenching machines in operation on construction projects. The machine illustrated in Fig. 8-29 casts the excavated earth onto a spoil pile along the trench, while the machine illustrated in Fig. 8-30 deposits the excavated earth onto a conveyor belt which deposits it into a tractor-pulled wagon.

LADDER-TYPE TRENCHING MACHINES

Figure 8-31 illustrates a ladder-type trenching machine. By installing extensions to the ladders or booms and by adding more buckets and chain links, it is possible to dig trenches in excess of 30 ft deep with the large machines. Trench widths in excess of 12 ft may be dug. Most of these machines have booms whose lengths may be varied, thereby permitting a single machine to be used on trenches varying considerably in depth. This eliminates the need of owning a different machine for each depth range. A machine may have 30 or more digging speeds to suit the needs of any given job.

The excavating part of the machine consists of two endless chains, which travel along the boom, to which there are attached cutter buckets equipped with teeth. In addition, shaft-mounted side cutters may be installed on each side of the boom to increase the width of a trench. As the buckets travel up the underside of the boom, they bring out earth and deposit it on a belt conveyor, which discharges it along either side of the trench. As a machine moves over uneven ground, it is possible to vary the depth of cut by adjusting the position, but not the length, of the boom.

Table 8-17 gives representative specifications for ladder-type trenching machines. The prospective purchaser of a machine should check the manufacturer's specifications for the particular machine under consideration. The various trench widths for a given machine are obtained by using different bucket widths and installing side cutters.

Figure 8-31 Ladder-type trenching machine.

Table 8-17 Representative specifications for ladder-type trenching machines

Max trench depth, ft (m)	Trench width, in. (mm)	Engine power, hp (kW)	Bucket speed, fpm (m/sec)	Travel speed, mph (km/hr)	Digging speed, fpm (m/min)
4.5 (1.37)	6–8 (152–203)	47 (35)	245–538 (1.24–2.72)	0.7–3.4 (1.1–5.5)	2.2–21.8 (0.67–6.6)
8.5 (2.58)	16–36 (407–920)	55 (41)	96–225 (0.48–1.14)	1.4–3.2 (2.2–5.1)	0.5–13.8 (0.15–4.2)
12.5 (3.81)	16–42 (407–1,070)	74 (55)	135–542 (0.68–2.74)	1.4–3.2 (2.2–5.1)	0.3–9.7 (0.09–2.95)
15.0 (4.57)	18–54 (457–1,370)	90 (67)	103–168 (0.52–0.85)	1.7 (2.7)	0.7–15.5 (0.21–4.75)

As can be seen from Table 8-17, ladder-type trenching machines have considerable flexibility with regard to trench depths and widths. However, the machines are not suitable for excavating trenches in rock or where large quantities of ground water, combined with unstable soil, prevent the walls of a trench from remaining in place. If the soil, such as loose sand or mud, tends to flow into the trench, it may be desirable to adopt some other method of excavating the trench. Usually, the trench is lined on both sides with sheet piling, lumber or steel, prior to excavating with a clamshell bucket.

SELECTING THE MOST SUITABLE EQUIPMENT FOR EXCAVATING TRENCHES

The choice of equipment to be used in excavating a trench will depend on the job conditions, the depth and width of the trench, the class of soil, the extent to which ground water is present, the width of the right of way for the disposal of excavated earth, and the type of equipment already owned by a contractor.

If a relatively shallow and narrow trench is to be excavated in firm soil, the wheel-type machine is probably the most suitable. However, if the soil is rock, which requires blasting, the most suitable excavator will be a hoe; a less desirable substitute could be a dragline. If the soil is an unstable, water-saturated material, it may be necessary to use a dragline, hoe, or clamshell and let the walls establish a stable slope. If it is necessary to install solid sheeting to hold the walls in place, neither a hoe nor a dragline will work satisfactorily. A clamshell, which can excavate between the trench braces that hold the sheeting in place, will probably be the best equipment for the job.

Consider the selection of a machine to excavate a trench 24 ft deep and 10 ft wide in soil which is sufficiently firm to require only shoring to hold the walls in place. A trench of this size can be excavated with a ladder-type machine, provided that the length and height of the conveyor belt are adequate to dispose of the earth along one side of the trench. The cross-sectional area of the trench will be 240 sq ft. If the loose earth has a 30 percent swell, the cross-section area of the spoil pile will be

$$240 \times 1.3 = 312 \text{ sq ft}$$

If the excavated earth will repose with 1:1 side slopes, the pile will have a height of 17.6 ft and a base width of 35.2 ft. If a minimum of 4 ft of clearance is required along the side of the trench, the end of the conveyor must have a height clearance of 17.6 ft and a length of approximately 27 ft, measured from the center of the trench. The casting effect on the earth, as it leaves the end of the conveyor belt, may permit the use of a shorter conveyor. Unless the machine under consideration satisfies these clearances, it is probable that difficulties will be experienced in disposing of the earth. A dragline would have no difficulty in disposing of the excavated earth. Also, a dragline can be used to backfill the trench if more suitable equipment is not available.

PRODUCTION RATES OF TRENCHING MACHINES

Many factors will influence the production rates of trenching machines. These include the class of soil, depth and width of the trench, extent of shoring required, topography, climatic conditions, extent of vegetation such as trees, stumps, and roots, physical obstructions such as buried pipes, sidewalks, paved streets, buildings, etc., and the speed with which the pipe can be placed in the trench. Any factors that may affect the progress on a project should be considered in estimating the probable digging speed of a trenching machine.

In laying oil and gas pipes through open, level country, with no physical obstructions to interfere with the progress, it is possible to install in excess of 6,000 ft of pipe in an 8-hr day. This is equivalent to approximately 800 ft per hr, which is not excessive for a wheel-type machine. However, if a trench must be excavated into rock over rough terrain covered with heavy timber, it may not be possible to excavate more than a few hundred feet per day.

If a trench is dug for the installation of sewer pipe, under favorable conditions, it is possible that the machine could dig 300 ft of trench per hour. However, an experienced pipe-laying crew may not be able to lay more than 25 joints of small-diameter pipe, 3 ft long, in an hour. Thus the speed of the machine will be limited to about 75 ft per hr regardless of its ability to dig more trench. In estimating the probable rate of digging a trench, an appropriate operating factor must be applied to the speed at which the machine could dig if there were no interruptions.

Example 8-11 Estimate the probable average production rate, in feet per hour, in excavating a trench 36 in. wide, with a maximum depth of 12 ft, in hard, tough clay. The trench will be dug for the installation of a 21-in.-diameter sewer pipe, which can be laid at a rate of approximately 30 ft per hr. An examination of the site along the trench reveals that there are obstructions which will reduce the digging speed to approximately 60 percent of the theoretically possible speed. This will require the application of an operating factor of 0.6 to the speed of the machine.

An examination of Table 8-17 indicates a ladder-type machine with a maximum digging depth of 12.5 ft. Considering the class of soil and the depth and width of the trench, the maximum possible digging speed should be about 1 fpm, or 60 ft per hr. The application of the operating factor will reduce the average speed to 36 ft per hr. However, since only 30 ft of pipe can be laid per hour, this will be the controlling speed.

The probable cost per linear foot of trench, for excavating only, should be as follows:

Trenching machine	= $29.60
Operator	= 14.20
Helpers, 3 men @ $8.00 per hr	= 24.00
Foreman, one-half time charged to excavating	= 9.00
Total cost	= $76.80
Cost per lin ft, $76.80 ÷ 30 ft per hr	= $ 2.56

EARTH-AND-ROCK SAWS

As illustrated in Fig. 8-32, this machine consists of a vertical wheel with a horizontal shaft which is supported on an adjustable boom mounted on the

Figure 8-32 Crawler-mounted cutting wheel. *(Vermeer Manufacturing Company.)*

rear of a track-type or wheel-type power unit such as a tractor. The machine is used to cut narrow trenches, up to about 6 in. in width and 30 in. or more in depth, in frozen earth, caliche, coral, other rocks, and concrete.

The sawing is performed by round carbide-tipped rotating cutters attached to the wheel, with the type selected depending on the properties of the material to be sawed. Their teeth, which are replaceable, rotate freely in their mounting pockets to maintain even tooth wear.

WHEEL EXCAVATORS

Figure 8-33 illustrates a wheel excavator that has been used to produce up to 1,750 cu yd bank measure per hour. The milling action of the wheel permits the excavator to cut almost any material, including weathered and broken rock. The depth of cut, up to 13 ft or more, assures a blending of the materials for the full depth excavated, and the pulverizing action facilitates the placement and compaction of the material on the fill.

As illustrated in the figure, the wheel deposits the excavated material onto a variable-speed conveyor belt, which discharges it into either of two hauling units.

PROBLEMS

8-1 Select the minimum size power shovel that will excavate 120,000 cu yd bank measure of good common earth in 130 working days of 8 hr each. The average depth of excavation will be 10 ft, and the average angle of swing will be 120°. The job and management factors will be good. The operating factor will be 0.83.

Figure 8-33 Wheel excavator loading bottom-dump wagons. *(Barber-Greene Company.)*

Figure 8-34 Bulldozers feeding a trap loader. *(Kolman Division, Athey Products Corp.)*

8-2 For each of the stated conditions determine the probable output of a $1\frac{1}{2}$-cu-yd power shovel expressed in cubic yards per hour bank measure. Use an operating factor of 0.83.

	Class of earth			
Condition	Moist loam	Moist loam	Common earth	Hard clay
Depth of dig, ft	12	8	12	10
Angle of swing, degrees	90	120	60	130
Job conditions	Fair	Good	Excellent	Poor
Management conditions	Fair	Good	Good	Poor

8-3 A 2-cu-yd power shovel whose cost per hour, including the wages to an operator and an oiler, is $96.00, is assumed to excavate good common earth under each of the stated conditions. Determine the cost per cubic yard for each condition. Use an operating factor of 0.83.

Condition	(1)	(2)	(3)	(4)
Depth of dig, ft	10	12	11	8
Angle of swing, degrees	60	90	120	150
Job conditions	Good	Fair	Excellent	Poor
Management conditions	Good	Fair	Excellent	Poor

8-4 Determine the probable production in cubic yard bank measure for a 2-cu-yd dragline when excavating good common earth under good job and management conditions. The average depth of dig will be 10 ft, and the average angle of swing will be 120°. The operating factor will be 0.75.

8-5 Determine the largest capacity heavy-duty dragline bucket that can be used with a dragline equipped with an 80-ft boom when the boom is operating at an angle of 45°. The earth will weigh 98 lb per cu ft loose measure.

8-6 Select the minimum size crane required to unload pipe weighing 24,000 lb per joint and lower it into a trench when the distance from the center line of the crane to the trench is 32 ft.

8-7 Select the minimum size crane and the minimum length boom required to hoist a load of 32,000 lb from a truck at ground level and place it on a platform 36 ft above the ground. The vertical distance from the bottom of the load to the boom point of the crane will be 10 ft. The maximum horizontal distance from the center of rotation of the crane to the hoist line of the crane when lifting the load will be 32 ft.

REFERENCES

1. Power Crane and Shovel Association, A Bureau of Construction Industry Manufactuers Association, 111 East Wisconsin Avenue, Milwaukee, WI 53202.
1a. "Man the Builder, The Functional Design and Job Application of Power Cranes and Excavators," *Technical Bulletin* 1, 1971.
1b. "Operating Cost Guide For Estimating Costs of Owning and Operating Power Cranes, Draglines, Clamshells, Backhoes and Shovels, 3/8 thru 5 cu yd—5 thru 125 tons," *Technical Bulletin* 2, 1965.
1c. "Proper Sizing of Excavators, Draglines, Clamshells, Backhoes, Shovels and Hauling Equipment," *Technical Bulletin* 3, 1966.
1d. "Cable-controlled Power Cranes, Draglines, Hoes, Shovels, Clamshells, Mountings, Attachments, Applications," *Technical Bulletin* 4, 1968.
1e. "Power Crane Applications In Industrial Plants," *Technical Bulletin* 5, 1954.
1f. "Hydraulic Excavators and Telescoping-boom Cranes," 1974.
1g. "Hydraulic Excavator User's Safety Manual," 1975.
1h. "Operating Cost Guide For Estimating Costs of Owning and Operating Cranes and Excavators," 1976.
1i. "Mobile Power Crane and Excavator Standards," *PCSA Standard* 1, 1968.
1j. "Mobile Hydraulic Crane Standards," *PCSA Standard* 2, 1968.
1k. "Mobile Hydraulic Excavator Standards," *PCSA Standard* 3, 1969.
2. "Crane Load Stability Test Code—SAE J765," SAE Recommended Practice Handbook, Society of Automotive Engineers, Inc., 1967.
3. Construction Industry Manufacturers Association, 111 East Wisconsin Avenue, Milwaukee, WI 53202.
4. Drott Manufacturing Division of J. I. Case, P.O. Box 1087, Wausau, WI 54401.
5. Northwest Engineering Company, 201 West Walnut Street, Green Bay, WI 54305.
6. Vermeer Manufacturing Company, P. O. Box 200, Pella, IA 50219.
7. Barber-Greene Company, 400 North Highland Avenue, Aurora, Il 60507.

CHAPTER

NINE

TRUCKS AND WAGONS

TRUCKS

In handling earth, aggregate, rock, ore, coal, and other materials, trucks serve one purpose. They are hauling units which, because of their high speeds when operating on suitable roads, have high capacities and provide relatively low hauling costs. They provide a high degree of flexibility, as the number in service may be increased or decreased easily to permit modifications in the total hauling capacity of a fleet. Most trucks may be operated over any haul road for which the surface is sufficiently firm and smooth and on which the grades are not excessively steep. Some units now in use are designated as off-highway trucks because their sizes and total loads are larger than are permitted on highways. These trucks are used for hauling materials on large projects, where the sizes and costs are justified.

Trucks may be classified according to a great many factors, including the following:

1. Size and type of engine—gasoline, diesel, butane, propane
2. Number of gears
3. Kind of drive—two-wheel, four-wheel, six-wheel, etc.
4. Number of wheels and axles and arrangement of driving wheels
5. Method of dumping the load—rear-dump, side-dump
6. Class of material hauled—earth, rock, coal, ore, etc.
7. Capacity, in tons or cubic yards
8. Method of dumping the load for rear dumps—hydraulic or cable

If trucks are to be purchased for general material hauling, the purchaser should select units that are adaptable to the purposes for which they will be used. However, if trucks are to be used on a given project for a given purpose, the purchaser should select trucks that most nearly fit the requirements of the project.

REAR-DUMP TRUCKS

Rear-dump trucks are suitable for use in hauling many types of materials. The shape of the body, such as the extent of sharp angles, corners, and the contour of the rear, through which the materials must flow during dumping, will affect the ease or difficulty of dumping. The bodies of trucks that will be used to haul wet clay and similar materials should be free of sharp angles and corners. Dry sand and gravel will flow easily from almost any shape of body. If quarry rock is to be hauled, bodies should be shallow with sloping sideboards. Figure 9-1 shows a power shovel loading a rear-dump truck.

Figure 9-2 shows a 22-ton single-axle dual-wheel rear-dump truck dumping its load. The body of this truck is approximately 15 ft 3 in. long, 8 ft 4 in. wide,

Figure 9-1 A 100-cu-yd struck capacity rear-dump truck.

Figure 9-2 Hydraulically operated rear-dump truck.

and 3 ft 6 in. deep, inside dimensions. The struck capacity is 14.8 cu yd. It is equipped with 14.00 by 24, 20-ply front tires and 18.00 by 24, 24-ply rear tires.

BOTTOM-DUMP WAGONS

If units are to be used to haul materials, such as sand, gravel, reasonably dry earth, coal, etc., which flow easily, the use of bottom-dump wagons will reduce the time required to unload the units (see Figs. 9-3, 9-4, and 9-5). Such units are particularly suitable for use where the materials are distributed in layers on a fill or are discharged through grizzlies into hoppers. When discharging the loads onto fills, the wagons can dump their loads while moving. When discharging through grizzlies, they will need to stop for only a few seconds. The rapid rate of discharging the load gives the wagons a time advantage over rear-dump trucks.

As the doors through which these units discharge their loads have limited openings, difficulties may be experienced in discharging such materials as wet, sticky clay, especially if they are in large lumps.

Figure 9-3 Bottom-dump wagon, 150-ton payload, 940-hp diesel engine.

These wagons are satisfactory hauling units on projects such as earthen dams, levees, highways, and airports, where large quantities of materials are to be transported and haul roads can be kept in reasonably good condition. They may be loaded by power shovels, draglines, or portable belt loaders (see Fig. 9-4).

CAPACITIES OF TRUCKS AND WAGONS

There are at least three methods of expressing the capacities of trucks and wagons: by the load which it will carry, expressed in tons; by its struck volume; and by its heaped volume, the latter two expressed in cubic yards.

Figure 9-4 Bottom-dump wagon being loaded by a dragline.

Figure 9-5 Bottom-dump wagon approaching the dump.

The struck capacity of a truck is the volume of material which it will haul when it is filled to the top of the sides, with no material above the sides. The heaped capacity is the volume of material which it will haul when the load is heaped above the sides. The capacity should be expressed in cubic yards. While the struck capacity remains fixed for any given unit, the heaped capacity will vary with the height to which the material may extend above the sides and with the length and width of the body. Wet earth or sandy clay may be hauled with a slope of 1:1, while dry sand or gravel may not permit a slope greater than 3:1. In order to determine the probable heaped capacity of a unit, it is necessary to know the struck capacity, the length and width of the body, and the slope at which the material will remain stable while the unit is moving. Smooth haul roads will permit a larger heaped capacity than rough haul roads. Because of variations in the heaping capacities of units it may be better to compare them on the basis of their struck capacities. In any event the capacities should be determined or compared in a realistic manner.

The weight capacity may limit the volume of the load when a unit is used to haul heavy material, such as iron ore. However, when the specific gravity of the material is such that the safe load is not exceeded, a unit may be filled to its heaped capacity.

In some instances it is possible to add sideboards to increase the depth of the body of a truck or wagon, thereby permitting it to haul a larger load. This practice probably will increase the hourly cost of operating a unit, because of higher fuel consumption, reduced tire life, more frequent failures of parts, such as axles, gears, brakes, and clutches, and higher maintenance costs. However, if the value of the extra material hauled is greater than the total increase in the cost of operating a vehicle, the overloading is justified. In considering hauling larger volumes of materials, the maximum safe loads on the tires should be checked to prevent excessive overloading, which might result in considerable lost time due to tire failures.

PERFORMANCE CAPABILITIES OF TRUCKS AND WAGONS

The productive capacity of a truck or wagon depends on the size of its load and the number of trips it can make in an hour. The size of the load can be determined from the specifications furnished by the manufacturer. The number of trips per hour will depend on the weight of the vehicle, the horsepower of the engine, the haul distance, and the condition of the haul road.

The productive capacity may be determined as illustrated in the example beginning on page 264 using the rimpulls of the vehicle, if this information is available, the weight of the vehicle, and the condition of the haul road.

Another method of determining the production is to use the performance chart furnished by most manufacturers for their vehicles. Such a chart is illustrated in Fig. 9-6, for a 22-ton rear-dump truck.

Example 9-1 The specifications for the truck are as follows:

Engine, 225 fwhp
Capacity
    Struck, 14.7 cu yd
    Heaped, 2:1, 18.3 cu yd

Net weight empty	= 36,860 lb
Payload	= 44,000 lb
Gross vehicle weight	= 80,860 lb

Determine the maximum speed for the truck when it is hauling a load of 22 tons up a 6 percent grade on a haul road having a rolling resistance of 60 lb per ton, equivalent to a 3 percent adverse grade. Because the chart is based on zero rolling resistance, it is necessary to combine the grade and rolling resistance, which gives an equivalent total resistance equal to $6 + 3 = 9$ percent of the vehicle weight.

The steps in using the chart are as follows:

1. Find the vehicle weight on the lower left horizontal scale.
2. Read up the weight line to the intersection with the slanted total resistance line.
3. From this intersection read horizontally to the right to the intersection with the performance curve.
4. From this intersection read down to find the vehicle speed.

If these four steps are followed, it will be determined that the truck will operate in the second speed range, and that its maximum speed will be 6.5 mph.

The chart should be used to determine the maximum speed for each section of a haul road having a significant difference in grade or rolling resistance.

While a performance chart indicates the maximum speed at which a vehicle

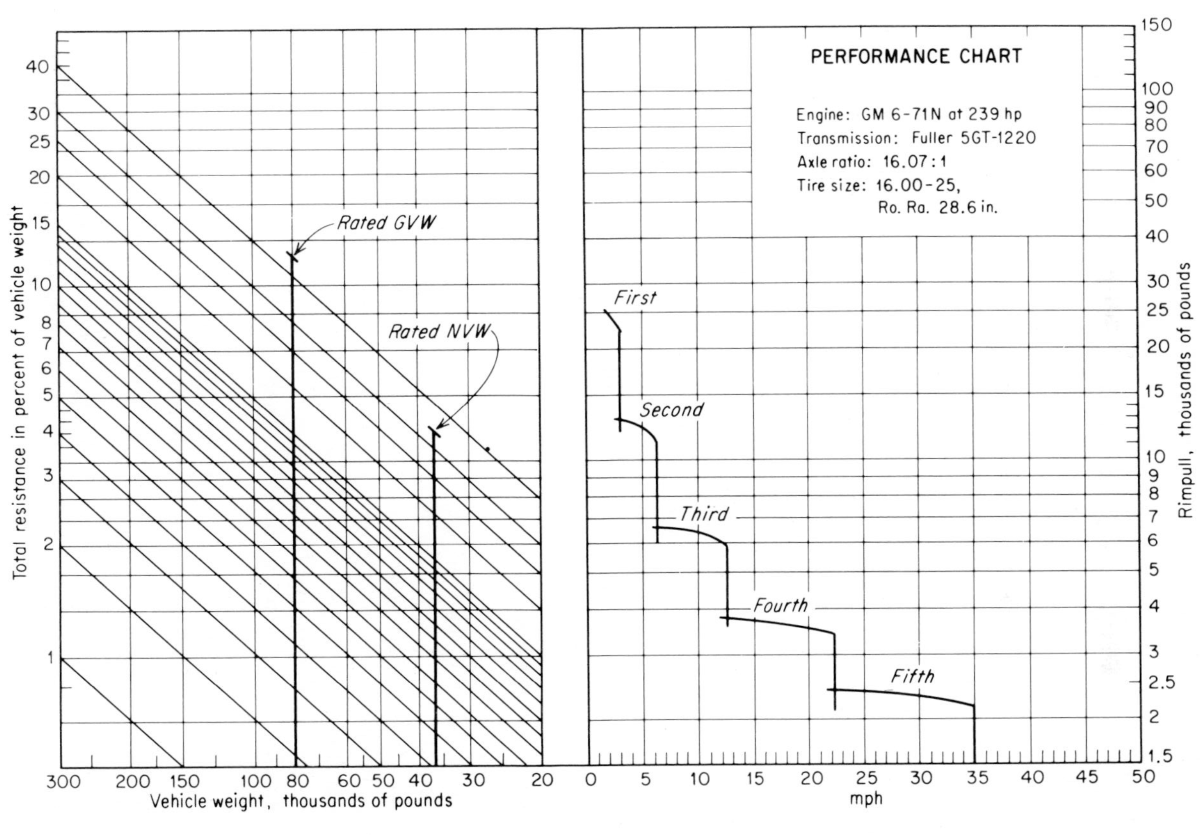

Figure 9-6 Performance chart for truck.

can travel, the vehicle will not necessarily travel at this speed. If conditions other than total resistance limit the speed to less than the value given in the chart, the anticipated effective speed should be used.

BALANCING THE CAPACITIES OF HAULING UNITS WITH THE SIZE OF EXCAVATOR

In loading with power shovels, draglines, or belt loaders, it is desirable to use units whose capacities balance the output of the excavator. If this is not done, operating difficulties will develop and the combined cost of excavating and hauling material may be higher than when balanced units are used. For example, when an excavator is used to load earth into trucks, the size of the trucks may introduce several factors which will affect the production rate and the cost of handling earth.

1. Advantages of small compared with large trucks:
   a. They are more flexible in maneuvering, which may be an advantage on short hauls.
   b. They may have higher speeds.
   c. There is less loss in production when one truck in a fleet breaks down.
   d. It is easier to balance the number of trucks with the output of the excavator, which will reduce the time lost by the trucks or the excavator.
2. Disadvantages of small compared with large trucks:
   a. It is more difficult for the excavator to load owing to small target for depositing earth.
   b. More total time is lost in spotting the trucks because of the larger number required.
   c. More drivers are required to haul a given output of material.
   d. The greater number of trucks required increases the danger of bunching up at the pit, along the haul road, or at the dump.
   e. The greater number of trucks required may increase the total investment in hauling equipment, with more expensive maintenance and repairs and more parts to stock.
3. Advantages of large compared with small trucks:
   a. Fewer trucks are required, which may reduce the total investment in hauling units and the cost of maintenance and repairs.
   b. Fewer drivers are required.
   c. The smaller number of trucks facilitates synchronizing the equipment and reduces the danger of bunching up by the trucks. This is especially true for long hauls.
   d. They give a larger target for the excavator during loading.
   e. They reduce the frequency of spotting trucks under the excavator.
   f. There are fewer trucks to maintain and repair and fewer parts to stock.
   g. The engines ordinarily use cheaper fuels.

4. Disadvantages of large compared with small trucks:
   a. The cost of truck time at loading is greater, especially with small excavators.
   b. The heavier loads may cause more damage to the haul roads, thus increasing the cost of maintaining haul roads.
   c. It is more difficult to balance the number of trucks with the output of the excavator.
   d. Repair parts may be more difficult to obtain.
   e. The largest sizes may not be permitted to haul on highways.

A rule-of-thumb practice which is frequently used in selecting the size of trucks is to use trucks with a minimum capacity of four to five times the capacity of the excavator bucket or dipper, when loading with a dragline or shovel. The dependability of this practice is discussed in the following analysis.

Example 9-2 Consider a $\frac{3}{4}$-cu-yd shovel excavating good common earth, with a 90° swing, with no delays waiting for hauling units, and with a 21-sec cycle time. If the dipper and the trucks are operated at their heaped capacities, the swelling effect of the earth should permit each to carry its rated or struck capacity, expressed in cubic yards bank measure. Assume that the number of dippers required to fill a truck will equal the capacity of the truck divided by the size of the dipper, both expressed in cubic yards. The sizes of the trucks considered are based on the struck capacities. Assume that the time for a travel cycle, excluding the time for loading, will be the same for the several sizes of trucks considered. If this is not true, an appropriate travel cycle should be determined for each truck. The time for a travel cycle, which includes traveling to the dump, dumping, and returning to the shovel, will be 6 min.

If 3-cu-yd trucks are used, it will require four dippers to fill a truck. With a shovel cycle of 21 sec it will be necessary to provide a new truck every 84 sec, or 1.4 min. The minimum round-trip cycle for a truck will be 7.4 min. The minimum number of trucks required to keep the shovel busy will be the round-trip time divided by the loading time $= 7.4 \div 1.4 = 5.3$. Thus it will be necessary to use six trucks to keep the shovel busy or else permit the shovel to idle between trucks. Since the time required to load six trucks will be $6 \times 1.4 = 8.4$ min, the lost time per truck cycle will be $8.4 - 7.4 = 1$ min per truck. This will produce an operating factor of

$$\frac{7.4}{8.4} \times 100 = 88 \text{ percent for the trucks}$$

If 6-cu-yd trucks are used, it will require eight dippers to fill a truck. The time required to load a truck will be 168 sec, or 2.8 min. The minimum round-trip cycle for a truck will be 8.8 min. The minimum number of trucks required to keep the shovel busy will be $8.8 \div 2.8 = 3.15$. For this condition it probably will be cheaper to provide three trucks and let the shovel idle a short time between trucks. The time required to load three trucks will be $3 \times 2.8 = 8.4$ min. Thus the shovel will lose $8.8 - 8.4 = 0.4$ min in loading three trucks. The time lost will be $0.4/8.8 \times 100 = 4.5$ percent, which is not serious. If four trucks are used, the time required to load them will be $4 \times 2.8 = 11.2$ min. As this will increase the round-trip cycle of each truck from 8.8 to 11.2 min, the lost time per truck cycle will be 2.4 min per truck. This will result in a loss of

$$\frac{2.4}{11.2} \times 100 = 21.4 \text{ percent of the truck time}$$

which is equivalent to an operating factor of 78.6 percent for the trucks.

If 15-cu-yd trucks are used, it will require 20 dippers to fill a truck. The time required to load a truck will be 420 sec, or 7 min. The minimum round-trip cycle for a truck will be

13 min. The minimum number of trucks required to keep the shovel busy will be $13 \div 7 = 1.85$. Use two trucks. Since the time required to load two trucks will be $2 \times 7 = 14$ min, the lost time per truck cycle will be $14 - 13 = 1$ min per truck. This will produce an operating factor of $\frac{13}{14} \times 100 = 93$ percent for the trucks.

In the previous example note that the production of the shovel is based on a 60-min hour. This policy should be followed when balancing a servicing unit with the units being served because at times both types of units will operate at maximum capacity if the number of units is properly balanced. However, the average production of a unit, shovel or truck, for a sustained period of time, should be based on applying an appropriate efficiency or operating factor to the maximum productive capacity.

THE EFFECT OF THE SIZE OF TRUCKS ON THE COST OF HAULING EARTH

A comparison of the cost of hauling earth with each of several sizes of trucks based on the previous analysis is illustrated in Table 9-1. The information appearing in the table is obtained as illustrated in the following example.

Example 9-3 Assume that the shovel operates at 80 percent efficiency while it is excavating, with no lost time waiting for trucks.

No. cycles per min, $60 \div 21 = 2.86$
No. cycles per hr, $60 \times 2.86 = 171.6$
Ideal output per hr, $171.6 \times \frac{3}{4} = 128$ cu yd
Output at 80 percent efficiency, $0.8 \times 128 = 102$ cu yd per hr
Travel cycle for each truck, 6 min

If 6-cu-yd trucks are used, the ideal number will be 3.15, as previously determined. If three trucks are used, the output will be $(3.0/3.15) \times 102 = 97$ cu yd per hr.

Cost per hr for a truck and driver $= \$22.00$
Total cost per hr for trucks, $3 \times \$22.00 = \$66.00$

$$\text{Truck cost while loading, } \frac{2.8 \times \$22.00}{60} = \$\ 1.05$$

$$\text{Truck cost per cu yd of earth loaded, } \frac{\$1.05}{6} = \$0.175$$

The hauling cost per cu yd equals the total truck cost per hour divided by the output per hour, $\$66.00 \div 97 = \$0.684$.

The information given in Table 9-1 indicates that, for the given power shovel and project, the lowest hauling cost will be obtained if three 6-cu-yd trucks are used. For other sizes of power shovels and truck travel cycles the comparative costs given in the table will not necessarily hold true. If the travel cycle for the larger trucks is greater than for the 3-cu-yd trucks, namely, 6 min,

Table 9-1 Comparison of the cost of hauling common earth with various sizes of trucks, using a 3/4-cu-yd power shovel for loading

				Truck cost				
			Load-ing	Per hr		At loading		Hauling cost
Size truck, cu yd	No. of trucks	Output, cu yd per hr	time min	Per truck	Total	Per Truck	Per cu yd	per cu yd
3	5	96	1.4	$16.75	$ 83.75	$ 0.39	$0.130	$0.875
3	6	102	1.4	16.75	100.50	0.39	0.130	0.987
6	3	97	2.8	22.00	66.00	0.99	0.165	0.684
6	4	102	2.8	22.00	88.00	0.99	0.165	0.865
10	2	89	4.6	31.80	63.60	2.43	0.243	0.714
10	3	102	4.6	31.80	95.40	2.43	0.243	0.936
15	2	102	7.0	48.50	97.00	5.67	0.378	0.954
20	2	102	9.3	68.30	136.60	10.60	0.530	1.344

the actual time should be used in preparing information similar to that given in the table.

If the size of the excavator is increased, the time lost by the larger trucks at loading will be reduced, which will reduce the hauling cost per cubic yard. One disadvantage in using large trucks, for which costs are paid by the hour, is that the cost of the trucks while they are being loaded will be higher than for smaller trucks. This results from two factors, the longer time required to load and the higher hourly cost of the larger trucks. Since it is desirable to have a truck under the excavator at all times the total hourly truck cost while loading 15-cu-yd trucks will be $48.50, compared with $22.00 for 6-cu-yd trucks, regardless of the size of the excavator. Unless this higher cost for the larger trucks can be recovered by more economical performance during the travel cycle, the use of the larger trucks will not be justified.

THE EFFECT OF THE SIZE OF THE EXCAVATOR ON THE COST OF EXCAVATING AND HAULING EARTH

If the size of the excavator is increased, while the size of trucks remains constant, the resulting increase in the output of the shovel will reduce the time required to load a truck. This will reduce the truck cost per cubic yard during loading. The effect which the size of a power shovel has on the truck cost at loading and the hauling cost is illustrated in Table 9-2. The material will be good common earth, the depth of cut will be optimum, and the angle of swing will be 90°. The operating factor for the shovel will be 80 percent, with no lost time waiting for trucks. Trucks with a heaped capacity of 15 cu yd bank measure will be used to haul the earth. The travel cycle for the trucks will be 8 min. The cost per hour for a truck and

Table 9-2 The effect of the size of the power shovel on the cost of hauling earth with 15-cu-yd trucks

		Truck time				Truck cost at loading		
Size shovel, cu yd	Output per hr, cu yd	Load-ing, min	Round trip, min	No.of trucks	Truck cost per hr	Per truck	per cu yd	cost per cu yd
$\frac{1}{2}$	76	11.8	19.8	2	\$ 97.20	\$9.57	\$0.636	\$1.281
$\frac{3}{4}$	108	8.3	16.3	2	97.20	6.72	0.450	0.903
1	125*	6.4	14.4	2	97.20	5.19	0.345	0.780
1	140	6.4	14.4	3	145.80	5.19	0.345	1.041
$1\frac{1}{2}$	191	4.7	12.7	3	145.80	3.81	0.255	0.765
2	231*	3.8	11.8	3	145.80	3.09	0.207	0.630
2	240	3.8	11.8	4	194.40	3.09	0.207	0.810
$2\frac{1}{2}$	280	3.2	11.2	4	194.40	2.58	0.171	0.693
3	312	2.9	10.9	4	194.40	2.37	0.159	0.624

* These values are reduced because the hauling capacities of the trucks limit the outputs.

driver will be \$48.60.

Sample calculations using a 1-cu-yd shovel are as follows:

Ideal output of the shovel, 175 cu yd per hr
Output at 80 percent efficiency, $0.80 \times 175 = 140$ cu yd per hr
Time required to load a truck, $\frac{15 \times 60}{140} = 6.4$ min
Round-trip time per truck, with no delays waiting for the shovel, $6.4 + 8.0 = 14.4$ min
No. trucks needed, $14.4 \div 6.4 = 2.25$
Output using 2 trucks, $\frac{2.0 \times 140}{2.25} = 125$ cu yd per hr
Output using 3 trucks, 140 cu yd per hr
Cost per hr for 2 trucks, $2 \times \$48.60$ = \$ 97.20
Cost per hr for 3 trucks, $3 \times \$48.60$ = \$145.80
Cost per truck during loading, $\frac{6.4}{60} \times \$48.60$ = \$ 5.19
Truck cost during loading per cu yd, \$5.19 ÷ 15 cu yd = \$ 0.345
Hauling cost per cu yd, using 2 trucks, \$97.20 ÷ 125 cu yd = \$ 0.777
Hauling cost per cu yd, using 3 trucks, \$145.80 ÷ 140 cu yd = \$ 1.041

While the information given in Table 9-2 indicates that the cost of hauling earth is reduced as the size of the shovel is increased, the job planner is concerned with the combined cost of excavating and hauling earth. This cost may be obtained by adding the cost of operating the shovel, including labor, to the cost of the

Table 9-3 The cost of loading and hauling earth, using various sizes of power shovels and 15-cu-yd trucks

Size shovel, cu yd	Output, cu yd per hr,	Shovel cost per hr	No. of trucks	Truck cost per hr	Excavating cost per cu yd	Hauling cost per cu yd	Total cost per cu yd
$\frac{1}{2}$	76	$ 36.90	2	$ 97.20	$0.486	$1.281	$1.767
$\frac{3}{4}$	108	41.70	2	97.20	0.387	0.903	1.290
1	125*	43.20	2	97.20	0.345	0.780	1.125
1	140	43.20	3	145.80	0.309	1.041	1.350
$1\frac{1}{2}$	191	64.20	3	145.80	0.336	0.765	1.101
2	231*	89.70	3	145.80	0.390	0.630	1.020
2	240	89.70	4	194.40	0.375	0.810	1.185
$2\frac{1}{2}$	280	101.10	4	194.40	0.363	0.693	1.056
3	312	121.50	4	194.40	0.390	0.624	1.164

* These values are reduced because the hauling capacities of the trucks limit the outputs.

trucks. Table 9-3 gives this information. The costs given in the table do not include the cost of moving the equipment to the project and setting it up. The cost of a shovel is based on the cost of owning and operating, with an allowance for the operator and an oiler.

THE EFFECT OF GRADE ON THE COST OF HAULING EARTH WITH TRUCKS

In constructing a fill it frequently is possible to obtain the earth from a borrow pit located either above or below the fill. If the borrow pit is above the fill, the effect of the favorable grade on the loaded truck is to reduce the required rimpull by 20 lb per gross ton for each 1 percent of grade. If the borrow pit is below the fill, the effect of the adverse grade on the loaded truck is to increase the required rimpull by 20 lb per gross ton for each 1 percent of grade. Obviously the grade of the haul road will affect the hauling capacity of a truck, its performance, and the cost of hauling earth. It may be more economical to obtain earth from a borrow pit above, instead of below, the fill, even though the haul distance from the higher pit is greater than from the lower pit. This is an item which should be given consideration in locating borrow pits.

If earth is hauled downhill, it may be possible to add sideboards to the vehicle to increase the hauling capacity, up to the maximum load which the tires can carry. In some instances it will be desirable to use larger tires to permit the trucks to haul greater loads. If the earth is hauled uphill, it may be necessary to reduce the size of the load or the travel speed of the truck, either of which will increase the cost of hauling earth.

Example 9-4 The following example will illustrate the effect of grade on the cost of hauling earth.

The project requires 1,000,000 cu yd of earth, bank measure.

The material will be good common earth, weighing 2,700 lb per cu yd bank measure, with a swell of 25 percent.

Borrow pit 1 will require an average haul of 0.66 mile up an average grade of 2.2 percent.

Borrow pit 2 will require an average haul of 0.78 mile down an average slope of 1.4 percent.

Both borrow pits are easily accessible to the trucks, which will permit spotting on either side of the shovel, whose angle of swing will not exceed 90°. Excavating can be done at optimum depth.

Job conditions will be excellent, and management conditions will be good. The job-management factor should be not less than 0.80.

The earth will be excavated with a 3-cu-yd power shovel, with a probable output of $0.80 \times 390 = 312$ cu yd per hr bank measure.

The average rolling resistance of the haul road is estimated to be 60 lb per ton.

The coefficient of traction between the truck tires and the haul road will average 0.60.

The earth will be hauled with bottom-dump wagons, whose estimated heaped capacity will be 15 cu yd bank measure.

The average elevation will be 600 ft above sea level.

The specifications for the trucks are as follows:

Pay-load capacity, 40,000 lb
Engine, diesel, 200 hp
Empty weight, 36,800 lb
Gross weight, loaded, 76,800 lb
Gross weight distribution
  Front axle, 12,000 lb
  Drive axle, 32,400 lb
  Trailer axle, 32,400 lb
Size tires on drive and trailer axles, 24.00 × 25

Gear	Speed, mph	Rimpull, lb
1st	3.2	19,900
2d	6.3	10,100
3d	11.9	5,350
4th	20.8	3,060
5th	32.7	1,945

The maximum usable rimpull of a loaded truck, as limited by the coefficient of traction, will be $32{,}400 \times 0.6 = 19{,}440$ lb. This is sufficiently high to eliminate the danger of tire slippage, except possibly in first gear.

The cost of hauling earth from borrow pit 1 is determined as follows:

The combined effect of rolling resistance and grade on a loaded truck will be

Rolling resistance	=	60 lb per ton
Grade resistance, 2.2 × 20	=	44 lb per ton
Total resistance	=	104 lb per ton

Gross weight of truck, 76,800 ÷ 2,000 = 38.4 tons
Required rimpull, 38.4 × 104 = 3,994 lb
Maximum speed of loaded truck, 11.9 mph

The combined effect of rolling resistance and grade on an empty truck will be

Rolling resistance = 60 lb per ton
Grade resistance, 2.2×20 = −44 lb per ton
Total resistance = 16 lb per ton
Weight of empty truck, $36{,}800 \div 2{,}000 = 18.4$ tons
Required rimpull, $18.4 \times 16 = 294$ lb
Maximum speed of an empty truck, 32.7 mph

The time required for each operation in a round-tip cycle should be about as follows:

Loading, 15 cu yd ÷ 312 cu yd per hr = 0.0482 hr
Lost time in pit and accelerating, 1.5 min = 0.0250 hr
Travel to the fill, 0.66 mile ÷ 11.9 mph = 0.0555 hr
Dumping, turning, and accelerating, 1 min = 0.0167 hr
Travel to pit, 0.66 mile ÷ 32.7 mph = 0.0202 hr
Round-trip time = 0.1656 hr

Assume that the trucks will operate an average of 50 min per hr.

No. trips per hr, $\frac{1}{0.1656} \times \frac{50}{60} = 5.02$
Volume of earth hauled per truck, $15 \times 5.02 = 75.3$ cu yd per hr
No. trucks required, $312 \div 75.3 = 4.15$

Use four trucks, which will reduce the output of the shovel slightly. If a truck and driver cost $32.40 per hour, the cost of hauling earth will be

$32.40 ÷ 75.3 = $0.429 per cu yd

The cost of hauling earth from borrow pit 2 is determined as follows:
The combined effect of rolling resistance and grade on a loaded truck will be

Rolling resistance = 60 lb per ton
Grade resistance, 1.4×20 = − 28 lb per ton
Total resistance = 32 lb per ton
Gross weight of truck, 38.4 tons
Required rimpull, $38.4 \times 32 = 1{,}229$ lb

The available rimpull in fifth gear is 1,945 lb, which is more than will be required by the truck. Sideboards can be installed to increase the hauling capacity of the truck. The gross load should be limited to a weight that can be pulled by not over 80 percent of the rimpull, with the remaining rimpull reserved to accelerate the truck and to be used on sections of the haul road having higher rolling resistance or less steep grades.

Net available rimpull, $0.8 \times 1{,}945 = 1{,}556$ lb
Required rimpull for 15 cu yd = 1,229 lb
Surplus rimpull = 327 lb
Possible additional load, $327 \div 32 = 10.2$ tons
Possible additional volume, $\frac{10.2 \times 2{,}000}{2{,}700} = 7.55$ cu yd

In order to compensate for the additional weight of the sideboards, the volume of the earth should be increased by not more than 7 cu yd. This will give a total volume of 22 cu yd per load.

The combined effect of rolling resistance and grade on the empty truck will be

Rolling resistance = 60 lb per ton
Grade resistance, 1.4 × 20 = 28 lb per ton
Total resistance = 88 lb per ton
Weight of empty truck, including sideboards, 19 tons
Required rimpull, 19 × 88 = 1,672 lb
Maximum speed of an empty truck, 32.7 mph

The time required for each operation in a round-trip cycle should be about as follows:

Loading, 22 cu yd ÷ 312 cu yd per hr = 0.0707 hr
Lost time in pit and accelerating, 2 min = 0.0333 hr
Travel to fill, 0.78 mile ÷ 32.7 mph = 0.0238 hr
Dumping, turning, and accelerating, 1.5 min = 0.0250 hr
Travel to pit, 0.78 mile ÷ 32.7 mph = 0.0238 hr
Round-trip time = 0.1766 hr

Assume that the trucks will operate an average of 50 min per hour.

No. trips per hr, $\frac{1}{0.1766} \times \frac{50}{60} = 4.72$
Volume of earth hauled per truck, 22 × 4.72 = 103.8 cu yd per hr
No. trucks required, 312 ÷ 103.8 = 3.01

Use three trucks.

If a truck and driver cost $32.40 per hr, the cost of hauling the earth will be $32.40 ÷ 103.8 = $0.312 per cu yd.

A comparison of the cost of hauling the earth from the two pits will reveal the extent of savings that may be effected by using pit 2.

Hauling cost from pit 1 = $0.429 per cu yd
Hauling cost from pit 2 = 0.311 per cu yd
Reduction in hauling cost = $0.118 per cu yd

Another item that is favorable to pit 2 is the reduction in the number of trucks from four to three units.

THE EFFECT OF ROLLING RESISTANCE ON THE COST OF HAULING EARTH

An important factor which affects the production capacity of a truck or a tractor-pulled wagon is the rolling resistance of the haul road. Rolling resistance is determined primarily by two factors, the physical condition of the road and the tires used on the hauling unit. A great deal can be done to reduce rolling resistance by properly maintaining the road and by selecting proper sizes of tires and then keeping them inflated to the correct pressure. Money spent for these purposes may return dividends, through reduced hauling costs, far in excess of the expenditures. This is one field where the application of engineering knowledge will yield excellent returns.

An earth haul road which is given little or no maintenance will soon become rough, loose, and soft and may develop a rolling resistance of 150 lb per ton or more, depending on the type of soil and weather conditions. If a road is properly maintained with a patrol grader, sprinkled with water, and compacted as required, it may be possible to reduce the rolling resistance to 50 lb per ton or less. Also, sprinkling the road will reduce the damage to hauling equipment by eliminating dust, will reduce the danger of vehicular collision by improving visibility, and will prolong the life of tires because of the cooling effect which the moisture has on the tires.

The selection of proper tire sizes and the practice of maintaining correct air pressure in the tires will reduce that portion of the rolling resistance that is due to tires. A tire supports its load by deforming where it contacts the road surface until the area in contact with the road will, considering the air pressure in the tires, produce a total force on the road equal to the load on the tire. If the load on a tire is 5,000 lb and the air pressure is 50 psi, the area of contact will be 100 sq in. This neglects any supporting resistance furnished by the side walls of the tire. If, for the same tire, the air pressure is permitted to drop to 40 psi, the area of contact will be increased to 125 sq in. The additional area of contact will be produced by additional deformation of the tire. This will increase the rolling resistance because the tire will be continually climbing a steeper grade as it rotates. The size tire selected and the inflated pressure should be based on the resistance which the surface of the road offers to penetration by the tire. For rigid road surfaces, such as concrete, small-diameter high-pressure tires will give lower rolling resistance, while, for soft road surfaces, large-diameter low-pressure tires will give lower rolling resistance because the larger areas of contact will reduce the depth of penetration by the tires.

Example 9-5 This example illustrates the effect which rolling resistance has on the cost of hauling earth.

A project requires a contractor to excavate and haul 1,900,000 cu yd of common earth. The contract must be completed within 1 year. By operating three shifts, with 7 hr actual working time per shift, 6 days per week, it is estimated that there will be 5,600 working hours, allowing for lost time due to bad weather. This will require an output of approximately 350 cu yd per hr bank measure, which should be obtained with a 4-cu-yd power shovel.

The job conditions are as follows:

Length of haul, 1 way, 3.5 miles
Slope of haul road, minus 0.5% from borrow pit to the fill
Weight of earth in place, 2,600 lb per cu yd
Swell, 30%
Weight of loose earth, 2,600 ÷ 1.3 = 2,000 lb per cu yd
Elevation, 800 ft above sea level

For hauling the earth the contractor considers using rubber-tire-equipped tractor-pulled bottom-dump wagons, which may be purchased with standard or optional gears. The optional gears will permit the unit to operate at a higher speed. Specifications and performance data are as follows:

	Standard tractor	Optional tractor
Tractor engine	150 bhp	150 bhp
Max speed	19.8 mph	27.4 mph
Mechanical efficiency	82%	82%
Rimpull at max speed	2,330 lb	1,685 lb

Heaped capacity of standard wagon, 32,000 lb or 16 cu yd loose measure, based on 3:1 slope
Inside length of wagon, 14 ft 2 in.
Average inside width of wagon, 7 ft 1 in.
Heaped capacity of wagon with sideboard extensions, 2 ft 0 in. high, 46,800 lb, or 23.4 cu yd loose measure, based on 3:1 slope

	Standard equipment	Optional equipment
Gross weight:		
Tractor and wagon	29,400 lb	29,400 lb
Sideboards	......	1,600 lb
Pay load	32,000 lb	46,800 lb
Total weight	61,400 lb	77,800 lb
Gross weight, tons	30.7	38.9
Delivered cost	$36,200	$36,900
Cost per hr, including driver	$27.40	$28.80*

*The higher cost per hour for the optional equipment is allowed because of the more severe conditions to which it will be subjected.

An analysis of the performance of the standard equipment, operating on a haul road with an estimated rolling resistance of 80 lb per ton, will give the probable hauling cost per cubic yard. This rolling resistance is representative of haul roads which are not carefully maintained.

The combined effect of rolling resistance and grade on a loaded unit will be

Rolling resistance = 80 lb per ton
Grade, 0.5×20 = −10 lb per ton
Total = 70 lb per ton
Gross weight of vehicle, 30.7 tons
Required rimpull, $30.7 \times 70 = 2{,}149$ lb
Available rimpull = 2,330 lb

The tractor can pull the loaded wagon, with a surplus rimpull for acceleration. The rimpull required for the return trip to the shovel will be

14.7 tons × 90 lb per ton = 1,323 lb

which will permit travel at maximum speed.

The time required for each operation in a round-trip cycle should be about as follows:

Volume of earth per load, $16 \div 1.30 = 12.3$ cu yd bm
Loading, 12.3 cu yd ÷ 350 cu yd per hr = 0.0351 hr
Lost time in pit and accelerating, 1.5 min = 0.0250 hr
Travel to the fill, 3.5 miles ÷ 19.8 mph = 0.1770 hr
Dumping, turning, and accelerating, 1.0 min = 0.0167 hr
Travel to pit, 3.5 miles ÷ 19.8 mph = 0.1770 hr
Round-trip time = 0.4308 hr

Assume that the wagons will operate an average of 45 min per hr.

No. trips per hr, $\frac{1}{0.4308} \times \frac{45}{60} = 1.74$
Volume of earth hauled per wagon, $12.3 \times 1.74 = 21.4$ cu yd per hr
No. wagons required, $350 \div 21.4 = 16.4$

It will be necessary to provide 17 wagons if the specified output is to be maintained. The actual volume of earth hauled per wagon will be $350 \div 17 = 20.6$ cu yd per hr.

Hauling cost per cu yd, \$27.40 ÷ 20.6 = \$1.330

Let us analyze the performance of the optional equipment to determine whether it will operate at the maximum possible speed while hauling 23.4 cu yd loose measure. It will be necessary to reduce the rolling resistance of the haul road by providing continuous maintenance. While it is possible to reduce the rolling resistance to 40 lb per ton during most of the time the project is in operation, a value of 50 lb per ton will be used in order to provide a margin of safety.

The combined effect of rolling resistance and grade on a loaded unit will be

Rolling resistance = 50 lb per ton
Grade, 0.5×20 = −10 lb per ton
Total = 40 lb per ton
Gross weight of vehicle, 38.9 tons
Required rimpull, $38.9 \times 40 = 1{,}556$ lb
Available rimpull at 27.4 mph = 1,685 lb

The tractor can pull the load at the maximum speed, with a surplus for acceleration. The rimpull required for the return trip to the shovel will be

15.5 tons × 60 lb per ton = 930 lb

which will permit travel at maximum speed.

The time required for each operation in a round-trip cycle should be about as follows:

Volume of earth per load, $23.4 \div 1.30 = 18.0$ cu yd bm
Loading, 18 cu yd ÷ 350 cu yd per hr = 0.0515 hr
Lost time in pit and accelerating, 2 min = 0.0333 hr
Travel to the fill, 3.5 miles ÷ 27.4 mph = 0.1277 hr
Dumping, turning, and accelerating, 1.5 min = 0.0250 hr
Travel to pit, 3.5 miles ÷ 27.4 mph = 0.1277 hr
Round-trip time = 0.3652 hr

Assume that the wagons will operate an average of 45 min per hr.

No. trips per hr, $\frac{1}{0.3652} \times \frac{45}{60} = 2.05$

Volume of earth hauled per wagon, $18 \times 2.05 = 36.9$ cu yd per hr

No. wagons required, $350 \div 36.9 = 9.5$

It will be necessary to provide 10 wagons if the specified output is to be maintained. The actual volume of earth hauled per hour per wagon will be $350 \div 10 = 35$ cu yd.

Hauling cost per cu yd, $\$28.80 \div 35 = \$0.825$

The reduction in the cost of hauling the earth with the optional equipment will be

Cost using standard equipment	= \$1.330 per cu yd
Cost using optional equipment	= 0.825 per cu yd
Reduction in cost	= \$0.505 per cu yd

Total reduction for project, $1{,}900{,}000 \times \$0.505 = \$959{,}500$

The reduction in the cost of hauling earth and in the amount of money invested in hauling equipment resulting from the improvement in the rolling resistance of the haul road illustrates the value of analyzing a project. Although the reduction may appear to be unreasonably large, it is possible to produce similar results for many projects involving the hauling of earth. Even the cost of paving the haul road would be justified if this were the only method of reducing the rolling resistance.

Most manufacturers of trucks and tractor-pulled wagons can furnish units with standard or optional gears. For equipment already in service the standard gears may be replaced with optional gears at reasonable costs. Sideboards may be purchased from the equipment manufacturer, or they may be made locally in a machine shop.

Example 9-6 The effect of rolling resistance on the performance of equipment and the cost of hauling earth is further illustrated in Table 9-4. The information given in the table is based on using the optional tractor-pulled wagons of the previous analysis, an output of 350 cu yd of earth per hour bank measure, a one-way haul distance of 3.5 miles, and a level haul road. If the haul road is not level, similar information may be obtained by combining the effect of rolling resistance and grade.

The speeds and rimpulls of the hauling units are as follows:

Gear	Speed, mph	Rimpull, lb
1st	4.1	11,250
2d	6.5	7,120
3d	10.6	4,360
4th	17.0	2,720
5th	27.4	1,685

The following sample calculations will show how the information given in the table is obtained. Consider a haul road with a rolling resistance of 100 lb per ton.

Table 9-4 The effect of rolling resistance on the cost of hauling earth

	Rolling resistance, lb per ton			
Item	40	60	100	150
Maximum speed loaded, mph	27.4	17.0	10.6	6.5
Maximum speed empty, mph	27.4	27.4	27.4	17.0
Number of trucks required	10	12	15	22
Cost of trucks per hr	$288.00	$345.60	$432.00	$633.60
Volume of earth hauled per hr, cu yd	350	350	350	350
Hauling cost per cu yd	$0.822	$0.985	$1.234	$1.808

Gross weight of loaded unit, 38.9 tons
Weight of empty unit, 15.5 tons
Required rimpull for loaded unit, $38.9 \times 100 = 3{,}890$ lb
Maximum speed, 10.6 mph
Required rimpull for empty unit, $15.5 \times 100 = 1{,}550$ lb
Maximum speed, 27.4 mph

The round-trip time will include fixed time, which should be reasonably constant regardless of the condition of the haul road, plus the travel time to and from the fill.

The fixed time will be

Loading, 18 cu yd ÷ 350 cu yd per hr	= 0.0515 hr
Lost time in pit and accelerating, 2 min	= 0.0333 hr
Dumping, turning, and accelerating, 1.5 min	= 0.0250 hr
Total fixed time	= 0.1098 hr
Travel to the fill, 3.5 miles ÷ 10.6 mph	= 0.3310 hr
Travel to shovel, 3.5 miles ÷ 27.4	= 0.1277 hr
Round-trip time	= 0.5685 hr

Trips per 45-min hour, $\frac{1}{0.5685} \times \frac{45}{60} = 1.32$
Volume per wagon, $18 \times 1.32 = 23.75$ cu yd per hr
No. wagons required, $350 \div 23.75 = 14.7$ (Use 15)
Actual volume per wagon, $350 \div 15 = 23.3$ cu yd per hr
Hauling cost per cu yd $28.80 ÷ 23.3 = $1.234

THE EFFECT OF ALTITUDE ON THE PERFORMANCE OF HAULING EQUIPMENT

Contractors who have established satisfactory production rates for earth-hauling equipment at one altitude frequently find it desirable to bid on a project located at a different altitude. Unless an adjustment is made for the performance of the equipment at the higher altitude, it is possible that a substantial error may be made in estimating the cost of hauling the earth. As previously discussed, the effect of altitude is to reduce the sea-level power of a four-cycle internal-combustion engine by approximately 3 percent for each additional 1,000 ft of altitude above 1,000 ft unless a supercharger is installed on the engine. Power losses of this magnitude are too large to ignore in analyzing a project for bid purposes.

Example 9-7 This example will illustrate the effect of altitude on the performance of hauling equipment and the cost of hauling earth. The hauling units are commonly used in the construction industry.

The job conditions are as follows:

Weight of earth, 2,700 lb per cu yd bm
Swell, 25%
Weight of loose earth, 2,700 ÷ 1.25 = 2,160 lb per cu yd
Haul distance, 1.5 miles, over level road
Rolling resistance, 50 lb per ton

The earth will be excavated with a power shovel, whose output will be 280 cu yd per hr.

The specifications for the hauling units are as follows:

Type, tractor-pulled bottom-dump wagons
Tractor engine, 200 bhp
Wagon capacity, 16 cu yd heaped volume
Wagon capacity, 16 ÷ 1.25 = 12.8 cu yd bm
Weight of tractor and wagon = 36,800 lb
Weight of load, 16 cu yd @ 2,160 lb = 34,560 lb
Gross loaded weight = 71,360 lb, or 35.68 tons
Cost per hr, including operator, $31.60

Tractor-performance data at sea level

Gear	Speed, mph	Rimpull, lb
1st	3.0	20,250
2d	5.8	10,450
3d	11.1	5,520
4th	19.4	3,130
5th	30.5	1,990

Compare the performance of a hauling unit at sea level with its performance at 5,000 ft above sea level, all other conditions remaining constant.

Performance at sea level:

Required rimpull for loaded unit, 35.68 × 50 = 1,784 lb
Maximum speed loaded, 30.5 mph
Maximum speed empty, 30.5 mph

The probable round-trip time should be as follows:

Loading, 12.8 cu yd ÷ 280 cu yd per hr	= 0.0458 hr
Lost time in pit and accelerating, 1.5 min	= 0.0250 hr
Travel to the fill, 1.5 miles ÷ 30.5 mph	= 0.0493 hr
Dumping, turning, and accelerating, 1.5 min	= 0.0250 hr
Travel to pit, 1.5 miles ÷ 30.5 mph	= 0.0493 hr
Round-trip time	= 0.1944 hr

Assume that units will operate an average of 45 min per hr.

No. trips per hr, $\frac{1}{0.1944} \times \frac{45}{60} = 3.86$

Volume per hr, 12.8 × 3.86 = 49.5 cu yd bm

No. units required, 280 ÷ 49.5 = 5.7

It will be necessary to use six units.

Volume hauled per unit, 280 ÷ 6 = 46.7 cu yd per hr

Hauling cost per cu yd, \$31.60 ÷ 46.7 = \$0.677

Performance at 5,000-ft elevation:

Loss in available rimpull, $\frac{0.03(5{,}000 - 1{,}000)}{1{,}000} \times 100 = 12\%$

Correction factor for rimpull at 5,000 ft, 0.88

Available rimpull

Gear	Speed, mph	Rimpull at sea level, lb	Rimpull at 5,000 ft, lb
1st	3.0	20,250	17,820
2d	5.0	10,450	9,196
3d	11.1	5,250	4,620
4th	19.4	3,150	2,772
5th	30.5	1,990	1,751

Required rimpull for loaded unit, 1,784 lb

Maximum speed loaded, 19.4 mph

Required rimpull empty, 15.5 × 50 = 775 lb

Maximum speed empty, 30.5 mph

The probable round-trip time should be as follows:

Loading, 12.8 cu yd ÷ 280 cu yd per hr	= 0.0458 hr
Lost time in pit and accelerating, 1.75 min	= 0.0290 hr
Travel to the fill, 1.5 miles ÷ 19.4 mph	= 0.0773 hr
Dumping, turning, and accelerating, 1.75 min	= 0.0290 hr
Travel to pit, 1.5 miles ÷ 30.5 mph	= 0.0493 hr
Round-trip time	= 0.2304 hr

No. trips per hr, $\frac{1}{0.2304} \times \frac{45}{60} = 3.25$

Volume per hr, 12.8 × 3.25 = 41.6 cu yd

No. units required, 280 ÷ 41.6 = 6.7

It will be necessary to use 7 units.

Volume hauled per unit, 280 ÷ 7 = 40 cu yd per hr

Hauling cost per cu yd, \$31.60 ÷ 40 = \$0.790

In the calculations for the 5,000-ft altitude the time lost by a unit in the pit and at the dump was increased by 0.25 min to allow for the effect of the loss in power at this altitude.

PROBLEMS

9-1 A truck for which the information in Fig. 9-6 applies operates over a haul road with a +4 percent slope and a rolling resistance of 90 lb per ton. If the gross vehicle weight is 70,000 lb, determine the maximum speed of the truck.

9-2 If the truck of Prob. 9-1 operates on a haul road having a −4 percent slope, determine the maximum speed.

9-3 Prepare a table similar to Table 9-1, using a $1\frac{1}{2}$-cu-yd shovel whose adjusted production will be 160 cu yd per hr.

9-4 A 2-cu-yd power shovel will be used to load common earth into trucks whose capacities are 15.0 cu yd bank measure. Determine the number of trucks required to haul the earth for the following conditions:

For the power shovel:
- Depth of dig, 12 ft
- Angle of swing, 90°
- Job conditions, good
- Management conditions, good

For the trucks:
- Weight of earth, 2,900 lb per cu yd bank measure
- Empty weight of truck, 34,820 lb
- Performance chart of Fig. 9-6 applies.
- Assume that the time at the dump will be 1.75 min.
- Rolling resistance of the haul road, 80 lb per ton
- Distance to dump, 1 mile of −2 percent slope and $\frac{3}{4}$ mile of +4 percent slope.

Assume that operating conditions limit the average speed of the trucks to 75 percent of the maximum possible speed. Note that the number of trucks should be based on the ideal production rate of the shovel and the no-delay cycle time of the trucks.

CHAPTER

TEN

OPERATION ANALYSES

GENERAL INFORMATION

In examples appearing earlier in this book methods of determining the probable production rates for various types of equipment are illustrated. One might be justified in questioning the accuracy of the results, especially if they are based on conditions that are assumed to represent those that will occur on a given project. To what extent do actual job conditions conform with the conditions that were assumed when planning a project? It is certain that there will be some variations.

Consider a truck which is loaded by a power shovel. As illustrated in Fig. 10-1, the truck cycle includes at least the following elements:

1. Load
2. Haul
3. Dump
4. Return
5. Spot at the shovel

Obviously, there will be a range of times required to perform these five functions, depending upon such factors as operator efficiency, weather, equipment condition, and the like. The usual method of analyzing the times is to use the average time, realizing there will be rates of production above and below the average.

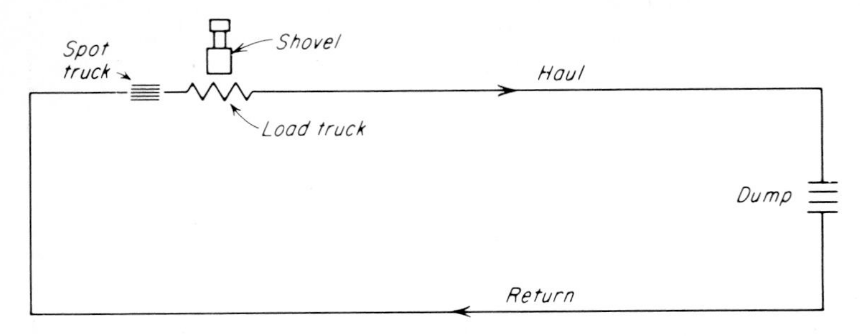

Figure 10-1 Elements of a truck cycle.

Thus, with varying rates of loading trucks and varying truck cycle times, it will not be possible to synchronize the loading and hauling operations for a sustained period of time without experiencing some delays by the shovel waiting for a truck, or by a truck waiting to be loaded.

When a contractor undertakes an analysis of the operations of his personnel or equipment, he should carefully determine the range of times he can expect to be required to perform each element of the operation and to perform the entire cycle. Then he should always strive to reduce the cycle time by eliminating or at least decreasing unnecessary delays.

In order to effectively utilize the information gathered on the operations, the powerful tools of probability and statistics should be employed. In the following section, the basics of probability are presented. For a more complete treatment of probability the reader is referred to the many excellent texts on the subject, a few of which are referenced at the close of this chapter [1–3].

PROBABILITY BASICS

Probability may be defined as the relative frequency of events in the long run. For example, a 50 percent probability of tossing tails with a coin means that if many tosses are made, one-half of the tosses will be tails. Probability is founded upon four fundamental axioms. They are:

Axiom 1 The probability (p) of an event (E) occurring $\langle p(E) \rangle$ is a number between 0 and 1.0.

$$0 \leqslant p(E) \leqslant 1.0 \tag{10-1}$$

Axiom 2 The sum of the probabilities of all mutually exclusive outcomes for a given event equals 1.0. Another way of saying this axiom is that the probability of an event occurring (p) plus the probability of that same event *not* occurring (q) is equal to 1.0.

$$p + q = 1.0 \tag{10-2}$$

Axiom 3 The probability of either of two mutually exclusive events occurring is equal to the sum of the probabilities of each of the two events.

$$p(E_1 + E_2) = p(E_1) + p(E_2) \tag{10-3}$$

Axiom 4 The probability of two independent events occurring simultaneously is equal to the product of the probabilities of the two independent events.

$$p(E_1E_2) = p(E_1) \times p(E_2) \tag{10-4}$$

To illustrate the meaning of these four axioms, consider the probabilities associated with the toss of a six-sided die.

1. What is the probability of tossing a "two" on one throw of a die? *Answer*: $p(2) = \frac{1}{6}$ (Axiom 1), because there is only one out of six possible outcomes from the toss which will result in a "two" appearing on top.
2. What is the probability of *not* throwing a "two" on any one throw of a die? *Answer*: $q(2) = \frac{5}{6}$ (Axiom 2), because there are six possible outcomes and only one would be a "two." Thus $\frac{6}{6} - \frac{1}{6} = \frac{5}{6}$.
3. What is the probability of throwing either a "two" *or* a "five" on one throw of a die? *Answer*: $p(2 + 5) = \frac{1}{6} + \frac{1}{6} = \frac{2}{6}$ (Axiom 3).
4. What is the probability of throwing two "twos" in succession with two throws of a die? *Answer*: $p(2,2) = p(2) \times p(2) = \frac{1}{6} \times \frac{1}{6} = \frac{1}{36}$ (Axiom 4).

   Note that the throw of two single die in succession is the same as the throw of a pair of dice simultaneously.

EXPECTED VALUE

Using the basics of probability, the concept of expected value can be explained. Expected value (EV) is the *weighted* average of all possible outcomes from a given situation in which probabilities can be assigned or assumed. The mathematical expression for EV is:

$$\text{EV} = \sum_{m}^{i=1} p_i O_i \tag{10-5}$$

where p_i = independent probability for event i
O_i = outcome for event i if the event actually occurs
i = 1,2,3, . . . , m

Note: The sum of all the probabilities, p_i's, must equal 1.0 (Axiom 2).

To illustrate, suppose that the probability of rain tomorrow is estimated to be 30 percent. If it rains, a contractor will be forced to stop work on a job, losing $2,000. If it does not rain, the contractor expects to earn $3,500. What is the expected value of his earnings tomorrow?

Using Eq. (10-5), we have

$$\text{EV} = 0.3 \times (-2{,}000) + 0.7 \times (+3{,}500) = \$1{,}850$$

What does the value $1,850 really represent? The contractor is either going to lose $2,000 or make $3,500. Remember the probability is the relative frequency of events in the long run. Therefore, if the contractor has a large number of jobs going on tomorrow, each one with the expected losses and earnings as described, on those jobs experiencing rain he will lose money and on those jobs which do not experience rain he will make money, *On the average*, he can expect to earn $1,850 *per job* that day.

CONSTRUCTION EXAMPLES USING PROBABILITY

Example 10-1 In observing a truck loading cycle, a contractor notes the following times and frequencies of occurrence:

Time to complete the cycle, min	Frequency
7	4
8	7
9	10
10	11
11	8
12	7
13	3
14	1
16	1

(*a*) What is the probability of any one cycle time taking 12 min?

SOLUTION: Since there were 7 out of 52 occurrences in which the cycle time was 12 min, the probability of any one cycle time being 12 min is 7/52 or 13.5 percent.

(*b*) What is the probability of any one cycle time taking 17 min?

SOLUTION: Zero (at least theoretically), because none of the observed times were that long.

(*c*) What is the expected value of cycle time in minutes?

SOLUTION: Using Eq. (10-5), the expected value would be:

$$\text{EV} = \tfrac{4}{52}\times 7 + \tfrac{7}{52}\times 8 + \tfrac{10}{52}\times 9 + \tfrac{11}{52}\times 10 + \tfrac{8}{52}\times 11 + \tfrac{7}{52}\times 12$$
$$+ \tfrac{3}{52}\times 13 + \tfrac{1}{52}\times 14 + \tfrac{1}{52}\times 16 = 10.1 \text{ min}$$

Example 10-2 A contractor's equipment records reveal that during the past year, her welders were idle 5.5 percent of the time because of welding equipment maintenance and repair. During this idle time, she figures she is losing $230 per hour, and while the welders are working she figures they are making her company $40 per hour. If a spare welding machine can be rented for $12 per hour, should it be rented?

SOLUTION: The expected value of her profits without the standby equipment is

$$\$40\times 0.945 - \$230\times 0.055 = \$25.15$$

The expected net profits with the standby equipment is

$$\$40 \times 1.00 - \$12 = \$28.00$$

The analysis indicates she should rent the standby welding machine as her "expected profit" would be higher.

Example 10-3 Assume the same conditions as given in Example 10-2, except that the standby welding equipment is not new and thus can be expected to have a downtime of about 20 percent itself. Should the standby equipment be utilized?

SOLUTION: In this case, there is the probability of two independent events occurring simultaneously (both pieces of equipment down at the same time). The probability of both pieces being down at the same time is (Axiom 4)

$$p(\text{both down}) = 0.055 \times 0.20 = 0.011$$

Without the standby, the expected values would be \$25.15 (from Example 10-2). With the standby, the expected value would be

$$\$40(1.00 - 0.011) - \$230(0.011) - \$12.00 = \$25.3$$

In this case, it would not be advantageous to rent the standby welding equipment.

MOTION AND TIME STUDIES

These are studies made by an observer using a stop watch and a clip board, with forms, on which appropriate time elements may be recorded, as they are observed. Consider a truck cycle. The truck is loaded; then it hauls to the dump site, backs into position, dumps its load, and returns to the shovel for another load, where it may have to wait to be served by the shovel, defined here as spotting.

A break point is selected for the beginning and the end of each element. For example, loading time starts when the shovel begins serving the truck, and it ends when the truck begins moving away from the shovel, at which time hauling time starts. Hauling time continues until the truck stops at the dump site, preparatory to backing into dumping position. Dumping time may include turning, backing, and dumping the load. Returning time starts when the truck begins moving away from the dump, and ends when the truck returns to the shovel site. Spotting time is the time required to maneuver into a position for loading. If it is necessary for a truck to wait for position at the dump or at the shovel, additional elements may be included in the cycle, and the time for each of these additional elements should be recorded. Also, on the form used for recording the time required for each element, there should be a space for recording any delays, such as stopping for fuel, water, or tires, and personal delays, etc.

The stop watch should be calibrated to read time in minutes and hundredths of a minute. Once a time study is started, the watch should operate continuously, with the observer simply recording the time reading at the beginning of the study and thereafter at the end of each element, which will be the beginning of the following element, until the study is finished. The actual time required for each element can be determined later.

If a study is to be made for a power shovel, the observer can select one location and remain there. However, if a study is to be made for equipment that travels beyond his view, such as a truck or a scraper, the observer should accompany the unit during the full period of study.

Figure 10-2 illustrates a form or observation sheet that can be used to record the time for each element. The entries opposite *R* are observed on the watch and recorded immediately. The entries opposite *T* represent the times required for the specified elements, which times are calculated later. This figure is a record of a study made for a power shovel, whose cycle is divided into the elements load, swing, dump, and return, with a further provision for recording any delays.

Dividing the cycle into four elements may or may not be justified. If the average time and the range in times per element are not desired, the study may be limited to determining the cycle times only.

Figure 10-3 illustrates the information obtained from a motion and time study for a wheel-type tractor-pulled scraper. The cycle is divided into five elements, with a further provision for recording any delays. It may be desirable to classify the causes of delays and to indicate them on the sheet, such as

1. Personal
2. Mechanical
3. Service, fuel, oil, water, etc.
4. Other, as applicable

DURATION OF A TIME STUDY

The duration of a time study should permit the observer to record enough cycles to assure results having the desired accuracy. One observation may have some value, but it is highly improbable that it will give a reliable estimate of the average time of continuing operations. Because of the cost of making the observations and the subsequent calculations, it is desirable to limit the observations to the minimum number required to produce a reliable estimate of the range of times to be expected.

There are time-recording meters that may be attached to construction equipment, which will record the time for each element of a cycle, either automatically or semiautomatically.

When a comprehensive time study is being planned, consideration should be given to making the observations at intermittent intervals, separated by several hours or days to allow for the effects of varying conditions that may affect the cycle time.

MOTION AND TIME STUDY												Sheet No. 1	
Project No. 162		Operation: Power shovel No. 6										Date: 6/14/83	
Start timing 9:15 am End timing 9:19 am		Operator Jim Brown							Observer J. G. Smith				
Element		Cycles, time in min										Summary	
		1	2	3	4	5	6	7	8	9	10	ΣT	$\bar{T}$
Start		0.00											
Fill dipper	T	0.15	0.12	0.14	0.13	0.14	0.15	0.12	0.17	0.12	0.16	1.40	0.14
	R	0.15	0.59	1.02	1.44	1.91	2.34	2.75	3.22	3.65	4.07		
Swing	T	0.12	0.12	0.10	0.12	0.11	0.10	0.12	0.09	0.09	0.10	1.07	0.11
	R	0.27	0.71	1.12	1.56	2.02	2.44	2.87	3.31	3.74	4.17		
Dump	T	0.09	0.07	0.08	0.09	0.07	0.08	0.08	0.09	0.07	0.08	0.80	0.08
	R	0.36	0.78	1.20	1.65	2.09	2.52	2.95	3.40	3.81	4.25		
Return	T	0.11	0.10	0.11	0.12	0.10	0.11	0.10	0.13	0.10	0.12	1.10	0.11
	R	0.47	0.88	1.31	1.77	2.19	2.63	3.05	3.53	3.91	4.37		
Delay	T												
	R												
Cycle	T	0.47	0.41	0.43	0.46	0.42	0.44	0.42	0.48	0.38	0.46	4.37	0.44

ΣT = sum of element times $\bar{T}$ = average time for element = $\Sigma T/N$

Figure 10-2 Time study for a power shovel.

MOTION AND TIME STUDY													Sheet No. 1
Project No. 158				Operation: Scraper No. 4									Date: 6/14/83
Start timing 2:14 pm End timing 4:10 pm				Operator Gus Weaver						Observer J. G. Smith			
Element	Load		Haul		Dump		Return		Wait		Cycle	Delay	
Cycle	T	R	T	R	T	R	T	R	T	R	time	T	Type
1	0.86	0.86	3.88	4.74	0.42	5.16	3.16	8.32	0.36	8.68	8.68	0	
2	0.93	9.61	4.24	13.85	0.36	14.21	3.34	17.55	0.24	17.79	9.11		
3	0.78	18.57	4.08	22.65	0.47	23.12	2.86	25.98	0.44	26.42	8.63		
4	0.98	27.40	4.16	31.56	0.34	31.90	2.98	34.88	0.42	35.30	8.88	3.18	6
5	0.82	39.30	4.18	43.48	0.31	43.79	3.04	46.83	0.28	47.11	8.63		
6	0.80	47.91	3.96	51.87	0.38	52.25	2.92	55.17	0.37	55.54	8.43		
7	0.88	56.42	4.22	60.64	0.41	61.05	2.94	63.99	0.46	64.45	8.91		
8	0.96	65.41	4.38	69.79	0.44	70.23	3.18	73.41	0.42	73.83	9.38	2.86	7
9	1.04	77.73	4.14	81.87	0.37	82.24	2.80	85.04	0.52	85.56	8.87		
10	0.92	86.48	4.04	90.52	0.33	90.85	2.84	93.69	0.60	94.29	8.73		
11	0.87	95.16	4.19	99.35	0.34	99.69	2.94	102.63	0.32	102.95	8.66	4.21	6
12	0.92	108.08	4.26	112.34	0.36	112.70	2.86	115.56	0.42	115.98	8.82		
ΣT	10.76		49.73		4.53		35.86		4.85		105.73	10.25	
$\bar{T}$	0.90		4.14		0.38		2.99		0.40		8.81		

ΣT = sum of element times $\bar{T}$ = average time for element

Figure 10-3 Time study for a wheel-type scraper.

STATISTICAL METHODS OF DETERMINING THE NUMBER OF OBSERVATIONS NEEDED

Statistics involves the application of probability theory to the analysis of data scatter. There are two general types of data scatter—systematic and random. Statistics can be used to analyze random scatter only. It is of little use in analyzing systematic scatter. If the scatter is more or less random, then there are several parameters of importance in analyzing the variability of observations. They include:

1. Measures of central tendencies, which include:
   a. The arithmetic mean or average, which is equal to the sum of the individual values divided by the total number of values.
   b. The median, which is the middle value when the values are arranged in ascending or descending order.
   c. The mode, which is the most frequently occurring value.
2. Measures of dispersion, which include:
   a. The range, which equals the maximum value minus the minimum value.
   b. The variance, which is a mathematically determined quantity whose value depends on the shape of the frequency distribution of the data. In mathematical terms it is equal to the root mean square of the deviations of all the values for their average value.
   c. The standard deviation, which is the square root of the variance.

In order to determine what kind of frequency distribution the data exhibit, it is usually necessary first to obtain at least 30 individual values, then group the data into classes of data (usually about 10 to 12 classes are sufficient) and plot the frequency of occurrence of each data class versus class interval (see Fig. 10-4). Fortunately, many real-life data exhibit random scatter, and when plotted in such a manner, can be mathematically represented by the Normal, or Gaussian distribution function (Fig. 10-5). This function, which can be completely expressed using only two parameters, the mean and the variance, allows us to perform a number of important statistical calculations. In particular, as discussed in Chap. 2, the area under any part of the curve can be equated to the probability of occurrence of any given value between the bounds of the area under consideration. Furthermore, if it is known or can be assumed that the data scatter is approximately normally distributed, you can obtain reasonable estimates of the mean and variance (or standard deviation) even if you do not have 30 values. These estimates make use of the Student's t distribution, which has a similar shape to the normal distribution except that it has a wider scatter (see Fig. 10-5). One other important point should be made here. In real-life conditions you will only be able to obtain a sample of the data you are analyzing. Fortunately, if the sample is collected in a random fashion, then the mean of the sample very closely approximates the mean of the entire data set. Unfortunately, the same cannot be said for the variance. The sample variance

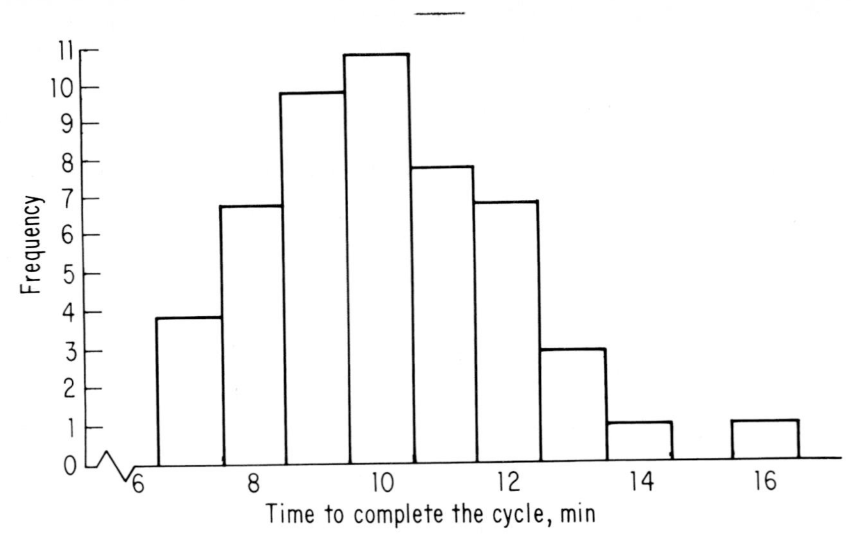

Figure 10-4 Frequency histogram for the data from Example 10-1.

is not the same as the variance of the entire data set. But, we can still use the sample variance as long as we recognize its limitations.

Knowing the power of statistics, we can use it to determine a number of important factors. For example, suppose we wish to determine the number of observations required to produce results having a specific accuracy (i.e., a

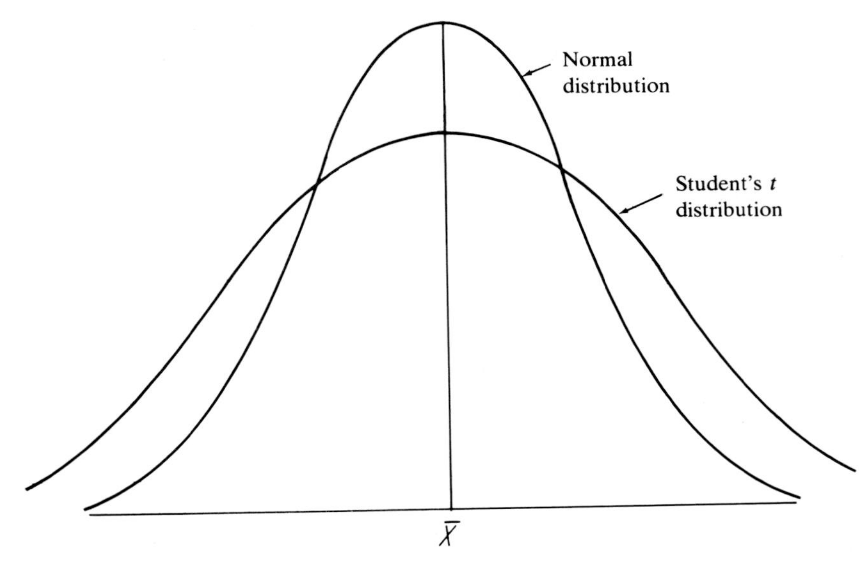

Figure 10-5 Normal and Student's t frequency distribution curves.

probability level or confidence that our prediction is correct). Assuming normally distributed data scatter, we can use the following procedure.

1. Specify a confidence interval I, which is a time interval, that conforms with the desired accuracy of the study.
2. Specify a confidence coefficient C, which indicates the probability that the results will conform with the desired accuracy.
3. Observe M cycles of the operation being studied.
4. Calculate the sample standard deviation s from Eq. (10-6),

$$s = \sqrt{\frac{\Sigma T^2 - (\Sigma T)^2/M}{M-1}} \tag{10-6}$$

Assume that one wishes to obtain an average cycle time with a 90 percent probability that the determined value will be accurate within the specified time interval or tolerance I. Calculate the confidence interval I_m provided by this sample of M observations, using Eq. (10-7)

$$I_m = 2t_{0.90}\left(\frac{s}{\sqrt{M}}\right) \tag{10-7}$$

where t is the value of Student's t distribution, as given in Table 10-1 for $C = 0.90$ and $M - 1$ degrees of freedom. For M equal to 10 observations, $t_{0.90} = 1.83$. Substituting these values of M and t into Eq. (10-7)

$$I_M = 2 \times 1.83\left(\frac{s}{\sqrt{10}}\right) = 1.16s \qquad \text{for 10 observations} \tag{10-8}$$

If I_M, as determined in Eq. (10-3), is equal to or less than the specified

Table 10-1 Values of t for Student's t distribution for $C = 0.90$

M	t	M	t
5	2.13	18	1.75
6	2.02	19	1.74
7	1.94	20	1.73
8	1.90	21	1.73
9	1.86	22	1.72
10	1.83	23	1.72
11	1.81	24	1.71
12	1.80	25	1.71
13	1.78	26	1.71
14	1.77	27	1.71
15	1.76	28	1.70
16	1.76	29	1.70
17	1.75	30	1.70
		Above 30	1.65

value of I, the number of observations is sufficient. If I_M is greater than the specified value of I, additional observations are required. The total number of observations required N can be determined from Eq. (10-9).

$$N = \frac{4(t)^2 s^2}{I^2} \tag{10-9}$$

If we use the previously selected value of $t = 1.83$, Eq. (10-9) becomes

$$N = \frac{4(1.83)^2 s^2}{I^2} = \frac{13.4 s^2}{I^2} \quad \text{for 10 observations} \tag{10-10}$$

The use of Eqs. (10-6) through (10-10) may be illustrated by applying them to the time study for the power shovel illustrated in Fig. 10-2. Note that 10 observations were made. Assume that it is desired to be 90 percent certain that the average cycle time obtained from the time study is within ± 0.02 min of the true cycle time. For this condition the confidence interval I will be $2 \times 0.02 = 0.04$ min.

Use Eq. (10-6) to determine the standard deviation s.

$$\begin{aligned} \Sigma T^2 &= T_1^2 + T_2^2 + T_3^2 + \cdots + T_{10}^2 \\ &= (0.47)^2 + (0.41)^2 + (0.43)^2 + \cdots + (0.46)^2 \\ &= 1.9183 \end{aligned}$$

$$\Sigma T = 0.47 + 0.41 + 0.43 + \cdots + 0.46 = 4.37$$

$$s = \sqrt{\frac{\Sigma T^2 - (\Sigma T)^2/M}{M - 1}} = \sqrt{\frac{1.9183 - (4.37)^2/10}{10 - 1}} = 0.030$$

Using Eq. (10-8),

$$I_M = 1.16s = 1.16 \times 0.030 = 0.035$$

Because the value of $I_M = 0.035$ is less than the permissible value of $I = 0.04$, no additional observations are required. Thus the determined average cycle time of 0.44 min is sufficiently accurate for the specified conditions.

If the average cycle time for the scraper study illustrated in Fig. 10-3 is desired for a confidence interval of 0.2 min, the solutions of Eqs. (10-6) and (10-7) reveal that $M = 22$. Thus it will be necessary to observe at least $22 - 12 = 10$ more cycles to attain the desired accuracy.

The solution of Eqs. (10-6) through (10-10) at the project are very time-consuming. An alternate method of determining the number of observations required uses Eq. (10-11) to calculate the value of s, which gives reasonable accuracy.

$$s = \frac{R}{d} \tag{10-11}$$

where R = the difference between the maximum and the minimum values of the cycle time

d = a conversion factor whose value depends on M, as given in Table 10-2

If the value of s in Eq. (10-11) is substituted in Eq. (10-9), the resulting equation will be

$$N = \frac{4t^2R^2}{I^2d^2} \tag{10-12}$$

If Eqs. (10-11) and (10-12) are applied to the cycle times in Fig. 10-2, and the value of d is obtained from Table 10-2 for $M = 10$, the results will be

$$R = 0.48 - 0.38 = 0.10$$

and

$$N = \frac{4 \times (1.83)^2 \times (0.10)^2}{(0.04)^2 \times (3.078)^2} = 8.95 \text{ or } 9$$

which agrees with the previous findings.

If Eq. (10-12) is applied to the cycle times for the scraper in Fig. 10-3, under the previously specified conditions,

$$C = 0.90$$
$$I = 0.20$$
$$M = 12$$
$$t = 1.80$$
$$R = 9.38 - 8.43 = 0.95$$
$$d = 3.258$$

$$N = \frac{4(1.80)^2(0.95)^2}{(0.2)^2(3.258)^2} = 27.6 \text{ or } 28$$

Table 10-2 Values of d factors for time studies using $s = R/d$

M	d	M	d
5	2.326	19	3.689
6	2.534	20	3.735
7	2.704	21	3.778
8	2.847	22	3.818
9	2.970	23	3.856
10	3.078	24	3.891
11	3.173	25	3.925
12	3.258	26	3.956
13	3.336	27	3.985
14	3.407	28	4.012
15	3.472	29	4.038
16	3.532	30	4.053
17	3.588		
18	3.640		

An examination of Eq. (10-12) discloses that, with other factors constant, the number of observations required varies inversely with the square of the confidence interval. For example, the confidence interval for the previous study of the scraper is 0.2 min. This means that there is a 90 percent probability that 28 observations, using the alternate method, or 22 observations, using the more exact method, will give an average cycle time that is within plus or minus 0.1 min of the true value. Because this accuracy might justifiably be considered unrealistic, the interval could be increased to a higher value, such as 0.3 min, for a result within plus or minus 0.15 min of the true value. If I is increased to 0.3, Eq. (10-12) gives a value of N equal to 12.3. Thus the 12 observations already made are adequate.

TIME-LAPSE PHOTOGRAPHY

Time-lapse photography has emerged recently as an extremely useful and important tool in measuring productivity of construction operations. It consists of a motion picture camera (usually super-8 mm) which can expose picture frames at specific rates, from the standard movie rate of 18 frames per second to rates as slow as one frame every 99 secs [4]. This allows a time compression of up to 5,346 to 1, which means that one 50-ft roll of film normally recording about $3\frac{1}{2}$ min of activity at 18 frames a second, can summarize up to 4 hr of activity on the same 50-ft roll at one frame every 4 sec. Of course the operation, if viewed at normal running speeds with a projector, are vastly speeded up and of little value. But if the time-lapse camera is coupled with a movie projector capable of advancing frames at any desired speed from manual to normal running speeds, the film can be viewed and analyzed for productivity in much the same way as with a stop watch and a clip board.

The equipment required for time-lapse photography studies is not expensive, as the entire set usually will cost less than $5,000. While black and white film may be used, color film usually is employed because it is easier to distinguish between different workers.

The most efficient use of time-lapse photography is in areas where activity is concentrated. The camera can be set up adjacent to the work area and the camera not moved during the filming operation. Thus about the only operations which do not readily lend themselves for time-lapse analysis are those which are spread out over a large area (such as highway construction on flat terrain with no overpasses upon which to set the camera).

As the camera is taking pictures at precise time intervals and the projector can display the film at precise time intervals for each frame, an accurate time log of the activities can be determined. And as all activities within the range of the camera eye are recorded, the chances for omission of important data is drastically reduced (when compared to an observer, who usually can watch only one major activity at a time).

Interestingly, one potential disadvantage of the time-lapse camera method has been shown *not* to exist. The presence of the camera, if left alone in an unobtrusive spot, does not affect the productivity of the work. It is quickly forgotten by the workers, and an accurate record of the work is captured for later analysis.

The method of analysis is straightforward and relatively easy. There are a number of excellent texts and technical reports showing the method of analyzing various construction operations and the resulting increases in productivity resulting from such an analysis (see references [5] and [6] at the close of the chapter). Following are a few examples of the use of time-lapse photography in analyzing construction operations.

TIME-LAPSE STUDY OF STRUCTURAL STEEL COLUMN AND BEAM ERECTION

A procedure that may be used in performing a methods-improvement study is illustrated by a time-lapse photography study of the erection of structural steel columns and beams on a small single-story shopping mall [7]. The entire project involved the erection of 55 columns and 44 beams for a single-story shopping mall consisting of approximately 35,000 sq ft. The contractor chose to have a crew of seven people performing the erection of both elements, as the process was the same for both, except for the exchange of welding for bolting in columns and beams, respectively. The process involved tieing, hoisting, placing, aligning, bolting (or welding), untieing, and returning. As such, the erection operation was largely repetitive and ideal for study to find possible ways to increase productivity. Time-lapse pictures were taken after the contractor had planned the erection sequence, laid out the construction area, selected the crew, commenced work, and optimized the procedure through a trial-and-error process. The construction site layout is shown in Fig. 10-6. The entire operation was recorded on time-lapse film taken over an extended period of time. The film was analyzed to identify the elements and their associated durations for each cycle of operation for each crew member. A crew balance diagram was drawn for both the column and beam erector tasks using the average durations for each element (see Figs. 10-7 and 10-8). For the column erection task, there were eight tasks identified (six operations and two transportations). Examination of Figs. 10-7 and 10-8 reveals significant "idle" times for several of the crew while they were waiting for some other element to be completed. Then, through a redesign of the construction site (Fig. 10-9), it was estimated that the cycle times could be reduced by $2\frac{1}{4}$ min for the column erection (a 22 percent reduction) and 3 min for the beam erection (a 37 percent reduction). Furthermore, it was found that these two activities could be accomplished with one less person (six instead of seven). The revised average durations for each cycle of operation for each crew member is given in Fig. 10-10 (column erection) and Fig. 10-11 (beam erection).

By the use of time-lapse photography, a number of other analyses can be

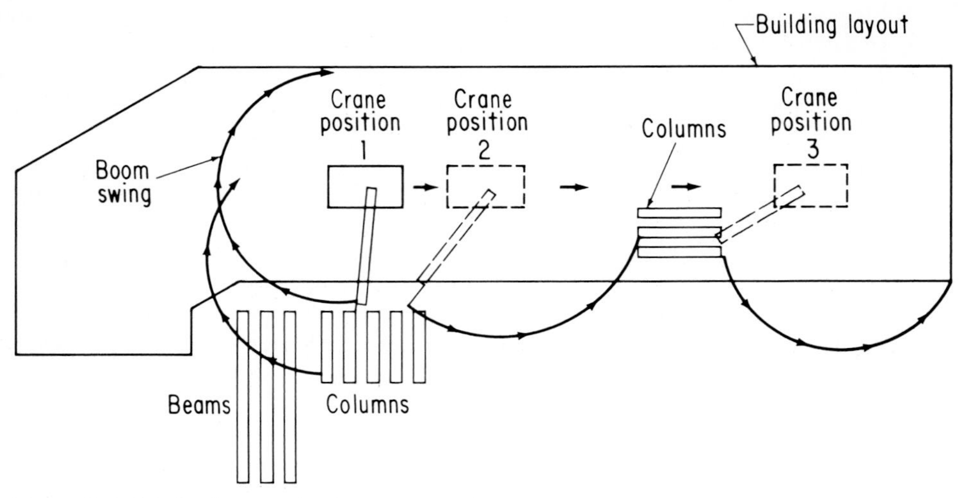

Figure 10-6 Existing beam and column erection layout.

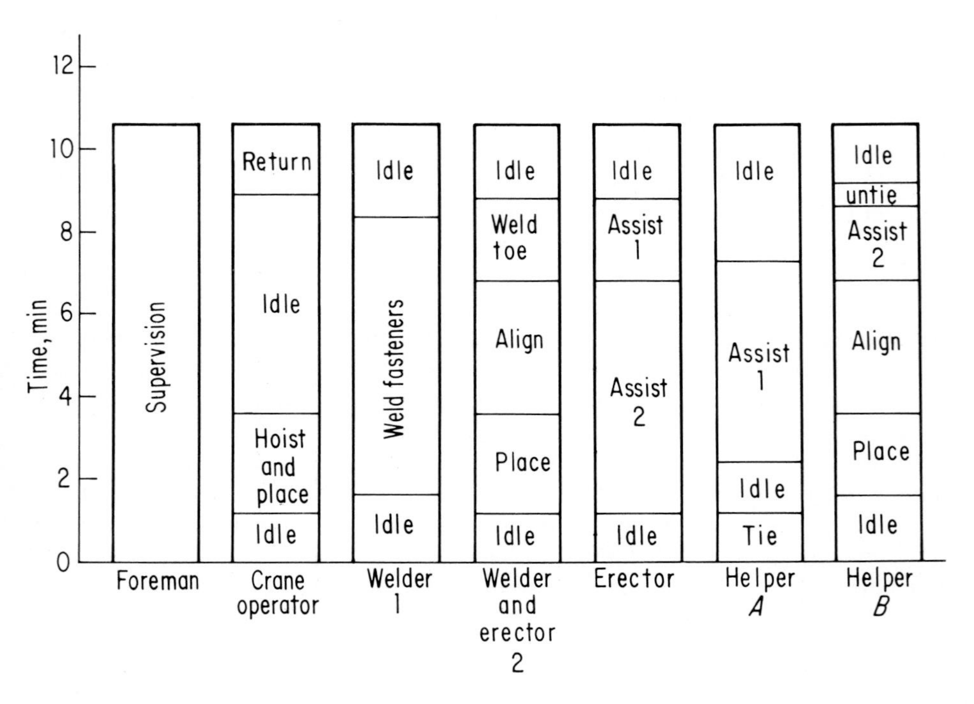

Figure 10-7 Existing crew balance—columns.

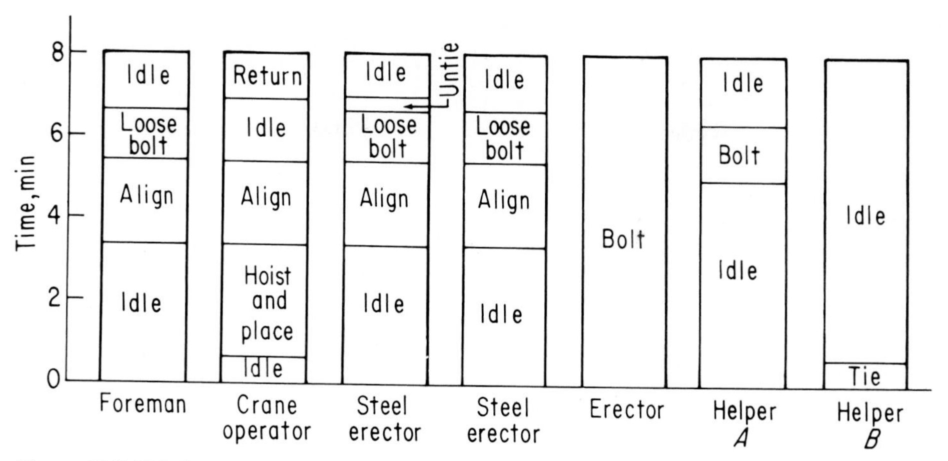

Figure 10-8 Existing crew balance—beams.

made and appropriate recommendations given to the contractor to improve his overall productivity. These include such important factors as identification of some of the causes of excessive idle time, identification of constricted areas within the construction site, identification of crew activities which can be eliminated without loss of efficiency, and identification of some quality management items (for example, concrete remaining in ready-mix trucks too long). Through such studies, significant improvements in productivity are often achieved, quickly recovering the costs involved in the study and resulting in construction cost reductions.

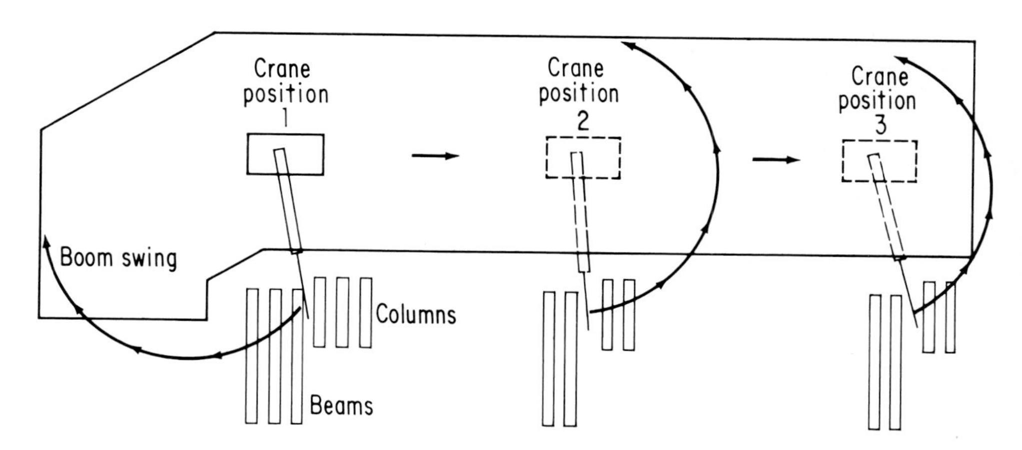

Figure 10-9 Proposed beam and column erection layout.

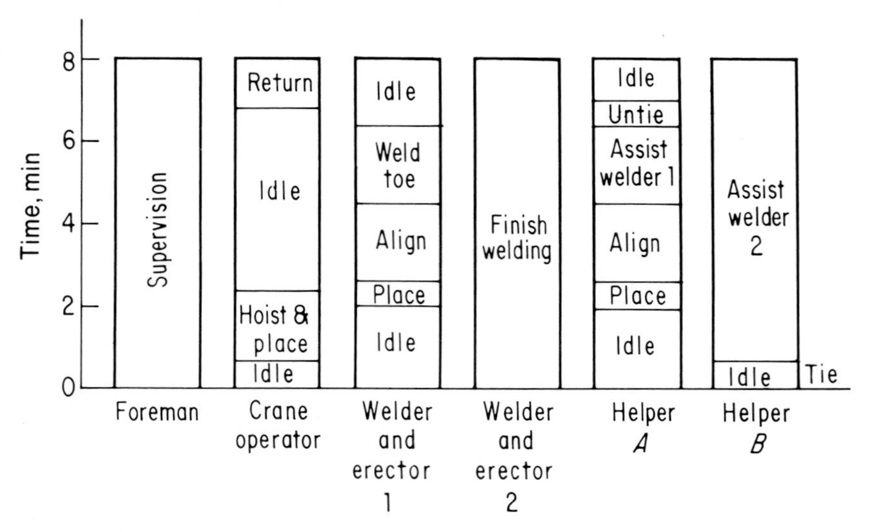

Figure 10-10 Proposed crew balance—columns.

EXAMPLE OF JOB PLANNING IN CONSTRUCTING ROOF TRUSSES

On a multimillion-dollar housing project the contractor chose to use prefabricated lightweight wood roof trusses instead of erecting them in place on each building. Figure 10-12 illustrates the type of truss used. As noted, each

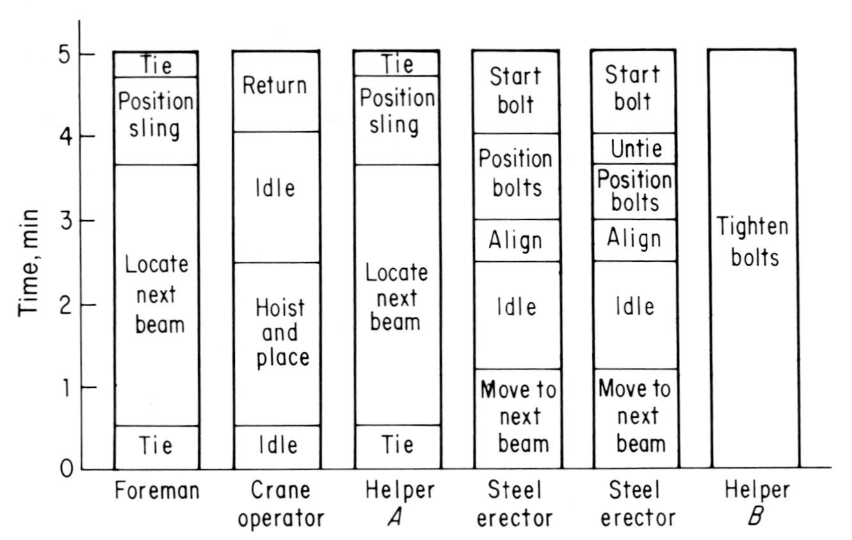

Figure 10-11 Proposed crew balance—beams.

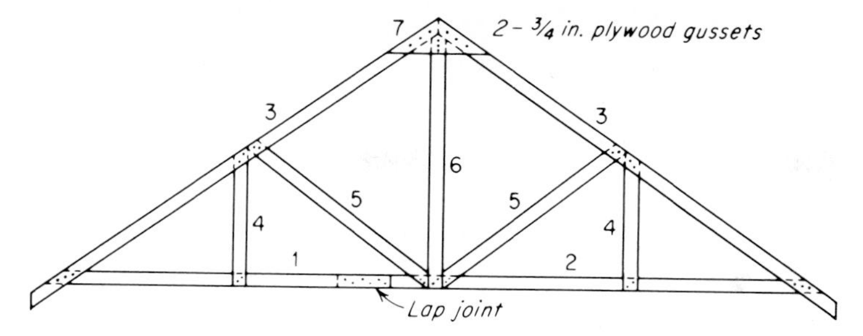

Figure 10-12 Roof truss.

truss required seven different sizes and shapes of members, including the $\frac{3}{4}$-in.-thick plywood gusset. Because several hundred identical trusses were required, the fabricating and erecting operations were highly repetitive, thereby permitting economy by job planning.

The contractor selected a site near the center of the project for the storage of stock sizes of lumber, the installation of a bench-type electric-motor-driven radial saw, a 36-in.-high jig-equipped work table, large enough to support a roof truss lying flat, and the spotting of two wheel-mounted flat-bed trailers to receive the finished trusses. Figure 10-13 illustrates the flow diagram for fabricating the trusses.

Stock lumber of appropriate sizes and lengths was stored in a compartmentalized rack at 1. When a given truss member was cut in quantity, a stop was set on the bench saw to enable the carpenter to saw the members without having to measure them for length or shape. The cut members were stored by size in the rack 3, from which they were transferred to work table 4 as needed. Two carpenters, using only hammers and nails, completed the fabrication of the trusses. Jigs attached to the table permitted the members to be fitted into position quickly and accurately. When a truss was fabricated, it was loaded onto a flat-bed trailer. When a trailer was loaded, it was attached to a farm tractor which pulled it to a building, where the trusses were hoisted into

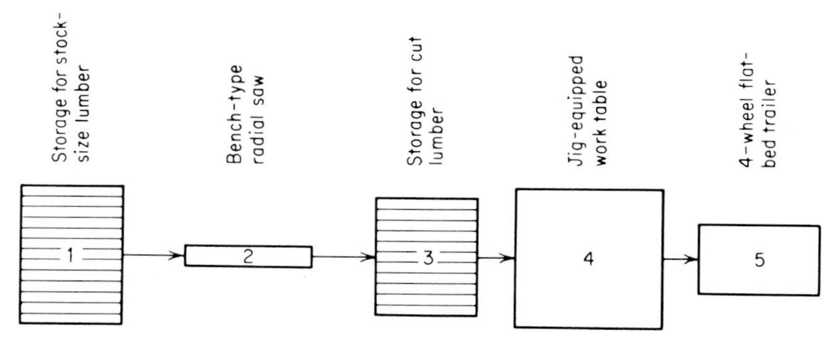

Figure 10-13 Flow diagram for fabricating roof truss.

position. The tractor was disconnected from the loaded trailer and attached to an empty trailer for a return trip to the fabricating site.

By preplanning the members to be cut from each length of stock lumber the waste was kept to a minimum.

APPLYING MOTION AND TIME STUDIES TO BUILDING HOUSES

During the early stages of building several hundred homes to only three plans the contractor conducted motion and time studies of all operations as a means of increasing the efficiency of and reducing the cost of construction. The houses were erected on concrete slabs, using wood framing for all walls and the flat roof, metal lath and stucco on the outside walls, Sheetrock on the inside walls and ceilings, a composition roof, and steel-sash windows.

He used a central shop to precut all lumber and to preassemble most of the rough-in plumbing. The lumber for a given house was assembled into a single bundle, secured with steel bands, and then delivered to the house site.

When two carpenters and a helper arrived at the slab, they cut the steel bands and proceeded with the erection of the wall frames, assembling panels on the slab, tilting them to vertical positions, and plumbing, bracing, and attaching the bottom plates to the concrete slab, using gun-driven studs. Because the lumber was assembled in the bundle in the order that it was used, no time was lost looking for a given member. The only tools used by the carpenters were nail aprons, hammers, carpenter's levels, and a gun. At the end of the first day the framing, including the ceiling joists, was completed.

The delivery of materials and the arrival of craftsmen for each succeeding activity were scheduled to assure efficient operations with minimum lost time.

As a result of preplanning and scheduling, a house was completed in a maximum of 2 weeks and at a substantial reduction in cost when compared with the cost of a single custom-built house of equal size and quality.

APPLYING THE THEORY OF QUEUES TO DETERMINE THE MOST ECONOMICAL NUMBER OF HAULING UNITS

On a given project involving earthwork the most economical number of hauling units is the number that will produce the lowest cost per unit of earth, considering the combined cost of the excavator and the hauling units. If the production rate of the excavator were constant, and if the loads and cycle times of the hauling units were constant, it would be fairly simple to determine the most economical number of hauling units to use on any given project, as illustrated in previous examples appearing in Chap. 9. However, it is well known that truck cycle times are not constant even though the hauling conditions and the number of trucks operating remain constant. There may be times when several trucks are waiting in a queue to be loaded; then later, for no apparent reason, the excavator may have to wait for a truck, resulting in a loss in production. If additional trucks are added to the fleet, to reduce or eliminate the lost production by the excavator, the rate of production will likely be increased, but often not enough to compensate for the increased cost of the extra truck or trucks.

The theory of queues can be applied to a situation involving an earth loader and hauling units to analyze statistically the cost of excavating and hauling earth when using varying numbers of hauling units; from this the optimum number of units can be determined. Actual observations and cost determinations made on operating projects have verified the accuracy of this theory [8].

The application of this theory will be illustrated by considering a power shovel used to load trucks, which will haul the earth to a dump site, dump the earth, and then return to the shovel for additional loads, as illustrated in Fig. 10-1. The symbols used in developing and applying the equations are as follows:

Q = output of shovel in cu yd per hr bank measure
f = operating factor for the shovel, such as a 45-min hr = 0.75
q = capacity of trucks in cu yd bank measure
n = number of trucks in the fleet
P_0 = probability of no trucks in the queue
r = mean arrival rate of truck per hr, excluding loading time, with no delays
$T_a = \frac{1}{r}$, cycle time for a truck, excluding loading time, hr
m = number of trucks loaded per hr
$x = \frac{m}{r}$, number of trucks needed in the fleet
$T_s = \frac{1}{m}$, time to load a truck, hr
C = total cost per hour for shovel and trucks

The production of the shovel in cubic yards per hour will be

$$Q = fmq \tag{10-13}$$

Equation (10-13) gives the ideal rate of production for the shovel. If it must wait for trucks at times, the rate will be reduced, as indicated by Eq. (10-14)

$$Q = (1 - P_0)fmq \tag{10-14}$$

The total cost per hour for the shovel and trucks will be

$$C = nC_t + C_s \tag{10-15}$$

where C_t = cost per hr per truck
C_s = cost per hr for shovel

The cost per cubic yard will be

$$c = \frac{nC_t + C_s}{Q} \tag{10-16}$$

In order to determine the actual production of the shovel from Eq. (10-14) it is necessary to determine the value of P_0 when varying numbers of trucks are used. The reader is referred to books on the theory of probability for a complete treatment of this subject. The probability of there being no truck in the queue, resulting in the shovel's having to wait until a truck arrives, is given by the equation

$$P_0(n, x) = \frac{e^{-x}x^n/n!}{\sum_{j=0}^{n} (e^{-x}x^j/j!)} = \frac{p(n, x)}{P(n, x)} \tag{10-17}$$

This is a cumulative Poisson expression, whose values can be determined by using Poisson distribution functions obtained from tables appearing in some handbooks and books treating the theory of probability [9]. Such tables give the values for the numerator and denominator expressions for values n and x [10].

Example 10-4 Determine the optimum number of trucks, the probable production, and the minimum cost per cubic yard to excavate and haul earth using the following equipment and operating conditions, with volumes in bank measure.

Power shovel, rate of production 300 cu yd per hr when loading
Operating factor, 50-min hour, 0.833
Production per hr, 0.833 × 300 = 250 cu yd
Capacity of trucks, 15 cu yd
Average cycle time for trucks, 0.2040 hr, excluding loading time
Cost per hr for shovel, including operator and oiler, $62.40
Cost per hr for truck and driver, $21.00

From this information the following values are determined:

$Q' = 300$ cu yd per hr
$f = 0.833$
$Q = 0.833 \times 300 = 250$ cu yd per hr
$T_a = 0.2040$ hr
$r = 1 \div 0.2040 = 4.91$

$$T_s = \frac{15}{300} = 0.050 \text{ hr}$$

$$m = \frac{1}{0.050} = 20 \text{ trucks loaded per hr}$$

$$x = \frac{20.0}{4.91} = 4.1, \text{ number of trucks needed}$$

Table 10-3 gives the value of the functions required to evaluate Eq. (10-17) with x

Table 10-3 Poisson distribution functions

(1) n	(2) $p(n, 4.1)$	(3) $1-p(n, 4.1)$	(4) $p(n, 4.1)$
0	0.0166	1.0000	0.0166
1	0.0679	0.9834	0.0845
2	0.1393	0.9155	0.2238
3	0.1904	0.7762	0.4142
4	0.1951	0.5858	0.6093
5	0.1600	0.3907	0.7693
6	0.1093	0.2307	0.8786
7	0.0640	0.1214	0.9427
8	0.0328	0.0573	0.9755
9	0.0150	0.0245	0.9905
10	0.0061	0.0095	0.9966

replaced by its value 4.1. The equation may be rewritten as

$$P_0(n, 4.1) = \frac{p(n, 4.1)}{P(n, 4.1)}$$

Most tables give the functions appearing in columns 2 and 3 only. In order to obtain the functions in column 4 it is necessary to subtract from 1 the function appearing in column 3 for the next larger value of n. For example, the function in column 4 for $n = 3$ is obtained by subtracting the function for n equals 4, appearing in column 3. Thus $1 - 0.5858 = 0.4142$.

Applying the values of the functions appearing in Table 10-3 in Eq. (10-17) gives the values listed for P_0 appearing in Table 10-4. The values of P_0 are the probabilities that there will be no truck in the queue or at the shovel for loading, thus causing a delay for the shovel and a reduction in its rate of production.

Table 10-5 gives the probable production in cubic yards per hour based on using the indicated number of trucks and the probability factors appearing in Table 10-4.

Table 10-6 gives the variations in the cost per cubic yard based on using varying numbers of trucks and Eq. (10-16). For example, the cost using four trucks is

$$c = \frac{4 \times 21.00 + 62.40}{170} = \$0.81 \text{ per cu yd}$$

Table 10-4 Values of P_0

n	$p(n, 4.1) \div P(n, 4.1)$	P_0
1	0.0679 ÷ 0.0845	0.804
2	0.1393 ÷ 0.2238	0.622
3	0.1904 ÷ 0.4142	0.460
4	0.1951 ÷ 0.6093	0.320
5	0.1600 ÷ 0.7693	0.208
6	0.1093 ÷ 0.8786	0.124
7	0.0640 ÷ 0.9427	0.068
8	0.0328 ÷ 0.9755	0.034
9	0.0150 ÷ 0.9905	0.015
10	0.0061 ÷ 0.9966	0.006

Table 10-5 Variation in the probable production with the number of trucks used

No. of trucks	$1 - P_0$	Normal production, cu yd per hr	Probable production, cu yd per hr
1	0.196	250	49.0
2	0.378	250	94.5
3	0.540	250	135.0
4	0.680	250	170.0
5	0.792	250	198.0
6	0.876	250	219.5
7	0.932	250	233.4
8	0.966	250	242.0
9	0.985	250	246.5
10	0.994	250	248.5

Example 10-5 This example presents a graphical solution for the optimization of an excavator-truck earth-moving system by considering it as a cyclic queueing system. Two different situations are analyzed with reference to the variability of the service time of the excavator and the transit time of the trucks. They are: (1) constant service time and constant transit time; (2) random service time and random transit time. Other conditions may exist and may be analyzed in a similar manner [11].

An earth-moving system composed of one excavator and N trucks may be considered as a queueing system that is described as follows: a truck is loaded, travels to the dump site, dumps, and returns to the back of the queue or, if there is no queue, begins loading immediately. (Fig. 10-14). If there is a continuous queue of trucks, the excavator will excavate an average of X cubic yards of earth per hour. If the excavator is idle for a proportion P_0 of the time, the cost per cubic yard of earth moved will be

$$C_N = \frac{K_1 + NK_2}{X(1 - P_0)} = \frac{K_2}{X} F_N\left(\frac{K_1}{K_2} + N\right) \tag{10-18}$$

Table 10-6 Variations in the cost of excavating and hauling earth with the number of trucks used

No. of trucks	Total cost per hr	Production, cu yd per hr	Cost per cu yd
1	$83.40	49.0	$1.71
2	104.40	94.5	1.05
3	125.40	135.0	0.93
4	146.40	170.0	0.81
5	167.40	198.0	0.84
6	188.40	219.5	0.86
7	209.40	233.4	0.89
8	230.40	242.0	0.96
9	251.40	246.5	1.02
10	272.40	248.5	1.10

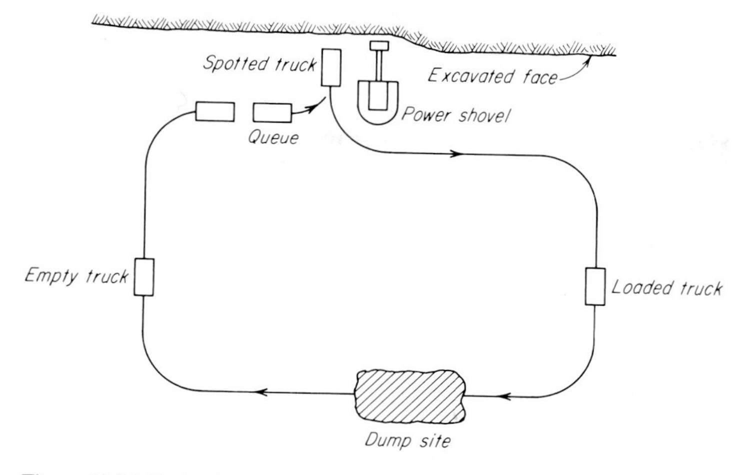

Figure 10-14 Basic elements of an earth-moving operation.

where K_1 = cost per hr for the excavator, including labor
K_2 = cost per hr for a truck, including labor
$F_N = 1/(1 - P_0)$

The problem is to determine P_0 and hence F_N for any particular N and any particular set of assumptions about the service time and the transit time.

The service time is defined as the time that elapses from the start of loading one truck until the excavator is ready to start loading another truck. The transit time is the time required for a truck to leave the excavator and arrive at the back of the queue. Both of these times, in general, will be subject to random fluctuations.

If the average service time for a truck is T_s and the average transit time is T_t, R is defined as the ratio T_t/T_s. The standard deviations are c_sT_s and c_tT_t. Then c_s and c_t are coefficients of variation.

In the simplest theory, a completely deterministic one, $c_s = c_t = 0$, whereas in the queueing theory approach the probability distributions are negative exponential functions and thus $c_s = c_t = 1$. These two situations may be considered as extremes between which any practical situation will lie.

The following analyses illustrate how this subject may be examined.

Constant service time and constant transit time When $c_s = c_t = 0$, the optimum value of N in this deterministic analysis is either the integer immediately below $R + 1$ or the integer immediately above. If R_0 is the highest integer that is less than R, the choice lies between $N = (R_0 + 1)$ and $N + 1$.

When $N = R_0 + 1$, the shovel is idle a fraction $(R - R_0)/(R + 1)$ of the time, so that the F_N value is $(R + 1)/(R_0 + 1)$. With $N + 1$ trucks, the shovel is never idle, and therefore $F_{N+1} = 1$.

From Eq. (10-18) it can be seen that the two systems are equally good if $C_N = C_{N+1}$, or

$$\frac{R+1}{R_0+1}\left(\frac{K_1}{K_2}+N\right)=\frac{K_1}{K_2}+N+1$$

or

$$\frac{K_1}{K_2}=\frac{1-E}{E}(1+R_0) \tag{10-19}$$

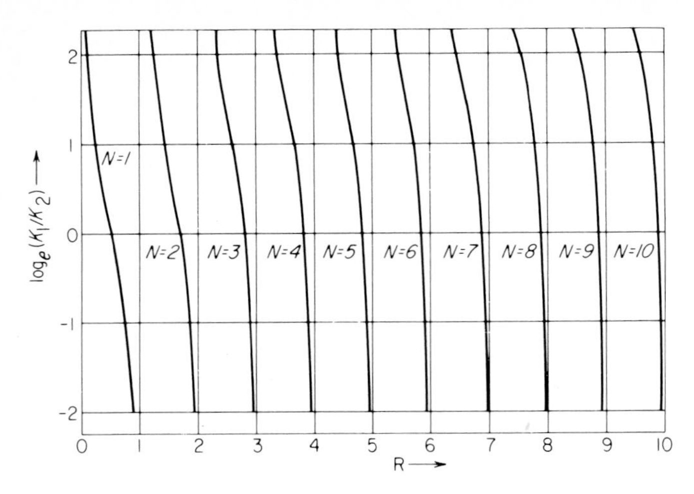

Figure 10-15 Regions of optimal N in the parameter space $\log_e(K_1/K_2)$-R for the condition $c_s = c_t = 0$.

where $E = R - R_0$

The regions of optimal N are shown in Fig. 10-15 in the parameter space which has axes R and K_1/K_2 at right angles.

RANDOM SERVICE TIME AND RANDOM TRANSIT TIME When service time and transit time are random, the distributions of both times are negative exponential functions, and the system is the cyclic queueing system analyzed by Griffis [10]:

$$P_0 = \frac{1}{\sum_{i=0}^{N} \frac{N!}{(N-i)!} \frac{1}{R^i}} \tag{10-20}$$

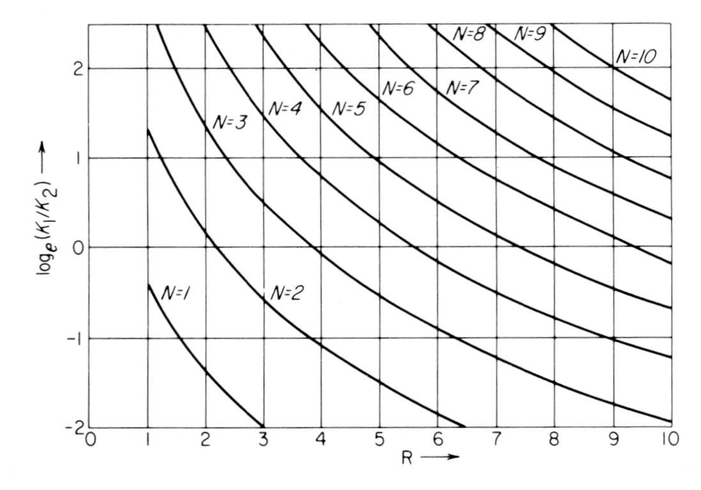

Figure 10-16 Regions of optimal N in the parameter space $\log_e(K_1/K_2)$-R for the condition $c_s = c_t = 0$.

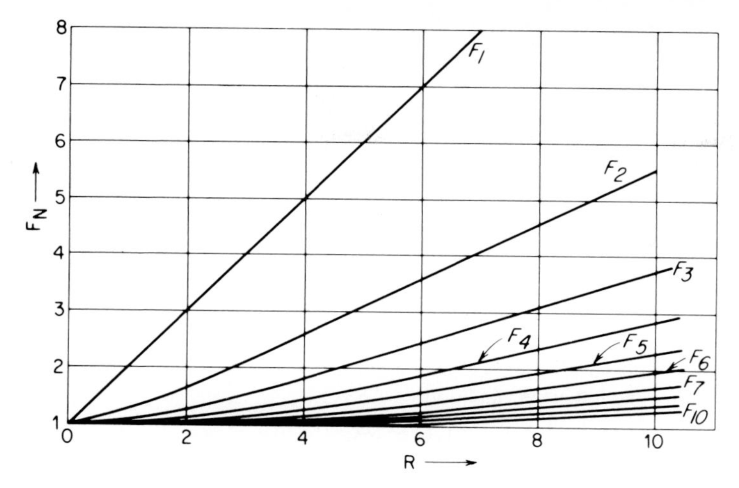

Figure 10-17 Relationship between F_N and R for different values of N_{opt}.

In order to avoid numerical calculations using tables of the cumulative Poisson distribution, the values of F_N have been calculated for various values of R. The critical values of K_1/K_2 have been calculated by the procedure of the previous section. Below a critical value of K_1/K_2, N is the optimal number of trucks, whereas immediately above it, $N + 1$ is better. The results of this are shown in Fig. 10-16.

This representation has the advantage of convenience for an analyst at the site, who does not need to perform any calculations other than those needed to find K_1/K_2 and $R = (T_t/T_s)$. For example, if the ratio of average transit time to average service time is 7 while the ratio of hourly costs is 1.5 ($\log K_1/K_2 = 0.405$), the optimum value of N is read from Fig. 10-16 as 6. From Fig. 10-15 in the deterministic analysis, the choice would have been eight trucks.

Figure 10-17 shows a plot of F_N against R, which reveals that for $R = 7$, $F_6 = 1.50$, whereas for $R = 8$, $F_8 = 1.22$. Because from Eq. (10-18) C is proportional to $F_n(K_1/K_2 + N)$, the difference between C_6 and C_8 (according to the queueing theory calculations) is about 3 percent.

Persons wishing additional information on other conditions related to service time and transit time may refer to the paper presented by Cabrera and Maher [11].

PROBLEMS

10-1 During a motion and time study of the operations of a power shovel, 15 cycle times were observed, with the following results in minutes:

0.52	0.38	0.43	0.46	0.49
0.41	0.36	0.46	0.48	0.43
0.50	0.39	0.41	0.45	0.47

Determine if the average cycle time obtained from these observations is sufficiently accurate for a confidence factor of 0.9 and a confidence interval of 0.03 min.

Use the exact method to determine if sufficient observations have been made. If not, how many additional observations are required?

Use the approximate method to determine the number of observations required.

10-2 Use the observed cycle times listed in Prob. 10-1 and the exact method to determine the number of observations required for each of the following conditions:

$C = 0.9$

$I = 0.02, 0.03, 0.04, 0.05,$ and 0.06 min

10-3 During a motion and time study of trucks hauling earth from a power shovel, the cycle times in minutes for the trucks were as follows:

4.62	4.10	4.96	4.82
4.24	5.12	4.50	4.74
4.68	4.36	4.68	4.88

Use the exact method to determine the number of observations required to obtain an average cycle time for a confidence factor of 0.9 and confidence intervals of 0.15 min, 0.20 min, and 0.30 min.

Make the same determinations using the approximate method.

10-4 Use the theory of queues to determine the optimum number of trucks to use in hauling earth from a power shovel to a fill for the following conditions:

Ideal production for the power shovel, 225 cu yd per hr bm
Operating factor for the power shovel, 0.75
Capacity of trucks, 11 cu yd bm
Average cycle time for trucks, excluding loading time, 0.260 hr
Cost per hr for shovel, including labor, $62.00
Cost per hr for truck and driver, $20.50

Because Table 10-3 gives the values of the Poisson distribution functions for 4.1 trucks only, it will be necessary to use a table of values that apply for the number of trucks needed for this project.

10-5 If the average cycle time for the trucks of Prob. 10-4 is increased to 0.360 hr with no other changes, determine the optimum number of trucks needed.

REFERENCES

1. Ang, A. H. S., and W. H. Tang: *Probability Concepts in Engineering Planning and Design*, Wiley, New York, 1975, 409p.
2. Benjamin, Jack R., and C. Allin Cornell: *Probability, Statistics, and Decision for Civil Engineers*, McGraw-Hill, New York, 1970, 684p.
3. Taha, Hamdy A.: *Operations Research, An Introduction*, Macmillan, New York, 1976, 648p.
4. Taha, Hamdy A.: *Owners Manual for Timelapse Model 1210 Camera*, Timelapse, Inc., 250 Polaris Ave., Mountainview, CA 94043.
5. Parker, Henry C., and Clarkson H. Oglesby: *Methods Improvement for Construction Managers*, McGraw-Hill, New York, 1972.
6. Thomas, H. Randolph, and Jeffrey Daily: Crew Performance Measurement via Activity Sampling, *The Journal of Construction Engineering and Management*, vol. 109, no. 3, September 1983, ASCE.
7. Graduate student project at Texas A&M University, College Station, Tex., 1983, headed by Walter L. Heme.
8. O'Shea, J. B., G. N. Slutkin, and L. R. Shaffer: "An Application of the Theory of Queues to the Forecasting of Shovel-truck Fleet Productions," Department of Civil Engineering, University of Illinois, Urbana, Ill., February 1964.
9. Burington, R. S., and D. C. May, Jr.: *Handbook of Probability and Statistics with Tables*. 1st ed., Handbook Publishers, Inc., Sandusky, Ohio, 1953.
10. Griffis, Fletcher H.: Optimizing Haul Fleet Size Using Queueing Theory, *Proceedings ASCE*, vol. 5753, no. CO1, pp. 75–88, January 1968.
11. Cabrera, J. G., and M. J. Maher: "Optimizing Earthmoving Plant: Solution for the Excavator-Truck System," Highway Research Board No. 454, pp. 7–15, 1973.

CHAPTER
ELEVEN
BELT-CONVEYOR SYSTEMS

GENERAL INFORMATION

Belt-conveyor systems are used extensively in the field of construction, where they frequently provide the most satisfactory and economical method of handling and transporting materials, such as earth, sand, gravel, crushed stone, mine ores, cement, concrete, etc. Because of the continuous flow of materials at relatively high speeds, belt conveyors have high capacities.

The essential parts of a belt-conveyor system include a continuous belt, idlers, a driving unit, driving and tail pulleys, take-up equipment, and a supporting structure. Additional accessories, as described later, may be included when desirable or necessary.

A conveyor for transporting materials a short distance may be a portable unit or a fixed installation. Figure 11-1 illustrates a portable conveyor used to stockpile aggregate which is delivered by trucks. This machine is available in lengths of 33 to 60 ft, with belt widths of 18, 24, and 30 in. It is self-powered with a gasoline-engine drive through a shaft and gearbox to the driving pulley. The operating features include swivel wheels, V-type truck, hydraulic hoist, low-mast height, and antifriction bearings throughout.

When a belt-conveyor system is used to transport materials a considerable distance, up to several miles in some instances, the system should consist of a number of different flights, as there is a limit to the maximum length of a belt. Each flight is a complete conveyor unit which discharges its load onto the tail end of the succeeding unit. Such a system will operate over any terrain

Figure 11-1 Portable belt conveyor.

Figure 11-2 (*a*) Troughing idlers installed. (*b*) Belt transporting aggregate.

provided the slopes do not exceed those for which the given material may be transported.

The limestone rock for the Bull Shoals Dam, whose maximum size was 6 in., was transported 7 miles from the primary crushing plant at the quarry to the dam site. The conveyor system consisted of 21 flights, varying in length from 600 to 2,800 ft, each powered with a 100-hp electric motor. The belts, which were 30 in. wide, were operated at a speed of 525 fpm to deliver 350 cu yd of material per hour. The entire system required 14,000 idlers, which were supported primarily by wood structures. Figure 11-2 illustrates troughing idlers installed on a project and an operating belt transporting aggregate.

THE ECONOMY OF TRANSPORTING MATERIALS WITH A BELT CONVEYOR

One of the first questions that arises in considering the use of a belt conveyor is whether this method of transportation is the most dependable and economical when compared with other methods. The proper way to answer this question is to estimate the cost of transporting the material by each method under consideration using the methods described in Chap. 3. Assume that a belt conveyor is to be compared with trucks for hauling aggregate for a large concrete project.

The net total cost of the conveyor system will include the installed cost of the system, an access road for installing and servicing the system, maintenance, replacements, and repairs, fuel, or electrical energy, and labor, less the net salvage value of the system upon completion of its use. Interest on the investment, plus taxes and insurance, if they apply, should be included. Likewise, any cost of obtaining a right of way for the system should be included. The unit cost of moving the material, per ton or cubic yard, may be obtained by dividing the net total cost of the system by the number of units to be transported.

The cost of transporting the materials by truck will include the cost of constructing and maintaining a haul road, plus the cost of operating the trucks. The unit cost of moving the materials may be obtained by dividing the net total cost by the number of units to be transported.

If either method requires additional handling costs at the source or at the destination, these costs should be included prior to determining the unit cost of moving the materials.

In the construction of the Bull Shoals Dam more than 4,500,000 tons of aggregate was transported on belt conveyors at a reported cost of $0.045 per ton-mile. It was estimated that the contractors saved $560,000 on the purchase and installation of the conveyor system compared with a fleet of trucks, plus a haul road and incidentals required for the trucks. In addition, it was estimated that there was a saving of $375,000 on labor operating the system compared with trucks.

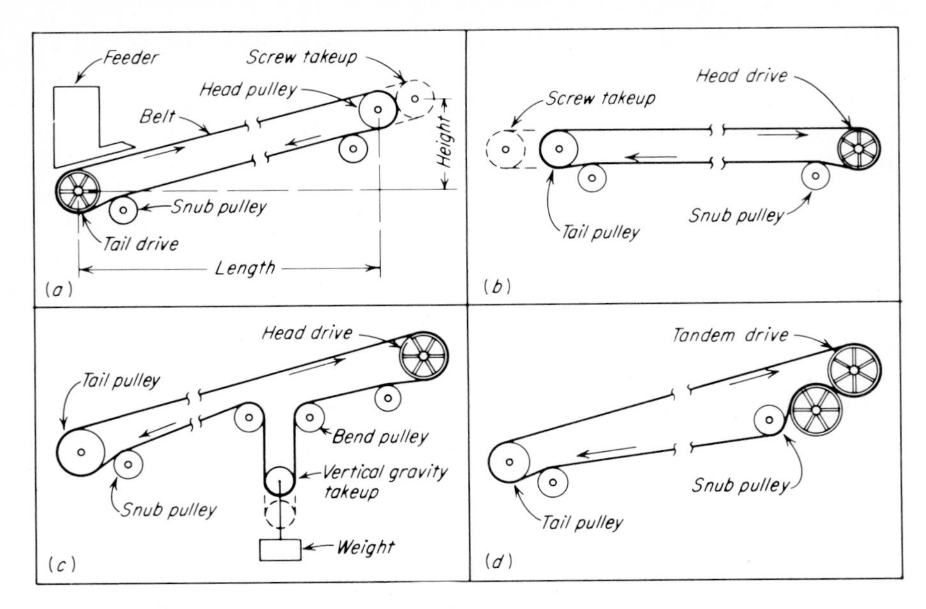

Figure 11-3 Representative belt-conveyor systems.

REPRESENTATIVE BELT-CONVEYOR SYSTEMS

Figure 11-3 illustrates four belt-conveyor systems based on the location of the drive pulley, the number of drive pulleys, and the take-up method of maintaining the necessary tension in the belt.

CONVEYOR BELTS

The belt is the moving and supporting surface on which the material is transported. Many types, sizes, and grades are available, from which the most suitable belt for a given service may be selected.

Belts are manufactured by joining several layers or plies of woven cotton duck into a carcass which provides the necessary strength to resist the tension in the belt. The layers are covered with an adhesive which combines them into a unified structure. Special types of reinforcing, such as rayon, nylon, and steel cables, are employed sometimes to increase the strength of a belt. A measure of the strength of a belt is indicated by the number and weight of the several layers of fabric. The number of layers is expressed as 4-, 6-, 7-, 8-, etc., ply. The weight of each layer of fabric is expressed as 28-, 32-, 36-, 42-, etc., oz, the number indicating the weight of a piece of duck 42 in. wide and 36 in. long. The width of a belt is expressed in inches. Thus, a belt might be specified as a 36-in.-wide 6-ply 42-oz belt.

The top and bottom surfaces of a belt are covered with rubber to protect the carcass from abrasion and injury from the impact at loading. Various thicknesses of covers may be specified. Figure 11-4 illustrates cross sections of belts having different types of construction.

It is necessary to select a belt with sufficient strength to resist the maximum tension to which it will be subjected, as determined by methods which will be developed later.

Also, it is necessary to select a belt that is wide enough to transport the material at the required rate. Most belts used on construction projects travel over troughing rollers to increase the carrying capacities. The number of tons that can be transported in an hour is determined by using Eq. (11-1).

$$T = \frac{60ASW}{2,000} \tag{11-1}$$

where T = weight of material, tons per hr
A = cross-sectional area of material, sq ft
S = speed of the belt, ft per min
W = weight of material, lb per cu ft

The area of the cross section will depend on the width of the belt, the depth of troughing, the angle of repose for the material, and the extent to which the belt is loaded to capacity. Figure 11-5 illustrates how the cross-section area may vary with the width of a belt and the angle of repose for the material. In the figure the troughing idlers are set at an angle of 20° above the

Figure 11-4 Types of conveyor-belt construction. (*a*) Standard. (*b*) Shock pad. (*c*) Stepped pad. (*d*) Stepped ply.

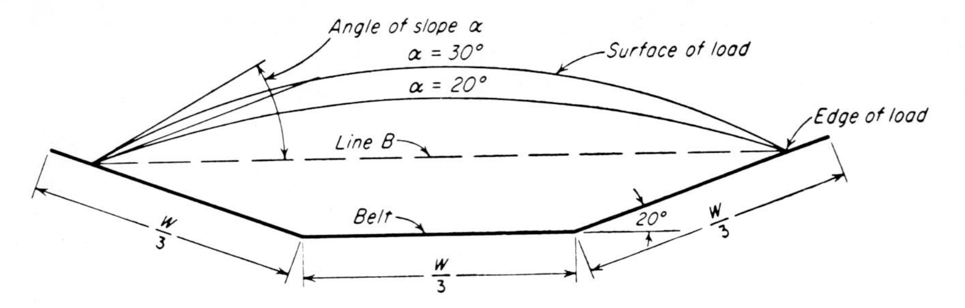

Figure 11-5 Cross-section area of a load on a conveyor belt.

horizontal. In order to eliminate side spillage, it is assumed that materials will not be placed closer than $0.05W + 1$ in. from the sides of the belt, where W is the width of the belt in inches. It is assumed that the top surface of the material will be an arc of a circle. Table 11-1 gives the cross-sectional areas for various belt widths and loading conditions. These areas are subject to variation and should not be considered as exact unless the loading conditions are as stated. The area of surcharge is the area above line B of Fig. 11-5.

The carrying capacity of a 42-in. belt, moving 100 fpm, loaded with sand weighing 100 lb per cu ft, with a 20° angle of repose, will be 100 fpm × 100 lb × 1.115 sq ft × 60 min ÷ 2,000 lb per ton = 334.5 tons per hr. The carrying capacity of this belt for other speeds may be obtained by multiplying 334.5 by the ratio of the speed of the two belts.

Table 11-2 gives the approximate carrying capacities of troughed conveyor belts, in tons per hour, for various widths and materials for a speed of 100 fpm. Table 11-3 gives the suggested maximum speeds which are considered good practice for conveyor belts of different widths when handling various kinds of materials. Table 11-4 gives representative allowable working tensions in duck

Table 11-1 Areas of cross sections of materials for loaded belts

Width of belt, in.	$0.05W + 1$, in.	Area of level load, sq ft	Area of surcharge, sq ft, for angle of repose, deg			Total area, sq ft, for angle of repose, deg		
			10	20	30	10	20	30
16	1.8	0.072	0.029	0.059	0.090	0.101	0.131	0.162
18	1.9	0.096	0.038	0.078	0.118	0.134	0.174	0.214
20	2.0	0.122	0.048	0.098	0.150	0.170	0.220	0.272
24	2.2	0.185	0.072	0.146	0.225	0.257	0.331	0.410
30	2.5	0.303	0.118	0.238	0.365	0.421	0.541	0.668
36	2.8	0.450	0.174	0.351	0.540	0.624	0.801	0.990
42	3.1	0.627	0.241	0.488	0.749	0.868	1.115	1.376
48	3.4	0.833	0.321	0.649	0.992	1.154	1.482	1.825
54	3.7	1.068	0.408	0.826	1.264	1.476	1.894	2.332
60	4.0	1.333	0.510	1.027	1.575	1.843	2.360	2.908

Table 11-2 Carrying capacities of troughed conveyor belts, in tons per hour for a speed of 100 fpm*

Width of belt, in.	Max lumps: Sized, in.	Max lumps: Unsized, in.	Weight of material, lb per cu ft: 30	50	90	100	125	150	160	180	200
14	2	$2\frac{1}{2}$	9	15	28	31	39	46	49	56	62
16	$2\frac{1}{2}$	3	13	21	38	42	52	63	67	75	83
18	3	4	16	27	48	54	67	81	86	97	107
20	$3\frac{1}{2}$	5	20	33	60	67	83	100	107	120	133
24	$4\frac{1}{2}$	8	30	50	90	100	125	150	160	180	200
30	7	14	47	79	142	158	197	236	252	284	315
36	9	18	70	117	210	234	292	351	374	421	467
42	11	20	100	167	300	333	417	500	534	600	667
48	14	24	138	230	414	460	575	690	736	828	920
54	15	28	178	297	534	593	741	890	948	1,070	1,190
60	16	30	222	369	664	738	922	1,110	1,180	1,330	1,480

* Courtesy Hewitt-Robins.

belts for various thicknesses and widths. The pulley diameter is the minimum size that should be used for the indicated service.

IDLERS

Idlers provide the supports for a belt conveyor. For the load-carrying portion of a belt the idlers are designed to provide the necessary troughing, while for

Table 11-3 Maximum speeds of conveyor belts, in fpm*

Kind and condition of material handled	Width of belt, in.: 14	16	18	20	24	30	36	42	48	54	60
Unsized coal, gravel, stone, ashes, ore, or similar material	300	300	350	350	400	450	500	550	600	600	600
Sized coal, coke, or other breakable material	250	250	250	300	300	350	350	400	400	400	400
Wet or dry sand	400	400	500	600	600	700	800	800	800	800	800
Crushed coke, crushed slag, or other fine abrasive material	250	250	300	400	400	500	500	500	500	500	500
Large lump ore, rock, slag, or other large abrasive material	...	...	...	...	350	350	400	400	400	400	400

* Courtesy Hewitt-Robins.

Table 11-4 Allowable working tension and pulley diameter for conveyor belts*

No. plies	Weight per ply, oz	Width of belt, in. 16	18	20	24	30	36	42	48	Diameter of pulley, in. Head, drive, tripper	Tail, take-up, snub	Bend
3	32	1,440	1,620	.....	.....	.....	.....	.....	.....	16	12	12
3	36	.....	.....	1,800	2,160	.....	.....	.....	.....	20	16	12
3	42	.....	.....	2,200	2,640	3,300	.....	.....	.....	20	16	12
3	48	.....	.....	.....	3,840	.....	.....	.....	.....	24	20	16
4	28	1,600	1,800	2,000	2,400	3,000	.....	.....	.....	20	16	12
4	32	1,920	2,160	2,400	2,880	3,600	4,320	.....	.....	20	16	12
4	36	.....	.....	2,600	3,120	3,900	4,680	.....	.....	24	20	16
4	42	.....	.....	.....	.....	4,800	5,760	6,720	.....	24	20	20
4	48	.....	.....	.....	.....	6,450	7,750	9,020	.....	30	24	20
5	28	2,000	2,250	2,500	3,000	3,750	4,500	.....	.....	24	20	16
5	32	.....	2,700	3,000	3,480	4,500	5,400	.....	.....	24	20	16
5	36	.....	.....	3,400	4,080	5,100	6,120	7,140	.....	30	24	20
5	42	.....	.....	.....	.....	6,600	7,920	9,240	10,560	30	24	20
5	48	.....	.....	.....	.....	8,700	10,400	12,180	13,920	36	30	24
6	28	.....	.....	3,000	3,600	4,500	5,400	.....	.....	30	24	20
6	32	.....	.....	.....	4,320	5,400	6,480	7,560	.....	30	24	20
6	36	.....	.....	.....	.....	6,300	7,560	8,820	10,080	36	30	24
6	42	.....	.....	.....	.....	.....	9,720	11,340	12,900	36	30	24
6	48	.....	.....	.....	.....	.....	13,000	15,120	17,300	42	36	30
7	28	.....	.....	.....	.....	5,250	6,300	.....	.....	36	30	24
7	32	.....	.....	.....	.....	6,300	7,560	8,820	10,080	36	30	24
7	36	.....	.....	.....	.....	.....	8,820	10,300	11,780	42	36	30
7	42	.....	.....	.....	.....	.....	.....	13,200	15,140	42	36	30
7	48	.....	.....	.....	.....	.....	.....	17,640	20,180	48	42	36
8	32	.....	.....	.....	.....	.....	8,640	10,080	11,520	42	30	24
8	36	.....	.....	.....	.....	.....	.....	11,760	13,450	48	42	30
8	42	.....	.....	.....	.....	.....	.....	.....	17,300	48	42	30
8	48	.....	.....	.....	.....	.....	.....	.....	23,050	54	48	42
9	32	.....	.....	.....	.....	.....	.....	11,340	12,900	48	36	30
9	36	.....	.....	.....	.....	.....	.....	13,200	15,140	54	48	36

* Courtesy Hewitt-Robins.

the return portion of a belt the idlers provide flat supports. The essential parts of a troughing idler include the rolls, brackets, and base. Antifriction bearings are generally used in idlers, with high-pressure grease fittings to permit periodic lubrication of the bearings. The rolls may be made of steel tubing or cast iron, either plain or covered with a composition, such as rubber, where it is necessary to protect a belt against damage due to impact. The diameters of the rolls most commonly used are 4, 5, 6, and 7 in. Large-diameter rolls give lower friction and better belt protection, especially when the load includes large lumps of material. Figure 11-6 illustrates troughing and return idlers.

Figure 11-6 Belt idlers. (*a*) Heavy-duty troughing, (*b*) Return.

Spacing of idlers. Troughing idlers should be spaced close enough to prevent excessive deflection of the loaded belt between the idlers. As indicated in Table 11-5, the maximum spacing will vary with the width of the belt and the weight of the load carried. The idler spacing should be reduced at the point where the load is fed onto the belt.

Table 11-5 Recommended maximum spacing of troughing idlers*

Width of belt, in.	Weight of material, lb per cu ft		
	30–70	70–120	120–150
14	5 ft 6 in.	5 ft 0 in.	4 ft 9 in.
16	5 ft 6 in.	5 ft 0 in.	4 ft 9 in.
18	5 ft 6 in.	5 ft 0 in.	4 ft 9 in.
20	5 ft 6 in.	5 ft 0 in.	4 ft 9 in.
24	5 ft 6 in.	5 ft 0 in.	4 ft 9 in.
30	5 ft 0 in.	4 ft 6 in.	4 ft 3 in.
36	5 ft 0 in.	4 ft 6 in.	4 ft 3 in.
42	4 ft 6 in.	4 ft 0 in.	3 ft 9 in.
48	4 ft 0 in.	3 ft 3 in.	3 ft 0 in.
24	4 ft 0 in.	2 ft 9 in.	2 ft 6 in.
60	4 ft 0 in.	2 ft 3 in.	2 ft 0 in.

* Courtesy of Hewitt-Robins.

As the sole function of the return idlers is to support the empty belt, the spacing can be increased to approximately 10 ft.

Training idlers. Sometimes a conveyor belt is operated under conditions which make it difficult to keep the belt centered on the troughing idlers. If the conditions cannot be corrected sufficiently to keep the belt centered, it may be necessary to install training idlers, spaced 50 to 60 ft apart. Figure 11-7 illustrates a set of training idlers.

Idler friction. In analyzing a belt conveyor to determine the horsepower required, it is necessary to include the power needed by the idlers. This power will depend on the type and size idler, the kind of bearings, the weight of the revolving parts, the weight of the belt, and the weight of the load. Table 11-6 gives representative friction factors for idlers equipped with antifriction bearings. Manufacturers of idlers will furnish information giving the weights of the revolving parts of their idlers.

The information in Table 11-6 is used as in the following example.

Example 11-1 Consider a conveyor 100 ft long, with a 5-ply 32-oz 30-in.-wide belt weighing 6.8 lb per ft. The load will weigh 100 lb per cu ft, or 54 lb per foot of conveyor. The revolving

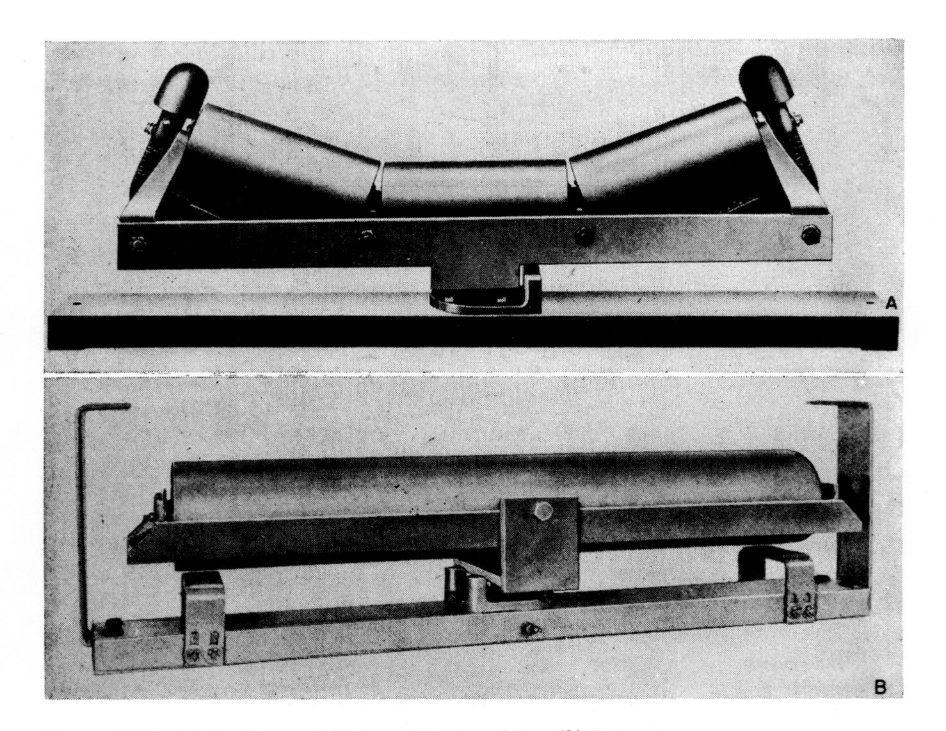

Figure 11-7 Training idlers. (*a*) Reversible troughing. (*b*) Return.

Table 11-6 Friction factors for conveyor-belt idlers equipped with antifriction bearings*

Diameter of idler pulley, in.	Friction factor
4	0.0375
5	0.036
6	0.030
7	0.025

* Courtesy Hewitt-Robins.

parts will weigh 50 lb for a troughing idler and 31 lb for a return idler. Both idlers are 6 in. in diameter.

From Table 11-6 the idler friction factor is 0.030.

No. troughing idlers required, $100 \div 4.5$	= 22
Add extra idlers at loading point	= 3
Total no. troughing idlers	= 25
No. of return idlers, $100 \div 10$	= 10

Total weight of the revolving parts of idlers will be

Troughing, 25×50	= 1,250 lb
Return, 10×31	= 310 lb
Weight of belt, 200×6.8	= 1,360 lb
Weight of load, 100×54	= 5,400 lb
Total weight	= 8,320 lb

The force required to overcome idler friction, $8{,}320 \times 0.03 = 249.6$ lb.
For a belt speed of 100 fpm the energy required per min will be

$$100 \times 249.6 = 24{,}960 \text{ ft-lb}$$

The horsepower required to overcome idler friction will be

$$P = \frac{24{,}960 \text{ ft-lb per min}}{33{,}000 \text{ ft-lb per min per hp}} = 0.76$$

For other belt speeds the required horsepower will be

$$P = \frac{0.76 \times \text{speed, fpm}}{100}$$

POWER REQUIRED TO DRIVE A BELT CONVEYOR

The total external power required to drive a loaded belt conveyor is the algebraic sum of the power required by each of the following:

1. To move the empty belt over the idlers
2. To move the load horizontally

3. To lift or lower the load vertically
4. To turn all pulleys
5. To compensate for drive losses
6. To operate a tripper, if one is used

The power required for each of these operations can be determined with reasonable accuracy for any given conveyor system, as explained later.

POWER REQUIRED TO MOVE AN EMPTY BELT

The power required to move an empty conveyor belt over the idlers will vary with the type of idler bearings, the diameter and spacing of the idlers, and the length, weight, and speed of the belt. The energy required to move an empty belt is given by the equation

$$E = LSCQ \tag{11-2}$$

where E = energy, ft-lb per min
L = length of conveyor, ft
S = belt speed, fpm
C = idler-friction factor, from Table 11-6
Q = weight of moving parts per foot of conveyor

Table 11-7 Representative values of Q**

	Idlers, 5-in.-diameter, steel pulleys					Weight of conveyor, lb per ft			
	Troughing		Return			Idlers			
Width of belt, in.	Weight of revolving parts, lb	Spacing	Weight of revolving parts, lb	Spacing	Weight of belt, lb per ft	Troughing	Return	Belt	Q, lb per ft
14	18	5′0″	9	10′0″	2.8	3.6	0.9	5.6	10.1
16	20	5′0″	11	10′0″	3.3	4.0	1.1	6.6	11.7
18	22	5′0″	12	10′0″	4.1	4.4	1.2	8.2	13.8
20	24	5′0″	14	10′0″	4.6	4.8	1.4	9.2	15.4
24	26	5′0″	17	10′0″	7.0	5.2	1.7	14.0	20.9
30	31	4′6″	21	10′0″	8.5	6.9	2.1	17.0	26.0
36	36	4′6″	25	10′0″	11.3	8.0	2.5	22.6	33.1
42	40	4′0″	29	10′0″	17.0	10.0	2.9	34.0	46.0
48	45	3′3″	34	10′0″	23.8	13.8	3.4	47.6	64.8
54	74	2′9″	54	10′0″	29.2	26.9	5.4	73.2	105.5
60	80	2′3″	60	10′0″	32.5	35.6	6.0	74.0	115.6

* Courtesy Hewitt-Robins.

Equation (11-2) may be expressed as horsepower by dividing by 33,000, to give

$$P = \frac{LSCQ}{33{,}000} \tag{11-3}$$

Representative values of Q are given in Table 11-7. If more accurate values are desired for a given conveyor, they may be determined from the design of the particular conveyor and the weight of the belt used.

Example 11-2 The use of Eq. (11-3) is illustrated by determining the horsepower required to move a 30-in.-wide belt on a conveyor whose length is 1,800 ft, equipped with 5-in.-diameter idler pulleys, with antifriction bearings. Assume a belt speed of 100 fpm.

From Table 11-6 the value of C will be 0.036.

Table 11-8 Horsepower required to move empty conveyor belts for a speed of 100 fpm*

Length of conveyor, ft	Width of belt, in.										
	14	16	18	20	24	30	36	42	48	54	60
50	0.05	0.06	0.07	0.08	0.11	0.14	0.18	0.25	0.35	0.54	0.63
100	0.11	0.13	0.15	0.17	0.23	0.28	0.36	0.51	0.70	1.14	1.25
150	0.16	0.19	0.22	0.25	0.34	0.42	0.53	0.76	1.05	1.71	1.88
200	0.22	0.25	0.30	0.33	0.45	0.56	0.71	1.01	1.40	2.28	2.50
250	0.27	0.32	0.37	0.42	0.56	0.70	0.89	1.27	1.75	2.85	3.13
300	0.33	0.38	0.45	0.50	0.68	0.84	1.07	1.52	2.10	3.42	3.76
400	....	....	0.60	0.66	0.90	1.12	1.43	2.03	2.80	4.56	5.01
500	....	....	....	0.83	1.13	1.40	1.79	2.53	3.50	5.70	6.26
600	....	....	....	1.00	1.35	1.68	2.14	3.04	4.20	6.84	7.51
800	....	....	....	....	1.80	2.25	2.86	4.05	5.60	9.12	10.00
1,000	....	....	....	....	2.26	2.81	3.57	5.07	7.00	11.40	12.50
1,200	....	....	....	....	....	3.37	4.29	6.08	8.40	13.70	15.00
1,400	....	....	....	....	....	3.93	5.00	7.09	9.80	16.00	17.50
1,600	....	....	....	....	....	4.49	5.72	8.10	11.20	18.30	20.10
1,800	....	....	....	....	....	5.05	6.43	9.12	12.60	20.50	22.60
2,000	....	....	....	....	....	5.62	7.15	10.10	14.00	22.80	24.90
2,200	....	....	....	....	....	....	7.86	11.10	15.40	25.10	27.60
2,400	....	....	....	....	....	....	8.58	12.20	16.80	27.40	30.10
2,600	....	....	....	....	....	....	9.29	13.20	18.20	29.60	32.60
2,800	....	....	....	....	....	....	10.00	14.20	19.60	31.90	35.00
3,000	....	....	....	....	....	....	10.70	15.20	21.00	34.20	37.60

* Courtesy Hewitt-Robins.

The power values given in this table are based on the use of 5-in.-diameter idlers. For 4-in.-diameter idlers increase the values by 4 percent. For 6-in.-diameter idlers decrease the values by 17 percent.

From Table 11-7 the value of Q will be 26 lb per foot of conveyor length. The power required to move the empty belt will be

$$P = \frac{1{,}800 \times 100 \times 0.036 \times 26}{33{,}000} = 5.10 \text{ hp}$$

Table 11-8 gives representative values for the horsepower required to move empty conveyor belts. The values are based on using 5-in.-diameter idlers with antifriction bearings, and the belt widths given in Table 11-7.

POWER REQUIRED TO MOVE A LOAD HORIZONTALLY

The power required to move a load horizontally may be expressed by Eq. (11-4) if Q is replaced by W, the weight of the load in pounds per foot of belt.

$$P = \frac{LSCW}{33{,}000} \tag{11-4}$$

Table 11-9 Horsepower required to move loads horizontally on conveyor belts*

Length of conveyor, ft	Load, tons per hr													
	50	100	150	200	250	300	350	400	500	600	700	800	900	1,000
50	0.09	0.18	0.27	0.36	0.46	0.55	0.64	0.73	0.91	1.1	1.3	1.5	1.6	1.8
100	0.18	0.36	0.55	0.74	0.91	1.1	1.3	1.5	1.8	2.2	2.6	2.9	3.3	3.6
150	0.27	0.55	0.82	1.1	1.4	1.6	1.9	2.2	2.7	3.3	3.8	4.4	4.9	5.5
200	0.36	0.73	1.1	1.5	1.8	2.2	2.6	2.9	3.6	4.4	5.1	5.8	6.6	7.3
250	0.46	0.91	1.4	1.8	2.3	2.7	3.2	3.6	4.6	5.5	6.4	7.3	8.2	9.1
300	0.55	1.1	1.6	2.2	2.7	3.3	3.8	4.4	5.5	6.6	7.7	8.8	9.9	10.9
400	0.73	1.5	2.2	2.9	3.6	4.4	5.1	5.8	7.3	8.7	10.2	11.6	13.1	14.6
500	0.91	1.8	2.7	3.6	4.6	5.5	6.4	7.3	9.1	10.9	12.7	14.5	16.4	18.2
600	1.10	2.1	3.2	4.2	5.3	6.4	7.4	8.5	10.6	12.7	14.8	17.0	19.1	21.0
800	1.40	2.7	4.1	5.5	6.8	8.2	9.5	10.8	13.7	16.4	19.1	22.0	25.0	27.0
1,000	1.70	3.3	5.0	6.7	8.3	10.0	11.7	13.3	16.7	20.0	23.0	27.0	30.0	33.0
1,200	2.0	3.9	5.9	7.9	9.8	11.8	13.8	15.7	19.8	24.0	28.0	32.0	36.0	39.0
1,400	2.3	4.5	6.8	9.1	11.4	13.7	15.9	18.1	23.0	27.0	32.0	36.0	41.0	45.0
1,600	2.6	5.2	7.7	10.3	12.9	15.5	18	21	26	31	36	41	46	52
1,800	2.9	5.8	8.7	11.5	14.4	17.3	20	23	28	35	40	46	52	58
2,000	3.2	6.4	9.6	12.7	15.9	19.1	22	25	32	38	45	51	57	64
2,200	3.5	7.0	10.5	13.9	17.4	21.0	24	28	35	42	49	56	63	70
2,400	3.9	7.6	11.4	15.2	18.9	23.0	27	30	38	46	53	61	68	76
2,600	4.1	8.2	12.3	16.4	20.0	25.0	29	33	41	49	57	65	74	82
2,800	4.4	8.8	13.2	17.6	22.0	26.0	31	35	44	53	62	70	79	88
3,000	4.7	9.4	14.1	18.8	23.0	28.0	33	37	47	56	66	75	85	94

* Courtesy Hewitt-Robins.

The power values given in this table are based on the use of 5-in.-diameter idlers. For 4-in.-diameter idlers increase the values by 4 percent.

This equation may be expressed in terms of the load moved in tons per hour. Let

$$T = \text{tons material moved per hr}$$
$$SW = \text{lb material moved per min}$$
$$60SW = \text{lb material moved per hr}$$

$$T = \frac{60SW}{2{,}000} = \frac{3SW}{100}$$

Solving, we get

$$SW = \frac{100T}{3} \qquad (11\text{-}5)$$

Substituting this value of SW in Eq. (11-4), we determine that the horsepower required to move a load horizontally is

$$P = \frac{100LCT}{3 \times 33{,}000} = \frac{LCT}{990} \qquad (11\text{-}6)$$

Table 11-9 gives values for the horsepower required to move loads horizontally on conveyor belts. The values are based on using 5-in.-diameter idlers with antifriction bearings. For 6-in.-diameter idlers, decrease the values by 17 percent.

POWER REQUIRED TO MOVE A LOAD UP AN INCLINED BELT CONVEYOR

When a load is moved up an inclined belt conveyor, the power required may be divided into two components: the power required to move the load horizontally and the power required to lift the load through the net change in elevation. The power required to move the load horizontally may be determined from Eq. (11-6). The power required to lift the load through the net change in elevation may be determined as follows: Let

$$H = \text{net change in elevation, ft}$$
$$T = \text{tons material per hr}$$

From Eq. (11-5),

$$\frac{100T}{3} = \text{lb material per min}$$

$$\frac{100TH}{3} = \text{energy, ft-lb per min}$$

Dividing by 33,000 gives the horsepower,

$$P = \frac{100TH}{3 \times 33{,}000} = \frac{TH}{990} \qquad (11\text{-}7)$$

Table 11-10 Horsepower required to lift a load*

Net lift, ft	Load, tons per hr											
	50	100	150	200	250	300	350	400	500	600	800	1,000
5	0.3	0.5	0.8	1.0	1.3	1.5	1.8	2.0	2.5	3.0	4.0	5.1
10	0.5	1.0	1.5	2.0	2.5	3.0	3.5	4.0	5.1	6.1	8.1	10.0
15	0.8	1.5	2.3	3.0	3.8	4.5	5.3	6.1	7.6	9.1	12.0	15.0
20	1.0	2.0	3.0	4.0	5.1	6.1	7.1	8.1	10.0	12.0	16.0	20.0
25	1.3	2.5	3.8	5.1	6.3	7.6	8.8	10.0	13.0	15.0	20.0	25.0
30	1.5	3.0	4.5	6.1	7.6	9.1	11.0	12.0	15.0	18.0	24.0	30.0
40	2.0	4.0	6.1	8.1	10.0	12.0	14.0	16.0	20.0	24.0	32.0	40.0
50	2.5	5.1	7.6	10.0	13.0	15.0	18.0	20.0	25.0	30.0	40.0	51.0
75	3.8	7.6	11.0	15.0	19.0	23.0	27.0	30.0	38.0	45.0	61.0	76.0
100	5.1	10.0	15.0	20.0	25.0	30.0	35.0	40.0	51.0	61.0	81.0	101
125	6.3	13.0	19.0	25.0	32.0	38.0	44.0	51.0	63.0	76.0	101	126
150	7.6	15.0	23.0	30.0	38.0	45.0	53.0	61.0	76.0	91.0	121	152
200	10.0	20.0	30.0	40.0	51.0	61.0	71.0	81.0	101	121	162	202
300	15.0	30.0	45.0	61.0	76.0	91.0	106	121	152	185	242	303
400	20.0	40.0	61.0	81.0	101	121	141	162	202	242	323	404
500	25.0	51.0	76.0	101	126	151	177	202	252	303	404	505

* Courtesy Hewitt-Robins.

If the load is moved up an inclined conveyor, the power given in Eq. (11-7) must be supplied from an outside source. If the load is moved down an inclined conveyor, the power will be supplied to the belt by the load.

DRIVING EQUIPMENT

A belt conveyor may be driven through the head or tail pulley or through an intermediate pulley. In the event high driving forces are required, it may be necessary to use more than one pulley, with the pulleys arranged in tandem to increase the areas of contact with the belt. Smooth-faced or lagged pulleys may be used, depending on the desired coefficient of friction between the belt and the pulley surface. The pulley may be driven by an electric motor or a gasoline or diesel engine. It is usually necessary to install a suitable speed reducer, such as gears, chain drives, or belt drives, between the power unit and the driving pulley. The power loss in the speed reducer should be included in determining the total power required to drive a belt conveyor. This loss may amount to 5 to 10 percent or more, depending on the type of speed reducer.

The coefficient of friction between a steel shaft and babitted bearings will be approximately 0.10.

When the power is transmitted from a driving pulley to a belt, the effective driving force, which is transmitted to the belt, is equal to the tension in the tight side less the tension in the slack side of the belt, expressed in pounds.

$$T_e = T_1 - T_2 \tag{11-8}$$

where T_e = effective tension or driving force between pulley and belt
T_1 = tension in tight side of belt
T_2 = tension in slack side of belt

The coefficient of friction between a rubber belt and a bare steel or cast-iron pulley is approximately 0.25. If the surface of a pulley is lagged with a rubberized fabric, the coefficient of friction will be increased to approximately 0.35.

When power is transmitted from a pulley to a belt, the tension in the slack side of the belt should not exceed the amount required to prevent slippage between the pulley and the belt. For a driving pulley with a given diameter and speed, the effective tension T_e required to transmit a given horsepower to the belt may be determined from the following equation,

$$P = \frac{\pi D T_e N}{33{,}000} \tag{11-9}$$

where P = hp transmitted to belt
D = diameter of pulley, ft
T_e = effective force between pulley and belt, lb
N = rpm

The equation may be rewritten as

Table 11-11 Tension factors for driving pulleys*

Arc of contact, deg	Bare pulley	Lagged pulley
Single-pulley drive		
200	1.72	1.42
210	1.70	1.40
215	1.65	1.38
220	1.62	1.35
240	1.54	1.30
Tandem drive		
360	1.26	1.13
380	1.23	1.11
400	1.21	1.10
450	1.18	1.09
500	1.14	1.06

* Courtesy Hewitt-Robins.

$$T_e = \frac{33{,}000P}{\pi DN} \tag{11-10}$$

The ratio T_1/T_e is defined as the pulley tension factor. This factor varies with the type of pulley surface, bare or lagged, and the arc of contact between the belt and the pulley. Values for the factor are given in Table 11-11. The factor may be expressed as

$$F = \frac{T_1}{T_e} \tag{11-11}$$

If the required effective force T_e between a pulley and a belt whose arc of contact is 210° is 3,000 lb, the minimum tension in the tight side of the belt may be determined from Eq. (11-11) and Table 11-11. From Table 11-11, $F = 1.70$ for a bare pulley.

$$T_1 = FT_e$$
$$= 1.70 \times 3{,}000 = 5{,}100 \text{ lb}$$

If the same pulley is lagged, the value of F will be 1.40 and

$$T_1 = 1.40 \times 3{,}000 = 4{,}200 \text{ lb}$$

For these conditions the minimum values of T_1, T_2, and T_e will be as follows:

	Bare pulley	Lagged pulley
T_1	5,100 lb	4,200 lb
T_e	3,000 lb	3,000 lb
T_2	2,100 lb	1,200 lb

Thus, it is evident that by lagging a drive pulley the tension in a belt may be reduced, possibly enough to permit the use of a lighter and less expensive belt.

POWER REQUIRED TO TURN PULLEYS

A belt conveyor includes several pulleys, around which the belt is bent. For the shaft of each pulley there is a bearing friction that requires the consumption of power. The power required will vary with the tension in the belt, the weight of the pulley and shaft, and the type of bearing, babbitted or antifriction. For a given conveyor the friction factors for each pulley may be determined reasonably accurately, and from this information the additional power required to

Table 11-12 Percent of shaft horsepower required to overcome pulley friction for conveyors with head drive and babbitted bearings*

Length of conveyor, ft	Slope of conveyor, %				
	0	2-10	10-19	19-29	29-36
20	112	93	53	35	28
30	76	63	36	25	19
50	45	38	22	15	13
75	30	25	15	12	9
100	22	19	11	8	7
150	15	14	9	7	6
200	14	11	8	6	5
250	12	10	7	5	5
300	11	8	6	5	4
400	9	6	5	4	4
500	7	6	5	4	3
600	6	5	4	3	3
700	5	4	4	3	3
800	4	4	3	3	3
1,000	4	4	3	3	3
2,000	4	4	3		
3,000	4	3	3		

* Courtesy Hewitt-Robins.
For antifriction bearings use one-half of the above percentages.

compensate for the loss due to pulley friction may be obtained. Table 11-12 gives the percent of the power delivered to a conveyor required to overcome pulley friction for conveyors with head drive and babbitted bearings for all pulley shafts.

CONVEYOR-BELT TAKE-UPS

Because of the tendency of a conveyor belt to elongate after it is put into operation, it is necessary to provide a method of adjusting for the increase in length.

A screw take-up may be used to increase the length of the conveyor by moving the head or tail pulley. This adjustment may be sufficient for a short belt but not for a long belt (see Fig. 11-3).

Another take-up, which is more satisfactory, depends on forcing the returning belt to travel under a weighted pulley, which provides a uniform tension in the belt regardless of the variation in length.

HOLDBACKS

If a belt conveyor is operated on an incline, it is advisable to install a holdback on the driving pulley to prevent the load from causing the belt to run backward in the event of a power failure. A holdback is a mechanical device which permits a driving pulley to rotate in the normal direction but prevents it from rotating in the opposite direction. The operation of a holdback should be automatic. At least three types are available. They are the roller, ratchet, and differential band brake, all of which operate automatically.

A holdback must be strong enough to resist the force produced by the load less the sum of the forces required to move the empty belt, move the load horizontally, turn the pulleys, drive the tripper, and to overcome drive losses.

If a belt conveyor is operated on a decline, the effect of the load is to move the belt forward. If this effect exceeds the total forces of friction, it will be necessary to install a suitable braking unit to regulate the speed of the belt. To overcome this difficulty, an electric motor or generator may be used as the driving unit. In starting an empty belt, the unit will act as a motor, but when the effect of the load is sufficient to overcome all resistances, the unit will act as a generator to regulate the belt speed.

FEEDERS

The purpose of a feeder is to deliver material to a belt at a uniform rate. A feeder may discharge directly onto a belt, or it may discharge the material through a chute in order to reduce the impact of the falling material on the belt. Several types of feeders are available, each of which has advantages and disadvantages when compared with another type. Among the more popular types are the following:

1. Apron
2. Reciprocating
3. Rotary vane
4. Rotary plow

An apron feeder usually receives the material from a gated hopper, which regulates the flow onto the feeder. The feeder consists of a moving, flat, rubber-covered belt or a number of flat steel plates connected to two moving chains. This feeder moves the material from under the hopper and discharges it through a receiving unit onto the conveyor belt. A belt feeder is suitable for handling material consisting of relatively small pieces. If the material contains large pieces of highly abrasive rock or stone, a steel-plate-type feeder will usually prove more satisfactory than a belt type.

A reciprocating feeder consists of steel plate placed under a hopper. The plate is operated through an eccentric drive to produce the reciprocating effect, which moves the material onto the conveyor belt.

A rotary-vane feeder consists of a number of vanes mounted on a horizontal shaft. As the material flows down an inclined plane, the rotating vanes deliver measured amounts to the conveyor. The rate of feeding may be regulated by varying the speed of the rotating vanes.

A rotary-plow feeder consists of a number of plows, or vanes, mounted on a vertical shaft. The plows rotate over a horizontal table onto which the material is allowed to flow. The rate of feeding may be regulated by varying the speed of the plows.

TRIPPERS

When it is necessary to remove material from a belt conveyor before it reaches the end of the belt, a tripper should be installed on the conveyor. A tripper consists of a pair of pulleys which are so located that the loaded belt must pass over one pulley and under the other. As the belt passes over the top pulley, the load will be discharged from the belt into an auxiliary hopper or chute.

A tripper may be stationary or a traveling type. The latter type may be propelled by a hand-operated crank, a separate motor, or the conveyor belt. If a tripper is installed on a conveyor, additional power should be provided to operate it. Figure 11-8 illustrates a belt-propelled automatically controlled tripper.

Figure 11-8 Belt-propelled automatically controlled tripper.

Example 11-3 Belt-conveyor design Design a belt conveyor to transport unsized crushed limestone. The essential information is as follows:

Capacity, 300 tons per hr
Horizontal distance, 360 ft
Vertical lift, 40 ft
Maximum size stone, 6 in.
Weight of stone, 100 lb per cu ft
Required belt width, from Table 11-2, 24 in.
Maximum speed, from Table 11-3, 400 fpm
Capacity at 400 fpm, from Table 11-2, $4 \times 100 = 400$ tons per hr
Required belt speed, $\dfrac{400 \times 300}{400} = 300$ fpm

This speed, 300 fpm, will be satisfactory provided the feeder supplies material at a uniform rate. If the rate of feeding is irregular, it may be necessary to increase the speed to assure the specified rate of delivery. The design will be based on a speed of 350 fpm to provide a margin of safety.

The power required to operate the loaded conveyor will be as follows:

To drive the empty belt, Table 11-8, $0.81 \times 350/100$	= 2.84 hp*
To move the load horizontally, Table 11-9	= 3.92 hp*
To lift the load, Table 11-10	= 12.00 hp
Subtotal	= 18.76 hp
For pulley friction, Table 11-12, 4% of 18.76	= 0.75 hp
Subtotal required by belt	= 19.51 hp
For drive losses, 10% of 19.51	= 1.95 hp
Total power required	= 21.46 hp

Determine the type, size, and number of driving pulleys required to operate the belt. The belt will be driven through a head pulley. When a belt is driven by a pulley, the effective driving force transmitted to the belt is equal to the difference in the belt tensions on the tight side and the slack side, expressed in pounds. This difference is referred to as the effective driving force or tension. Let

T_1 = tight-side tension
T_2 = slack-side tension
T_e = effective tension
$T_e = T_1 - T_2$

The value of T_e can be determined from the horsepower transmitted to the belt and the belt speed in fpm.

$$T_e = \frac{\text{hp} \times 33{,}000}{\text{belt speed, fpm}} = \frac{19.5 \times 33{,}000}{350} = 1{,}838 \text{ lb}$$

$$T_1 - T_2 = 1{,}838 \text{ lb}$$

It is desirable to operate the belt at the lowest practical tight-side and slack-side tensions. The necessary tensions are maintained by the take-ups. The maximum slack-side tension will occur as the belt leaves the driving pulley. This tension will equal the tension at the tail pulley plus the weight of the vertical component of the belt. Field observations indicate that the tension in the belt at the loading point should be not less than 20 lb per in. of belt width. This tension will be transmitted to the slack side of the belt at the tail pulley. The

*These values are obtained by interpolating in the tables.

minimum possible slack-side tension will be

Tension at tail pulley, 20×24 = 480 lb
Weight of belt, 40 ft × 6 lb per ft = 240 lb
Total tension = 720 lb

The values of T_1 and T_2 for each of three driving arrangements will be as follows:

Arc of contact, deg	T_1, lb	T_2, lb
	Single bare drive	
215	1,838 × 1.65 = 3,035	3,035 − 1,838 = 1,197
220	1,838 × 1.62 = 2,980	2,980 − 1,838 = 1,142
	Single lagged drive	
215	1,838 × 1.38 = 2,538	2,538 − 1,838 = 700
220	1,838 × 1.35 = 2,480	2,480 − 1,838 = 642
	Tandem bare drive	
400	1,838 × 1.21 = 2,225	2,225 − 1,838 = 387
450	1,838 × 1.18 = 2,170	2,170 − 1,838 = 332

Regardless of the type of drive selected, the minimum slack-side tension in the belt just as it leaves the head pulley will be 720 lb. Adding the effective tension T_e gives a minimum tight-side tension of 720 + 1,838 = 2,558 lb. If, for a single lagged drive, with an arc of contact of 215°, T_2 is increased to 720 lb, T_1 will be 2,558 lb, which satisfies the tension requirements. Thus, a single lagged pulley will be used to drive the belt.

Reference to Table 11-4 indicates that a 3-ply 42-oz belt has a safe working stress of 2,640 lb, which is satisfactory. The thickness of this belt will permit it to trough satisfactorily. Reference to Table 11-4 indicates that the minimum pulley diameters should be head, 20 in.; tail, take-up, and snub, 16 in.; bend, 12 in.

The troughing idlers should be 5 in. in diameter, spaced 5 ft 0 in. apart, with a maximum spacing of 1 ft 6 in. at the loading point. The return idlers should be 5 in. in diameter, spaced 10 ft 0 in. apart.

Examples illustrating the use of belt-conveyor systems. The following two examples demonstrate some of the advantages of using belt-conveyors on construction projects.

Example 11-4 When the Los Angeles County Flood Control District prepared an estimate covering the cost of removing 7,300,000 tons of silt, sand, and gravel from the lake bed above the San Gabriel Dam in California, the estimate was based on use of conventional equipment for excavating and hauling the material to a dump site some 800 ft above the lake. It was estimated that the project would cost $7,977,000 and that it would require 857 days to complete the job. The project was awarded to a contractor at a cost of $4,593,000. He completed the project in some 200 days less than the allotted time.

The contractor installed a belt-conveyor system consisting of 165 flights of belts, each 105 ft long and 48 in. wide, to transport the material from the lake to the disposal area. The electric motor selected to power each flight varied from 40 to 100 hp, depending on the slope of the flight. In order to ensure the same rate of flow of the material on each flight all belts operated at the same speed.

A feature of the system was continuous weighing of the material as it moved along on the belt, giving a metered readout of the total tonnage measured from the start of a shift, with another meter showing the cumulative total from the beginning of the week. Still another meter indicated continuously the percent of maximum capacity the belt was carrying at any given time.

The conveyor was loaded at two separate portable stations by tractor-mounted front-end loaders. The method permitted one station to be moved to another location as necessary, while loading was continued at the other station. Grizzlies were installed at the loader stations to remove any boulders that were too large for the conveyor to handle [1].

Example 11-5 When the earth-filled Portage Dam on the Peace River in British Columbia, Canada, was constructed, it was necessary to transport 57,500,000 cu yd of earth over a distance of approximately 3 miles. The earth was excavated from moraine deposits by bulldozers, which fed the material downhill to trap loaders similar to the ones illustrated in Fig. 8-34. The trap loaders fed portable shuttle conveyors, which transferred the soil to two fixed gathering conveyors. The gathering conveyors then transferred the soil to the permanently installed plant-feed conveyor.

When material in a given location became depleted or an excessively long dozer haul was required, the trap loaders and the shuttle conveyors were moved to new locations nearer the available material.

Because the specifications required a stated grading for the soil, it was necessary to blend different classes of soil before placing it in the dam. Part of the blending was accomplished by the bulldozers before the loads were deposited on the trap loaders. Final blending was performed at the blender, as illustrated in Fig. 11-9.

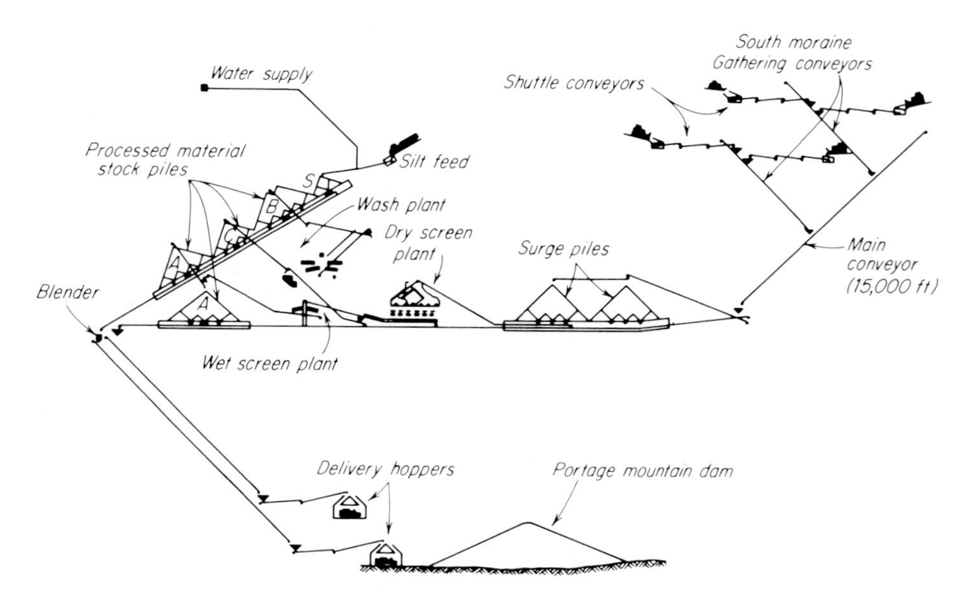

Figure 11-9 Portage Mountain Dam conveying and processing system.

Table 11-13. Details of Belt-Conveyor Equipment for the Portage Mountain Dam

Item	Elevation difference, ft	Belt width, in.	Length of belt, ft	Capacity, tons per hr	Speed, ft per min	Motor horsepower	Location
Belt loaders	------	60	25	3,500	450	1 × 100	Moraine
Shuttle conveyors	Variable	48	100	3,500	600	1 × 100	Moraine
Gathering conveyor	146F*	60	2,000	6,000	800	2 × 500	Moraine
Gathering conveyor	180F	60	1,000	6,000	800	2 × 500	Moraine
Plant feed conveyor	350F	66	15,000	12,000	1,150	4 × 850	Moraine to plant
Conveyor No. 1	98R*	72	655	12,000	900	3 × 500	Surge pile to feed
Conveyor No. 2	75F	72	2,136	12,000	900	2 × 125	Plant feed to blender
Conveyor No. 3	75R	60	540	6,300	700	1 × 500 1 × 200	Surge pile to dry screen
Conveyor No. 4	25R	48	241	2,100	400	1 × 125	$-\frac{3}{8}$-in. material dry screen to stockpile
Conveyor No. 5	14R	48	199	2,100	400	1 × 75	$-\frac{3}{8}$-in. material dry screen to stockpile
Conveyor No. 6	62R	48	364	2,700	600	1 × 200 1 × 100	$-\frac{3}{8}$-in. material dry screen to wet screen
Conveyor No. 7	97R	48	466	3,000	600	2 × 200	Wet screen to stockpile
Conveyor No. 7A	0	48	123	3,000	600	1 × 30	Wet screen to stockpile
Conveyor No. 8	100R	48	502	4,200	700	1 × 500 1 × 200	$-\frac{3}{8}$-in. material to stockpile to wash plant
Conveyor No. 9	16R	48	211	1,800	500	1 × 75	Wash plant to stockpile
Conveyor No. 9A	60R	48	220	1,800	550	2 × 75	Wash plant to stockpile
Conveyor No. 9B	17R	48	94	1,800	550	1 × 75	Wash pile to stockpile
Conveyor No. 10	7F	60	778	6,000	600	1 × 200	Reclaim from stockpile
Conveyor No. A	312F	60	1,907	6,000	800	3 × 500	Blender to dam
Conveyor No. B	303F	60	1,874	6,000	800	3 × 500	Blender to dam
Conveyor A_1 and A_2	Variable	60	122	6,000	800	1 × 200 1 × 75	Blender to dam
Conveyor B_1 and B_2	Variable	60	122	6,000	800	1 × 200 1 × 75	Blender to dam

F designates a fall; R designates a rise.

The soil was hauled by trucks from the delivery hoppers to the dam.

Table 11-13 lists pertinent information concerning the details of the conveyor equipment for the project [2].

PROBLEMS

11-1 Determine the minimum belt width and the minimum belt speed, in fpm, required to transport 300 tph of unsized crushed stone weighing 100 lb per cu ft. The maximum size of the stone is 4 in.

11-2 What is the capacity of a 36-in.-wide belt, in tph, when the belt is transporting gravel weighing 100 lb per cu ft and is moving at the maximum recommended speed.

11-3 A conveyor 500-ft long with a 30-in.-wide 5-ply 36-oz belt is used to transport material weighing 125 lb per cu ft. The angle of repose for the material is 20°. The revolving parts will weigh 36 lb for each troughing idler and 25 lb for each return idler. Both idlers are 5 in. in diameter and are equipped with antifriction bearings. Determine the horsepower required to move this belt at a speed of 300 fpm.

11-4 Using the information given in Table 11-7, determine the horsepower required to move an empty conveyor belt 30 in. wide on a 900-ft-long conveyor equipped with 5-in.-diameter idlers with antifriction bearings when the belt is moving at a speed of 300 fpm.

11-5 If the belt of Prob. 11-4 transports 300 tph up a 5 percent slope, determine the horsepower required when the belt is moving at the minimum speed necessary to transport the material. The material will weigh 125 lb per cu ft.

11-6 A 4-ply 36-in.-wide belt on a conveyor 500-ft long will be used to transport its maximum capacity of unsized gravel weighing 125 lb per cu ft up a 4 percent slope, using 5-in. antifriction idlers. Determine the power required to operate the belt when it is traveling at its maximum recommended speed.

11-7 Design a conveyor belt system to transport 300 tph of crushed stone under the following conditions:

Weight of stone, 125 lb per cu ft
Horizontal distance, 600 ft
Vertical lift, 60 ft

Use 5-in.-diameter idlers with antifriction bearings.

The design should furnish the following information:
The required width of the belt
The number of plies required for the belt
The required belt speed
The number of 5-in. diameter troughing and return idlers required
The spacing of the idlers
The minimum diameters of the head, tail, take-up, snub, and bend pulleys

REFERENCES

1. Belt-conveyor System Wins Cleanout Job, *Construction Methods and Equipment*, vol. 51, pp. 154–157, March 1969.
2. Low, W. Irvine: Portage Mountain Dam Conveyor System, *Journal of the Construction Division, American Society of Civil Engineers*, vol. 93, pp. 33–51, September 1967.
3. Golz, Alfred R.: Portage Mountain Dam Conveyor System, *Journal of the Construction Division, American Society of Civil Engineers*, vol. 94, pp. 258–259, October 1968.
4. Low, W. Irvine: Portage Mountain Dam Conveyor System, *Journal of the Construction Division, American Society of Civil Engineers*, vol. 95, pp. 126–127, July 1969.

CHAPTER

TWELVE

COMPRESSED AIR

GENERAL INFORMATION

Compressed air is used extensively on construction projects for drilling rock or other hard formations, loosening earth, operating air motors, hand tools, pile drivers, pumps, mucking equipment, cleaning, etc. In many instances the energy supplied by compressed air is the most convenient method of operating equipment and tools.

When air is compressed, it receives energy from the compressor. This energy is transmitted through a pipe or hose to the operating equipment, where a portion of the energy is converted into mechanical work. The operations of compressing, transmitting, and using air will always results in a loss of energy, which will give an overall efficiency less than 100 percent, sometimes considerably less.

FUNDAMENTAL GAS LAWS

As air is a gas, it obeys, within reason, the fundamental laws which apply to gases. The laws with which we are concerned are related to the pressure, volume, temperature, and transmission of air.

DEFINITIONS OF GAS-LAW TERMS

In order to understand the laws which relate to compressed air, it is necessary to define certain terms which are used in developing and applying these laws.

The essential definitions are as follows:

Gauge pressure This is the pressure exerted by the air in excess of atmospheric pressure. It is usually expressed in psi or inches of mercury and is measured by a pressure gauge or a mercury manometer.

Absolute pressure This is the total pressure measured from absolute zero. It is equal to the sum of the gauge and the atmospheric pressure, corresponding to the barometric reading. The absolute pressure should be used in dealing with the laws of gases.

Psi is the abbreviation for pounds per square inch of pressure.

Psf is the abbreviation for pounds per square foot of pressure.

Vacuum This is a measure of the extent to which pressure is less than atmospheric pressure. For example, a vacuum of 5 psi is equivalent to an absolute pressure of $14.7 - 5 = 9.7$ psi.

Standard conditions Because of the variations in the volume of air with pressure and temperature, it is necessary to express the volume at standard conditions if it is to have a definite meaning. Standard conditions are an absolute pressure of 14.7 psi and a temperature of 60°F.

Temperature Temperature is a measure of the amount of heat contained by a unit quantity of gas. It is measured with a thermometer or some other suitable temperature-indicating device.

Fahrenheit temperature This is the temperature indicated by a thermometer calibrated according to the Fahrenheit scale. For this thermometer pure water freezes at 32°F and boils at 212°F, at a pressure of 14.7 psi. Thus, the number of degrees between freezing and boiling water is 180.

Celsius temperature This is the temperature indicated by a thermometer calibrated according to the Celsius scale. For this thermometer pure water freezes at 0°C and boils at 100°C, at a pressure of 14.7 psi.

Relation between Fahrenheit and Celsius temperatures As 180 degrees on the Fahrenheit scale equals 100 degrees on the Celsius scale, 1°C equals 1.8°F. A Fahrenheit thermometer will read 32° when a Celsius thermometer reads 0°.

Let T_F = Fahrenheit temperature and T_C = Celsius temperature. For any given temperature the thermometer readings are expressed by the following equation:

$$T_F = 32 + 1.8T_C \tag{12-1}$$

Absolute temperature This is the temperature of a gas measured above absolute zero. It equals degrees Fahrenheit plus 459.6 or, as more commonly used, 460.

ISOTHERMAL COMPRESSION

When a gas undergoes a change in volume without any change in temperature, this is referred to as isothermal expansion or compression.

ADIABATIC COMPRESSION

When a gas undergoes a change in volume without gaining or losing heat, this is referred to as adiabatic expansion or compression.

Boyle's law states that when a gas is subjected to a change in volume due to a change in pressure, at a constant temperature, the product of the pressure times the volume will remain constant. This relation is expressed by the equation

$$P_1V_1 = P_2V_2 = K \tag{12-2}$$

where P_1 = initial absolute pressure
V_1 = initial volume
P_2 = final absolute pressure
V_2 = final volume
K = a constant

Example Determine the final volume of 1,000 cu ft of air when the gauge pressure is increased from 20 to 120 psi, with no change in temperature. The barometer indicates an atmospheric pressure of 14.7 psi.

$P_1 = 20 + 14.7 = 34.7$ psi
$P_2 = 120 + 14.7 = 134.7$ psi
$V_1 = 1{,}000$ cu ft

From Eq. (12-2)

$$V_2 = \frac{P_1V_1}{P_2} = \frac{34.7 \times 1{,}000}{134.7} = 257.8 \text{ cu ft}$$

BOYLE'S AND CHARLES' LAWS

When a gas undergoes a change in volume or pressure with a change in temperature, Boyle's law will not apply. Charles' law introduces the effect of absolute temperature on the volume of a gas when the pressure is maintained constant. It states that the volume of a given weight of gas at constant pressure varies in direct proportion to its absolute temperature. It may be expressed mathematically by the equation

$$\frac{V_1}{T_1} = \frac{V_2}{T_2} = C \tag{12-3}$$

where V_1 = initial volume
T_1 = initial absolute temperature
V_2 = final volume
T_2 = final absolute temperature
C = a constant

The laws of Boyle and Charles may be combined to give the equation

$$\frac{P_1V_1}{T_1} = \frac{P_2V_2}{T_2} = \text{a constant} \tag{12-4}$$

Equation (12-4) may be used to express the relations between pressure, volume, and temperature for any given gas, such as air. It is illustrated by the following example.

Example 12-1 One thousand cubic feet of air, at an initial gauge pressure of 40 psi and temperature of 50°F, is compressed to a volume of 200 cu ft at a first temperature of 110°F. Determine the final gauge pressure. The atmospheric pressure is 14.46 psi.

$P_1 = 40 + 14.46 = 54.46$ psi
$V_1 = 1{,}000$ cu ft
$T_1 = 460 + 50 = 510°F$
$V_2 = 200$ cu ft
$T_2 = 460 + 110 = 570°F$

Rewriting Eq. (12-4) and substituting these values, we get

$$P_2 = \frac{P_1 V_1}{T_1} \frac{T_2}{V_2} = \frac{54.46 \times 1{,}000}{510} \frac{570}{200} = 304 \text{ psi}$$

Final gauge pressure = 304 − 14.46 = 289.54 psi.

ENERGY REQUIRED TO COMPRESS AIR

Equation (12-2) may be expressed as $PV = K$, where K is a constant so long as the temperature remains constant. However, in actual practice the temperature usually will not remain constant, and the equation must be modified to provide for the effect of changes in temperature. The effect of temperature may be provided for by introducing an exponent n to V. Thus, Eq. (12-2) may be rewritten as

$$P_1 V_1^n = P_2 V_2^n = K \tag{12-5}$$

For air the values of n will vary from 1.0 for isothermal compression to 1.4 for adiabatic compression. The actual value for any compression condition may be determined experimentally from an indicator card obtained from a given compressor.

When the pressure of a given volume of air is increased by an air compressor, it is necessary to furnish energy to the air. Consider a single compression cycle for an air compressor, as indicated in Fig. 12-1. Air is drawn into the cylinder at pressure P_1 and is discharged at pressure P_2. P_1 does not need to be atmospheric pressure. The initial volume is V_1. As the piston compresses the air, the pressure-volume will follow the curve CD. At D, when the pressure is P_2, the discharge valve will open and the pressure will remain constant while the volume decreases to V_2, as indicated by line DE. Point E represents the end of the piston stroke. At point E the discharge valve will close, and as the piston begins its return stroke, the pressure will decrease along line EB to a value of P_1, when the intake valve will open and allow additional air to enter the cylinder. This will establish line BC.

The work done along the line CD may be obtained by integrating the equation $dW = V\,dP$.

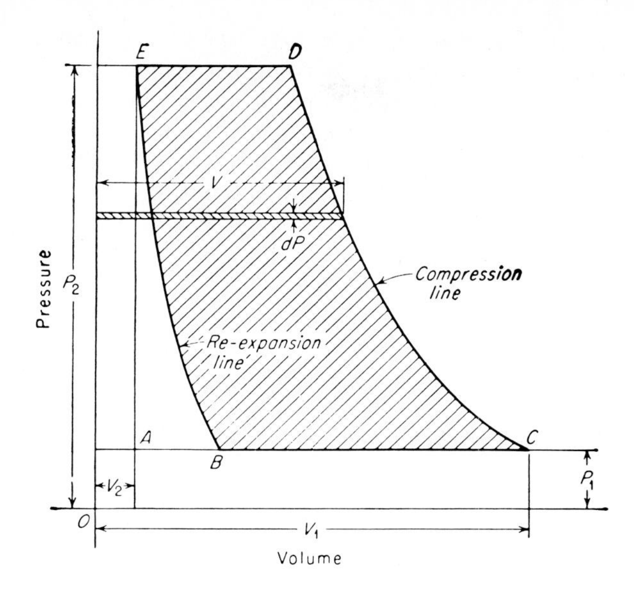

Figure 12-1 Cycle for isothermal compression of air.

From Eq. (12-5), $V^n = K/P$. If both sides of the equation are raised to the $1/n$ power, the equation will be

$$V = \left(\frac{K}{P}\right)^{1/n}$$

Substituting this value of V gives

$$dW = \left(\frac{K}{P}\right)^{1/n} dP$$

Integrating gives

$$W = K^{1/n} \int_1^2 \frac{dP}{P^{1/n}} \tag{12-6}$$

For isothermal compression $n = 1$. Substituting this value in Eq. (12-6) gives

$$W = K \int_1^2 \frac{dP}{P} = -K \log_e \frac{P_2}{P_1} + C$$

When $P_2 = P_1$ and no work is done, the constant of integration is equal to zero. The minus sign may be disregarded. Thus, for isothermal compression of air, the equation may be written as

$$W = K \log_e \frac{P_2}{P_1} \tag{12-7}$$

If it is desired to convert from natural to common logarithms, $\log_e (P_2/P_1)$ may be replaced by $2.302 \log (P_2/P_1)$. For the given compression conditions $n = 1$, and $K = P_1V_1$. If the compression is started on air at standard conditions, P_1 will be 14.7 psi at 60°F. Since work is commonly expressed in foot-pounds, it is necessary to express P_1 in psf. This is done by multiplying P_1 by 144. When these substitutions are made, Eq. (12-7) may be written as

$$W = 14.7 \times 144\ V_1 \times 2.302 \log \frac{P_2}{P_1}$$

$$= 4{,}873\ V_1 \log \frac{P_2}{P_1} \tag{12-8}$$

The value of W is in foot-pounds per cycle. One horsepower is equivalent to 33,000 ft-lb per min. If V_1 in Eq. (12-8) is replaced by V, the volume of free air per minute at standard conditions, the horsepower required to compress V cu ft of air from an absolute pressure of P_1 to P_2 psi will be

$$\text{hp} = \frac{4{,}873\ V \log(P_2/P_1)}{33{,}000}$$

$$\text{hp} = 0.1477\ V \log \frac{P_2}{P_1} \tag{12-9}$$

Example 12-2 Determine the theoretical horsepower required to compress 100 cu ft of free air per minute, measured at standard conditions, from atmospheric pressure to 100 psi gauge pressure. Substituting in Eq. (12-9), we get

$$\text{hp} = 0.1477 \times 100 \times \log \frac{114.7}{14.7}$$

$$= 14.79 \times \log 7.8$$
$$= 14.79 \times 0.892$$
$$= 13.2$$

If air is compressed under other than isothermal conditions, the equation for the required horsepower may be derived in a similar manner. However, since n will not equal 1, it must appear as an exponent in the equation. Equation (12-10) gives the horsepower for nonisothermal conditions.

$$\text{hp} = \frac{n}{n-1} 0.0643\ V\left[\left(\frac{P_2}{P_1}\right)^{(n-1)/n} - 1\right] \tag{12-10}$$

where the terms are the same as those used in Eq. (12-9)

Example 12-3 Determine the theoretical horsepower required to compress 100 cu ft of free air per minute, measured at standard conditions, from atmospheric pressure to 100 psi gauge pressure, under adiabatic conditions. The value of n will be 1.4 for air for adiabatic compression. Substituting in Eq. (12-10), we get

$$\begin{aligned}\text{hp} &= \frac{1.4}{1.4-1} \times 0.0643 \times 100(7.8^{0.4/1.4} - 1) \\ &= 22.5(7.8^{0.286} - 1) \\ &= 22.5 \times 0.79 \\ &= 17.8\end{aligned}$$

For air compressors used on construction projects the compression will be performed under conditions between isothermal and adiabatic. Thus, the theoretical horsepower will be between 13.2 and 17.8, the actual value depending on the extent to which the compressor is cooled during operation. The difference in the horsepowers required illustrates the importance of operating an air compressor at the lowest practical temperature.

EFFECT OF ALTITUDE ON THE POWER REQUIRED TO COMPRESS AIR

When a given volume of air, measured as free air prior to its entering a compressor, is compressed, the original pressure will average 14.7 psi absolute pressure at sea level. If the same volume of free air is compressed to the same gauge pressure at a higher altitude, the volume of the air after being compressed will be less than the volume compressed at sea level. The reason for this difference is that there is less weight in a cubic foot of free air at 5,000 ft than at sea level. Thus, while a compressor may compress air to the same discharge pressure at a higher altitude, the volume supplied in a given time interval will be less at the higher altitude. The use of Eq. (12-2) and the information in Table 4-4 (page 94) will demonstrate the correctness of this statement.

Because a compressor of a specified capacity actually supplies a smaller volume of air at a given discharge pressure at a higher altitude, it requires less power to operate a compressor at a higher altitude, as illustrated in Table 12-1.

AIR-COMPRESSOR DEFINITIONS AND TERMS

Many terms related to air compressors and compressed air have assumed uniform meanings. The essential terms are defined hereafter.

Air compressor This is a machine which is used to increase the pressure of air by reducing its volume.

Reciprocating compressor This is a machine which compresses air by means of a piston reciprocating in a cylinder.

Single-acting compressor This compressor is a machine which compresses air in only one end of a cylinder.

Table 12-1 Theoretical horsepower required to compress 100 cu ft of free air per min at different altitudes*

Altitude, ft	Isothermal compression Single- and two-stage gauge pressure				Adiabatic compression Single-stage gauge pressure			Adiabatic compression Two-stage gauge pressure			
	60	80	100	125	60	80	100	60	80	100	125
0	10.4	11.9	13.2	14.4	13.4	15.9	18.1	11.8	13.7	15.4	17.1
1,000	10.2	11.7	12.9	14.1	13.2	15.6	17.8	11.6	13.5	15.1	16.8
2,000	10.0	11.4	12.6	13.8	13.0	15.4	17.5	11.4	13.2	14.8	16.4
3,000	9.8	11.2	12.3	13.5	12.8	15.2	17.2	11.2	13.0	14.5	16.1
4,000	9.6	11.0	12.1	13.2	12.6	14.9	16.9	11.0	12.7	14.2	15.7
5,000	9.4	10.7	11.8	12.8	12.4	14.7	16.5	10.8	12.5	13.9	15.4
6,000	9.2	10.5	11.5	12.5	12.2	14.4	16.2	10.6	12.2	13.6	15.1
7,000	9.0	10.3	11.2	12.2	12.0	14.2	16.0	10.4	12.0	13.4	14.8
8,000	8.9	10.0	11.0	11.9	11.8	14.0	15.7	10.2	11.8	13.1	14.5
9,000	8.7	9.8	10.7	11.6	11.6	13.7	15.4	10.0	11.6	12.8	14.1
10,000	8.5	9.6	10.4	11.4	11.5	13.5	15.1	9.8	11.3	12.6	13.8

* Compressed Air and Gas Institute.

Double-acting compressor The double-acting compressor is a machine which compresses air in both ends of a cylinder.

Single-stage compressor This is a machine which compresses air from atmospheric pressure to the desired discharge pressure in a single operation.

Two-stage compressor This is a machine which compresses air in two separate operations. The first operation compresses the air to an intermediate pressure, while the second operation further compresses it to the desired final pressure.

Multistage compressor This is a compressor which produces the desired final pressure through two or more stages.

Rotary compressor The rotary compressor is a machine in which the compression is effected by the action of rotating elements.

Centrifugal compressor This compressor is a machine in which the compression is effected by a rotating vane or impeller that imparts velocity to the flowing air to give it the desired pressure.

Intercooler The intercooler is a heat exchanger which is placed between two compression stages to remove the heat of compression from the air.

Aftercooler This is a heat exchanger which cools the air after it is discharged from a compressor.

Inlet pressure This is the absolute pressure of the air at the inlet to a compressor.

Discharge pressure Discharge pressure is the absolute pressure of the air at the outlet from a compressor.

Compression ratio This is the ratio of the absolute discharge pressure to the absolute inlet pressure.

Free air Free air is air as it exists under atmospheric conditions at any given location.

Cfm Cfm is an abbreviation for cubic feet per minute.

Capacity Capacity is the volume of air delivered by a compressor, expressed in cfm of free air.

Theoretical horsepower This is the horsepower required to compress adiabatically the air delivered by a compressor through the specified pressure range, without any provision for lost energy.

Brake horsepower Brake horsepower is the actual horsepower input required by a compressor.

Compressor efficiency This is the ratio of the theoretical horsepower to the brake horsepower.

Volumetric efficiency This is the ratio of the capacity of a compressor to the piston displacement of the compressor.

Density of air This is the weight of a unit volume of air, usually expressed as pounds per cubic foot. Density varies with the pressure and temperature of the air. The weight of air at 60°F and 14.7 psi, absolute pressure, is 0.07658 lb per cu ft. The volume per pound is 13.059 cu ft.

Load factor The load factor is the ratio of the average load during a given period of time to the maximum rated load of a compressor.

Diversity factor This is the ratio of the actual quantity of air required for all uses to the sum of the individual quantities required for each use.

STATIONARY COMPRESSORS

Stationary compressors are generally used for installations where compressed air is required for a long period of time. The compressors may be reciprocating or rotary types, single-stage or multistage. The total quantity of air may be supplied by one or more compressors. The installed cost of a single compressor will usually be less than for several compressors having the same capacity. However, several compressors provide better flexibility for varying load demands, and, in the event of a shutdown for repairs, the entire plant does not need to be stopped.

Stationary compressors may be driven by steam, electric motors, or internal-combustion engines.

PORTABLE COMPRESSORS

Portable compressors are used when it is necessary to move the equipment frequently to meet job demands. The compressors may be mounted on rubber tires, steel wheels, or skids. They may be driven by gasoline or diesel engines.

Figure 12-2 Two stage diesel-engine-operated portable air compressor. *(Ingersoll-Rand Company.)*

They are available in single- or two-stage, reciprocating or rotary types. See Fig. 12-2.

Reciprocating compressors A reciprocating compressor depends on a piston, which moves back and forth in a cylinder, for the compressing action. The piston may compress air while moving in one or both directions. For the former it is defined as single-acting, while for the latter it is defined as double-acting. A compressor may have one or more cylinders.

Rotary compressors In recent years considerable effort has been directed toward the development of rotary compressors. These machines offer several advantages compared with reciprocating compressors, such as compactness, light weight, uniform flow, variable output, carefree operation, and long life.

Figure 12-3 illustrates a 600-cfm two-stage rotary compressor which has given excellent performance in the construction industry. Its operating weight is 9,500 lb, which is comparable with the weight of a 315-cfm portable reciprocating unit. The cost is approximately the same as for a 600-cfm reciprocating compressor.

Rotary screw compressors The working parts of a screw compressor are two helical rotors as illustrated in Fig. 12-4. The male rotor has four lobes and rotates 50 percent faster than the female rotor, which has six flutes, with which the male motor meshes. As the air enters and flows through the compressor it is compressed in the space between the lobes and the flutes. The inlet and outlet ports are automatically covered and uncovered by the shaped ends of the rotors as they turn.

Figure 12-3 Two-stage rotary air compressor. *(Ingersoll-Rand Company.)*

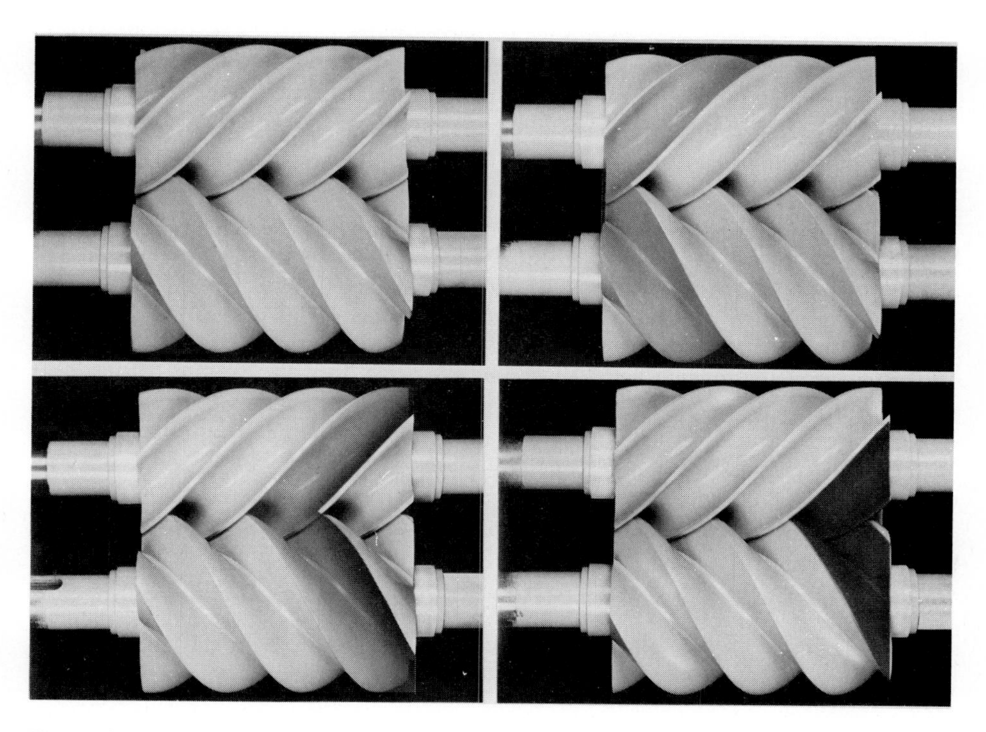

Figure 12-4 The operation of the helical rotors of a screw compressor. *(Atlas Copco, Inc.)*

Table 12-2 Representative specifications for rotary screw air compressors*

Capacity of free air		Normal operating pressure		Number of compression stages
cfm	cu m/min	psi	kp/cm^2	
125	3.5	102	7	1
170	4.8	102	7	1
250	7.1	102	7	1
335	9.5	102	7	1
365	10.3	100	7	2
425	12.0	100	7	2
600	17.0	100	7	2
700	19.8	100	7	2
900	25.5	102	7	2
1,200	34.0	102	7	2
1,500	42.5	102	7	2

* Atlas Copco, Inc.

These compressors are available in a relatively wide range of capacities, with single-stage or multistage compression and with rotors which operate under oil-lubricated conditions or with no oil, the latter to produce oil-free air.

They offer several advantages when compared with other types of compressors, including but not limited to the following:

1. Quiet operation to satisfy a wide range of legal requirements limiting the permissible volume of noise, with little or no loss in output
2. Few moving parts, with minimum mechanical wear and little maintenance requirements
3. Automatic controls actuated by the output pressure, which regulate the speed of the driving unit and the compressor to limit the output to only the demand required
4. Little or no pulsation in the flow of air and hence reduced vibrations

Table 12-2 lists information applicable to representative screw compressors operated under standard conditions, namely absolute inlet air pressure equal to 1 bar (1.02 kp/sq cm or 14.5 psi) and inlet air temperature and inlet coolant temperature equal to 15°C (60°F).

COMPRESSOR CAPACITY

Air compressors are rated by the piston displacement in cfm. However, the capacity of a compressor will be less than the piston displacement because of valve and piston leakage and the air left in the end-clearance spaces of the cylinders.

The capacity of a compressor is the actual volume of free air drawn into a compressor in a minute. It is expressed in cubic feet. For a reciprocating compressor in good mechanical condition the actual capacity should be 80 to 90 percent of the piston displacement. This is illustrated by an analysis of a 315-cfm two-stage portable compressor. The manufacturer's specifications give the following information:

No. low-pressure cylinders, 4
No. high-pressure cylinders, 2
Diameter of low-pressure cylinders, 7 in.
Diameter of high-pressure cylinders, $5\frac{3}{4}$ in.
Length of stroke, 5 in.
Rpm, 870

Consider the piston displacement of the low-pressure cylinders only as they determine the capacity of the unit.

Area of cylinder, $\frac{\pi \times 7^2}{4 \times 144} = 0.267$ sq ft
Displacement per cylinder per stroke, $0.267 \times \frac{5}{12} = 0.111$ cu ft
Displacement per minute, $4 \times 0.111 \times 870 = 386$ cu ft
Specified capacity, 315 cu ft
Volumetric efficiency, $315/386 \times 100 = 81.6\%$

EFFECT OF ALTITUDE ON CAPACITY OF COMPRESSORS

The capacity of an air compressor is rated on the basis of its performance at sea level, where the normal absolute barometric pressure is about 14.7 psi. If a compressor is operated at a higher altitude, such as 5,000 ft above sea level, the absolute barometric pressure will be about 12.2 psi. Thus, at the higher altitude there is less weight of air in a cubic foot of free volume than at sea level. If the air is discharged by the compressor at a given pressure, the compression ratio will be increased, and the capacity of the compressor will be reduced. This may be demonstrated by applying Eq. 12-2.

Assume that 100 cu ft of free air at sea level are compressed to 100 psi gauge with no change in temperature. Applying Eq. (12-2),

$$V_2 = \frac{P_1 V_1}{P_2}$$

where $V_1 = 100$ cu ft
$P_1 = 14.7$ psi absolute
$P_2 = 114.7$ psi absolute
$V_2 = \frac{14.7 \times 100}{114.7} = 12.85$ cu ft

At 5,000 ft above sea level
$V_1 = 100$ cu ft
$P_1 = 12.2$ psi absolute
$P_2 = 112.2$ psi absolute

$$V_2 = \frac{12.2 \times 100}{112.2} = 10.87 \text{ cu ft}$$

Table 12-3 lists the factors that should be applied to single-stage compressors to correct for the loss in capacity at various altitudes. For example, a compressor having a sea-level capacity of 600 cfm operating at a pressure of 100 psi gauge will have a capacity at 5,000 ft equal to $600 \times 0.925 = 555.0$ cfm if the operating pressure is 100 psi gauge. At an altitude of 10,000 the capacity will be further reduced to $600 \times 0.840 = 504$ cfm.

INTERCOOLERS

Intercoolers frequently are installed between the stages of a compressor to reduce the temperature of the air and to remove moisture from the air. The reduction in temperature prior to additional compression can reduce the total power required by as much as 10 to 15 percent. Unless an intercooler is installed, the power required by a two-stage compressor will be the same as for a single-stage compressor.

An intercooler requires a continuous supply of circulating cool water to remove the heat from the air. It will require 1.0 to 1.5 gal of water per minute for each 100 cfm of air compressed, the actual amount depending on the temperature of the water.

AFTERCOOLERS

Aftercoolers are installed sometimes at the discharge side of a compressor to cool the air to the desired temperature and to remove moisture from the air. It is highly desirable to remove excess moisture from the air, as it tends to freeze during expansion in air tools, and it washes the lubricating oil out of tools, thereby reducing the lubricating efficiency.

RECEIVERS

An air receiver should be installed on the discharge side of a compressor to equalize the compressor pulsations and to serve as a condensing chamber for the removal of water and oil vapors. A receiver should have a drain cock at its bottom to permit the removal of the condensate. Its volume should be one-tenth to one-sixth of the capacity of the compressor. A blowoff valve, to limit the maximum pressure, is desirable.

Table 12-3 The effect of altitude on the capacity of single-stage air compressors*

	Operating pressure, psi gauge, [psi absolute], (Pa)							
	80 [94.7] (6.53×10^5)		90 [104.7] (7.23×10^5)		100 [114.7] (7.91×10^5)		125 [139.7] (9.65×10^5)	
Altitude above sea level, ft (m)	Compressor ratio†	Factor‡	Compressor ratio	Factor	Compressor ratio	Factor	Compressor ratio	Factor
0	6.44	1.000	7.12	1.000	7.81	1.000	9.51	1.000
1,000	6.64	0.992	7.34	0.988	8.05	0.987	9.81	0.982
(305)	6.64	0.992	7.34	0.988	8.05	0.987	9.81	0.982
2,000	6.88	0.977	7.62	0.972	8.35	0.972	10.20	0.962
(610)	6.88	0.977	7.62	0.972	8.35	0.972	10.20	0.962
3,000	7.12	0.967	7.87	0.959	8.63	0.957	10.55	0.942
(915)	7.12	0.967	7.87	0.959	8.63	0.957	10.55	0.942
4,000	7.36	0.953	8.15	0.944	8.94	0.942	10.92	0.923
(1,220)	7.36	0.953	8.15	0.944	8.94	0.942	10.92	0.923
5,000	7.62	0.940	8.44	0.931	9.27	0.925	11.32	0.903
(1,525)	7.62	0.940	8.44	0.931	9.27	0.925	11.32	0.903
6,000	7.84	0.928	8.69	0.917	9.55	0.908	11.69	0.883
(1,830)	7.84	0.928	8.69	0.917	9.55	0.908	11.69	0.883
7,000	8.14	0.915	9.03	0.902	9.93	0.890	12.17	0.863
(2,135)	8.14	0.915	9.03	0.902	9.93	0.890	12.17	0.863
8,000	8.42	0.900	9.33	0.886	10.26	0.873	12.58	0.844
(2,440)	8.42	0.900	9.33	0.886	10.26	0.873	12.58	0.844
9,000	8.70	0.887	9.65	0.868	10.62	0.857	13.02	0.824
(2,745)	8.70	0.887	9.65	0.868	10.62	0.857	13.02	0.824
10,000	9.00	0.872	10.00	0.853	11.00	0.840	13.50	0.804
(3,050)	9.00	0.872	10.00	0.853	11.00	0.840	13.50	0.804
11,000	9.34	0.858	10.38	0.837	11.42	0.823	14.03	
(3,355)	9.34	0.858	10.38	0.837	11.42	0.823	14.03	
12,000	9.70	0.839	10.79	0.818	11.88	0.807	14.60	
(3,660)	9.70	0.839	10.79	0.818	11.88	0.807	14.60	
14,000	10.42	0.805	11.60		12.78		15.71	
(4,270)	10.42	0.805	11.60		12.78		15.71	
15,000	10.88	0.784	12.12		13.36		16.43	
(4,575)	10.88	0.784	12.12		13.36		16.43	

* Compressed Air and Gas Institute.

† The compressor ratio is the ratio of the volume of free air divided by the volume of the same air at the indicated pressure.

‡ When this factor is multiplied by the specified capacity of the compressor, at sea level, it will give the capacity at the indicated altitude and operating pressure.

LOSS OF AIR PRESSURE IN PIPE DUE TO FRICTION

The loss in pressure due to friction as air flows through a pipe or a hose is a factor which must be considered in selecting the size of a pipe or hose. Failure to use a sufficiently large line may cause the air pressure to drop so low that it will not satisfactorily perform the service for which it is provided.

The selection of the size of line is a problem in economy. The efficiency of most equipment operated by compressed air drops off rapidly as the pressure of the air is reduced. When the cost of lost efficiency exceeds the cost of providing a larger line, it is good economy to install a larger line. The manufacturers of pneumatic equipment generally specify the minimum air pressure at which the equipment will operate satisfactorily. However, these values should be considered as minimum and not desirable operating pressures. The actual pressure should be higher than the specified minimum.

Example 12-4 The cost of lost efficiency on a project resulting from the operation of pneumatic equipment at reduced pressure is estimated to be $1,000. The lost efficiency can be eliminated by installing a larger pipe line at an additional cost of $600. In this instance the contractor will save $400 by installing the larger pipe. Thus, it is good economy to use a larger pipe. However, it is not good economy to spend $1,000 to eliminate an operating loss of $600.

Several formulas are used to determine the loss of pressure in a pipe due to friction. The following equation has been used extensively [1]:

$$f = \frac{CL}{r}\frac{Q^2}{d^5} \tag{12-11}$$

where f = pressure drop, psi
L = length of pipe, ft
Q = cu ft of free air per sec
r = ratio of compression
d = actual ID of pipe, in.
C = experimental coefficient

For ordinary steel pipe the value of C has been found to equal $0.1025/d^{0.31}$. If this value is substituted in Eq. (12-11), we get

$$f = \frac{0.1025L}{r}\frac{Q^2}{d^{5.31}} \tag{12-12}$$

A chart for determining the loss in pressure in a pipe is given in Fig. 12-5.

Example 12-5 This example illustrates the use of the chart in Fig. 12.5. Determine the pressure loss per 100 ft of pipe resulting from transmitting 1,000 cfm of free air, at 100 psi gauge pressure, through a 4-in. standard-weight steel pipe. Enter the chart at the top at 100 psi; then proceed vertically downward to a point opposite 1,000 cfm; thence proceed parallel to the sloping guide lines to a point opposite the 4-in. pipe; thence proceed vertically downward to the bottom of the chart, where the pressure drop is indicated to be 0.225 psi.

Table 12-4 gives the loss of air pressure in 1,000 ft of standard-weight pipe

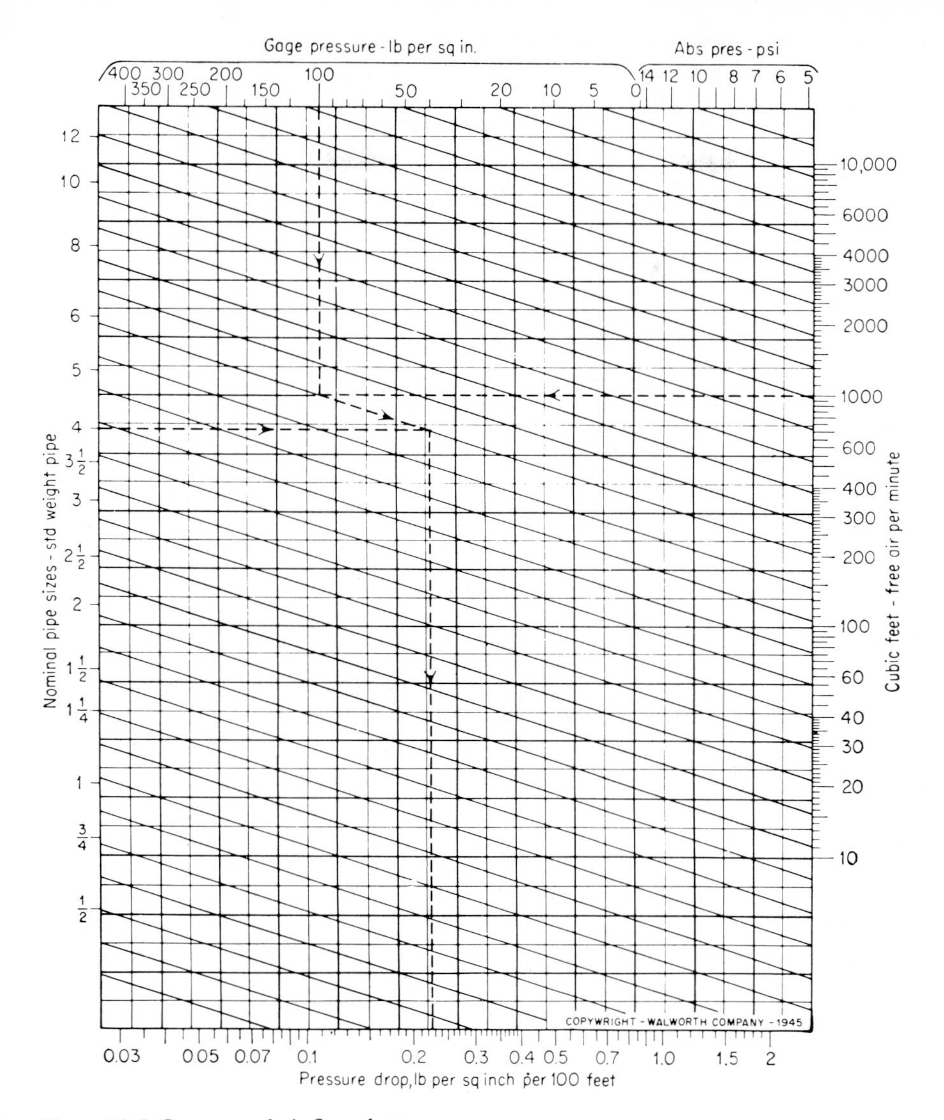

Figure 12-5 Compressed-air flow chart.

due to friction. For longer or shorter lengths of pipe the friction loss will be in proportion to the length. The losses given in the table are for an initial gauge pressure of 100 psi. If the initial pressure is other than 100 psi, the corresponding losses may be obtained by multiplying the values in Table 12-4 by a suitable factor. Reference to Eq. (12-12) reveals that for a given rate of flow through a given size pipe the only variable is r, which is the ratio of compression, based on absolute pressures. For a gauge pressure of 100 psi, $r = 114.7/14.7 = 7.8$,

Table 12-4 Loss of pressure in psi in 1,000 ft of standard-weight pipe due to friction for an inital gauge pressure of 100 psi

Free air per min., cu ft	Nominal diameter, in.												
	$\frac{1}{2}$	$\frac{3}{4}$	1	$1\frac{1}{4}$	$1\frac{1}{2}$	2	$2\frac{1}{2}$	3	$3\frac{1}{2}$	4	$4\frac{1}{2}$	5	6
10	6.50	0.99	0.28										
20	25.90	3.90	1.11	0.25	0.11								
30	68.50	9.01	2.51	0.57	0.26								
40	.....	16.00	4.45	1.03	0.46								
50	.....	25.10	6.96	1.61	0.71	0.19							
60	.....	36.20	10.00	2.32	1.02	0.28							
70	.....	49.30	13.70	3.16	1.40	0.37							
80	.....	64.50	17.80	4.14	1.83	0.49	0.19						
90	.....	82.80	22.60	5.23	2.32	0.62	0.24						
100	.....	.....	27.90	6.47	2.86	0.77	0.30						
125	.....	.....	48.60	10.20	4.49	1.19	0.46						
150	.....	.....	62.80	14.60	6.43	1.72	0.66	0.21					
175	.....	.....	.....	19.80	8.72	2.36	0.91	0.28					
200	.....	.....	.....	25.90	11.40	3.06	1.19	0.37	0.17				
250	.....	.....	.....	40.40	17.90	4.78	1.85	0.58	0.27				
300	.....	.....	.....	58.20	25.80	6.85	2.67	0.84	0.39	0.20			
350	.....	.....	.....	.....	35.10	9.36	3.64	1.14	0.53	0.27			
400	.....	.....	.....	.....	45.80	12.10	4.75	1.50	0.69	0.35	0.19		
450	.....	.....	.....	.....	58.00	15.40	5.98	1.89	0.88	0.46	0.25		
500	.....	.....	.....	.....	71.60	19.20	7.42	2.34	1.09	0.55	0.30		
600	.....	.....	.....	.....	.....	27.60	10.70	3.36	1.56	0.79	0.44		
700	.....	.....	.....	.....	.....	37.70	14.50	4.55	2.13	1.09	0.59		
800	.....	.....	.....	.....	.....	49.00	19.00	5.89	2.77	1.42	0.78		
900	.....	.....	.....	.....	.....	62.30	24.10	7.60	3.51	1.80	0.99		
1,000	.....	.....	.....	.....	.....	76.90	29.80	9.30	4.35	2.21	1.22		
1,500	.....	.....	.....	.....	.....	.....	67.00	21.00	9.80	4.90	2.73	1.51	0.57
2,000	.....	.....	.....	.....	.....	.....	.....	37.40	17.30	8.80	4.90	2.73	0.99
2,500	.....	.....	.....	.....	.....	.....	.....	58.40	27.20	13.80	8.30	4.20	1.57
3,000	.....	.....	.....	.....	.....	.....	.....	84.10	39.10	20.00	10.90	6.00	2.26
3,500	.....	.....	.....	.....	.....	.....	.....	.....	58.20	27.20	14.70	8.20	3.04
4,000	.....	.....	.....	.....	.....	.....	.....	.....	69.40	35.50	19.40	10.70	4.01
4,500	.....	.....	.....	.....	.....	.....	.....	.....	.....	45.00	24.50	13.50	5.10
5,000	.....	.....	.....	.....	.....	.....	.....	.....	.....	55.60	30.20	16.80	6.30
6,000	.....	.....	.....	.....	.....	.....	.....	.....	.....	80.00	43.70	24.10	9.10
7,000	.....	.....	.....	.....	.....	.....	.....	.....	.....	.....	59.50	32.80	12.20
8,000	.....	.....	.....	.....	.....	.....	.....	.....	.....	.....	77.50	42.90	16.10
9,000	.....	.....	.....	.....	.....	.....	.....	.....	.....	.....	.....	54.30	20.40
10,000	.....	.....	.....	.....	.....	.....	.....	.....	.....	.....	.....	.....	25.10
11,000	.....	.....	.....	.....	.....	.....	.....	.....	.....	.....	.....	.....	30.40
12,000	.....	.....	.....	.....	.....	.....	.....	.....	.....	.....	.....	.....	36.20
13,000	.....	.....	.....	.....	.....	.....	.....	.....	.....	.....	.....	.....	42.60
14,000	.....	.....	.....	.....	.....	.....	.....	.....	.....	.....	.....	.....	49.20
15,000	.....	.....	.....	.....	.....	.....	.....	.....	.....	.....	.....	.....	56.60

while, for a gauge pressure of 80 psi, $r = 94.7/14.7 = 6.44$. The ratio of these values of $r = 7.8/6.44 = 1.21$. Thus, the loss for an initial pressure of 80 psi will be 1.21 times the loss for an initial pressure of 100 psi. For other initial pressures the factors are given below.

Gauge pressure, psi	Factor
80	1.210
90	1.095
100	1.000
110	0.912
120	0.853
125	0.822

LOSS OF AIR PRESSURE THROUGH SCREW-PIPE FITTINGS

In order to provide for the loss of pressure resulting from the flow of air through fittings, it is common practice to convert a fitting to its equivalent length of pipe having the same nominal diameter. This equivalent length should be added to the actual length of the pipe in determining losses in pressure. Table 12-5 gives the equivalent length of standard weight pipe for computing pressure losses.

Table 12-5 Equivalent length in feet of standard-weight pipe having the same pressure losses as screwed fittings

Nominal pipe size, in.	Gate valve	Globe valve	Angle valve	Long-radius ell or on run of standard tee	Standard ell or on run of tee	Tee through side outlet
$\frac{1}{2}$	0.4	17.3	8.6	0.6	1.6	3.1
$\frac{3}{4}$	0.5	22.9	11.4	0.8	2.1	4.1
1	0.6	29.1	14.6	1.1	2.6	5.2
$1\frac{1}{4}$	0.8	38.3	19.1	1.4	3.5	6.9
$1\frac{1}{2}$	0.9	44.7	22.4	1.6	4.0	8.0
2	1.2	57.4	28.7	2.1	5.2	10.3
$2\frac{1}{2}$	1.4	68.5	34.3	2.5	6.2	12.3
3	1.8	85.2	42.6	3.1	6.2	15.3
4	2.4	112.0	56.0	4.0	7.7	20.2
5	2.9	140.0	70.0	5.0	10.1	25.2
6	3.5	168.0	84.1	6.1	15.2	30.4
8	4.7	222.0	111.0	8.0	20.0	40.0
10	5.9	278.0	139.0	10.0	25.0	50.0
12	7.0	332.0	166.0	11.0	29.8	59.6

Table 12-6 Loss of pressure, in psi, in 50 ft of hose and end couplings

Size of hose, in.	Gauge pressure at line, psi	Volume of free air through hose, cfm													
		20	30	40	50	60	70	80	90	100	110	120	130	140	150
$\frac{1}{2}$	50	1.8	5.0	10.1	18.1										
	60	1.3	4.0	8.4	14.8	23.5									
	70	1.0	3.4	7.0	12.4	20.0	28.4								
	80	0.9	2.8	6.0	10.8	17.4	25.2	34.6							
	90	0.8	2.4	5.4	9.5	14.8	22.0	30.5	41.0						
	100	0.7	2.3	4.8	8.4	13.3	19.3	27.2	36.6						
	110	0.6	2.0	4.3	7.6	12.0	17.6	24.6	33.3	44.5					
$\frac{3}{4}$	50	0.4	0.8	1.5	2.4	3.5	4.4	6.5	8.5	11.4	14.2				
	60	0.3	0.6	1.2	1.9	2.8	3.8	5.2	6.8	8.6	11.2				
	70	0.2	0.5	0.9	1.5	2.3	3.2	4.2	5.5	7.0	8.8	11.0			
	80	0.2	0.5	0.8	1.3	1.9	2.8	3.6	4.7	5.8	7.2	8.8	10.6		
	90	0.2	0.4	0.7	1.1	1.6	2.3	3.1	4.0	5.0	6.2	7.5	9.0		
	100	0.2	0.4	0.6	1.0	1.4	2.0	2.7	3.5	4.4	5.4	6.6	7.9	9.4	11.1
	110	0.1	0.3	0.5	0.9	1.3	1.8	2.4	3.1	3.9	4.9	5.9	7.1	8.4	9.9
1	50	0.1	0.2	0.3	0.5	0.8	1.1	1.5	2.0	2.6	3.5	4.8	7.0		
	60	0.1	0.2	0.3	0.4	0.6	0.8	1.2	1.5	2.0	2.6	3.3	4.2	5.5	7.2
	70	...	0.1	0.2	0.4	0.5	0.7	1.0	1.3	1.6	2.0	2.5	3.1	3.8	4.7
	80	...	0.1	0.2	0.3	0.5	0.7	0.8	1.1	1.4	1.7	2.0	2.4	2.7	3.5
	90	...	0.1	0.2	0.3	0.4	0.6	0.7	0.9	1.2	1.4	1.7	2.0	2.4	2.8
	100	...	0.1	0.2	0.2	0.4	0.5	0.6	0.8	1.0	1.2	1.5	1.8	2.1	2.4
	110	...	0.1	0.2	0.2	0.3	0.4	0.6	0.7	0.9	1.1	1.3	1.5	1.8	2.1
$1\frac{1}{4}$	50	...	...	0.2	0.2	0.2	0.3	0.4	0.5	0.7	1.1				
	60	...	...	....	0.1	0.2	0.3	0.3	0.5	0.6	0.8	1.0	1.2	1.5	
	70	...	...	....	0.1	0.2	0.2	0.3	0.4	0.4	0.5	0.7	0.8	1.0	1.3
	80	...	...	....	....	0.1	0.2	0.2	0.3	0.4	0.5	0.6	0.7	0.8	1.0
	90	...	...	....	....	0.1	0.2	0.2	0.3	0.3	0.4	0.5	0.6	0.7	0.8
	100	...	....	....	....	....	0.1	0.2	0.2	0.3	0.4	0.4	0.5	0.6	0.7
	110	...	...	....	....	....	0.1	0.2	0.2	0.3	0.3	0.4	0.5	0.5	0.6
$1\frac{1}{2}$	50	...	...	....	....	....	0.1	0.2	0.2	0.2	0.3	0.3	0.4	0.5	0.6
	60	...	...	....	....	....	....	0.1	0.2	0.2	0.2	0.3	0.3	0.4	0.5
	70	...	...	....	....	....	....	....	0.1	0.2	0.2	0.2	0.3	0.3	0.4
	80	...	...	....	....	....	....	....	....	0.1	0.2	0.2	0.2	0.3	0.4
	90	...	...	....	....	....	....	....	....	....	0.1	0.2	0.2	0.2	0.2
	100	...	...	....	....	....	....	....	....	....	....	0.1	0.2	0.2	0.2
	110	...	...	....	....	....	....	....	....	....	....	0.1	0.2	0.2	0.2

LOSS OF AIR PRESSURE IN HOSE

The loss of pressure resulting from the flow of air through hose is given in Table 12-6.

RECOMMENDED SIZES OF PIPE FOR TRANSMITTING COMPRESSED AIR

In transmitting air from a compressor to pneumatic equipment, it is necessary to limit the pressure drop along the line. If this precaution is not taken, the pressure may drop below that for which the equipment was designed and production will suffer.

At least two factors should be considered in determining the minimum size pipe. One is the necessity of supplying air at the required pressure. The other is the desirability of supplying energy, through compressed air, at the lowest total cost, considering the cost of the pipe and the cost of production obtained from the equipment. Considering the first factor, a smaller pipe may be used for a short run than for a long run. While this is possible, it may not be economical. For the latter factor, economy may dictate the use of a pipe larger than the minimum possible size. The cost of installing large pipe will be more fully justified for an installation that will be used for a long period of time than for one that will be used for a short period of time.

No book, table, or fixed data can give the correct size pipe for all installations. The correct method of determining the size pipe for a given installation is to make a complete engineering analysis of the particular installation.

Table 12-7 Recommended pipe sizes for transmitting compressed air at 80 to 125 psi gauge

Volume of air, cfm	Length of pipe, ft				
	50–200	200–500	500–1,000	1,000–2,500	2,500–5,000
	Nominal size pipe, in.				
30–60	1	1	$1\frac{1}{4}$	$1\frac{1}{2}$	$1\frac{1}{2}$
60–100	1	$1\frac{1}{4}$	$1\frac{1}{4}$	2	2
100–200	$1\frac{1}{4}$	$1\frac{1}{2}$	2	$2\frac{1}{2}$	$2\frac{1}{2}$
200–500	2	$2\frac{1}{2}$	3	$3\frac{1}{2}$	$3\frac{1}{2}$
500–1,000	$2\frac{1}{2}$	3	$3\frac{1}{2}$	4	$4\frac{1}{2}$
1,000–2,000	$2\frac{1}{2}$	4	$4\frac{1}{2}$	5	6
2,000–4,000	$3\frac{1}{2}$	5	6	8	8
4,000–8,000	6	8	8	10	10

Table 12-7 gives recommended sizes of pipe for transmitting compressed air for various lengths of run. This information is useful as a guide in selecting pipe sizes.

RECOMMENDED SIZES OF HOSE FOR TRANSMITTING COMPRESSED AIR

Most pneumatic equipment and tools require a length of flexible hose between the source of air and the equipment. As the loss of pressure in the hose is relatively high, the length should be no greater than is required for satisfactory operation.

Table 12-8 gives the recommended sizes of hose for transmitting various quantities of compressed air and for various types of pneumatic equipment and tools frequently used on construction projects.

DIVERSITY OR CAPACITY FACTOR

While it is necessary to provide as much compressed air as will be required to supply the needs of all operating equipment, it is unnecessarily extravagant to provide more air capacity than will be needed. It is probable that all equipment nominally used on a project will not be in operation at any given time. An analysis of the job should be made to determine the maximum actual need prior to designing the compressed-air system.

If 10 jackhammers are nominally drilling, it is probable that not more than 5 or 6 will be consuming air at a given time. The others will be out of use temporarily for changes in bits or drill steel or moving to new locations. Thus, the actual amount of air demand will be based on 5 or 6 drills instead of 10. The same condition will apply to other pneumatic tools.

Capacity factor is the ratio of the average load to the maximum mathematical load that would exist if all tools were operating at the same time. This ratio is also referred to as a diversity factor. For example, if a jackhammer required 90 cfm of air, 10 hammers would require a total of 900 cfm if they were all operated at the same time. However, with only 5 hammers operating at one time, the demand for air would be 450 cfm. Thus, the diversity factor would be $450 \div 900 = 0.5$.

Table12-9 illustrates a method of applying diversity factors to a project in which excavation is the primary operation.

AIR REQUIRED BY PNEUMATIC EQUIPMENT AND TOOLS

The approximate quantities of compressed air required by pneumatic equipment and tools are given in Table 12-10. The quantities are based on continuous operation at a pressure of 90 psi gauge.

Table 12-8 Recommended sizes of hose, in inches, for transmitting compressed air at 80- to 125-psi gauge

Volume of air, cfm	Types of air tools	Length of hose, ft		
		0–25	25–50	50–200
0–15	Spray guns $\frac{1}{4}$-in. drills Light chipping and scaling hammers $\frac{3}{8}$-in. impact wrenches	$\frac{5}{16}$	$\frac{3}{8}$	$\frac{1}{2}$
15–30	$\frac{5}{16}-\frac{1}{2}$-in. drills $\frac{5}{8}$-in. impact wrenches Chipping hammers 15-lb rock drills	$\frac{3}{8}$	$\frac{1}{2}$	$\frac{1}{2}$
30–60	$\frac{5}{8}$–1-in. drills $\frac{3}{4}$-in. impact wrenches Light grinders Rivet hammers Clay diggers Backfill tampers Small concrete vibrators Light and medium demolition tools 25-lb rock drills	$\frac{1}{2}$	$\frac{3}{4}$	$\frac{3}{4}$
60–100	1–2-in. drills $1\frac{1}{4}-1\frac{3}{4}$-in. impact wrenches Heavy grinders Large concrete vibrators Sump pumps 35–55-lb rock drills Heavy demolition tools	$\frac{3}{4}$	$\frac{3}{4}$	1
100–200	Winches and hoists Drifters Wagon drills 75-lb rock drills	1	1	$1\frac{1}{4}$

EFFECT OF ALTITUDE ON THE CONSUMPTION OF AIR BY ROCK DRILLS

As previously explained in this chapter, the capacity of an air compressor is the volume of free air that enters the compressor during a stated time, usually expressed in cfm. Because of the lower atmospheric pressure at higher altitudes, the quantity of air supplied by a compressor at a given gauge pressure will be less than at sea level. It is necessary to provide more compressor

Table 12-9 Illustration of the application of diversity factors in designing a compressed-air system

Equipment	Air required per unit, cfm	Number of units		Maximum air demand, cfm	Diversity factor	Probable air demand, cfm
		on job	working			
Wagon drills	200	6	4	1,200	0.67	800
Jackhammers	100	16	8	1,600	0.50	800
Drill sharpeners	160	2	1	320	0.50	160
Oil furnaces	80	2	2	160	1.00	160
Grinders	50	2	1	100	0.50	50
Sump pumps	160	3	2	480	0.67	320
Line loss	...	...	...	220	....	220
Total	...	...	...	4,080	....	2,510
Job diversity factor	...	...	...	....	0.80	
Total actual demand, 0.80 × 2,510	...	...	...	.....	....	2,008

Table 12-10 Quantities of compressed air required by pneumatic equipment and tools*

Equipment or tools	Capacity or size	Air consumption, cfm
Chipping hammers	Light	15–25
	Heavy	25–30
Clay diggers	Light, 20 lb	20–25
	Medium, 25 lb	25–30
	Heavy, 35 lb	30–35
Concrete vibrators	$2\frac{1}{2}$-in. tube diameter	20–30
	3-in. tube diameter	40–50
	4-in. tube diameter	45–55
	5-in. tube diameter	75–85
Drills or borers	1-in. diameter	35–40
	2-in. diameter	50–75
	4-in. diameter	50–75
Hoist	Single-drum, 2,000 lb pull	200–220
	Double-drum, 2,400 lb pull	250–260
Impact wrenches	$\frac{5}{8}$-in. bolt	15–20
	$\frac{3}{4}$-in. bolt	30–40
	$1\frac{1}{4}$-in. bolt	60–70
	$1\frac{1}{2}$-in. bolt	70–80
	$1\frac{3}{4}$-in. bolt	80–90

* Air pressure at 90 psi gauge.

Table 12-10 Quantities of compressed air required by pneumatic equipment and tools* (continued)

Equipment or tools	Capacity or size: Weight, lb	Capacity or size: Depth of hole, ft	Air consumption, cfm
Jackhammers	10	0–2	15–25
	15	0–2	20–35
	25	2–8	30–50
	35	8–12	55–75
	45	12–16	80–100
	55	16–24	90–110
	75	8–24	150–175
Paving breakers	35	.....	30–35
	60	.....	40–45
	80	.....	50–50
Riveting hammers	$\frac{5}{8}$-in. rivet		25–30
	$\frac{3}{4}$-in. rivet		30–35
	$\frac{7}{8}$-in. rivet		35–40
	$1\frac{1}{8}$-in. rivet		40–45
	$1\frac{1}{4}$-in. rivet		40–45
Saws:			
Circular	12-in. blade		40–60
Chain	18-30-in. blade		85–95
	36-in. blade		135–150
	48-in. blade		150–160
Reciprocating	20-in.		45–50
Spray guns	Light-duty		2–3
	Medium-duty		8–15
	Heavy duty		14–30
Sump pump	Single-stage, 10–40 ft head		80–90
	Single-stage, 100–150 ft head		150–170
	Two-stage, 100–150 ft head		160–180
Trampers, earth	35 lb		30–35
	60 lb		40–45
	80 lb		50–60
Wagon drills—drifters	3-in. piston		150–175
	$3\frac{1}{2}$-in. piston		180–210
	4-in. piston		225–275

* Air pressure at 90 psi gauge.

USE 100 psi

capacity at higher altitudes to assure an adequate supply of air at the specified pressure to rock drills.

Table 12-11 gives representative factors to be applied to specified compressor capacities to determine the required capacities at different altitudes. For example, if a single drill requires a capacity of 600 cfm of air at sea level, it will require a capacity of $600 \times 1.2 = 720$ cfm at an altitude of 5,000 ft, and $600 \times 1.3 = 780$ cfm at an altitude of 10,000 ft.

The values of the factors are adjusted to reflect representative diversity factors for the use of multidrills. Because these factors will not necessarily apply to all drills and projects, they should be used as a guide only.

THE COST OF COMPRESSED AIR

The cost of compressed air may be determined at the compressor or at the point of use. The former will include the cost of compressing, while the latter will include the cost of compressing plus transmitting, including line losses.

The cost of compressing should include the cost of the compressor, insurance, taxes, interest, maintenance, repair, fuel, lubrication, and labor. The cost is usually based on 1,000 cu ft of free air.

Eample 12-6 Determine the cost of compressing 1,000 cu ft of free air to a gauge pressure of 100 psi by using a 600-cfm two-stage portable compressor driven by a 180-hp diesel engine.

Table 12-11 Factors to be used in determining the capacities of compressed air required by rock drills at different altitudes*

	Number of drills									
	1	2	3	4	5	6	7	8	9	10
Altitude, ft	Factor									
0	1.0	1.8	2.7	3.4	4.1	4.8	5.4	6.0	6.5	7.1
1,000	1.0	1.9	2.8	3.5	4.2	4.9	5.6	6.2	6.7	7.3
2,000	1.1	1.9	2.9	3.6	4.4	5.1	5.8	6.4	7.0	7.6
3,000	1.1	2.0	3.0	3.7	4.5	5.3	5.9	6.6	7.2	7.8
5,000	1.1	2.1	3.1	3.9	4.7	5.5	6.1	6.8	7.4	8.1
5,000	1.2	2.1	3.2	4.0	4.8	5.6	6.3	7.0	7.6	8.3
6,000	1.2	2.2	3.2	4.1	4.9	5.8	6.5	7.2	7.8	8.5
7,000	1.2	2.2	3.3	4.2	5.0	5.9	6.6	7.4	8.0	8.7
8,000	1.3	2.3	3.4	4.3	5.2	6.1	6.8	7.6	8.2	9.0
9,000	1.3	2.3	3.5	4.4	5.3	6.2	7.0	7.7	8.4	9.2
10,000	1.3	2.4	3.6	4.5	5.4	6.3	7.1	7.9	8.6	9.4
12,000	1.4	2.5	3.7	4.6	5.6	6.6	7.4	8.2	8.9	9.7
15,000	1.4	2.6	3.9	4.7	5.9	6.9	7.7	8.6	9.3	1.02

* Compressed Air and Gas Institute.

The following information will apply:

Cost delivered to the project = $40,845
Life, 5 yr at 2,000 hr per yr
Average investment, 0.6 × $40,845 = $24,507
Fuel consumed per hr, 0.04 × 180 = 7.2 gal
Lubricating oil consumed per hr, 0.125 gal
The costs will be:
Annual costs:

Depreciation, $40,845 ÷ 5	= $ 8,169
Repairs, 75% of $8,169	= 6,126
Investment, 22% × $24,507	= 5,391
Total annual fixed cost	= $19,686

Hourly costs:

Fixed cost, $19,686 ÷ 2,000 hr	= $ 9.84
Fuel, 7.2 gal @ $1.00	= 7.20
Lubricating oil, 0.125 gal @ $3.20	= 0.40
Operator, 1/2 time @ $16.00 per hr	= 8.00*
Total cost per hr	= $25.44

*This cost will vary with location and union requirements. Volume of air compressed per hr, 60 × 600 = 36,000 cu ft.

Cost per 1,000 cu ft, $25.44 ÷ 36 = $0.71.

The cost per 1,000 cu ft of air for a compressor operating under various load factors might be as follows:

	Load factor, %		
	100	75	50
Hourly costs:			
Fixed cost*	$ 9.84	$ 9.84	$ 9.84
Fuel	7.20	5.38	4.28
Lubricating oil	0.40	0.30	0.24
Operator, ½ time†	8.00	8.00	8.00
Total cost per hr	$25.44	$23.52	$22.36
Volume of air per hour cu ft	36,000	27,000	18,000
Cost per 1,000 cu ft	$ 0.71	$ 0.87	$ 1.24

*This cost may vary slightly with the load factor.

†This cost may vary with the locations and the requirements of the local unions.

THE COST OF AIR LEAKS

The loss of air through leakage in a transmission line can be surprisingly large and costly. It results from poor pipe connections, loose valve stems, deteriorated hose, and loose hose connections. If the cost of such leaks were more fully known, most of them would be eliminated. The rate of leakage

Table 12-12 Cost per month for air leakage

Size of opening, in.	Cu ft or air lost per month at 100 psi	For indicated cost per 1,000 cu ft			
		$ 0.45	$ 0.60	$ 0.75	$ 1.20
$\frac{1}{32}$	45,500	$ 19.14	$ 27.30	34.14	$ 45.50
$\frac{1}{16}$	182,300	82.20	109.50	136.95	182.30
$\frac{1}{8}$	740,200	333.00	444.00	555.00	740.20
$\frac{1}{4}$	2,920,000	1,314.00	1,782.00	2,190.00	2,920.00

through an opening of known size can be determined by applying a formula for the flow of air through an orifice.

Table 12-12 illustrates the cost of air leakage for various sizes of openings and costs per 1,000 cu ft of air.

THE COST OF USING LOW AIR PRESSURE

The effect on the cost of production of operating pneumatic equipment at less than the recommended air pressure can be demonstrated by analyzing the performance of a group of jackhammers under different pressures. The hammers receive the air from a common header-type pipe line. Similar results would be obtained when using other kinds of pneumatic equipment.

Example 12-7 Determine the economy of using a 3-in. pipe instead of a $2\frac{1}{2}$-in. pipe to transmit compressed air to jackhammers. The air will be supplied to the entrance of each pipe at a pressure of 100 psi gauge. The stated conditions will apply.

Length of pipe, 1,000 ft
Installed cost of 3-in. pipe, $7,550
Installed cost of $2\frac{1}{2}$-in. pipe, $5,665
Extra cost of 3-in. pipe, $1,885
Extra cost chargeable to this project, considering the salvage value of the pipe, $1,200
Estimated length of project, 4 months
Hours worked per month, 180
No. jackhammers on the job, 16
No. jackhammers operating at one time, 8
Size of jackhammers, 55 lb
Air required per hammer at 90 psi, 100 cfm
Total air required, $8 \times 100 = 800$ cfm

Loss of pressure in 3-in. pipe, from Table 12-4, is $1.0 \times 5.89 = 5.9$ psi
Loss of pressure in 50 ft of $\frac{3}{4}$-in. hose, from Table 12-6, is 4.7 psi, interpolated for 100 cfm at a pressure of 94 psi
Total loss in pressure through pipe and hose, 10.6 psi
Air pressure at the hammer, $100.00 - 10.6 = 89.4$ psi
Loss in pressure through the $2\frac{1}{2}$-in. pipe, from Table 12-4, 19.0 psi

Pressure entering the $\frac{3}{4}$-in. hose, 81 psi
Loss of pressure in 50 ft of $\frac{3}{4}$-in. hose, 5.8 psi
Total loss in pressure through $2\frac{1}{2}$-in. pipe and hose, 24.8 psi
Pressure at the hammer, $100.00 - 24.8 = 75.2$ psi

If the pressure of air at the hammer is reduced to 75.2 psi, the quantity of air required to operate each hammer will be 90 cfm instead of 100 cfm.

The total quantity of air required will be $8 \times 90 = 720$ cfm.

The hammer efficiency at 75.2 psi will be about 80 percent of that at 90 psi.

Assume that the cost of air will be \$0.87 per 1,000 cu ft. Consider the effect of using each size pipe for 1 hr, as it applies to the rate of drilling rock, and the indicated costs. The results will be

Item	Size of pipe, in.	
	$2\frac{1}{2}$	3
Volume of air consumed, cu ft per hr		
by $2\frac{1}{2}$-in.-pipe, 60×720	43,200	
by 3-in. pipe		48,000
Cost of air @ \$0.87 per 1,000 cfm	\$ 37.58	\$ 41.76
Cost of labor, 20 men @ \$12.00 per hr	240.00	240.00
Cost of jackhammers, steel and bits*	36.00	42.00
Total cost per hr	\$313.58	\$323.76

*The cost per hour for jackhammers, steel and bits, is increased to compensate for the greater wear and reduced life of the units operated at the higher pressure provided by the 3-in. pipe.

Tests conducted on drilling equipment indicate that the drills operating at the lower pressure, namely 75.2 psi, will have an efficiency equal to about 80 percent of those operating at 89.4 psi. Thus, the increase in the depth of hole drilled at the higher pressure will be $100 - 80 = 20\%$

Value of increased production as related
to total cost per hr, $0.20 \times \$313.58 \quad = \$62.70$
Increased cost per hr with 3-in. pipe,
$\$323.76 - \$313.55 \quad = -10.18$
Net value of increased production per hr = \$52.52

The total value of the increased rate of production during the project will be:

Length of project, 4 months $\times$ 180 hr = 720 hr
Value of increased production, 720 hr @ \$52.52 = \$37,814.40

Thus, it is evident that spending an extra \$1,200 to provide the larger pipe is an excellent investment. If the length of the project is greater than 4 months, the value of using the larger pipe will be proportionally greater.

PROBLEMS

12-1 An air compressor draws in 1,000 cu ft of air at a gauge pressure of 0 psi and a temperature of 70°F. The air is compressed to a gauge pressure of 100 psi at a temperature of 140°F. The atmospheric pressure is 14.0 psi. Determine the volume of air after it is compressed.

12-2 An air compressor draws in 1,000 cu ft of free air at a gauge pressure of 0 psi and a temperature of 60°F. The air is compressed to a gauge pressure of 100 psi at a temperature of 130°F. The atmospheric pressure is 12.30 psi. Determine the volume of air after it is compressed.

12-3 Determine the theoretical horsepower required to compress 600 cfm of free air, measured at standard conditions, from atmospheric pressure to 100-psi gauge pressure when the compression is performed under isothermal conditions.

12-4 Solve Prob. 12-3 if the air is compressed under adiabatic conditions.

12-5 Determine the difference in horsepower required to compress 600 cfm of free air under adiabatic conditions for altitudes of 3,000 and 8,000 ft. The air will be compressed to 100-psi gauge pressure for each altitude.

12-6 A compressor has a capacity of 500 cfm of free air at 100-psi gauge pressure at zero altitude. If the compressor is operating at an altitude of 8,000 ft, determine the capacity when the air is compressed to 100-psi gauge pressure, with no change in temperature.

12-7 Compressors operating at zero altitude will supply enough air to operate eight drills. If the compressors and drills are operated at an altitude of 6,000 ft, how many drills can the compressor serve?

12-8 A 3-in. pipe with screwed fittings is used to transmit 1,000 cfm of free air at an initial pressure of 100-psi gauge pressure. The pipe line includes the following items:

900 ft of pipe
3 gate valves
8 on-run tees
6 standard ells

Determine the total loss of pressure in the pipe line.

12-9 If the air from the end of the pipe line of Prob. 12-8 is delivered through 50 ft of 1-in. hose to a rock drill that requires 160 cfm of air, determine the pressure at the drill.

REFERENCES

1. Harris, E. G.: *University of Missouri Bulletin*, vol. 1, no. 4, 1912.
2. Atlas Copco, Inc., 70 Demarest Drive, Wayne, NJ 07470.
3. Chicago Pneumatic Tool Company, 6 East 44th Street, New York, NY 10017.
4. Ingersoll-Rand Company, Phillipsburg, NJ 08865.

CHAPTER

THIRTEEN

DRILLING ROCK AND EARTH

INTRODUCTION

This chapter will deal with the equipment and methods used by the construction and mining industries to drill holes in both rock and earth. Although the same or similar equipment may in some instances be used for drilling both materials, they will be treated separately in this chapter.

Because the purposes for which drilling is performed vary a great deal from general to highly specialized, it is desirable to select the equipment and methods best suited to the specific service. For example, a contractor engaged in highway construction which requires drilling rock under varying conditions should select equipment that is suitable for various services. However, if equipment is selected to drill rock in a quarry where the material and conditions will not vary, specialized equipment should be considered. In some instances custom-made equipment, designed for use on that project only, may be justified.

More complete information related to the performance characteristics of drilling equipment will be presented in this chapter.

DEFINITIONS OF TERMS

Terms which are commonly used in describing drilling equipment and procedures are given below as a guide for the reader.

Percussion drill This is a drill which breaks rock into small particles by the impact from repeated blows.

Abrasion drill This drill grinds rock into small particles through the abrasive effect of a bit that rotates in the hole.

Cuttings Cuttings are the disintegrated rock particles that are removed from a hole.

Jackhammer, or sinker This device is an air-operated percussion-type drill that is small enough to be handled by one worker.

Drifter A drifter is an air-operated percussion-type drill, similar to a jackhammer, but so large that it requires mechanical mounting.

Wagon drill This is a drifter mounted on a mast supported by two or more wheels.

Stoper A stoper is an air-operated percussion-type drill, similar to a drifter, that is used for overhead drilling, as in a tunnel.

Churn drill The churn drill is a percussion-type drill consisting of a long steel bit that is mechanically lifted and dropped to disintegrate the rock. It is used to drill deep holes, usually 6 in. in diameter or larger.

Blast-hole drill This is a rotary drill consisting of a steel-pipe drill stem on the bottom of which is a roller bit that disintegrates the rock as it rotates over it. The cuttings are removed by a stream of compressed air.

Shot drill This is a rotary abrasive-type drill whose bit consists of a section of steel pipe with a roughened surface at the bottom. As the bit is rotated under pressure, chilled-steel shot are supplied under the bit to accomplish the disintegration of the rock. The cuttings are removed by water.

Diamond drill The diamond drill is a rotary abrasive-type drill whose bit consists of a metal matrix in which there are embedded a large number of diamonds. As the drill rotates, the diamonds disintegrate the rock. This drill is used extensively to obtain core samples.

Dry drill This is a drill which uses compressed air to remove the cuttings from a hole.

Wet drill A wet drill is one that uses water to remove the cuttings from a hole.

Core drilling Core drilling is the obtaining of core samples of rock from a hole, usually for exploratory purposes. The diamond and shot drills are used for core drilling.

Bit This is the portion of a drill which contacts the rock and disintegrates it. Many types are used.

Detachable bit This is a bit which may be attached to or removed from the drill steel or drill stem.

Forged bit This is a bit which is forged on drill steel.

Carbide-insert bit The carbide-insert bit is a detachable bit whose cutting edges consist of tungsten carbide embedded in a softer steel base.

Diamond bit The diamond bit is a detachable bit whose cutting elements consist of diamonds embedded in a metal matrix.

Depth per bit This is the depth of hole that can be drilled by a bit before it is replaced.

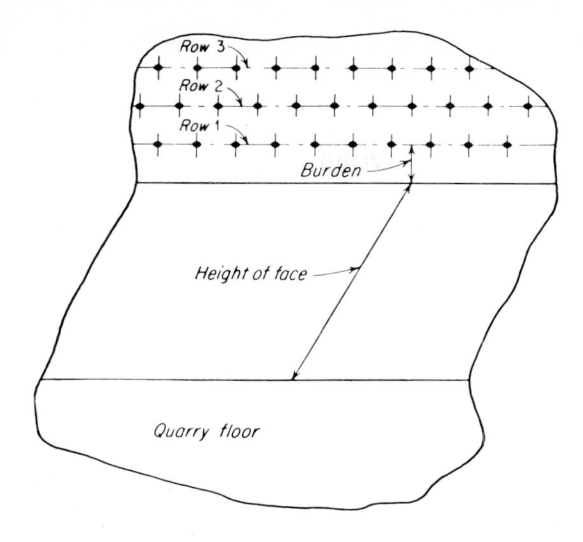

Figure 13-1 Dimensional terminology for drilling rock.

Drilling rate Drilling rate is the number of feet of hole drilled per hour per drill.
Face Face is the approximately vertical surface extending upward from the floor of a pit to the level at which drilling is being done.
Burden This is the horizontal distance from a face back to the first row of drill holes.
Drilling pattern Drilling pattern is the spacing of the drill holes.

Figure 13-1 illustrates the dimension terminology frequently used for drilling.

BITS

The bit is the essential part of a drill, as it is the part which must engage and disintegrate the rock. The success of a drilling operation depends on the ability of the bit to remain sharp under the impact of the drill. Many types and sizes are available.

Until recent years the bits for jackhammers and drifters were forged on one end of the drill steel. This practice has been pretty well discontinued in favor of detachable bits, which are screwed to the drill steel. Detachable bits have many advantages compared with forged bits. They are easily replaced and resharpened, are available in various sizes, shapes, and hardness, and are relatively inexpensive. They are usually resharpened on a grinder.

Steel bits for jackhammers and drifters are illustrated in Fig. 13-2. They are available in sizes from 1 to $4\frac{1}{2}$ in., the gauge size varying in steps of $\frac{1}{8}$ in. These bits may be resharpened two to six times.

The depth of hole that can be drilled with a steel bit will vary from a few inches to 30 or 40 ft or more, depending on the type of rock.

Carbide-insert bits Some types of rock are so abrasive that steel bits must be replaced after they have drilled only a few inches of hole. The cost of the bits and the time lost in changing are so great that it will usually be economical to use carbide-insert bits. This bit is illustrated in Figs. 13-2, 13-3, and 13-4. As noted in the figures, the actual drilling points consist of a very hard metal, tungsten carbide, which is embedded in steel. Although these bits are considerably more expensive than steel bits, the increased drilling rate and depth of hole obtained per bit will give an over-all economy in drilling hard rock.

A contractor on a highway project in Pennsylvania found that when drilling diabase rock the depth per steel bit was $\frac{1}{2}$ to 2 in. When he changed to carbide bits, he obtained an average depth per bit of 1,992 ft. The estimated saving by using carbide bits, in drilling 30,000 cu yd of rock, was in excess of $100,000. The cost analysis for this project is as follows:

Total quantity of rock, 300,000 cu yd
Depth of holes, 12 ft
Size of bits, $2\frac{1}{4}$ in.

Figure 13-2 Removable shoulder-drive-type rock bits. *(The Timken Company.)*

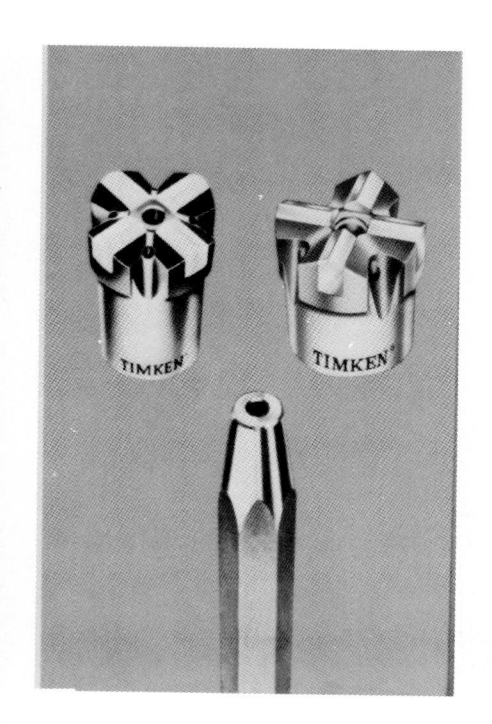

Figure 13-3 Removable tapered-socket-type rock bits. *(The Timken Company.)*

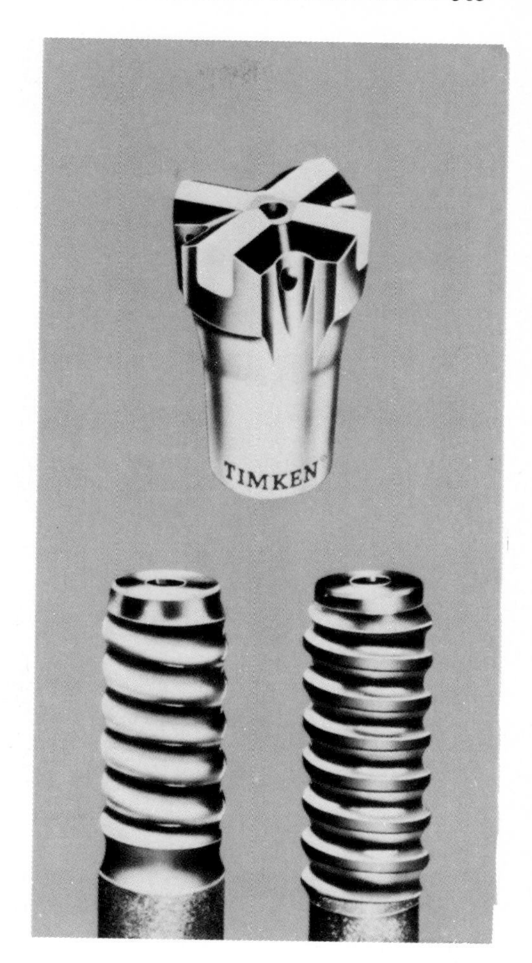

Figure 13-4 Removable bottom-drive-type carbide insert rock bit. *(The Timken Company.)*

Cu yd of rock per ft of hole, $2\frac{1}{2}$
Total depth of hole required, 300,000 ÷ 2.5 = 120,000 ft
Average depth of hole per steel bit, with resharpening, 0.48 ft
No. bits required, 120,000 ÷ 0.48 = 250,000
Average depth of hole per carbide bit, 1,992 ft
No. bits required, 120,000 ÷ 1,992 = 60

This result does not include the value of time saved in changing bits. This is an exceptional case, as such savings are not possible on all projects.

Tapered socket bits Figure 13-3 illustrates removable tapered socket bits, which are available in gauge sizes varying in $\frac{1}{8}$-in. (3.2-mm) steps from about 1 in. (25 mm) to 4 in. (102 mm) or more.

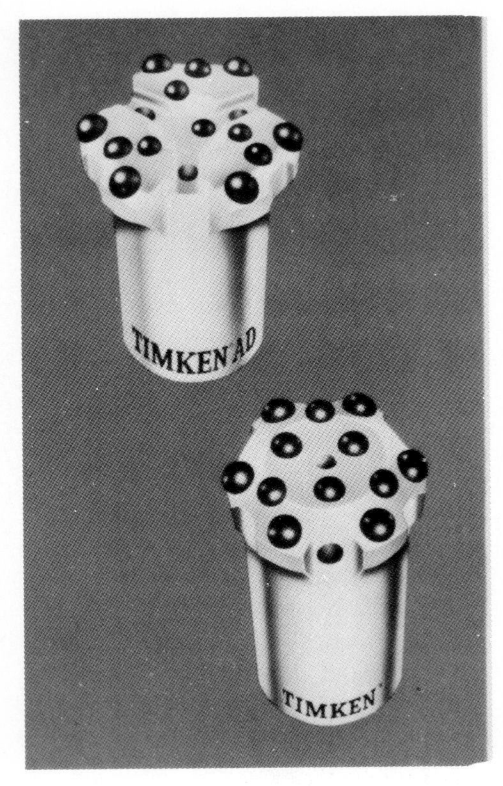

Figure 13-5 Removable button bits. *(The Timken Company.)*

Bottom-drive bits Figure 13-4 illustrates removable bottom-drive bits, which are available in gauge sizes varying from about $1\frac{1}{2}$ in. (38 mm) to 6 in. (152 mm) or more.

Button bits Figure 13-5 illustrates removable button bits, which are available in numerous sizes. These bits, which are available in different cutting face designs with a choice of insert grades, require no regrinding or sharpening. Their demonstrated performance has been superior to that of other types of bits when they are used to drill rocks that are more suitable for them, as indicated later in this chapter.

JACKHAMMERS

Jackhammers are hand-held air-operated percussion-type drills which are used primarily for drilling down holes. For this reason, they are frequently called sinkers. They are classified according to their weight, such as 45 or 55 lb. A complete drilling unit consists of a hammer, drill steel, and bit. As the compressed air flows through a hammer, it causes a piston to reciprocate at a speed up to 2,200 blows per minute, which produces the hammer effect. The energy of this piston is transmitted to a bit through the drill steel. Some of the air flows through a hole in the drill steel and the bit to remove the cuttings from the hole and to cool the bit. For wet drilling, water is used instead of air to remove the cuttings. Figure 13-6 shows a sectionalized jackhammer with the essential parts indicated. The drill steel is rotated slightly following each blow so that the points of the bit will not strike at the same spot each time.

Although jackhammers may be used to drill holes in excess of 20 ft deep, they seldom are used for holes exceeding 10 ft deep. The heavier hammers will drill holes up to $2\frac{1}{2}$ in. in diameter. Drill steel usually is supplied in 2-ft-length variations, but longer lengths are available. Representative specifications for jackhammers are listed in Table 13-1.

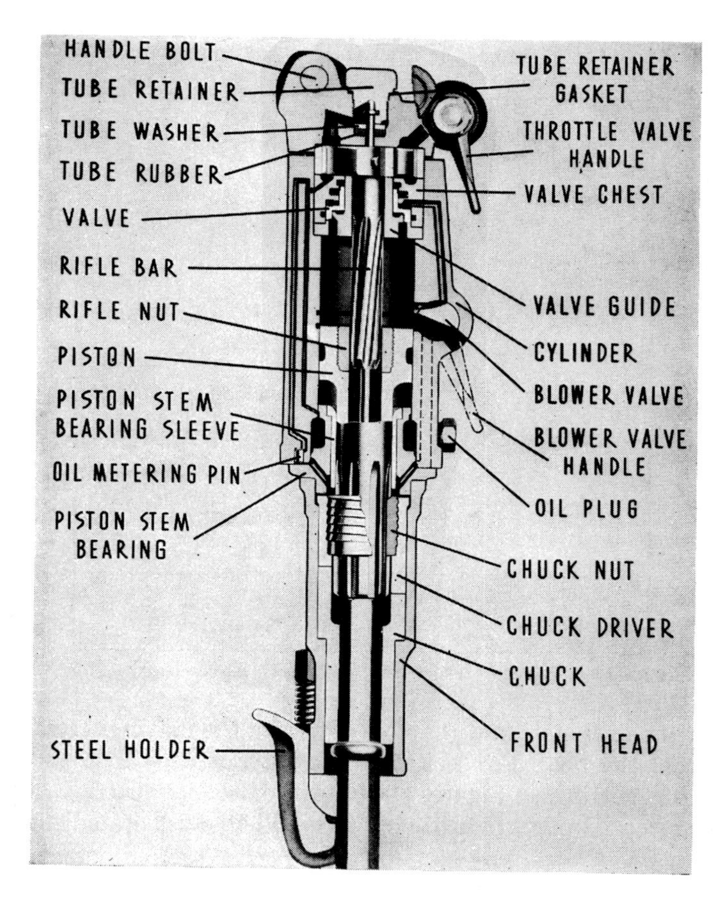

Figure 13-6 Section through a jackhammer.

Table 13-1 Representative specifications for jackhammers

Model	S33	S55	S73
Length overall, in.	$20\frac{1}{8}$	$23\frac{3}{8}$	25
Cylinder bore, in.	$2\frac{3}{8}$	$2\frac{5}{8}$	$2\frac{3}{4}$
Weight, lb	31	$56\frac{1}{2}$	67
Size steel recommended, in.	$\frac{7}{8}$	$\frac{7}{8}-1$	$\frac{7}{8}-1\frac{1}{4}$
Size air hose recommended, in.	$\frac{3}{4}$	$\frac{3}{4}-1$	$\frac{3}{4}-1$
Size water hose recommended, in.	$\frac{1}{2}$	$\frac{1}{2}$	$\frac{1}{2}$

DRIFTERS

Drifter drills are similar to jackhammers in operation, but they are larger and are used as mounted tools for drilling down, horizontal, or up holes. They vary in weight from 75 to 260 lb and are capable of drilling holes up to $4\frac{1}{2}$ in. in diameter. These tools are used extensively in mining and tunneling. Either air or water may be used to remove the cuttings.

When drifters are used for horizontal or up drilling, the feed pressure is supplied by a hand-operated screw or a pneumatic or hydraulic piston. The weight is usually sufficient to supply the necessary pressure for down drilling. Steel changes may be obtained in lengths of 24, 30, 36, 45, and 60 in. Representative specifications for automatic-feed drifters are listed in Table 13-2.

WAGON DRILLS

Wagon drills consist of drifters mounted on masts which are mounted on wheels to provide portability. They are used extensively to drill holes up to $4\frac{1}{2}$ in. in diameter and up to 30 ft or more in depth. They give better performance than jackhammers when used on terrain where it is possible for them to operate. They may be used to drill at any angle from down to slightly above

Table 13-2 Representative specifications for automatic-feed drifters

Model	79	89	93	99
Cylinder bore, in.	3	$3\frac{1}{2}$	$3\frac{1}{2}$	4
Size chuck available, in.	$\frac{7}{8}-1\frac{1}{4}$	$\frac{7}{8}-1\frac{1}{4}$	$\frac{7}{8}-1\frac{1}{4}$	$\frac{1}{4}-1\frac{1}{2}$
Size air hose recommended, in.	1	1	1	1
Size water hose recommended, in.	$\frac{1}{2}$	$\frac{1}{2}$	$\frac{1}{2}$	$\frac{1}{2}$
Weight of drill, less mounting, lb	111	134	140	181
Overall length, in.	$31\frac{3}{4}$	34	35	$35\frac{1}{8}$

Figure 13-7 Gasoline-engine-operated hand drill. *(Atlas Copco, Inc.)*

horizontal. The length of drill steel may be 6, 10, or 15 ft, or more, depending on the length of feed of the particular wagon drill.

TRACK-MOUNTED DRILLS

The track-mounted drills illustrated in Figs. 13-8, 13-9, and 13-10 have substantially replaced the wagon drill on construction work. Because of its ability to move quickly to a new location and, using the hydraulically operated boom, position the drill for resumption of drilling, its production rate may be three or more times that of a wagon drill. Holes can be drilled at any angle from under 15° back from vertical to above the horizontal, ahead, or on either side of the unit. All operations, including tramming, are powered by compressed air.

Depending on the size unit selected, these machines can drill holes up to about 6 in. in diameter, and to depths of 50 ft or more.

WHEEL-MOUNTED DRILLS

Figures 13-11 and 13-12 illustrate wheel-mounted drills. These drills are similar in sizes and capacities to the track-mounted drills. However, these drills require a more nearly level ground upon which to operate.

Figure 13-8 Track-mounted drill equipped with dust collector. *(Atlas Copco, Inc.)*

Figure 13-9 Track-mounted dual drills equipped with dust collector. *(Atlas Copco, Inc.)*

Figure 13-10 Track-mounted air compressor and drill. *(Ingersoll-Rand Company.)*

ROTARY-PERCUSSION DRILLS

These drills combine the hard-hitting reciprocal action of the percussion drill with the turning-under-pressure action of the rotary drill. Whereas the percussion drill only has a rotary action to reposition the bit's cutting edges, the rotation of this combination drill, with the bit under constant pressure, has demonstrated its ability to drill much faster than the regular percussion drill (Fig. 13-13). On the Smith Power tunnel near Eugene, Oregon, rotary-percussion drills are reported to have drilled blast holes three times as fast as regular percussion drills [1]. These drills require special carbide bits with the carbide inserts set at a different angle than those used with standard carbide bits.

In the Smith Power tunnel, four of these drills operating on a two-deck rail-mounted jumbo are reported to have drilled $1\frac{3}{4}$-in.-diameter holes at rates that varied from 5 to 10 fpm, depending on the hardness of the rock. A round of 40 to 48 holes, having an average depth of 8 ft, was drilled in as little as 15 min.

Figure 13-11 Wheel-mounted drill. *(Schramm, Inc.)*

Figure 13-12 Wheel-mounted drill. *(Joy Manufacturing Company.)*

PISTON DRILLS

These are percussion-type drills with the hollow drill tube attached to the piston. The stroke and rotation of the piston are adjustable to give the best performance for the particular type of rock being drilled. It is available with carbide-insert bits which are up to 6 in. in diameter. The drill has a practical depth limit of approximately 70 ft.

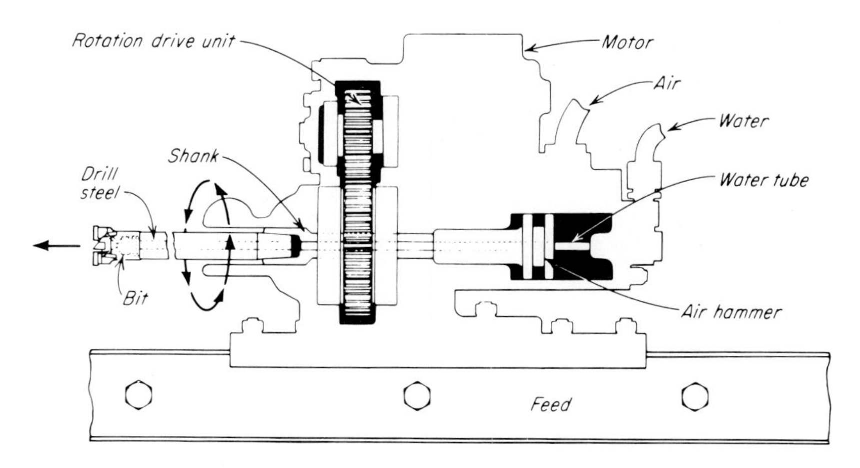

Figure 13-13 Rotary-percussion drill. *(Joy Manufacturing Company.)*

BLASTHOLE DRILLS

The blasthole drill is a self-propelled drill which is mounted on a truck or on crawler tracks (Fig. 13-14). Drilling is accomplished with a tri-cone roller-type bit attached to the lower end of a drill pipe. As the bit is rotated in the hole, a

Figure 13-14 Blasthole drill. *(Schramm, Inc.)*

Figure 13-15 Blasthole drill. *(Schramm, Inc.)*

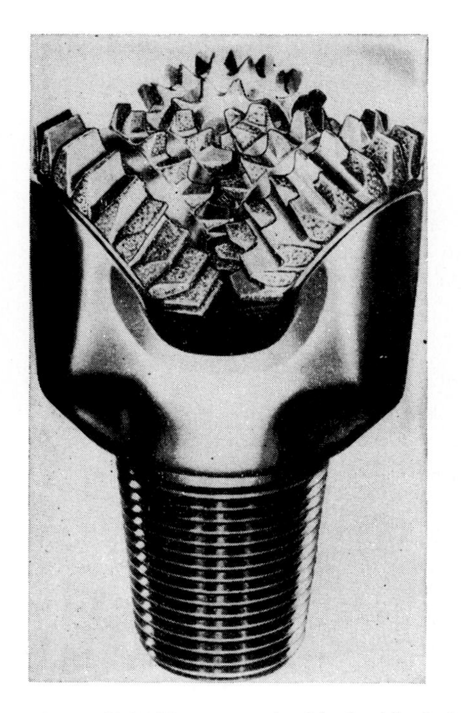

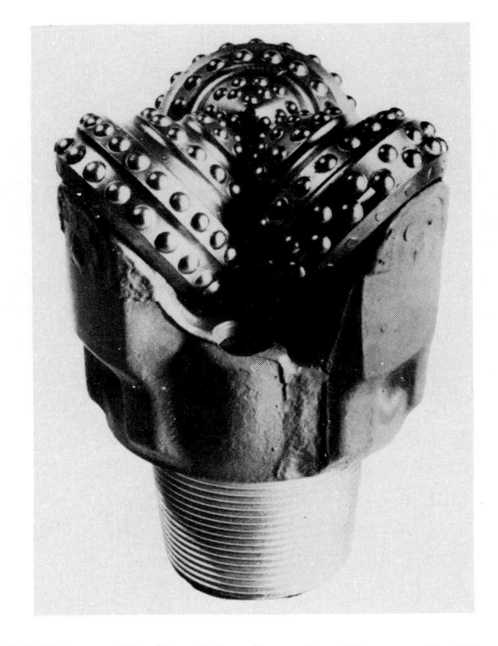

Figure 13-16 Representative bits for blastholes. (*a*) Tricone bit *(Joy Manufacturing Company)*. (*b*) Button-type bit. *(Reed Tool Company)*.

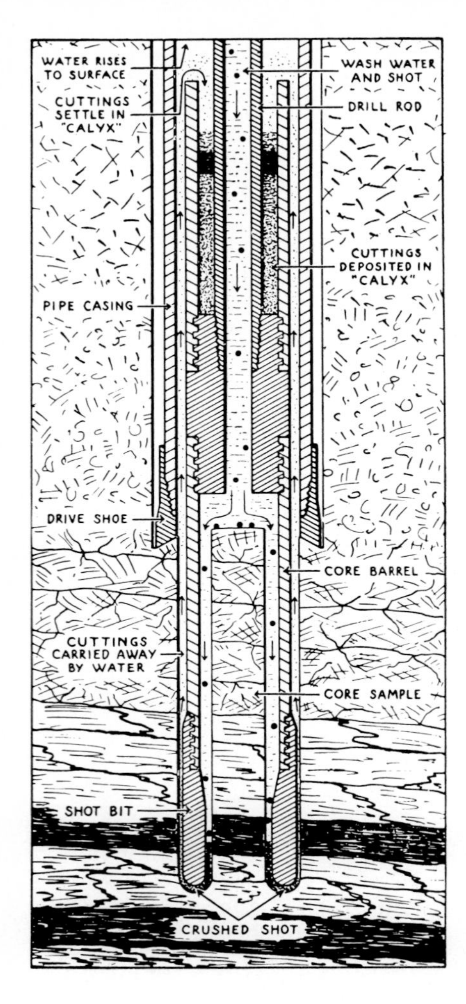

Figure 13-17 Shot or calyx core drill. *(Ingersoll-Rand Company.)*

continuous blast of compressed air is forced down through the pipe and the bit to remove the rock cuttings and cool the bit. Rigs are available to drill holes to different diameters and to depths up to approximately 300 ft. This drill is suitable for drilling soft to medium rock, such as hard dolomite and limestone, but is not suitable for drilling the harder igneous rocks.

In drilling dolomite for the Ontario Hydro-canal, heavyweight drills were used to drill 25- and 50-ft-deep holes. The 25-ft holes were drilled on 10- by 10- and 12- by 12-ft patterns, using $6\frac{1}{4}$- and $6\frac{3}{4}$-in. bits. The average drilling speed was approximately 30 ft per hr, including moving. The average life of the bits was 958 ft for the $6\frac{1}{4}$-in. and 1,374 ft for the $6\frac{3}{4}$-in. bits.

On other projects, drilling speeds have varied from $1\frac{1}{2}$ ft per hr in dense, hard dolomite to 50 ft per hr in limestone. The speed of drilling is regulated by pressure delivered through a twin-cylinder hydraulic feed.

Figure 13-18 Drilling unit for shot core drill. *(Acker Drill Company.)*

SHOT DRILLS

A shot drill is a tool which depends on the abrasive effect of chilled steel shot to penetrate the rock. The essential parts include a shot bit, core barrel, sludge barrel, drill rod, water pump, and power-driven rotation unit. The bit consists of a section of steel pipe, with a serrated lower end. As the bit is rotated, shot are fed to the lower end through the drill rod. Under the pressure of the bit these shot erode the rock to form a kerf around the core. Water, which is supplied through the drill rod, forces the rock cuttings up around the outside of the drill, where they settle in a sludge barrel, to be removed when the entire unit is pulled from the hole. Periodically it is necessary to break the core off and remove it from the hole in order that drilling may proceed.

Figure 13-17 illustrates a type of shot drill that has been used extensively. Figure 13-18 illustrates a drilling unit which is used to rotate the bit. The drive is through the spindle extending below the drill head.

Standard shot drills are capable of drilling holes up to 600 ft or more in depth, with diameters varying from $2\frac{1}{2}$ to 20 in. Special equipment has been used to drill up to 6 ft in diameter with depths in excess of 1,000 ft. Rock of any hardness may be drilled.

Although large holes are expensive, they permit a worker to be lowered into them for a thorough examination of the formation in place. For this purpose holes 30 in. in diameter or larger are sometimes drilled. Smaller holes provide continuous cores for examination for structural information.

The rate of drilling with a shot drill is relatively slow, sometimes less than 1 ft per hr, depending on the size of the drill and the hardness of the rock.

Example 13-1 On a project for an electric utility company near Oak Park, Ohio, the contractor used a shot drill to drill 100 large-diameter footings for columns in hard limestone rock. The holes varied from 30 in. (76 cm) to 60 in. (152 cm) in diameter and averaged about 12 ft (3.7 m) in depth. Holes 42 in. (107 cm) in diameter were drilled at an average rate of 1 to $1\frac{1}{2}$ ft (0.3 to 0.5 m) per hr, while holes 60 in. (152 cm) in diameter were drilled at an average rate of about 1 ft (0.3 m) per hr [2].

DIAMOND DRILLS

Diamond drills are used primarily for exploration drilling, where cores are desired for the purpose of studying the rock structure. The Diamond Core Drill Manufacturers' Association lists four sizes as standard—$1\frac{1}{2}$, $1\frac{7}{8}$, $2\frac{3}{8}$, and 3 in. Larger sizes are available, but the investment in diamonds increases so rapidly with an increase in size that shot drills may be more economical for larger-diameter holes.

A drilling rig consists of a diamond bit, a core barrel, a jointed driving tube, and a rotary head to supply the driving torque (see Fig. 13-19). Water is pumped through the driving tube to remove the cuttings. The pressure on the bit is regulated through a screw or hydraulic-feed swivel head. Core barrels are available in lengths varying from 5 to 15 ft. When the bit advances to a depth equal to the length of the core barrel, the core is broken off and the drill is removed from the hole. Diamond drills can drill in any desired direction from vertically downward to upward.

The selection of the size of diamonds depends on the nature of the formation to be drilled. Large stones are preferred for the softer formations and small stones for fine-grained solid formations.

Diamond drills are capable of drilling to depths in excess of 1,000 ft. Bit speeds may vary from approximately 200 to 1,200 rpm. The drilling rate will vary from less than a foot to several feet per hour, depending on the type of rock.

Table 13-3 gives information on the dimensions and diamond content of bits. The cost of diamonds varies with the quality and quantity used with the bit (see Figs. 13-20 and 13-21).

Figure 13-19 Drilling unit for diamond core drill. *(Acker Drill Company.)*

Figure 13-20 Diamond-point bits. *(Sprague & Henwood, Inc.)*

Table 13-3 Representative information for standard diamond coring bits

Size of bit, in.	Nominal		Net dimension		Minimum carat content
	Hole diameter, in.	Core diameter, in.	OD, in.	ID, in.	
EX	$1\frac{1}{2}$	$\frac{7}{8}$	1.460	0.845	6.75
AX	$1\frac{7}{8}$	$1\frac{1}{8}$	1.865	1.185	10.00
BX	$2\frac{3}{8}$	$1\frac{5}{8}$	2.330	1.655	14.00
NX	3	$2\frac{1}{8}$	2.945	2.155	18.00
$2\frac{3}{4}\times3\frac{7}{8}$	$3\frac{7}{8}$	$2\frac{3}{4}$	3.840	2.690	36.00
$4\times5\frac{1}{2}$	$5\frac{1}{2}$	4	5.435	3.970	60.00
$6\times7\frac{3}{4}$	$7\frac{3}{4}$	6	7.655	5.970	90.00

MANUFACTURERS' REPORT ON DRILLING EQUIPMENT AND TECHNIQUES [3]

In 1976 a magazine devoted primarily to construction methods and equipment published an article discussing and describing the contributions of drills to production profits. The article presented the views of representatives of drilling equipment manufacturers, with suggestions listed for selecting and using the equipment to achieve increased production at reduced costs. Their views are presented hereafter.

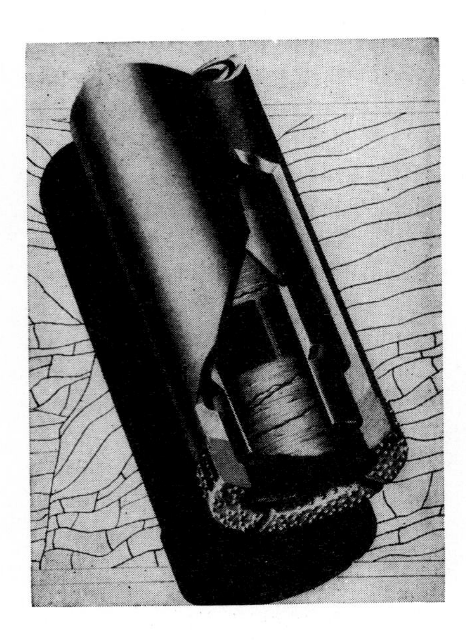

Figure 13-21 Diamond coring bit and double-tube core barrel. *(Sprague & Henwood, Inc.)*

The views of drilling equipment manufacturers The manufacturers' representatives who assessed the ingredients of drilling productivity included as important factors selection, maintainability, mobility, operator expertise, operability of equipment, and use of auxiliary attachments. The methods of drilling employed, the attitude of management toward this activity, and the interrelationship between the drill bit and the drilling rig also contribute to the ability to attain maximum production at minimum cost. Each user or prospective user of drilling equipment should examine these views and adopt the ones that seem to apply to his operations.

The views of the manufacturers are presented below.

The user of drills to be operated in a quarry should select equipment that will maximize productivity within the limits of his loading, hauling, and crushing equipment. The equipment which he selects must work under the most severe and grueling conditions of the given project. Before any equipment is selected, it should be tested for performance and the results of the tests should be made available to the prospective purchaser.

One representative suggested that use of increased air pressure and larger hammers and internal improvements such as valveless construction of air compressors, where the piston is its own valve (resulting in the reduction of moving parts), should contribute to better performance. It was also suggested that when drilling holes deeper than about 50 ft, down-the-hole drilling, i.e., the process of moving the cylinder and its percussive impact into the hole, should be considered. This move can increase the effective energy by as much as 10 percent by eliminating the loss of energy in extended lengths of drill rods.

Another suggestion was that more tests be conducted to determine the best sizes of blast holes, ratios of burden to spacing, density of explosives, and delay patterns to increase the yield of blasted rock per foot of hole, to determine the effect which such actions would have on the cost of production.

Greater care could perhaps be exercised by the drill operators as they add sections of steel rods to produce deeper holes to be certain that the abutting ends of drill rods bear tightly against each other to transmit the drilling forces to the bits. If these forces are transmitted by the couplings, they may cause excessive splitting of the couplings, with possible loss of bits, drill rods, and holes.

The use of button bits as a means of more evenly distributing the forces of the drills on the rock may be desirable on some projects. The superior geometric pattern of the buttons can help increase the penetration by as much as 10 to 20 percent, smooth the drilling operation, and reduce the stress on the drilling equipment when compared with conventional blade bits. However, button bits are not recommended for drilling long, close, parallel holes, as in presplitting, where holes drilled with button bits may drift or even cross each other. Moreover, their small contact surface may cause greater wear in drilling of hard, abrasive rock.

In some drilling operations productivity has been increased by increasing the pressure and volume of air supplied to the drill by maintaining a cleaner blast hole. At the same time, increasing the down pressure on the bit has improved the cutting action of the bit.

Maintenance and repair services for equipment can be improved, with less down time for repairs and replacement of parts, if the equipment is standardized to the extent permitted by the operations for which it is used. With fewer models of equipment to be serviced, an inventory of spare parts that fail most frequently can be maintained on the job.

Manufacturers have made drilling equipment to provide more service-free time by enclosing many feed, swing, and other assemblies in oil to reduce the wear.

Track drills have been improved by offering hydraulic or mechanical oscillation. Improved horsepower-to-weight ratios permit drills to navigate terrain that was considered inaccessible without winch assistance at one time. And automatic brakes assist in holding the drills in difficult positions.

The availability of extendable booms on track drills has increased the service ranges and performance of drills.

Also, the availability of drills with dust control systems permits use of drills in locations where environmental restrictions would otherwise preclude them.

Because of the high cost, complexity, and sophistication of much drilling equipment, some manufacturers have developed training programs to assist the users and operators of their equipment in increasing its efficiency.

SELECTING THE DRILLING METHOD AND EQUIPMENT

Holes are drilled for various purposes, such as to receive charges of explosives, for exploration, for the injection of grout, etc. Within practical limits the equipment which will produce the greatest overall economy for the particular project is the most satisfactory. Many factors affect the selection of equipment. Among them are the following:

1. The nature of the terrain. Rough surfaces may dictate jackhammers, regardless of other factors.
2. The required depth of holes.
3. The hardness of the rock.
4. The extent to which the formation is broken or fractured.
5. The size of the project.
6. The extent to which the rock is to be broken for handling or crushing.
7. The availability of water for drilling purposes. Lack of water favors dry drilling.
8. The purpose of the holes, such as blasting, exploration, or grout injection.
9. The size cores required for exploration. Small cores permit the use of diamond drills, while large cores suggest shot drills.

For small-diameter shallow blastholes, especially on rough surfaces where larger drills cannot operate, it is usually necessary to use jackhammers, even though the production rates will be low and the costs high.

For blastholes up to about 6 in. in diameter and up to about 50 ft deep, where crawler-mounted machines can operate, the choice may be between track-mounted, rotary-percussion, or piston drills.

If it is necessary to drill holes from 6 to 12 in. in diameter, from 50 to 300 ft deep, the blasthole or rotary drill is usually the best choice.

If cores up to 3 in. are desired, the diamond coring drill is the most satisfactory.

If intermediate-size cores, 3 to 8 in. outside diameter, are desired, the choice will be between a diamond drill and a shot drill. A diamond drill will usually drill faster than a shot drill; also, a diamond drill can drill holes in any direction, while a shot drill is limited to holes that are vertical, or nearly so.

SELECTING THE DRILLING PATTERN

The pattern selected for drilling holes to be loaded with explosives will vary with the type and size drill used, the depth of the holes, the kind of rock, the maximum size rocks permissible, and other factors.

If the holes are drilled to produce rock aggregate, the drilling pattern should be planned to produce rock pieces small enough to permit most of them to be handled by the excavator, such as a power shovel, or to pass into the crusher opening without secondary blasting. While this condition is possible, the cost of excess drilling and explosives to produce it may be so high that the production of some oversize rocks is permissible, in the interest of economy.

If small-diameter holes are spaced close together, the better distribution of the explosives will result in a more uniform rock breakage. However, if the added cost of drilling exceeds the value of the benefits resulting from better breakage, the close spacing is not justified.

As large-diameter holes permit greater explosive loading per hole, it is possible to increase the spacing between large holes and thereby reduce the cost of drilling.

In analyzing a job for drilling and blasting operations, there are three factors which should be considered. They are:

1. The cubic yards of rock per linear foot of hole
2. The number of pounds of explosive per cubic yard of rock
3. The number of pounds of explosive per linear foot of hole

The value of each of the three factors may be estimated in advance of drilling and blasting operations, but after experimental drilling operations are conducted, it probably will be desirable to modify the values to give better results.

The relationships between the three factors are illustrated in Table 13-4. The volumes of rock per linear foot of hole are based on the net depth of holes

Table 13-4 Drilling and blasting data

Size hole, in.	Hole pattern, ft	Area per hole, sq ft	Volume of rock per lin ft of hole, cu yd	Lb of explosive per lin ft of hole*	Lb of explosive per cu yd of rock* % of hole filled		
					100	75	50
$1\frac{1}{2}$	4 × 4	16	0.59	0.9	1.52	1.14	0.76
	5 × 5	25	0.93	0.9	0.97	0.73	0.48
	6 × 6	36	1.33	0.9	0.68	0.51	0.34
	7 × 7	49	1.81	0.9	0.50	0.38	0.25
2	5 × 5	25	0.93	1.7	1.83	1.37	0.92
	6 × 6	36	1.33	1.7	1.28	0.96	0.64
	7 × 7	49	1.81	1.7	0.94	0.71	0.47
	8 × 8	64	2.37	1.7	0.72	0.54	0.36
3	7 × 7	49	1.81	3.9	2.15	1.61	1.08
	8 × 8	64	2.37	3.9	1.65	1.24	0.83
	9 × 9	81	3.00	3.9	1.30	0.97	0.65
	10 × 10	100	3.70	3.9	1.05	0.79	0.53
	11 × 11	121	4.48	3.9	0.87	0.65	0.44
4	8 × 8	64	2.37	7.5	3.16	2.37	1.58
	10 × 10	100	3.70	7.5	2.03	1.52	1.02
	12 × 12	144	5.30	7.5	1.42	1.06	0.71
	14 × 14	196	7.25	7.5	1.03	0.77	0.52
	16 × 16	256	9.50	7.5	0.79	0.59	0.40
5	12 × 12	144	5.30	10.9	2.05	1.54	1.02
	14 × 14	196	7.25	10.9	1.50	1.13	0.75
	16 × 16	256	9.50	10.9	1.15	0.86	0.58
	18 × 18	324	12.00	10.9	0.91	0.68	0.46
	20 × 20	400	14.85	10.9	0.73	0.55	0.37
6	12 × 12	144	5.30	15.6	2.94	2.20	1.47
	14 × 14	196	7.25	15.6	2.05	1.54	1.02
	16 × 16	256	9.50	15.6	1.64	1.23	0.82
	18 × 18	324	12.00	15.6	1.30	0.97	0.65
	20 × 20	400	14.85	15.6	1.05	0.79	0.53
	24 × 24	576	21.35	15.6	0.73	0.55	0.37
9	20 × 20	400	14.85	35.0	2.36	1.77	1.18
	24 × 24	576	21.35	35.0	1.64	1.23	0.82
	28 × 28	784	29.00	35.0	1.21	0.91	0.61
	30 × 30	900	33.30	35.0	1.05	0.79	0.53
	32 × 32	1,024	37.90	35.0	0.92	0.69	0.46

* Based on using dynamite weighing 80 lb per cu ft.

and do not include subdrilling, which frequently is necessary. The pounds of explosive per linear foot of hole are based on filling the holes completely with 60 percent dynamite. The pounds of explosive per cubic yard of rock are based on filling each hole to 100, 75, and 50 percent of its total capacity with dynamite. When a hole is not filled completely with dynamite, the surplus volume is filled with stemming.

RATES OF DRILLING ROCK

The rates of drilling rock will vary with a number of factors, such as the type and size drill used, hardness of the rock, depth of holes, drilling pattern, time lost waiting for other operations, etc. Also, if pneumatic drills are used, the rate of drilling will vary considerably with the pressure of the air, as demonstrated in the section starting on page 384.

Another item that influences the rate of drilling is the availability factor. Because of the nature of the work that they do, drills are subjected to severe usage, which may result in frequent failures of critical parts, or a deterioration of the whole unit, entailing delays in drilling. The portion of time that a drill is operative is defined as the availability factor, which is usually expressed as a percent of the total time the drill is expected to be working.

Examples of rates and costs of drilling rock During 1964 a mining company operating an open-pit mine in British Columbia, Canada, made an analysis of extensive records related to the production rates of several types and makes of drills [4]. The results of the analysis appear in Table 13-5.

The ore in the pit consisted of low-grade chalcopyrite, which is associated with hematite and magnetite.

As indicated in the table, four drills were used during the tests. The types were

Table 13-5 Comparative production and cost of rock drills*

	Type drill			
Item	*A*	*B*	*C*	*D*
Diameter of hole, in.	9	9	$7\frac{7}{8}$	6
Availability, %	89	70	67	88
Tons drilled per shift	10,350	6,570	3,020	4,360
Operating cost per ton	$0.0085	$0.0154	$0.0210	$0.0126
Labor cost per ton	0.0034	0.0053	0.0113	0.0078
Total cost per ton	$0.0119	$0.0207	$0.0323	$0.0204

*The costs appearing in this table were applicable at the time the tests were conducted.

A. Electric-powered rotary
B. Diesel-engine-powered rotary
C. Diesel-engine-powered rotary
D. Electric-powered percussion

Because drill bits account for about 50 percent of the operating cost of the drills, cost tests were conducted using several types and makes of bits. The results of the tests indicated that tri-cone bits were the most economical.

Another example of the rates of drilling rock and the distribution of costs appears in Table 13-6. The project is an open-pit copper and zinc mine in Ontario, Canada [4]. The ores were greenstones and sulphides, the latter being hard and abrasive. The two units used were track-mounted rotary-percussion drills, which drilled 3-in.-diameter holes on patterns of 9 by 9 ft and 7 by 7 ft.

The effect of air pressure on the rate of drilling rock The cost of energy furnished by compressed air is high when compared with the cost of energy

Table 13-6 Rates and costs of drilling and blasting rock*

Item	Performance or cost
Drilling performance	
Tons drilled	368,250
Depth of holes drilled, ft	71,407
Tons per foot of hole	5.16
Depth of hole per bit, ft	174
Depth of hole per drill rod, ft	1,300
Depth of hole per coupling, ft	750
Depth of hole per drill shift, 8 hr, ft	238
Drilling costs per ton of rock	
Labor	$0.025
Supplies	0.059
Compressed air	0.017
Machine maintenance labor	0.002
Machine maintenance supplies	0.007
Miscellaneous	0.009
Total	$0.119
Blasting costs per ton of rock	
Average powder factor, lb/ton	0.429
Explosives	$0.084
Labor	0.009
Total cost	$0.093

*The costs appearing in this table were applicable at the time the tests were conducted.

supplied by electricity or diesel fuel. The ratio of costs may be as high as 6:1. For this reason it is essential that every reasonable effort be made to increase the efficiency of the compressed air system and the equipment which uses compressed air as a source of energy. One method of increasing the efficiency of pneumatic drills is to be certain that the specified air pressure at the drill is available.

It has been shown that the energy of a rock drill can be represented by the following equation [5, 6]

$$E \propto \frac{P^{1.5} A^{1.5} S^{0.5}}{W^{0.5}} \tag{13-1}$$

where E = energy per blow
P = air pressure
A = area of piston stroke
S = length of piston stroke
W = weight of piston

With all the factors constant in a given drill except the pressure of the air, this equation indicates that the energy delivered per blow varies with the 1.5 power of the pressure.

Factors which during the past have discouraged the use of higher pressures to increase the rate of drilling have been the limitations imposed by the design of the drills and reduced life spans of the drill steel and bits. However, these limitations have been overcome to a large extent in recent years.

The potential increase in production of a drill when operating at a higher pressure should not be the sole factor in a decision to use higher pressure. The value of the increased production should be compared with the probable increase in the cost of air and maintenance and repairs for the drill, including drill steel and bits. Also, any increase in maintenance and repairs may reduce the availability factor for the drill. The optimum pressure is the pressure that will result in the minimum cost of drilling a unit of hole depth, considering all factors related to the drilling including, but not limited to, the following:

1. The value of the increased production
2. The increased cost of providing air at a higher pressure
3. The cost of increased line leakage
4. The increased cost of maintenance and repairs for the drill
5. The adverse effect, if any, of increased noise in some instances such as tunneling
6. The effect which a higher pressure may have on the availability factor for the drill or air compressor

On most construction projects it is not practical to conduct studies to evaluate each of these factors. However, a limited number of studies have been made under conditions that did permit evaluation of the effects of varying the pressure of the air. The results of one such test are given in the next article.

Determining the optimum air pressure for drilling rock This is a report on the results of tests that were conducted in a mine in Ontario, Canada, recently [3]. The walls of the test station were marked off in panels, so that, by drilling in each of the panels at every stage of testing, variations in the drillability of the rock were minimized. Stop watches, an air flowmeter, pressure-reducing valves, pressure gauges, micrometer gauges, and tools were used to assure adequate controls and information.

Prior to starting the tests, seven new jackleg drills were obtained from five manufacturers and divided into two groups, as indicated in Table 13-7. Holes

Table 13-7 Variations in the rates of penetration by rock drills with varying air pressure

			Dynamic air pressure, psi gauge (Pa)*					
	Bore, in.	Stroke, in.	90 (7.2)†	100 (7.9)†	110 (8.6)†	120 (9.3)†	130 (10.0)†	140 (10.6)†
Drill group	(mm)	(mm)	Rate of penetration, in. per min (mm/min)					
A_1	$2\frac{21}{32}$	$2\frac{7}{8}$	12.47	14.31	16.10	20.60	21.84	24.49
	(67.4)	(73.0)	(317)	(364)	(409)	(524)	(555)	(623)
A_2	$2\frac{11}{16}$	$2\frac{9}{16}$	12.90	16.06	17.41	21.87	23.76	24.55
	(68.3)	(65.2)	(329)	(408)	(443)	(556)	(603)	(625)
A_3	$2\frac{3}{4}$	$2\frac{3}{4}$	13.90	17.49	16.78	22.88	23.58	26.63
	(69.7)	(75.0)	(355)	(444)	(426)	(582)	(600)	(678)
Average for group *A*			13.09	15.95	16.76	21.78	23.06	25.22
			(322)	(405)	(427)	(554)	(587)	(641)
Percent increase				21.85	28.04	66.38	76.16	92.67
B_1	$3\frac{1}{8}$	$2\frac{3}{8}$	14.48	18.49	16.08	21.75	21.01	25.96
	(79.5)	(60.5)	(367)	(469)	(408)	(553)	(535)	(659)
B_2	3	$2\frac{9}{16}$	14.15	19.04	19.72	22.94	21.32	22.89
	(76.0)	(65.2)	(360)	(484)	(500)	(582)	(540)	(582)
B_3	3	$2\frac{5}{8}$	14.58	15.82	15.23	21.95	18.93	20.87
	(76.0)	(66.6)	(370)	(402)	(387)	(558)	(481)	(531)
B_4	3	$1\frac{15}{16}$	10.32	12.59	17.77	20.28	21.97	24.01
	(76.0)	(49.2)	(263)	(320)	(452)	(515)	(583)	(610)
Average for group *B*			13.39	16.49	17.20	21.73	20.81	23.43
			(340)	(419)	(438)	(551)	(529)	(596)
Percent increase				23.15	28.45	62.28	55.41	74.97

* Psi pressures are gauge, while Pa pressures are absolute. Thus 14.7 must be added to the psi values before multiplying by the conversion to Pa units.

† Each of the listed values is multiplied by 10^5 to obtain the correct Pa units. Thus (7.2) is 7.2×10^5, (7.9) is 7.9×10^5, etc.

Table 13-8 Variations in the volume of air consumed with varying air pressures

	Dynamic air pressure, psi gauge (Pa),*					
Drill group	90 (7.2)†	100 (7.9)†	110 (8.6)†	120 (9.3)†	130 (10.0)†	140 (10.6)†
	Volume of air consumed, cfm (cu m/sec)					
I_1	132.2	154.2	160.5	178.5	185.0	210.0
	(0.062)	(0.073)	(0.076)	(0.084)	(0.087)	(0.099)
A_2	133.5	150.6	162.5	182.5	193.0	221.0
	(0.063)	(0.071)	(0.077)	(0.086)	(0.091)	(0.104)
A_3	159.2	171.0	193.2	214.0	228.5	255.7
	(0.075)	(0.081)	(0.091)	(0.102)	(0.108)	(0.121)
Average for group *A*	141.6	158.6	172.1	191.7	202.2	228.9
	(0.067)	(0.075)	(0.081)	(0.090)	(0.095)	(0.108)
Percent increase	.....	12.0	21.5	35.4	42.8	61.7
B_1	185.2	207.3	233.6	257.0	284.5	302.4
	(0.087)	(0.098)	(0.111)	(0.121)	(0.134)	(0.143)
B_2	165.1	188.2	208.5	241.3	252.5	273.0
	(0.078)	(0.089)	(0.098)	(0.114)	(0.119)	(0.129)
B_3	163.0	183.0	206.7	226.5	255.0	267.7
	(0.077)	(0.086)	(0.097)	(0.107)	(0.121)	(0.126)
B_4	182.2	195.2	213.8	241.2	269.5	290.5
	(0.086)	(0.092)	(0.102)	(0.114)	(0.127)	(0.137)
Average for group *B*	173.9	193.4	215.6	241.5	265.4	283.4
	(0.082)	(0.091)	(0.103)	(0.114)	(0.125)	(0.133)
Percent increase	......	11.21	24.0	38.9	52.6	63.0

* Psi pressures are gauge, while Pa pressures are absolute. Thus 14.7 must be added to the psi values before multiplying by the conversion factor to obtain Pa units.

† Each of the listed values is multiplied by 10^5 to obtain the correct Pa units. Thus (7.2) is 7.2×10^5, (7.9) is 7.9×10^5, etc.

were drilled at pressures of 90, 100, 110, 120, 130, and 140 psi. For each pressure increment a new carbide-insert bit was used on both 6-ft and 12-ft steel. All bits were $1\frac{1}{4}$ in. in diameter.

Table 13-7 lists the rates of penetration for each drill for each of the air pressures, while Table 13-8 lists the volume of air consumed, in cubic feet of free air per minute, for each drill and pressure.

Figure 13-22 shows the relationship between the average rate of penetration and the operating pressure for each group of drills. Figure 13-23 is a nomogram based on the information appearing in Fig. 13-22 which indicates the percent increase in penetration resulting from an increase in air pressure. For example, if the pressure is increased from 90 to 100 psi, the increase in penetration will be 38 percent.

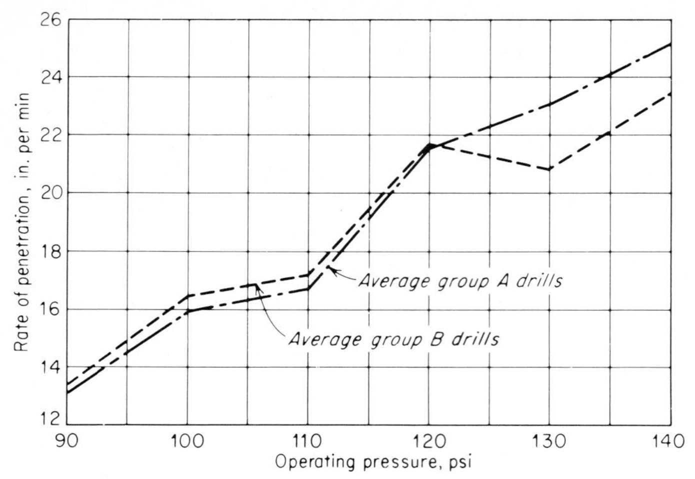

Figure 13-22 Variations in the rate of penetration with air pressure.

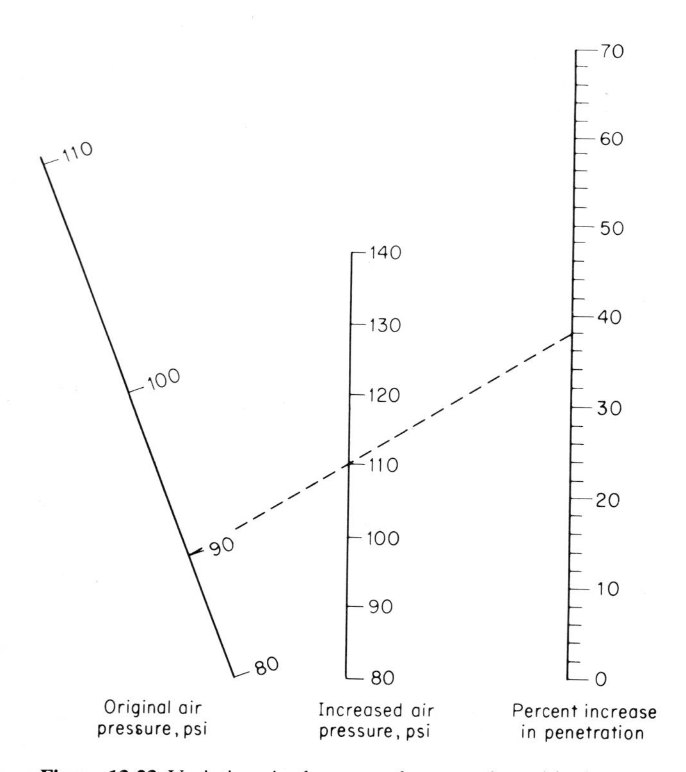

Figure 13-23 Variations in the rates of penetration with air pressure.

DETERMINING THE INCREASE IN PRODUCTION RESULTING FROM AN INCREASE IN AIR PRESSURE

If a drill is presently operating at a given air pressure, such as 90 psi, Fig. 13-23 indicates that if the pressure is increased to 110 psi, the rate of penetration of the drill will be increased 38 percent. This will not result in an increase of 38 percent in the production on the project. The increased rate of penetration is effective only during the time the drill is actually producing hole or drilling. Thus, the increase does not apply to the time that the drill is not actually drilling, which generally will remain the same, regardless of the rate of penetration.

Let us develop an equation which can be used to determine the increase in production resulting from an increase in the rate of penetration. The following symbols will be used:

T = elapsed time that the drill is on the job, hr

D = drilling factor, the portion of the elapsed time devoted to drilling $= \dfrac{T_1}{T}$

T_1 = time actually devoted to drilling, hr

Q_1 = total depth of hole drilled during T hr, ft

R_1 = average rate of drilling during T hr $= \dfrac{Q_1}{T}$, ft per hr

P = increase in rate of penetration resulting from increase in pressure, expressed as a fraction

R_2 = average rate of drilling resulting from increase in pressure $= R_1(1+P)$

T_2 = time required to drill Q_1 ft of hole at increased rate R_2, hr $= \dfrac{T_1}{1+P}$

T_s = time saved by increased rate of drilling

$$T_s = T_1 - T_2 = T_1 - \frac{T_1}{1+P}$$

$$= \frac{T_1(1+P) - T_1}{1+P} = \frac{T_1 + T_1P - T_1}{1+P} = \frac{T_1P}{1+P} \qquad (a)$$

But $T_1 = TD$. Thus,

$$T_s = \frac{TDP}{1+P} \qquad (b)$$

Let Q_2 = the increased depth of hole drilled at the increased rate of penetration during time T. Thus,

$$Q_2 = T_s R_2$$

$$= \frac{TDP}{1+P} \times R_1(1+P)$$

$$= TDPR_1 \qquad (c)$$

But $Q_1 = TR_1$. Thus

$$Q_2 = Q_1 DP$$

and

$$\frac{Q_2}{Q_1} = DP \tag{13-2}$$

which is the ratio of the increased production divided by the original production, expressed as a fraction

Example 13-2 Consider a 1,000-hr elapsed time for a drill on a project. During this time the drill actually penetrates rock 300 hr for a drilling factor of 0.3, for a total depth of hole equal to 10,000 ft. The initial operating air pressure at the drill is 90 psi. If the pressure is increased to 110 psi, what is the probable total depth of hole drilled in 1,000 hr, based on the information appearing in Fig. 13-23? Reference to this figure indicates an increased rate of penetration equal to 38 percent. Thus $P = 0.38$.

Applying equation (13-2),

$$\frac{Q_2}{Q_1} = DP = 0.3 \times 0.38 = 0.114$$

$$Q_2 = 0.114 Q_1 = 0.114 \times 10{,}000$$

$$= 1{,}140 \text{ ft additional depth of hole}$$

The total depth of hole will be $10{,}000 + 1{,}140 = 11{,}140$ ft. This should result in an increase of 11.4 percent in production if the increased depth of hole is reflected in increased production.

It should be emphasized that the information appearing in Fig. 13-23 does not necessarily apply to all drilling conditions. For other projects the increase in the rate of penetration may be more or less than the values obtained from this figure.

THE EFFECT OF INCREASED AIR PRESSURE ON THE COSTS OF MAINTENANCE AND REPAIRS OF DRILLS

During the time that tests were conducted by the mining company in Ontario, Canada, to evaluate the effect of increasing air pressure on the rate of penetration the company also determined the effect of increased pressure on the cost of providing compressed air and the cost of drills, bits, and drill steel. This information appears in Table 13-9.

CONDUCTING A STUDY TO DETERMINE THE ECONOMY OF INCREASING AIR PRESSURE

The decision to increase or not increase the air pressure at the drills should not be determined solely on the basis of the anticipated increase in production and

Table 13-9 Increase in drilling expense resulting from using increased air pressure

	Percent increase in expense Operating air pressure, psi			
Item	90	100	110	120
Compressor operation and maintenance	0	13.0	26.0	39.5
Drills	0	27.0	55.0	83.0
Bits	0	21.5	43.0	64.5
Steel	0	21.5	43.5	66.0

the increase in the cost of compressed air and drilling equipment. Drilling is only one item in a chain of operations, which may include drilling, blasting, loading, and hauling to a disposal area, or it may involve providing quarry rock for crushing into aggregate. The cost effect which operating at an increased pressure will have on the rate of production and also on cost of the related operations should be considered in reaching a decision. The objective is to provide rock at its disposal point, at a waste area, in a fill, or as crushed stone in stockpiles at the lowest practical cost per unit of material.

Figure 13-24 represents a curve that establishes the lowest total cost of producing the end product of a drilling operation. The curve is plotted to indicate this cost for varying air pressures. As noted, the optimum pressure is 102 psi.

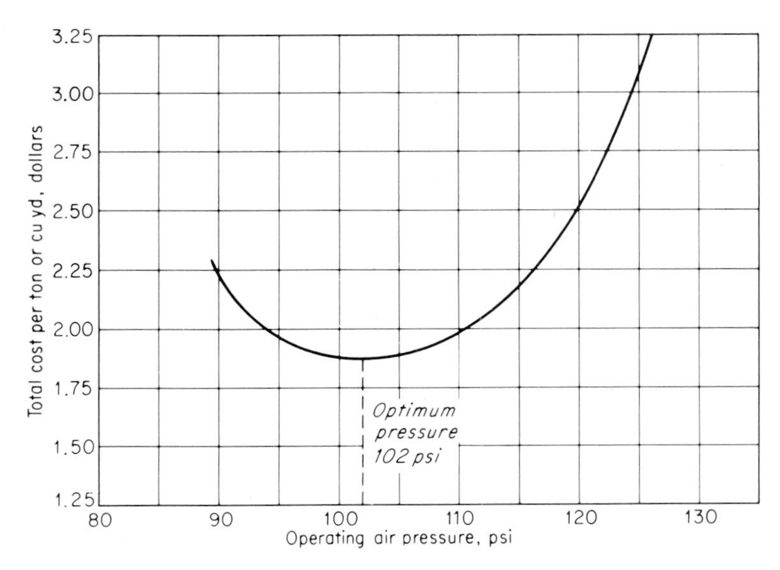

Figure 13-24 Variation in the total cost of rock product with air pressure.

TEST TO DETERMINE THE EFFECT OF AIR PRESSURE ON THE RATE OF PENETRATION OF A DRILL

During 1964 tests were conducted under the supervision of the senior author to determine the effect of air pressure on the rate of penetration of a track-mounted drill using 3-in.-diameter carbide-inset bits. The materials drilled were limestone in a commercial quarry and a large mass of homogeneous concrete, whose 28-day compressive strength averaged 6,790 psi, cast in a rectangular pit in the ground.

Adequate controls, such as valves, a pressure regulator, a recording pressure gauge, and an auxiliary 100-cu-ft air receiver in the line were used to assure the maintenance of the desired pressure. A constant down thrust was maintained on the drifter drill during all drilling operations.

The results of the tests are illustrated in Figs. 13-25 and 13-26. The maximum pressure of 105 psi was imposed by the inability of the 600-cfm compressor to supply air at a higher pressure.

It will be noted that, using values obtained from the curve in Fig. 13-25, the rate of penetration in concrete at 90 psi is 13.3 in. per min, while the rate at 100 psi is 14.4 in. per min. This is an increase of 8.3 percent in the rate of penetration. The corresponding increase for limestone is 6.7 percent. In each instance the percent increase in the rate of penetration was less than the percent increase in air pressure. Thus it appears that the use of an increased air pressure may not be economically justified.

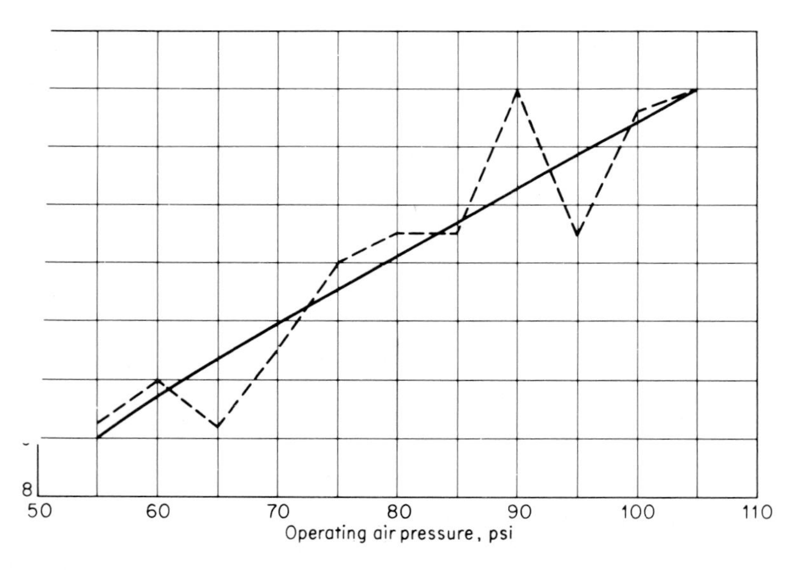

Figure 13-25 Variation in the rate of penetration of concrete with air pressure.

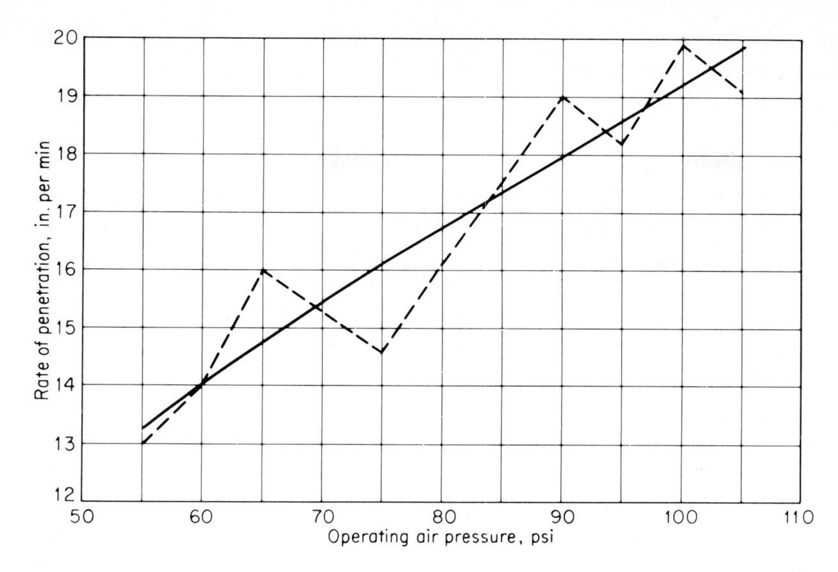

Figure 13-26 Variation in the rate of penetration of limestone with air pressure.

DRILLING EARTH

General information This section of the book will illustrate and discuss various types of equipment used to drill holes in earth, as distinguished from rock. Some equipment, such as that used for exploratory purposes in securing core samples and similar operations, may be used for drilling rock or earth.

Purposes for drilling holes in earth In the construction and mining industries holes are drilled into earth for many purposes, including, but not limited to, the following:

1. To obtain samples of soil for test purposes
2. To locate and evaluate deposits of aggregate suitable for construction purposes
3. To locate and evaluate deposits of minerals
4. To permit the installation of cast-in-place piles or shafts to support structures
5. To enable the driving of load-bearing piles into hard and tough formations
6. To provide wells for supplies of water or for deep drainage purposes
7. To provide shafts for ventilating mines, tunnels, and other underground facilities
8. To provide horizontal holes through embankments, such as those for highways, for the installation of unity conduits

Sizes and depths of holes drilled into earth As illustrated by the accompanying figures, most holes are drilled by rotating bits or heads attached to the lower end of a shaft called a kelly bar. This bar, which is supported by a truck or a tractor or another suitable mount, is rotated by an external motor or engine (Figs. 13-27 and 13-28).

The sizes of holes drilled may vary from a few inches to more than 12 ft (3.7 m). Drills may be equipped with a device attached to the lower end of the drill shaft, described as an underreamer, which will permit a gradual increase in the diameter of the hole, as illustrated in Fig. 13-29. This enlargement permits a substantial increase in the bearing area under a shaft-type concrete footing. Underream diameters as great as 144 in. or more have been drilled. Under favorable conditions it is possible to drill holes as deep as 200 ft for underreamed foundations.

The holes are drilled by a truck-mounted rig, whose essential parts include a power unit, cable drum, boom, rotary table, drill stem, and drill. The shaft is drilled first with a large earth auger or a bucket drill, equipped with cutting blades at the bottom; then the bottom portion of the hole is enlarged with a special drill known as an underreamer.

Figure 13-27 Tractor-mounted auger-type earth drill. *(Acker Drill Company.)*

Figure 13-28 Truck-mounted auger-type earth drill. *(Mobile Drilling Company.)*

Figure 13-29 illustrates a method used to drill these holes through unstable soils, such as mud, sand, or gravel, containing water. If it is possible to do so, the shaft is drilled entirely through the unstable soil; then a temporary steel casing is installed in the hole to eliminate ground water and caving. An alternate method is to add sections to the casing as drilling progresses until the full depth of bad soil is cased off. Then the hole is completed and filled with concrete, and the casing is pulled before the concrete sets.

This type of foundation has been used extensively in areas whose soils are subject to changes in moisture content to considerable depth. By placing the footings below the zone of moisture change, the effect of soil movements due to changes in moisture is eliminated.

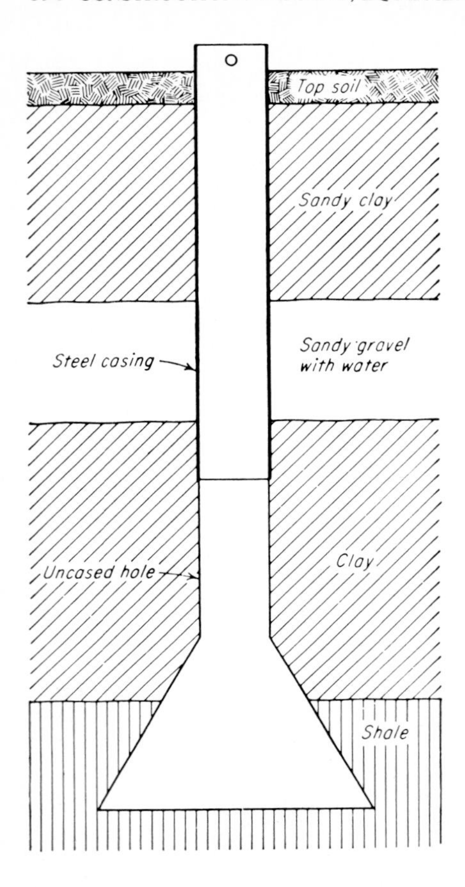

Figure 13-29 Steel casing used to permit footing to be drilled through unstable soil.

Among the advantages of drilled and underreamed foundations, compared with piles and conventional spread footings, are the following:

1. They are less expensive for some soils and projects.
2. They are easy to vary in length to adjust for soil conditions.
3. They permit inspection of soil prior to establishing depth or placing concrete.
4. They eliminate damage to adjacent structures due to vibration of the pile hammer.
5. They eliminate the use of forms for concrete.

Removal of cuttings Several methods are used to remove the cuttings from the holes.

One method of removing the cuttings is to attach the drill head, the actual cutting tool at the bottom of the drill shaft, to the lower end of an auger, as illustrated in Fig. 13-30, which extends from the drill head to above the surface of the ground. As the drill shaft and the auger rotate, the earth is forced to the

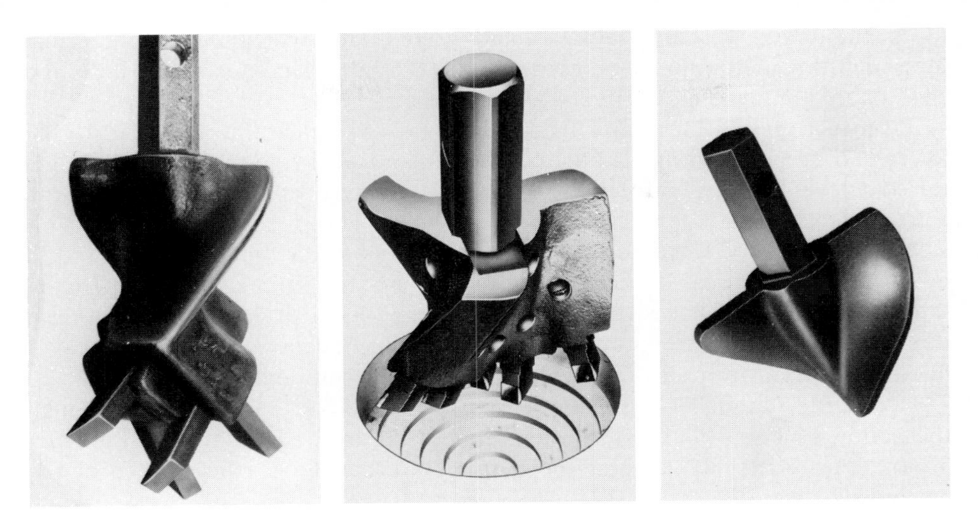

Figure 13-30 Drill heads or bits to be attached to the bottom of a drill kelly for drilling earth or soft rock. *(Mobile Drilling Company.)*

Figure 13-31 Gasoline-engine-powered auger-type boring machine. *(McLaughlin Manufacturing Company.)*

top of the hole, where it is removed and wasted. However, the depth of a hole for which this method may be used is limited by the diameter of the hole, the class of soil, and the moisture content of the soil.

Another method of removing the cuttings is to attach the drill head to the lower end of a section of the auger. When the auger is filled with cuttings, it is raised above the surface of the ground and rotated rapidly to free it of the cuttings.

A third method of removing the cuttings is to use a combination of a drill head with a cylindrical bucket, whose diameter is the same as the diameter of the hole. As the bucket is rotated, steel cutting blades attached to the bottom of the bucket force the cuttings up and into the bucket. When the bucket is filled, it is raised to the surface of the ground and emptied.

A fourth method of removing the cuttings is to force air or water through the hollow kelly bar and drill shaft to the bottom of the hole and then upward around the drill shaft, so that the cuttings are carried to the surface of the ground.

Drilling holes through unstable soils When holes are drilled through unstable soils, such as mud, silt, sand, or gravel-containing water, it may be necessary to install a temporary or permanent steel casing, as illustrated in Fig. 13-29, to prevent the flow of soil into the hole.

Figure 13-32 Gasoline-engine-powered auger-type boring machine with the auger enclosed in a steel casing. *(McLaughlin Manufacturing Company.)*

Earth-boring machines Figure 13-31 illustrates a self-cont engine-powered open-auger-type boring machine, which is u cased holes through earth. The machine illustrated will bore h 3 to 12 in. (76 to 304 mm) in diameter to depths up to 80 ft on the type and condition of the soil and the job conditions

Figure 13-32 illustrates a self-contained gasoline-engine-pow type boring machine with the auger enclosed in a steel pipe or casing, which is forced through the hole excavated by the auger. The machine illustrated is capable of boring holes for casing sizes varying from 4 to 30 in. (102 to 762 mm) or more to depths up to 200 ft (61 m), depending on the type and condition of the soil and the job conditions. As the boring advances, the machine automatically maintains a forward thrust on the casing and the auger.

These machines may also be powered hydraulically, by air, or by electric motors.

REFERENCES

1. Smith, Gordon R: Drilling, New Equipment, New Techniques, *Construction Methods and Equipment*, vol. 44, pp. 110–115, August 1962.
2. Shot-drill Cuts Hard Rock Sockets for Column Footings, *Construction Methods and Equipment*, vol. 53, pp. 84–85, May 1971.
3. Higgins, Lindley R: Drills Play Dramatic Role in Profit Production, *Construction Methods and Equipment*, vol. 58, pp. 54–61, September 1976.
4. The Cost of Drilling and Blasting Today's Pits, *Engineering and Mining Journal*, vol. 166, pp. 110–113, September 1965.
5. Pasieka, A. R., and J. C. Wilson: The Importance of High-Pressure Compressed Air to Mining Operations, *The Canadian Mining and Metallurgical Bulletin*, vol. 59, pp. 1093–1102, September 1966.
6. Knox, John: Factors Influence the Design and Application of Downhole Drills, *The Canadian Mining and Metallurgical Bulletin*, vol. 58, pp. 547–550, May 1965.
7. Acker Drill Company, Inc., P.O. Box 830, Scranton, PA 18501.
8. Atlas Copco, Inc., 70 Demarest Drive, Wayne, NJ 07470.
9. Ingersoll-Rand Company, Phillipsburg, NJ 08865.
10. Joy Manufacturing Company, Claremont, NH 03743.
11. McLaughlin Manufacturing Company, P.O. Box 303, Plainfield, IL 60544.
12. Mobile Drilling Company, Inc., 3807 Madison Avenue, Indianapolis, IN 46227.
13. Penn-Mar Mining Company, Route 5, Leitersburg Pike, Hagerstown, MD 21740.
14. Reed Tool Company, 12400 North Freeway, Houston, TX 77090.
15. Schramm, Inc., 800 East Virginia Avenue, West Chester, PA 19380.
16. Sprague & Henwood, Inc., Scranton, PA 18501.
17. The Timken Company, Canton, OH 44706.
18. Dick, Richard A., Larry R. Fletcher, and Dennis W. D'Andrea: "Explosives and Blasting Procedures," GPO No. 024-004-02115-6, Government Printing Office, Washington, D.C.
19. Nelmark, Jack D.: Large Diameter Blast Hole Drills, *Journal of the Mining Congress*, August 1980.

CHAPTER

FOURTEEN

BLASTING ROCK

BLASTING

The operation referred to as blasting is performed to loosen rock in order that it may be excavated or removed from its existing position. Blasting is accomplished by discharging an explosive that has been placed in a hole specially provided for this purpose. The energy associated with an explosion is the result of the pressure produced in the gases that are formed by the explosive.

There are many types of explosives and methods of using them. A full treatment of each explosive and method is too comprehensive for inclusion in this book. For more complete discussions of this subject the reader is referred to handbooks on blasting, published by manufacturers of explosives.

DEFINITIONS OF TERMS

The more common terms which are used in describing blasting operations are given below as a guide for the reader.

ANFO This is an explosive that is produced by mixing prilled ammonium nitrate with fuel oil.

Blasthole This is a hole that is drilled into rock to permit the placing of an explosive in it.

Blasting This is the detonation of an explosive to fracture the rock.

Blasting agent This is an explosive compound that is placed in a blasthole and detonated.

Blasting cap This is a hollow metal cap which is filled with a high explosive and detonated within or adjacent to the blasting agent as a means of detonating the agent.

Blasting machine This is a machine that is used to generate the electric current that detonates an electric blasting cap.

Blasting powder This is a slow-burning low explosive made from saltpeter, sulfur, and charcoal. It is seldom used for blasting rock.

Block holing This is the drilling of holes in oversize boulders to permit secondary blasting.

Booster This is a high explosive that is placed in a hole at desired spacings to assure that the explosives will detonate throughout the hole.

Borehole This is a blasthole.

Brisance This is an indication of the shattering effect shown by an explosive.

Burden This is the horizontal distance from the face, as in quarrying, to the line of blastholes nearest the face.

Cap-sensitive explosive This is an explosive which can be detonated by a No. 6 cap when the cap is detonated within or adjacent to the explosive.

Coyote tunnel This is a tunnel, several feet in diameter, into which a large quantity of explosive is placed for detonating purposes.

Crimping This is an operation to reduce the diameter of a cap near the open end to hold the fuse securely in the cap.

Cutoff This is the breaking of a fuse or electric circuit to the cap in a primer, usually by explosions in adjacent holes, before the cap in this hole is detonated.

Deck stemming This is the operation of placing inert material in a blasthole at spacings to separate explosive charges in the hole.

Density This is a measure of the energy of an explosive in a stated volume.

Detaline, *Hercudet*, and *Nonel* This is a nonelectric blast initiation system. It is immune to external disturbances such as electric storms, static electricity, or other stray electric currents. It may be used for underwater detonations. See also *Hercudet* and *Nonel*.

Detonation rate This is a measure of the speed at which an explosion travels from one location in an explosive to another location.

Downline This is a cord containing an enclosed explosive which extends from a trunkline into a blasthole where it is used to detonate an attached blasting cap.

Dynamite This is a high explosive whose primary constituent is nitroglycerin.

Electric blasting cap This is a small metal tube loaded with a charge of sensitive explosive. The cap is detonated by the heat produced by an electric current flowing through a wire bridge inside the cap.

Explosive This is a chemical compound which, under favorable conditions, will detonate quickly to produce a very high pressure.

Fuse primer This is a quantity of high explosive which is detonated by a fuse as a means of initiating the explosion of a main charge of explosive.

Gelatin dynamite This is a jellylike explosive made by dissolving nitrocotton in nitroglycerin. This explosive is entirely waterproof.

Hercudet This is a nonelectric blast initiation system. See *Detaline.*

High explosive This is an explosive that reacts to detonation at an extremely rapid rate.

Leading wires These are wires that are used to conduct the electric current from its source to the leg wires from electric blasting caps.

Leg wires These are wires that conduct the electric current from the lead wires to an electric cap.

Low explosive This is an explosive that produces pressure by progressive burning, thereby releasing energy relatively slowly.

LP delay cap This is a cap, electric or nonelectric, that delays the detonation for a long period of time as compared with MS caps.

MS delay cap This is a cap, electric or nonelectric, that delays the detonation of an explosive for a short period of time, measured in thousandths of a second.

Mud capping This is an operation in which an explosive is placed on an oversize boulder and covered with mud or earth, after which the explosive is detonated to fracture the boulder.

Nitroglycerin This is a colorless explosive liquid obtained by treating glycerol with a mixture of nitric and sulfuric acids.

Nonel This is a nonelectric blast initiation system. See *Detaline.*

Nonelectric delay blasting cap This is a blasting cap that is detonated by a fuse or a detonating cord.

Overbreak This is rock which is fractured outside of the desired space, as when tunneling.

Powder factor This is the quantity of explosive used to fracture a specified volume of rock, e.g., pounds of explosive per cubic yard of rock.

PETN This is the abbreviation for the chemical content of a high explosive with a very high rate of detonation.

Presplitting This operation involves drilling small holes at close intervals, loading them lightly with explosives, and detonating the explosives before the main charges to rupture the webs between the presplit holes.

Primacord This is a high-explosive detonating fuse or cord whose PETN core is contained in a waterproof covering of considerable strength. It is used to detonate high explosives and nonelectric blasting caps.

Primadet delay cap This nonelectric blasting cap is a small metal tube loaded with a charge of sensitive explosive, which is detonated by a detonating cord such as Primacord or Primaline.

Primaline This is a high-explosive detonating fuse or cord whose core is contained in a waterproof covering. It may be used as a downline with one end attached to a detonating cord trunkline and the other end inserted in a nonelectric blasting cap to detonate an explosive charge.

Primer This is the portion of a charge, consisting of a cap-sensitive explosive loaded with a firing device, which initiates the explosion.

Rounds This is a term which includes all the blastholes that are drilled, loaded, and exploded in one firing operation.

Safety fuse This is a fuse containing a low explosive enclosed in a suitable covering. When the fuse is ignited, it will burn at a predetermined speed. It is used to initiate explosions under certain conditions.

Secondary blasting This is an operation performed to reduce to desirable size the oversize boulders remaining after the primary explosion.

Slurry This is an explosive that is produced by mixing ammonium nitrate with TNT or with metals, such as aluminum, and water to form a gelatinlike mixture.

Stemming Stemming is the adding of inert material, such as rock dust or drill cuttings, in a blasthole on top of an explosive to confine the energy of the explosion.

Trunkline This is the main line of a detonating cord, extending from the ignition point to the blastholes containing explosives to be detonated. Secondary lines of detonating cords attached to the trunkline are used to detonate the blasting caps in the primers.

TNT This is a high explosive whose chemical content is trinitrotoluene.

DYNAMITE

Dynamite is available in many grades and sizes to meet the requirements of a particular job. The approximate strength is specified as a percentage, which is an indication of the ratio of the weight of nitroglycerin to the total weight of a cartridge. Individual cartridges vary in size from approximately 1 to 8 in. in diameter and 8 to 24 in. long.

Dynamite is used extensively for charging boreholes, especially for the smaller sizes. As it is placed in a hole, it is tamped sharply with a wooden pole to expand the cartridges so that they fill the hole. For this purpose it may be desirable to split the sides of a cartridge, or cartridges with perforated shells may be obtained. A charge may be fired by a blasting cap or a Primacord fuse. If a cap is used, it is placed in a hole made in one of the cartridges, which serves as a primer. Electric caps are supplied with two leg wires in lengths varying from 2 to 100 ft. These wires are connected with the wires from other holes to form a closed electric circuit for firing purposes.

AMMONIUM NITRATE EXPLOSIVES

This explosive is used extensively on construction projects for both above surface and underground blasting. The cost is only about one-fifth the cost of dynamite. Because this explosive must be detonated by special primers it is much safer than dynamite.

The explosive most commonly used is made by blending about 1 gal of diesel fuel with 100 lb of prilled ammonium nitrate fertilizer, which accounts for the common name used to identify it, ANFO. The mixture should be allowed to set for up to 24 hr to permit the oil to saturate the ammonium nitrate thoroughly.

Because the mixture is free-flowing, it can be poured directly into vertical holes, or it can be blown into horizontal holes using a suitable container, a hose, and compressed air at a pressure of approximately 10 psi.

Ammonium nitrate is not water-resistant. If it is to be used in wet holes, it should be enclosed in sealed plastic bags, or the holes should be prelined with plastic tubing, closed at the bottom, to exclude the water. The tubes, whose diameters should be slightly larger than holes, are installed in the holes by placing rocks or other weights in the bottoms of the tubes.

If holes are wet up to a certain depth and dry above that depth, the wet portions may be loaded with dynamite gel, ammonium nitrate in plastic bags, or slurry, and the dry portions loaded with bulk ammonium nitrate.

Ammonium nitrate is detonated by primers consisting of charges of dynamite placed at the bottoms of the holes and sometimes at intermediate depths. Electric blasting caps or Primacord may be used to detonate the dynamite.

In driving a 2,850-ft tunnel for the Blue Ridge Parkway the contractor used dynamite for one-half of the length, then changed to ammonium nitrate for the balance of the length, using the same size holes ($1\frac{7}{8}$ in. in diameter, depth, 10 ft) and the same pattern [1]. The ammonium nitrate gave better fragmentation plus about 1 ft more advance per round than the dynamite.

SLURRIES

This is a plastic water-resistant explosive that is made by blending several materials, such as inert gel, ammonium nitrate, and aluminum particles, with water to produce the desired consistency. It may be poured directly into the holes, or it may be packaged in plastic bags for placement in the holes. Because it is denser than water, it will sink to the bottom of the holes containing water. Also, its free-flowing properties assure that it will fill the holes completely, which improves its fragmentation effects.

Slurries are detonated by special primers, such as dynamite, TNT, or PETN, using electric blasting caps or Primacord.

Slurry is less expensive than dynamite and more expensive than ammonium nitrate. However, any cost comparison should be based on the total cost in the holes and the magnitude and degree of fragmentation produced by the explosive.

In its Eagle Mountain Mine in Southern California, Kaiser Steel Corporation conducted numerous tests of different explosives to develop one that best suited its needs [2]. The tests resulted in the selection of three types of metallic slurries for use in the 9-in. and $9\frac{7}{8}$-in. blastholes. The slurries had densities varying from 84.5 to 91 lb per cu ft, and bulk strengths varying from 1.4 to 2.5 times that of ammonium nitrate and fuel oil.

STEMMING

After a hole is filled with an explosive to the required depth, the balance of the hole should be filled with stemming. Stemming, which may consist of rock cuttings or other suitable inert material, confines the energy and increases the effectiveness of an explosion. If a continuous charge of explosive is not required from the bottom to the top of the charge, stemming may be placed between charges at predetermined intervals. When charges are separated with stemming, a separate primer should be provided for each charge. Figure 14-1 illustrates several methods of loading boreholes for firing with electric blasting caps.

FIRING CHARGES

It is common practice to fire several holes at one time, using either parallel or series circuits or a combination thereof. Before the final connection to the source of electric current is made, a circuit should be tested with a galvanometer in the line. Each circuit must be tested as a precaution to eliminate open breaks and misfires. Figure 14-2 illustrates three types of circuits.

In order to secure good breakage, with the desired degree of fragmentation, it is frequently necessary to place a higher concentration of explosive near the bottom of a hole than near the top. This may be done by

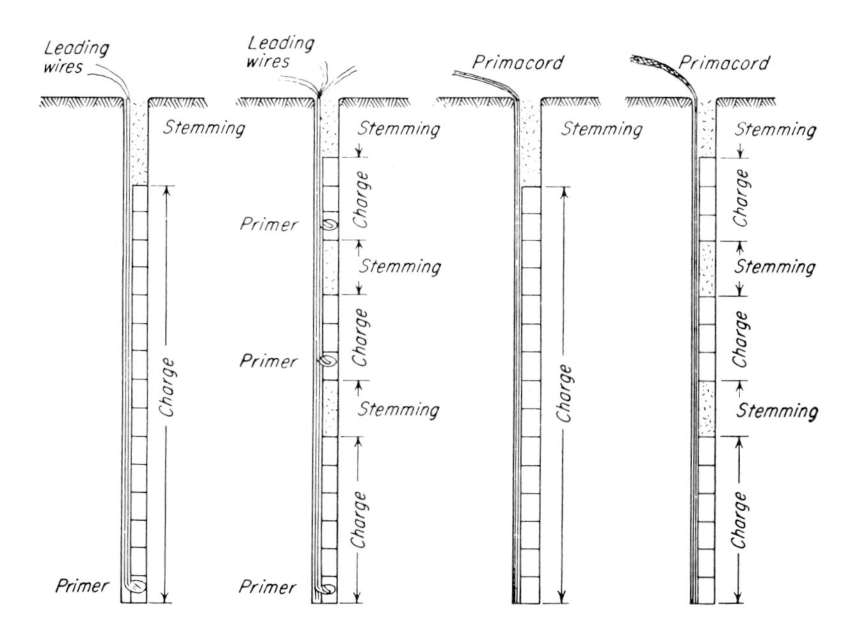

Figure 14-1 Methods of loading blastholes with explosives.

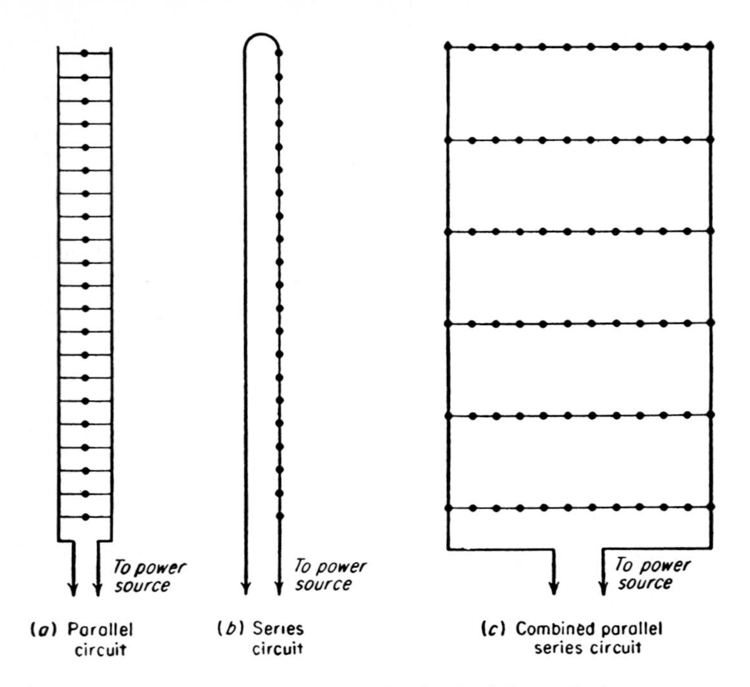

Figure 14-2 Representative types of circuits for firing explosives.

using a strong dynamite near the bottom and a less strong one near the top, or the same effect may be obtained by separating the charges near the top with stemming, provided the total charge in a hole is adequate.

SAFETY FUSE

This device is a continuous core of black powder enclosed in a covering of suitable material. When the core is ignited, it will convey a flame to an explosive attached to the opposite end. The flame will travel along the fuse at a predetermined uniform rate. Thus, the delay between lighting the fuse and the explosion is determined by selecting the proper length of fuse. Figure 14-3 illustrates the correct method of attaching the blasting cap to the end of the fuse.

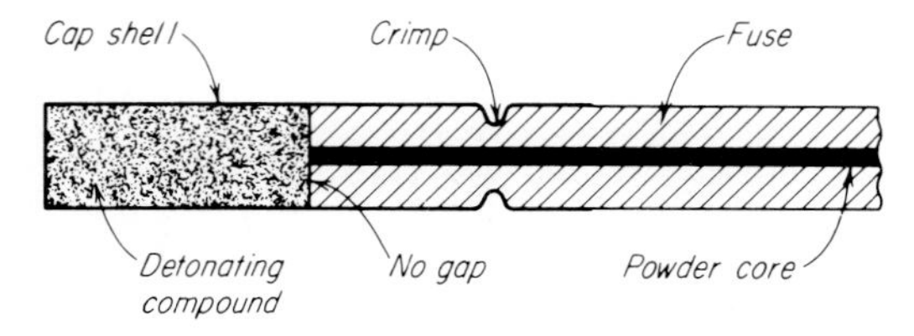

Figure 14-3 Method of seating square-cut fuse against detonating compound.

ELECTRIC BLASTING CAPS

Electric blasting caps are used to detonate charges of dynamite or Primacord fuse. A cap is exploded by passing an electric current through a wire bridge inside the cap. The current, which should be approximately 1.5 amp, heats the bridge, which detonates the explosive in the cap with sufficient violence to fire a charge of dynamite.

Regular-type electric blasting caps are supplied with leg wires whose lengths are indicated in Table 14-1. Number 22 gauge copper wires are used for leg lengths up to 24 ft and No. 20 gauge copper wires for lengths of 30 ft and longer.

In order to analyze an electric circuit used to fire blasting caps, it is necessary to know the resistance of the caps and the leading wires, which conduct the current to the caps. Table 14-2 gives the resistance of single-strand copper wire in the sizes most commonly used for firing electric caps.

Example 14-1 A total of 20 regular electric blasting caps, connected in a single series circuit, are to be fired. Determine the required voltage at the source of supply. The following information is available:

Current required to fire the caps, 1.5 amp
Length of leg wires per cap, 40 ft
Resistance per cap, 1.62 ohms
Distance from source of electricity to blast area, 400 ft
Length of leading wires, $2 \times 400 = 800$ ft
Size of leading wires, No. 20 gauge
Combined resistance of caps, $20 \times 1.62 = 32.4$ ohms
Resistance of leading wires, $0.8 \times 10.15 = 8.1$ ohms
Total resistance of circuit $= 40.5$ ohms

Table 14-1 Resistance of regular and delay blasting caps

Length of leg wires, ft (m)	Resistance, ohms per cap	
	Regular	Delay
4 (1.22)	0.94	1.45
6 (1.83)	1.00	1.51
8 (2.44)	1.07	1.58
10 (3.00)	1.13	1.64
12 (3.66)	1.20	1.71
16 (4.88)	1.32	1.84
20 (6.10)	1.45	1.97
24 (7.32)	1.58	2.10
30 (9.15)	1.41	1.93
40 (12.20)	1.62	2.13
50 (15.30)	1.82	2.33
60 (18.30)	2.02	2.53

Table 14-2 Resistance of copper wire

B. and S. gauge No.	Resistance, ohms per 1,000 ft (329 m)
8	0.628
10	0.999
12	1.588
14	2.525
16	4.015
18	6.385
20	10.150
22	16.140

From Ohm's law the voltage required is obtained from the equation

$$E = IR$$

where E = volts
I = current, amp
R = resistance, ohms

Thus,

$$E = 1.5 \times 40.5 = 60.7 \text{ volts}$$

Thus, any source of electricity that can supply at least 1.5 amp at 60.7 volts will be satisfactory. A 110-volt circuit is adequate.

Example 14-2 If the blasting caps of the previous example are fired in a parallel circuit, with other conditions the same except as noted, the required voltage and current may be determined as follows:

Current required per cap, 0.5 amp
Total current required, $20 \times 0.5 = 10$ amp
Leading wires, No. 14 gauge
Resistance of leading wires, $0.8 \times 2.525 = 2.0$ ohms
Resistance of caps, $1.62 \div 20 \quad = 0.08$ ohm
Total resistance $= 2.08$ ohms
Required voltage $E = IR$
$= 10 \times 2.08 = 20.8$ volts

Example 14-3 Determine the voltage required to fire the blasting caps in the circuit of Fig. 14-2(c) for the stated conditions.

Current required per cap, 1.5 amp
Current required by the 6 circuits, $6 \times 1.5 = 9.0$ amp
Resistance per parallel circuit, $10 \times 1.62 = 16.2$ ohms
Resistance of 6 circuits, $16.2 \div 6 = 2.7$ ohms
Distance from blast area to source of electricity, 400 ft
Leading wires, No. 14 gauge
Resistance of leading wires, $0.8 \times 2.525 = 2.0$ ohms
Resistance of caps $= 2.7$ ohms
Total resistance $= 4.7$ ohms
Required voltage $E = IR$
$= 9 \times 4.7 = 42.3$ volts

DELAY BLASTING CAPS

When the explosive charges in two or more rows of holes parallel to a face are fired at the same time, it is desirable to fire the charges in the holes nearest the face a short time ahead of those in the second row. This procedure will reduce the burden on the holes in the second row and thereby permit the explosive in the second row to break the rock more effectively. If there are more than two rows of holes, the detonations may progress in the order 1, 2, 3, 4, etc., where the numbers indicate the rows, starting with the row nearest the face.

Delay blasting caps are used to obtain the firing sequence. Such caps are available for delay intervals varying from a small fraction of a second to 10 or more seconds. For the shortest delay intervals the caps are called millisecond delay caps and are designated as MS-25, MS-50, MS-200, etc. The number indicates the period of delay in thousandths of a second.

Primacord Primacord [3] is a high-explosive fuse that is used to denote dynamite and other cap-sensitive explosives and sometimes the special primers that may be required to detonate ammonium nitrate explosives. The core of the fuse, which is the explosive PETN, is covered with a sheath for protection, tensile strength, waterproofing, and identification. The explosive has a detonation rate of about 22,000 ft per sec. When PETN is properly initiated, it explodes with great violence. When used as an explosive in Primacord, it is capable of initiating any cap-sensitive explosive with which it comes in contact.

The cord is manufactured in several types, grades, tensile strengths, and resistances to damage from external forces. Each type is specified as having the properties indicated below:

Reinforced Primacord

Core	Nominal grains per ft	Outside diameter, in.	Minimum tensile strength	Shipping weight per 2,000 ft
PETN	50	0.200 ± 0.008	200 lb	33 lb

When several blastholes are fired in a round, the cord is laid along the holes as a trunkline. At each hole, one end of a detonating cord serving as a downline is attached to the trunkline, while the other end extends into the blasthole. If it is necessary to use a blasting cap and/or a primer to initiate the blast in the hole, the bottom end of the downline may be cut square and securely inserted into a blasting cap as illustrated in Fig. 14-3.

Primadet delay blasting caps Primadet delays are nonelectric blasting caps of sufficient explosive strength to provide direct initiation of properly formulated ANFO mixtures when pneumatically loaded into blast holes up to $2\frac{1}{2}$ in. in diameter under normal conditions of density, confinement, and dryness. This

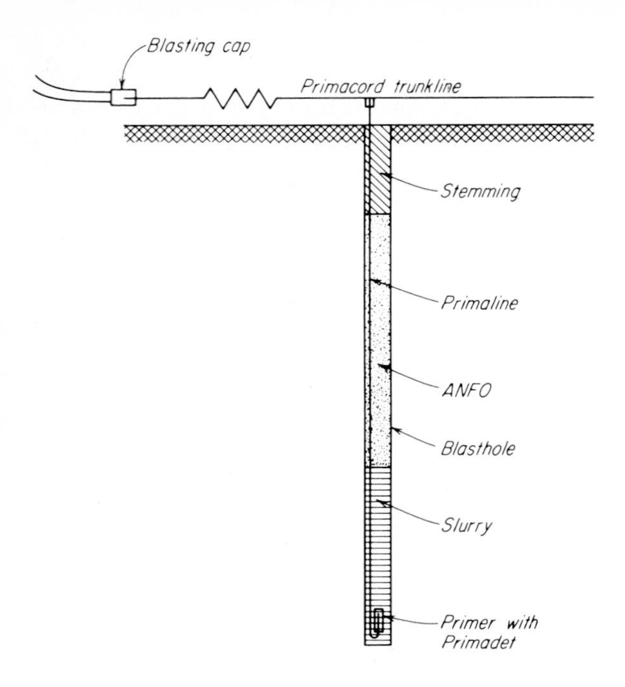

Figure 14-4 Loaded blasthole.

Table 14-3 Standard delay timings for MS and LP series Primadet delay caps

MS Primadet delays			LP Primadet delays		
Period	Average firing times, milliseconds	Average period interval, milliseconds	Period	Average firing times, seconds	Average period interval, second
0	Instant	Instant	1	0.2	0.2
1	25	25	2	0.4	0.2
2	50	25	3	0.6	0.2
3	75	25	4	1.0	0.4
4	100	25	5	1.4	0.4
5	125	25	6	1.8	0.4
6	150	25	7	2.4	0.6
7	175	25	8	3.0	0.6
8	200	25	9	3.8	0.8
9	250	50	10	4.6	0.8
10	300	50	11	5.5	0.9
11	350	50	12	6.4	0.9
12	400	50			
13	450	50			
14	500	50			
15	575	75			
16	650	75			

eliminates the need for a primer cartridge of explosives. These caps provide the precise timing of delay electric caps in both millisecond and long-period ranges but are immune to static electricity that may be generated during pneumatic loading operations, as well as other types of extraneous electricity that may be encountered. They are available with various delay intervals, as indicated in Table 14-3 [3].

A blasthole is loaded with a Primadet by attaching the cap to the lower end of a detonating cord, whose opposite end is attached to a trunkline detonating cord, as illustrated in Fig. 14-4. Also, the downline may be detonated by firing a blasting cap properly attached to the outer end of the line.

HANDLING MISFIRES

In shooting charges of explosives, it may be that one or more charges will fail to explode. This is referred to as a misfire. It is necessary to dispose of this explosive before excavating the loosened rock. The most satisfactory method is to shoot it if possible.

If electric blasting caps are used, the leading wires should be disconnected from the source of power prior to investigating the cause of the misfire. If the leg wires to the cap are available, test the cap circuit, and if the circuit is satisfactory, try again to set off the charge.

When it is necessary to remove the stemming to gain access to a charge in a hole, it should be removed with a wooden tool instead of a metal tool. If water or compressed air is available, either one may be used with a rubber hose to wash the stemming out of the hole. A new primer, set on top of or near the original charge, may be used to fire the charge.

PRESPLITTING ROCK

This is a technique of drilling and blasting which breaks rock along a relatively smooth surface, as illustrated in Fig. 14-5. Holes $2\frac{1}{2}$ to 3 in. in diameter are drilled along the desired surface at spacings varying from 18 to 36 in., or more in some instances, depending on the characteristics of the rock [4]. These holes are loaded with one or two sticks of dynamite at the bottoms, with smaller charges, such as $1\frac{1}{4}$- by 4-in. sticks spaced at 12-in. intervals to the top of the portion of the holes to be loaded. The sticks may be attached to Primacord with tape, or hollow sticks may be used, which permits the Primacord to pass through the sticks with cardboard tube spacers between the charges. After a hole is loaded, it should be stemmed to full depth with a free-flowing material.

When the explosives in these holes are detonated ahead of the production blast, the webs between the holes will fracture, leaving a surface joint which serves as a barrier to the shock waves from the production blast, thereby essentially eliminating breakage beyond the fractured surface.

Figure 14-5 An example of presplitting rock. *(E. I. du Pont de Nemours & Company.)*

Because of the variations in the characteristics of rocks the spacings of the holes and the quantity of explosive per hole should be determined by tests conducted at a given project [5].

This technique has been used successfully with vertical holes and with slanted holes whose slopes are not less than about 1 to 1. It has been used with limited success in tunnels.

INCREASING EFFICIENCY OF EXPLOSIVES WITH HOLES DRILLED AT AN ANGLE

Figure 14-6 illustrates two methods of drilling blastholes—one vertical, the other slanted. Studies conducted on the effectiveness of explosives in the two types of holes have demonstrated that the slanted holes allow greater efficiency, as well as offering other advantages such as [6]:

1. More uniform burden for full depth permits a uniform loading of slanted holes, with a reduced need for a heavy charge at the bottom.
2. An increased spacing of holes.
3. Less subdrilling.
4. Smoother faces and pit bottoms.
5. Generally better fragmentation of rock.

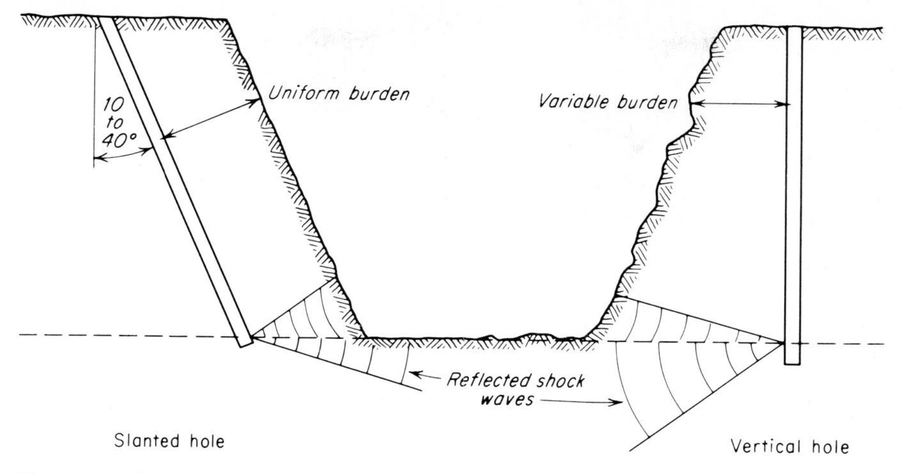

Figure 14-6 Slanted versus vertical blastholes.

When an explosive is detonated in a blasthole, shock waves are propagated in all directions. Only those waves that move toward a free face are highly effective in breaking rock. When a hole is slanted, more of the shock waves are directed upward, and thus they are more effective.

PROPER SPACING OF BLASTHOLES

Equation (14-1), which was developed and tested by Monsanto Chemical Company [7], may be used as a guide in determining the proper spacing of blastholes when using ammonium nitrate and fuel oil as an explosive. The equation indicates the maximum distance from the center of the hole that the explosive will fracture rock of a known or estimated tensile strength.

$$R = \frac{K}{12}\sqrt{\frac{P}{S}} \tag{14-1}$$

where R = critical radius to outer circle of fracture, ft
K = a coefficient whose average value is 0.8 for most rocks
P = maximum explosion pressure, psi
S = ultimate tensile strength of rock, psi

Table 14-4 gives the maximum spacings of blastholes per inch of diameter of hole.

Equation (14-1) and Table 14-4 assume that shock waves are propagated outward horizontally from the blastholes to form cylinders of fracture whose diameters are equal to $2R$. This assumption is more likely to be true with deep holes than with shallow ones. If cylinders of fracture are formed, as illustrated in Fig. 14-7, the staggering of holes in alternate rows should leave less

Table 14-4 Representative spacing of blastholes

Type of rock	Tensile strength of rock, psi (Pa)	Spacing of blastholes, ft per in. (mm/mm) of hole diameter
Anhydrite, strong	1,200 (8.27×10^6)	1.97 (23.6)
Anhydrite, weak	800 (5.52×10^6)	2.45 (29.5)
Granite, strong	1,298 (8.96×10^6)	1.92 (23.1)
Granite, average	888 (6.11×10^6)	2.32 (27.9)
Granite, weak	422 (2.90×10^6)	3.37 (40.5)
Graywacke	700 (4.82×10^6)	2.62 (31.5)
Greenstone	380 (2.62×10^6)	3.55 (42.7)
Limestone, strong	890 (6.13×10^6)	2.30 (27.6)
Limestone, average	480 (3.31×10^6)	3.15 (37.9)
Limestone, weak	280 (1.93×10^6)	4.12 (49.5)
Marble	860 (5.92×10^6)	2.37 (28.4)
Marlstone	480 (3.30×10^6)	3.15 (37.9)
Sandstone, strong	583 (4.01×10^6)	2.85 (34.3)
Sandstone, average	412 (2.83×10^6)	3.40 (40.8)
Sandstone, weak	280 (1.93×10^6)	4.12 (49.5)

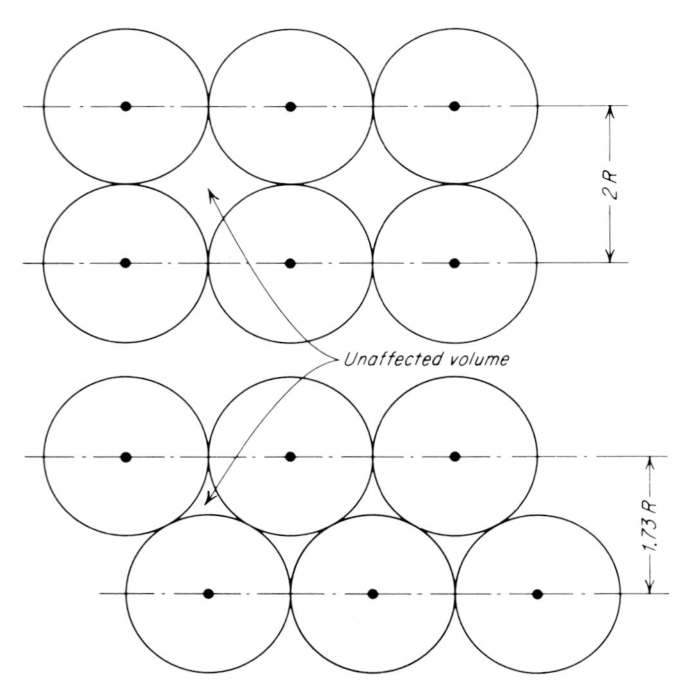

Figure 14-7 Reduction in unaffected volume resulting from staggered blastholes.

unaffected volumes than unstaggered holes of equal center-to-center spacing. However, staggering holes will reduce the spacing between the rows, as illustrated.

Because the spacing of holes determined from Eq. (14-1) may not apply under all conditions, it may be necessary to make some adjustments in the spacing of holes after observing the results of test blasting.

The equation may not be valid for small-diameter holes because the size of a hole may change appreciably before the blast attains maximum pressure. If all holes are blasted simultaneously, it may be possible to increase the spacing of the holes by as much as 50 percent of the values obtained from the equation [3, 8].

MONITORING THE SEISMIC EFFECT OF BLASTING

Because blasting operations may cause actual or alleged damages to buildings, structures, and other properties located in the vicinity of blasting operations, it may be desirable to examine, and possibly photograph, any structures for which charges of damage may be made later. Before beginning blasting operations, seismic recording instruments can be placed in the vicinity of the blasting to monitor the magnitudes of the effects of blasting. The monitoring may be conducted by the persons responsible for the blasting, or, if insurance covering this activity is carried by the company responsible for the blasting, a representative of the insurance carrier may provide this service [8, 9].

REFERENCES

1. ANFO Passes First Big Test as a Tunnel Explosive, *Construction Methods and Equipment*, vol. 44, pp. 90–93, September 1962.
2. Conger, H. M.: Metallized Slurry Blasting at Eagle Mountain, *Mining Engineering*, vol. 17, pp. 52–55, November 1965.
3. Ensign Bickford Company, P.O. Box 7, Simsbury, CT 06070.
4. Presplitting, What It Can Do, How It Works, *Construction Methods and Equipment*, vol. 46, pp. 136–141, June 1964.
5. Presplitting Done Under Sand Blanket for Suburban Freeway, *Roads and Streets*, vol. 116, pp. 31–33, February 1973.
6. Smith, Gordon R.: Drilling, New Equipment, New Techniques, *Construction Methods and Equipment*, vol. 44, pp. 110–115, August 1962.
7. Spaeth, G. L.: Formula for Proper Blasthole Spacing, *Engineering News-Record*, vol. 164, p. 53, April 1960.
8. Flying Long-reaching Drills Override Rugged Cliff Obstacles, *Construction Methods and Equipment*, vol. 52, pp. 66–72, March 1970.
9. Precision Blasting for Highway Cut Protects Old Rail Tunnel Nearby, *Construction Methods and Equipment*, vol. 59, pp. 53–55, February 1977.
10. "Do's and Don't's," Safety Library Publication No. 4, Institute of Makers of Explosives, 420 Lexington Avenue, New York, NY 10017.
11. "How to Destroy Explosives," Safety Library Publication No. 21, Institute of Makers of Explosives, 420 Lexington Avenue, New York, NY 10017.

12. "The American Table of Distances," Safety Library Publication No. 2, Institute of Makers of Explosives, 420 Lexington Avenue, New York, NY 10017, November 1971.
13. Dick, Richard A., Larry R. Fletcher, and Dennis V. D'Andrea: "Explosives and Blasting Procedures," GPO No. 024-004-02113-6, Government Printing Office, Washington, DC
14. Hemphill, Gary B.: "Blasting Operations," ABA Publishing Company, 406 West 32nd Street, Wilmington, DE 19802.
15. Nelmark, Jack D.: Large Diameter Blast Hole Drills, *Mining Congress Journal*, August 1980.
16. Roberts, A.: *Applied Geotechnology: A Text for Students and Engineers on Rock Excavation and Related Subjects*, Pergamon, Fairview Park, Elmsford, NY 10523.
17. Gustafeson, Rune: "Blasting Techniques," ABA Publishing Company, 406 West 32nd Street, Wilmington, DE 19802.
18. Burger, John R.: Nonelectric Blast Initiation, *Engineering and Mining Journal*, April 1982.

CHAPTER

FIFTEEN

TUNNELING

SCOPE OF THIS SUBJECT

The subject of tunneling is too broad to permit adequate coverage in this book. Therefore, only the fundamentals will be presented, with a limited number of examples to illustrate at least some of the construction methods used and current practices. Tunneling is an activity which is undergoing a great deal of study and development throughout a substantial portion of the world.

The references listed at the end of this chapter should assist readers who wish additional information on the subject in exploring it more fully. The list is representative of the types of information that are available. An examination of *Engineering Index*, available in many libraries, will assist readers in locating more sources of information.

Purposes of tunnels Tunnels are constructed for various purposes, including, but not limited to, provision of:

1. Passageways for railroads and automotive vehicles
2. Conduits for water and other liquids
3. Accesses to mines and underground spaces
4. Conduits for utility services
5. Passageways for persons

Types of earth excavated for tunnels Tunnels may be excavated in all types of soil, varying from loose earth, such as sand, gravel, clay, and shale, through the hardest rocks. This book will deal with tunnels driven through all these materials.

TYPES OF ROCK

The rocks which are encountered in tunneling operations can be divided into three major groups; igneous, sedimentary, and methamorphic. Each group can be subdivided according to origin, mineral content, physical condition, etc.

Igneous rocks Igneous rocks have cooled from molten masses which emerged through fissures from the interior of the earth. If a molten mass cooled prior to reaching the surface of the earth, the rock is defined as intrusive. Examples of intrusive rocks are granite and gabbro. If a molten mass cooled after reaching the surface of the earth, the rock is defined as extrusive. Examples of extrusive rocks are rhyolite and basalt.

Sedimentary rocks The sedimentary rocks with which the engineer is concerned include those which were deposited by flowing water, such as conglomerates, sandstones, shales, and clays, and those which were deposited by marine organisms, such as limestones and dolomites.

Metamorphic rocks If igneous or sedimentary rocks are subjected to high temperatures and pressures, they undergo changes in structure and texture. Rocks which have been subjected to such changes are described as metamorphic rocks.

Under the influence of moderate temperatures and pressures, clay and shales are transformed into slates and schists, which are low-grade metamorphic rocks. When subjected to high temperatures and pressures, slates and schists are metamorphosed into hard and dense gneiss. Limestone metamorphoses into marble and sandstone into quartzite.

PHYSICAL DEFECTS OF ROCKS

All rocks, regardless of the type, have physical or structural defects which have considerable effect on tunneling operations. These defects consist of fractures, whose magnitudes and spacings vary considerably. Simple fractures are defined as joints, whereas major fractures, associated with relatively large displacements, are defined as faults.

Joints Joints are surfaces of physical failure or separation with little or no displacement between the rock components on opposite sides of a joint. The joints may exist in two or three planes approximately at right angles with each other.

In driving a tunnel through a rock formation, the existence of joints will affect the extent to which the sides and roof must be supported during the tunneling operation. Also, joints provide passageways through which ground water may flow into a tunnel.

Faults A fault is a zone in a formation where a large displacement has occurred along the plane of failure. The displacement may be horizontal, vertical, or a combination thereof. A fault usually constitutes an undesirable hazard to tunnel driving. Because of the enormous forces that produce a fault the rock formation in the fault zone will be badly broken. The crushed material may vary in size from fine sand to large blocks, which tend to flow into a tunnel as it is driven through a fault zone. If ground water is present in the formation, the broken material within the fault zone will provide excellent passageways for the water to flow into the tunnel unless corrective steps are taken prior to excavating through the zone. It may be necessary to pressure-grout the formation ahead of the tunneling operation in order to eliminate the hazard of ground water.

PRELIMINARY EXPLORATIONS

While the approximate location of a tunnel is dictated by the service it is to provide, the final location should be based on the results of surface and subsurface explorations. Such explorations are made prior to selecting the exact location of a tunnel in order to determine the kinds of formation that exist and the extent to which ground water is present in the formations along the route of a proposed tunnel. The formations may include unconsolidated muck, sand, gravel, or clay, with or without ground water. There may be solid or badly broken rock, or there may be faults and folds to contend with. If a tunnel is driven through solid rock, little or no roof support may be required, whereas if it is driven through badly broken rock, it will be necessary to provide extensive wall and roof supports. If an exploration indicates the presence of significant quantities of ground water, it may be desirable to seek a more favorable location, or if this is not possible, it may be necessary to pressure-grout the formation ahead of excavation as a means of reducing the flow of water. Plans should be made to have adequate pumps available to remove the water.

Seismic exploratory methods have been used to obtain information on the characteristics of the formation along the proposed routes of tunnels. These studies are made by recording and analyzing the behavior of shock waves, generally propagated by explosives detonated in holes along the route, as the waves travel from the source to the recording instruments. Prior to driving the Musco Tunnel in Sweden intensive seismic studies were made. Where the studies indicated poor rock or fissures, borings were made with diamond core drills in order to obtain additional information.

Valuable information may be obtained from a surface exploration by a competent geologist who is reasonably familiar with the area. More definite information concerning a formation may be obtained by drilling holes along the proposed route and securing samples of the formation. The holes should be drilled at least to the bottom of the proposed tunnel and should be spaced sufficiently close to give representative samples of the formation. If the formation is free of severe structural irregularities and variations, the spacing of the holes may be greater than for a formation that contains faults, folds, or other structural irregularities.

If a formation is soft enough, the holes may be drilled with earth augers or split tubes, which will permit the recovery of undisturbed samples for examination. If the formation consists of unconsolidated material, such as sand or small gravel, holes may be jetted with water. For this purpose it will be necessary to supply a reasonably large quantity of water, under pressure, and enough pipe to permit holes to be jetted to the desired depth. However, the material recovered from jetted holes may not give true samples of the formation, and the information obtained from such samples may not be sufficiently dependable for selecting the location of a tunnel.

If a formation is rock, the holes may be drilled with wagon, churn, rotary, or other types of drills that produce cuttings. Since these drills produce cuttings instead of undisturbed samples or cores, the material recovered from the holes will not indicate whether the formation is solid or broken rock. As water must be added to holes drilled with churn drills in order to remove the cuttings, the cuttings will not indicate the extent to which ground water exists in a formation.

When cores from the exploratory holes are desired, they may be obtained with core or shot drills. Cores obtained with diamond bits usually vary in diameter from $\frac{7}{8}$ to 4 in., while cores obtained with shot drills usually vary in diameter from about 4 to 8 in. However, larger sizes may be obtained with either bit. Large-diameter cores will permit a more intelligent analysis of the structure of the formation. Cores should be assembled in the same order as they come from the hole. In general, the length of the core recovered from a hole will be less than the depth of the hole, the length varying with the kind of rock and the degree of solidity. Typical core recoveries should be about 80 to 90 percent for igneous rocks, 60 to 70 percent for limestone, 70 to 80 percent for sandstone, and 40 to 50 percent for shale.

After the preliminary explorations have been completed and the results analyzed, the location that will permit the construction of a satisfactory tunnel at the lowest practical cost can be selected.

NUMBER OF ENTRANCES

If a tunnel is relatively short, not more than a few hundred feet long, it may be driven from one entrance only. However, as the length is increased, conducting all operations from one entrance may result in excessive haul distances and high haulage costs, together with a general congestion between the portal and the head of the tunnel. Such a condition may be eliminated or alleviated by driving a tunnel from both ends. For long tunnels it may be advantageous to provide intermediate openings, such as shafts, to facilitate the removal of muck and water and the delivery of materials, supplies, air, and utilities. Intermediate shafts or openings permit operations at a greater number of headings, thus making possible an increase in the rate of driving a tunnel. This may be especially important for a project when an early completion is desirable.

SEQUENCE OF OPERATIONS FOR DRILL AND BLAST CONSTRUCTION

As soon as the construction of a tunnel is under way, the various operations should be carried on in a well-planned sequence. The actual operations will vary with the type and size tunnel, the method of attacking the heading, and the kind of formation encountered. The construction may be on the basis of one, two, or three shifts per day.

For a tunnel driven through rock the following operations might apply:

1. Setting up and drilling
2. Loading holes and shooting the explosives
3. Ventilating and removing the dust following an explosion
4. Loading and hauling muck
5. Removing ground water if necessary
6. Erecting supports for the roof and sides if necessary
7. Placing reinforcing steel
8. Placing the concrete lining

The first four operations are related to the driving of the tunnel and frequently establish the rate of progress in constructing a tunnel. Progress on the other operations should be coordinated with the rate of driving insofar as it is practical to do so.

A representative sequence of operations and the time required for each are given hereinafter for a railroad tunnel at the Conemaugh Damsite [1]. The tunnel, whose bore was 36 ft wide and 32 ft high, was driven through sandstone. In driving the main bore, artificial ventilation was not necessary because a pilot tunnel had been driven the full distance prior to starting excavation for the main bore. Each round* required 80 holes, 20 ft deep, which were drilled by nine drifters, mounted on a jumbo. The muck was loaded by a $1\frac{1}{4}$-cu-yd electric power shovel into narrow-gauge cars, whose capacity was 5 cu yd each. The rate of progress was approximately 20 ft per day for two shifts.

The time required for several operations in a cycle was as follows:

*A round involves drilling holes into the face of the tunnel, loading the holes with explosives, and detonating the explosives.

Shift	Operation	Time, hr Min	Time, hr Max
1	Drill the holes	5	6
	Load the holes	1	1
	Explode the dynamite		
2	Load and haul muck	9	9
Total time		15	16

The contractor who drove the power tunnel for the Kemano hydroelectric project in British Columbia chose to complete a drilling cycle in each 8-hr shift [2]. The tunnel was a 25-ft horseshoe bore. Each round required 87 to 96 holes, 13 to 15 ft deep, which were drilled with 15 drifters, mounted on a jumbo. The average rate of advance was about 12 ft per shift.

A typical time for each operation and for a cycle was as follows:

Operation	Time, hr
Drill the holes	$1\frac{3}{4}$
Load the holes	$\frac{3}{4}$
Explode the dynamite	
Ventilate during lunch	$\frac{3}{4}$
Load and haul muck	$4\frac{3}{4}$
Total time	8

DRIVING TUNNELS IN ROCK

There are several methods of attacking the faces of tunnels driven through rock. The method selected will depend on the size of the bore, the equipment available, the condition of the formation, and the extent to which timbering is required. The more common methods of attack are:

1. Full face
2. Heading and bench
3. Drift
4. Pilot tunnel

Each of these methods is described in the articles which follow.

Full-face attack When a tunnel is driven by the full-face attack method, the entire bore or face is drilled, the holes are loaded, and the explosives are discharged. Small tunnels whose dimensions do not exceed about 10 ft are

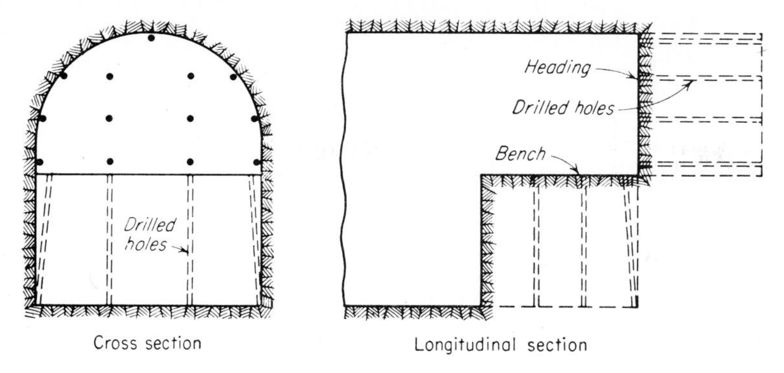

Figure 15-1 Bench method of driving a tunnel.

always driven by this method. Large-size tunnels in rock frequently are driven by the full-face method. With the development of the jumbo, or drill carriage, the use of this method has become increasingly more popular in driving large tunnels. A number of drills may be mounted on the front end of a jumbo and operated simultaneously with a high efficiency.

Heading and bench method The heading and bench method of driving a tunnel involves the driving of the top portion of the tunnel ahead of the bottom portion, as illustrated in Fig. 15-1. If the rock is firm enough to permit the roof to stand without supports, the top heading usually is advanced one round ahead of the bottom heading. If the rock is badly broken, the top heading may be driven well ahead of the bench and the bench used in installing the timbers to support the roof. The development of the jumbo has reduced the use of the heading and bench method of driving a tunnel.

Drift method In driving a large tunnel, it may be advantageous to drive a small tunnel, called a drift, through all or a portion of the length of the tunnel prior to excavating the full bore. A drift may be classified as center, bottom, side, or top, depending on its position relative to the main bore. Figure 15-2 illustrates the position of each type of drift.

The use of the drift method of driving a tunnel has several advantages and disadvantages.

Among the advantages are:

1. Any zone of bad rock or excessive water will be discovered prior to driving the full bore, thus permitting corrective steps to be taken early.
2. The drift will assist in ventilating the tunnel during later operations.
3. The quantity of explosives required may be reduced.
4. Side drifts may facilitate the installation of timbers to support the roof, especially for a tunnel driven through broken rock.

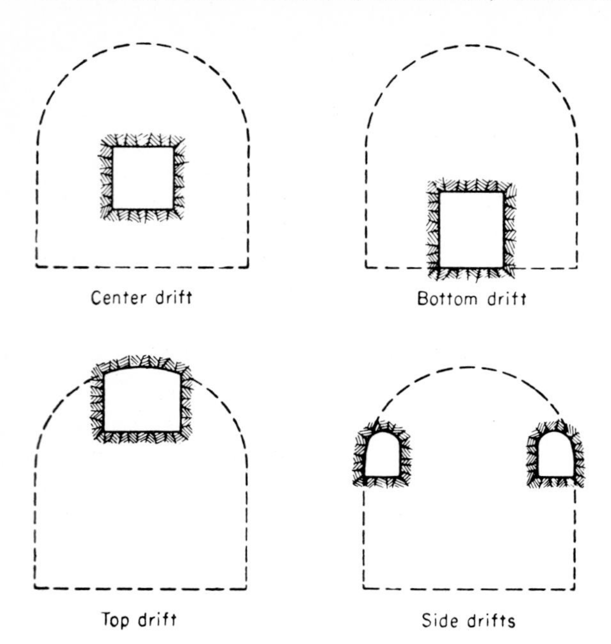

Figure 15-2 Types of drifts.

Among the disadvantages are:

1. Driving the main bore must be delayed until the drift is finished.
2. The cost of drilling and handling muck in a small drift will be high because much of the work must be performed by hand instead of by power-operated equipment.

DRILLING ROCK

In driving a tunnel through rock, it is necessary to drill holes for the explosives that loosen the rock. The most commonly used drill is a drifter, equipped with drill steel and detachable bits, either steel or carbide-insert. Drills and bits are of the types described in Chap. 13. Water frequently is used instead of compressed air to remove the cuttings from the holes, as a means of reducing the amount of dust in the air.

For any given project the best depth and spacing of holes over the face of the tunnel should be determined experimentally. The depth of holes will vary with the size and shape of the tunnel, the kind of rock, and the drilling equipment used. The depth advanced during one drilling and shooting operation is called a round. This distance frequently varies from 5 to 20 ft. It will be necessary to drill holes deeper than the advance per round because of loss in depth resulting in blasting. For example, it may be necessary to drill holes 14 ft deep in order to pull 12 ft, the latter value being the effective depth per round.

DRILL MOUNTINGS FOR SMALL TUNNELS

Drills used in small tunnels and drifts usually are mounted on bars or columns, which are made from sections of steel pipe, equipped with a screw jack at one or both ends. Bars are installed horizontally in a tunnel whose width is less than the height, while columns are installed vertically in a tunnel whose height is less than the width. Installation consists in placing the bar or column in position and extending the jack until the bar or column is securely wedged in position. Figure 15-3 illustrates the use of bars, while Fig. 15-4 illustrates the use of columns.

The drill may be mounted directly on the bar through an adjustable clamp, which permits movement along the length of the bar. When a column is used, the drill is mounted on an arm, which in turn is mounted on the column through an adjustable clamp. The drill may be moved along the arm or the column.

While bars and columns are satisfactory for use in small tunnels, the excess

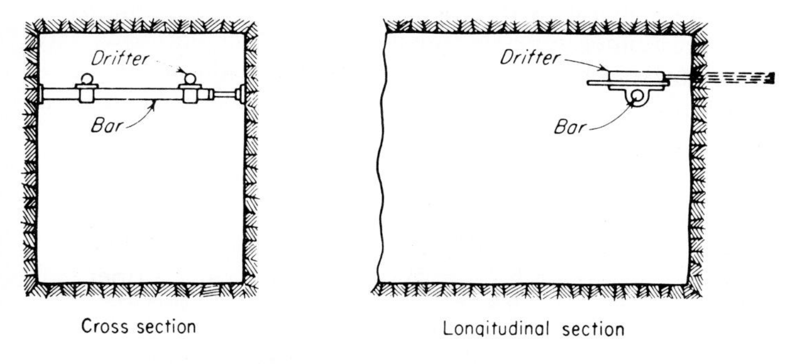

Figure 15-3 Drifter supported by a bar.

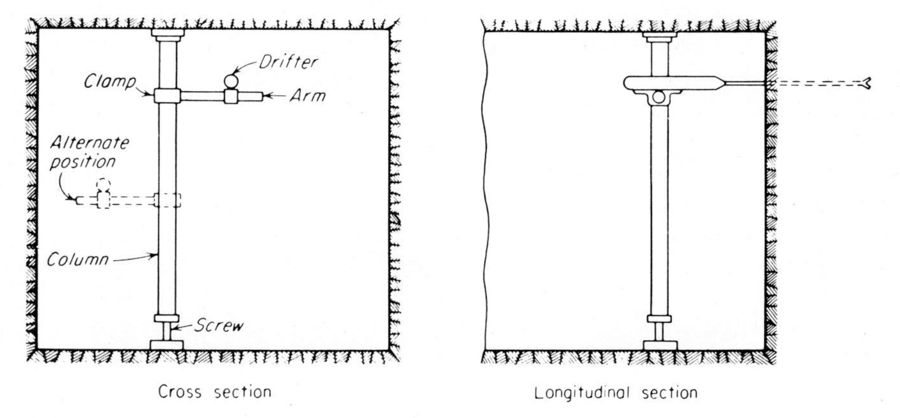

Figure 15-4 Drill mounted on a column.

lengths and weight required for use in large tunnels make them too difficult to handle. In drilling in large tunnels, it is more satisfactory to mount the drills on jumbos.

DRILL JUMBOS

A drill jumbo is a portable carriage with one or more working platforms, equipped with bars, columns, or booms to support the drills. The supports are designed to permit the drills to be spaced to any desired pattern. The main members of some jumbos have been constructed from welded steel pipe designed to transmit compressed air to the drills.

A recent improvement in drilling equipment is a hydraulic or airpowered boom to support rock drills. This boom, which is mounted on a jumbo, is equipped with controls that permit the operator to spot a drill in any desired position in a few seconds. Figure 15-5 illustrates a powered boom mounting in operation.

A jumbo may be constructed with one or more working platforms, depending on the size of the tunnel in which it will be used. The platforms may be connected to the jumbo structure with hinges which permit them to be raised or lowered to allow other equipment, such as a mucker or cars, to pass under the jumbo. Several drills may be operated from each platform.

Figure 15-5 Jumbo mounting 16 drifters, with hinged centers to permit passage of trucks. *(Joy Manufacturing Company.)*

A jumbo may be mounted on skids, on wheels for traveling on rails, or on pneumatic tires. Tire mounting gives a jumbo considerable freedom of movement, which facilitates spotting it in position for drilling operations, since it is not restricted to movement on rails.

Self-propelled jumbos, with one or more working platforms, have been constructed on trucks and tractors. When an air compressor is mounted on or attached to the same vehicle, it provides a highly mobile and versatile drilling machine. If the machine is powered with an internal combustion engine, the engine should be diesel driven and equipped with an exhaust scrubber to eliminate the discharge of carbon monoxide gas in the tunnel.

DRILLING PATTERNS

A drilling pattern represents the positions of the holes drilled into the face of a tunnel in advancing one round. The pattern that will produce the most economical and satisfactory breakage of rock for a given tunnel should be determined by conducting tests, using different patterns and quantities of explosives.

Figure 15-6 illustrates a pattern of holes 12 ft deep and $1\frac{3}{4}$ in. in diameter

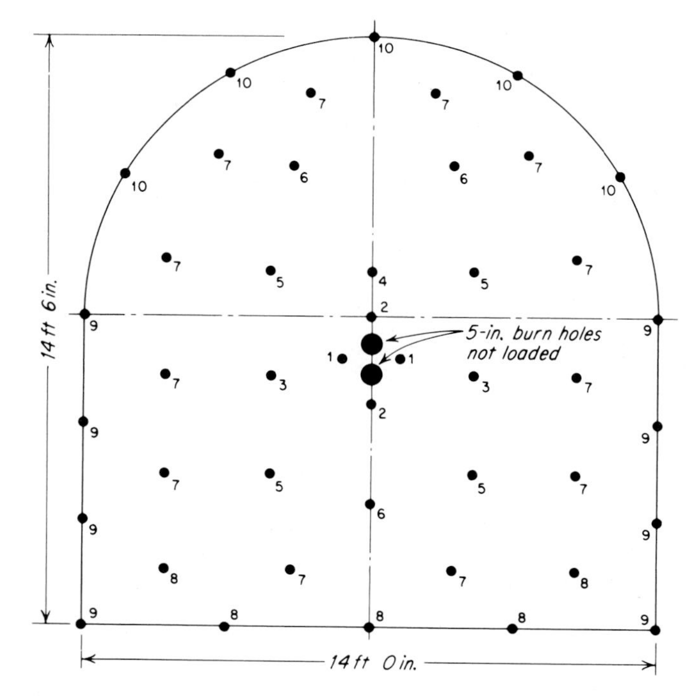

Figure 15-6 Firing pattern for blastholes in a tunnel.

used in driving a tunnel for water in California. The two 5-in.-diameter burn holes, located near the center of the face, were drilled to increase the effectiveness of the blasts in holes 1 and 2, which were detonated in the sequence indicated by the numbers. The numbers adjacent to the other holes indicate the sequence of firing the holes, using MS electric blasting caps. The explosives produced an average advance of 11 ft per round.

Contracts for driving tunnels usually provide that the contractor will be paid a given price per linear foot of tunnel, or per cubic yard of excavation lying within a specified payline. Also, if the tunnel is lined with concrete, it will be necessary to increase the quantity of concrete to replace the excess rock removed beyond the payline, perhaps at the contractor's expense. Therefore, it is desirable to keep overbreak to a minimum. It is good practice to use a template to locate the payline and all holes to be drilled. A daub of paint at each hole location will assist in spotting the drills more quickly.

LOADING AND SHOOTING HOLES

In general the information concerning explosives appearing in Chap. 14 will apply to the explosives used in driving tunnels. However, when selecting explosives for use in tunnels, consideration should be given to their fume properties.

Ammonium nitrate blended with fuel oil has proven to be effective and economical for use in tunnels. In driving the Canyon Tunnel in California the contractor used ANFO to load $1\frac{3}{4}$-in.-diameter holes 12 ft deep with considerable success [3, 4]. Among the advantages resulting from the use of ANFO compared with dynamite were the following:

1. Fragmentation was 40 percent better with the same hole pattern.
2. Loading the blasting time was reduced 15 to 20 percent.
3. Cost of ANFO was about 5 cents per lb versus 22 cents per lb for dynamite, and the overall cost of explosives for the job ran about one-half the cost of using dynamite only.
4. ANFO produced appreciably less toxic blast fumes than dynamite produced.
5. ANFO was simpler and safer to store and handle.

DRIVING TUNNELS WITH TUNNEL-BORING MACHINES

A recently developed technique in driving tunnels through both earth and rock is the use of tunnel-boring machines, frequently identified in the literature as TBM's or mechanical moles. Several types of moles are illustrated in Figs. 15-7, 15-8, 15-9, 15-10.

The function of a mole is to loosen the earth or break the rock to be

Figure 15-7 Disc cutterhead for semihard rock formations. *(Caldwell Division of Smith Industries International.)*

Figure 15-8 Mechanical mole with oscillating arms. *(Caldwell Division of Smith Industries International.)*

removed from a tunnel into cuttings, which can be conveyed to the rear of the machine, where they can be loaded into muck cars or trucks or onto conveyor belts to be transported to the ultimate disposal site.

The essential parts of a mode Depending on the manufacturer and the type of service to be provided, the essential parts of a mole might include the following items, beginning at the front end and continuing to the rear:

Figure 15-9 Mechanical mole with a rotating head. *(Reed Tool Company.)*

1. A rotating cutterhead, mounting teeth for excavating earth or discs for excavating rock
2. Muck buckets mounted on a rotating muck ring to elevate the muck and to discharge it into or onto a primary conveyor, which transports it to the rear of the machine
3. A cylindrical metal shield, whose diameter is essentially the same as the diameter of the tunnel
4. Extendable and retractable clamp legs, with shoes to engage the inner surface of the tunnel bore to prevent the mole from rotating while in operation
5. Thrust cylinders or rams, usually hydraulically operated, to maintain a forward pressure on the cutting head
6. A control console
7. Rear support legs
8. Auxiliary sprag legs (may be optional)
9. Hydraulically operated extendable and retractable rams, which bear against the tunnel lining members or the tunnel surfaces, to provide the thrust needed to advance the mole

Figure 15-10 Mechanical mole equipped with disc-type cutter head. *(Robbins Company.)*

10. The motor or motors, usually electric, required to operate all energy-requiring components of the mole
11. Possibly a laser guidance system

The operation of a mode When a mole is in position to operate, the cutterhead and discs are pressed against the face of the tunnel. As this thrust is maintained, the cutterhead is rotated and the muck is delivered to the rear of the mole. When the cutterhead reaches the end of the advance stroke, the rear legs are extended downward to support the machine. The clamp legs are retracted to free the main frame and the thrust rams move the frame forward. The clamp legs are then extended to anchor the mole in position for the beginning of the next cycle.

Methods of transporting muck Figure 15-11 illustrates a mechanical mole equipped with a self-contained belt conveyor, which transports the cuttings from the face of the tunnel to the rear of the machine, where it is discharged into hauling equipment for removal from the tunnel.

Figure 15-11 Mechanical mole equipped with belt conveyor to transport muck to hauling units. *(Robbins Company.)*

Other methods of removing the cuttings are illustrated and described in the following articles.

Sizes of moles There are no theoretical limits on the sizes of moles, either minimum or maximum. In actual practice they have been used to drive tunnels whose diameters vary from less than 5 to 40 ft or more. While many moles have been designed and manufactured to be used on a given project and have a specified diameter, some machines have been equipped to permit increases or decreases in their diameters, so that they may be used to drive tunnels having different diameters.

Limits on the types of earth rock that can be excavated by moles Moles have been used to drive tunnels through all types of earth and solid rock with compressive strength as high as 40,000 psi [5]. However, when driving a tunnel through hard rock, the rate of advance may be slow, and the cost of replacing the cutter teeth or discs may be so high that alternate methods of driving the tunnel are more desirable.

Rates of driving tunnels with moles The rate of advance of a mole when driving a tunnel will depend on several factors, including but not limited to the following:

1. Type of earth or rock excavated
2. Amount of ground water present, if any
3. Power of the driving equipment
4. Skill of the operator
5. Method of removing the muck
6. Extent of ground support required
7. Type of ground support provided

The following examples indicate production rates attained by moles.

Example 15-1 This example concerns the Azotea tunnel in New Mexico [6].

Diameter of bore, 13 ft 6 in., max
Type of rock, shale and sandstone
Hardness of shale, 2.5 Moh's scale
Compression strength of shale, 1,380 to 5,890 psi
Hardness of sandstone, 5.0 Moh's scale
Compression strength of sandstone, 3,015 to 8,500 psi
Maximum advance during a three-shift day, 241.5 ft
Average advance while mole was operating, 12.1 ft per hr

Example 15-2 This following information applies to the Blanco tunnel in New Mexico [6].

Diameter of bore, 10 ft
Type of rock, shale and sandstone
Maximum advance per month, three shifts, 6,713 ft
Maximum advance per day, 375 ft
Advance during 12 months, 41,179 ft
Average advance per month, 3,432 ft

While these examples report rates of advance that have been attained, they do not represent rates attained on other projects, some of which have been considerably less.

Figure 15-12 demonstrates graphically the progress in rates of drilling tunnels during more than 100 years [4]. It will be noted that the rates have increased significantly since introduction of the mechanical mole.

Table 15-1 gives information on the rates of advance for tunnels that have been driven with moles.

PRODUCTION EXPERIENCES WITH MOLES

The following examples furnish information illustrating and describing experiences and results of using moles to drive tunnels under various conditions. The references appearing at the end of this chapter list publications containing additional information on this subject.

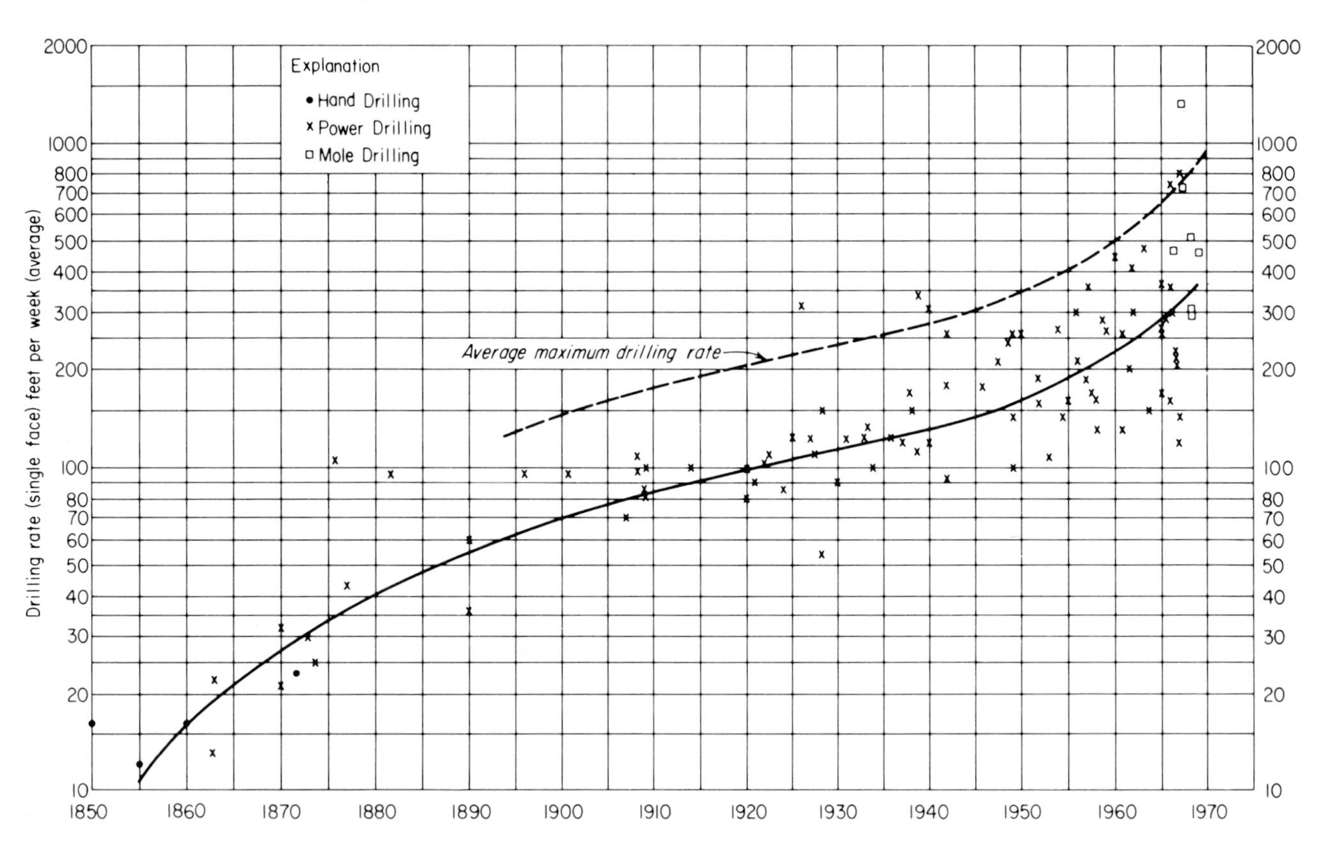

Figure 15-12 Progress in driving tunnels. *(Journal of the Construction Division, American Society of Civil Engineers.)* [4]

Table 15-1 Rates of progress on tunnels driven by moles [4]

Date	Location	External diameter, ft-in.	Type of rock	Rate of advance, ft per 24 hr: Maximum	Rate of advance, ft per 24 hr: Average
1954	Mittry mole, Oahe dam diversion tunnels	25-9	Soft shale, faulted squeezing	96	46
1955	Oahe miner Oahe dam	25-9	Clay and soft shale, faulted	120	38
1958	M-K miner Oahe dam	29-6	Soft shale, bentonite	135	32
1959	Prairie miner Oahe dam	29-6	Soft shale	153	63
1959	Humber River sewer at Toronto	10-9	Shale and limestone, hardness, 2.5	94	*
1961–63	Poatina tunnel, Australia	16-0	Sandstone—hardness 3.5 to 5.0, compressive strength 13,000 to 16,000 psi	150	84
1963–64	Vancouver interceptor sewer	7-6	Sandstone, shale, and coal	200	*
1965	Philadelphia sewer	13-8	Mica schist, hornblende, 6,000 to 25,000 psi	*	41
1965–66	St. Louis metropolitan sewer	8-0	Limestone, compressive strength 14,000 to 17,000 psi	*	48
1965–66	Navaho No. 1, New Mexico	20-10	Layered sandstone and shale, 5,000 to 6,000 psi	160	61
1964–66	Azotea, New Mexico	13-5	Sandstone and shale	241	55
1966–67	Oso, New Mexico	10-2	Shale	414	157
1965–67	Blanco, New Mexico	10-0	Shale	375	107

* Information not available.

Mangla Dam Tunnels [7] This project in West Pakistan involved driving five diversion and power tunnels, using what was then the world's largest full-face tunnel-boring machine. The diameter of the bore was 36 ft 8 in. The tunnels were driven through soft rock. An extensible rope belt conveyor was employed, which continuously increased in length as the machine advanced. An overhead monorail crane was used to carry steel ribs to the mole and place them on the machine's upper deck ring-beam conveyor. This was the second machine to employ a shield of flexible steel fingers to protect against caving rock at the point where the steel ribs, wire mesh, and lagging were placed. This shield provided continuous support without relaxation from a point close behind the cutters to the final emplacement of the primary lining. The machine established a record by excavating 4,160 cu yd of *in situ* rock in a 24-hr day while advancing 106 lin ft.

Mersey River Tunnels, Liverpool, England [7] When the mole used to drive the tunnels for the Mangla Dam finished that assignment, it was transported to

Liverpool, England, where it was modified in size and used to drive two highway tunnels under the Mersey River. The mole was reduced in size from a diameter of 36 ft 8 in. to 33 ft 11 in. Also, the contractor redesigned the machine's propulsion and tunnel support erection system to permit installation of precast concrete segments for the final lining of the tunnel. Figure 15-13 illustrates longitudinal and cross-sectional views of this machine, showing the working platform and the precast concrete segments installed at a location immediately behind the cutter head.

New developments in tunneling [8] Tunnel-boring machines, or moles, exemplify the ingenuity of the manufacturing industry in developing new equipment for rapid excavation of tunnels. The machines have been used with considerable success to set new records in rapid excavation of several tunnels on Bureau of Reclamation projects. In excavating the 2-mile-long, 20-ft-diameter tunnel No. 1 on the Navajo Indian irrigation project in northern New Mexico, the contractor completed the excavation in 9 months, from June 23, 1965 to March 19, 1966, averaging 61 ft of advance each operating day. By comparison, the rate of excavation of a comparable tunnel using conventional methods would have averaged only 45 ft per day. Three tunnels of the San Juan-Chama project in New Mexico, the 8.6-mile Blanco Tunnel, the 5-mile Oso Tunnel,

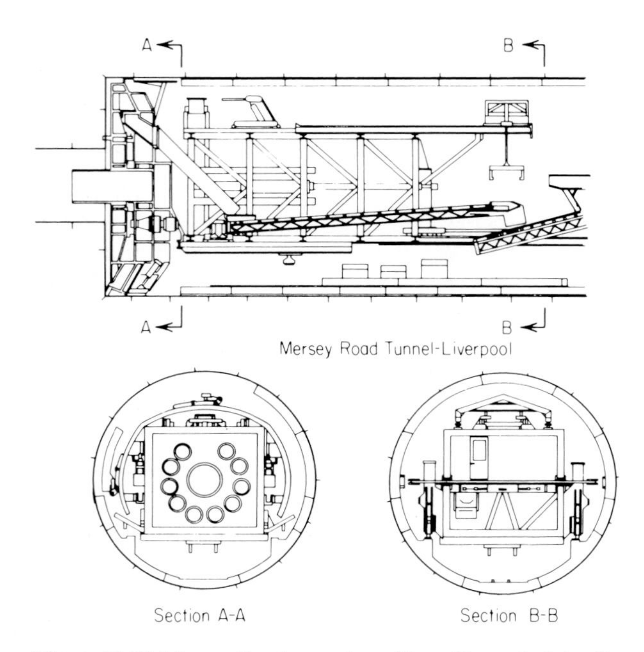

Figure 15-13 Mersey Road tunnel machine. *(Journal of the Construction Division, American Society of Civil Engineers.)*

and the 12.7-mile Azotea Tunnel, were also excavated by boring machines. Excavation of the Blanco and Oso Tunnels set world records for progress. Using a mole, the crews for the 8-ft-7-in.-diameter Blanco Tunnel advanced 367 ft in one day and 375 ft in another day. During the month of March 1967 they advanced 6,713 ft. The crews for the 8-ft-7-in.-diameter Oso Tunnel excavated 403 ft in one day. During the month of March 1967 they advanced 6,851 ft.

The geological formations through which these boring machines were used were relatively soft sandstones and shales. However, the industry is continuing to develop boring machines for rapid excavation through much harder rock. For example, the 10-ft-diameter, 4-mile-long River Mountain Tunnel on the Southern Nevada Water Project was largely being driven through hard volcanic rock. Using tungsten carbide insert-type bits and hard steel kerf-type cutters on the mole's rotating head, the rate of advance was as great as 250 ft per day, with an average rate of 6.5 ft per shift hour.

San Francisco Bay Area Rapid Transit system [9] In driving the tunnels for this transit system, which was started in 1968, two types of soft-ground tunneling machines were used.

The Memco machine employed a cutting wheel working the entire face, scraping the ground and dropping the cuttings onto a conveyor belt, from which they were carried out by the muck trains. The wheel was turned clockwise or counterclockwise by planetary gears driven by hydraulic motors having 2,000,000 ft-lb of torque. Fifty jacks of 115 tons capacity each shoved the cutter wheel against the face during excavation. The advance for each shovel was about 30 in. One machine excavated 19,056 ft of tunnel in clayey sand, 1,200 ft of which required compressed air with outside dewatering. Another machine excavated 7,600 ft in fairly graded sand and clay, also under compressed air.

Another type of mole with oscillating cutter arms was a Caldwell machine, similar to the one illustrated in Fig. 15-10, which employed four independently activated cutter blades, each operating in its own quadrant, sweeping back and forth to scrape the ground away at the face. When operated under 6,000 psi of hydraulic pressure, each arm could produce a torque of 972,000 ft-lb. A bulkhead mounted behind the cutting arm channeled the excavated material downward onto a 36-in. drag-chain conveyor belt and thence to the muck train. The machine was advanced by means of 20 hydraulic jacks, with a total capacity of 3,020 tons, in increments of 30 in.

The Memco machine was 17 ft 1 in. in outside diameter and 12 ft 5 in. long, with a 163-ft-long trailing conveyor and power pack and a total of 1,350 hp.

The Caldwell machine was 18 ft 0 in. in outside diameter and 15 ft 0 in. long, with 24 hydraulic jacks, each having a capacity of 125 tons and an advance stroke of 36 in.

Three different mucking methods were used to move the excavated material out of the shield:

1. A backhoe mounted on the framing of the shield dumped the muck onto a conveyor belt.
2. An Eimco model 620 rail-mounted and pneumatically operated loader cast the material into muck cars behind the shield.
3. Diesel-engine-powered rubber-tired loaders carried the muck back through the tunnel to an outside stockpile.

Japan, tunnel mole [10] Figure 15-14 illustrates a full-face mole with a supporting system controlled almost entirely from above the ground, that was used to drive an 11-ft-diameter water tunnel under a railroad yard in Osaka, Japan. As the highly automated mole bored through sandy gravel soil containing cobblestones, it supported the face of the tunnel with a pressurized clay-water slurry between its cutting head and a watertight steel bulkhead.

Excavated muck entering the mole shield through the cutting wheel joined the slurry in a mix, which was forced through a discharge pipe to a trommel, or rotary screen, for treatment. Some muck was extracted and delivered to the surface of the ground, while clay-water slurry was recirculated to the head of the mole, where it replaced the mix being pumped out. The controlled pressure of the slurry at the face counterbalanced the ground-water pressure, which permitted the space to the rear of the bulkhead to be maintained at atmospheric pressure.

The 15-ft-long mole was driven forward by ten 80-ton jacks with 42-in. strokes, which pushed against the tunnel lining.

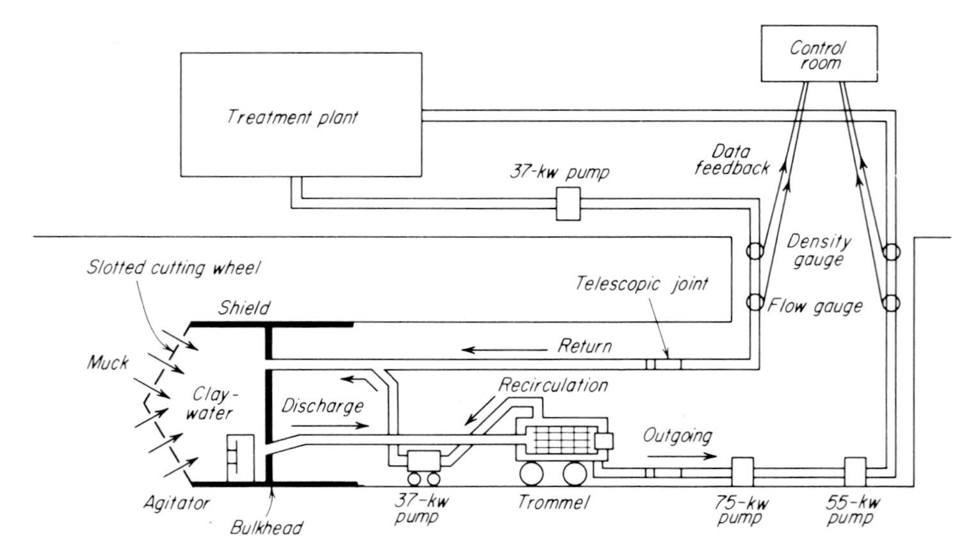

Figure 15-14 Automated mole's pressurized-mix support system. *(Construction Methods and Equipment.)*

A data feedback system linked to a station on the surface of the ground monitored and controlled the tubehead pressure, the mucking conditions, and the pipeline flows. Information from this monitoring system was relayed to the mole operator to enable him to perform the operations that would produce the maximum production.

The slurry pressure in front of the bulkhead was continuously measured by a diaphragm gauge that activated the pressure-adjusting instrument. In turn, this instrument activated the incoming pipeline pump to vary its speed and to adjust the pressure in front of the bulkhead.

Readings from gauges on incoming and outgoing pipelines, plus the measured weights of stones removed by the trommel, were instantly and continuously processed by a computer. Thus, the exact volume of muck removed was determined. The results appeared on digital indicators and level gauges in the surface control room. From that room, the mole operator was directed to adjust the rotation of the cutting wheel and the thrust of the machine when necessary to maintain the desired production rate.

When the slurry-muck flowed to the trommel, all cobblestones were removed ahead of the screen and hauled out of the tunnel. A portion of the

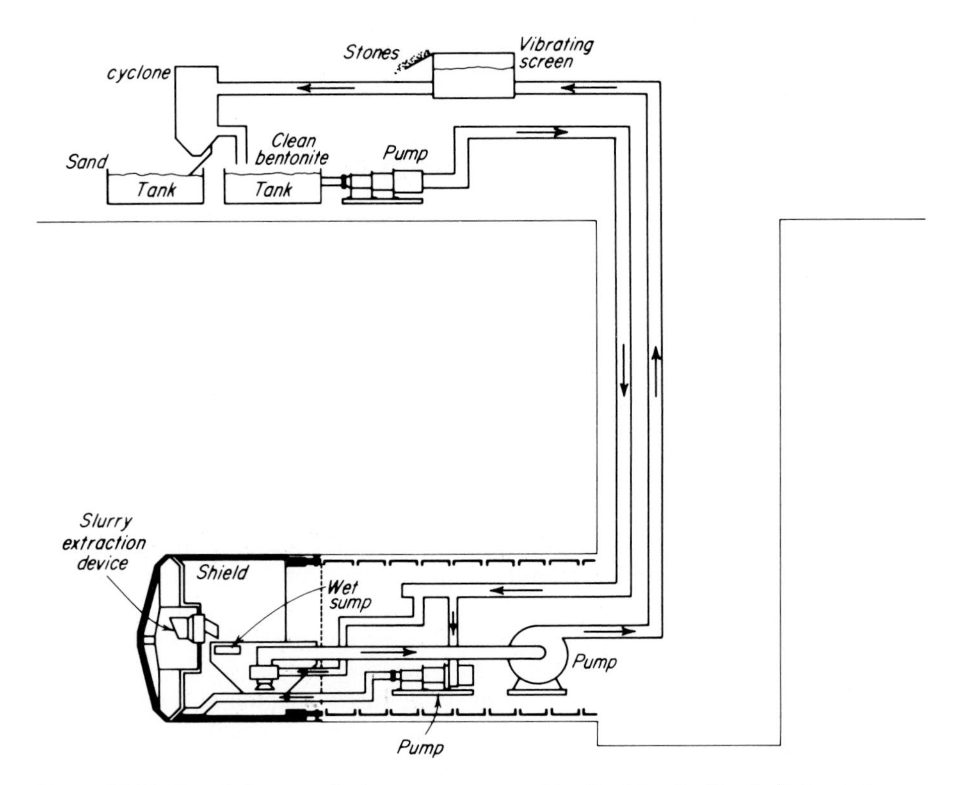

Figure 15-15 Use of slurry method to excavate tunnel in alluvial soil. *(Roads & Streets.)*

muck flowed through a pipeline to the treatment plant at the surface of the ground, where it was processed. That which was returned to the mole was mixed in the proportions of about 25 percent clay and 75 percent water, plus 0.04 percent of an additive to improve viscosity.

Tunnel in alluvial soil makes fast advance at low cost [11] A tunnel-boring machine with a bentonite-slurry shield was used to drive a 13-ft-6-in.-diameter tunnel through wet sand and gravel for the Fleet Underground Railway Line in London, England. The methods used were similar to those used in the previous example.

As the machine moved forward a distance equal to one advance stroke, a complete ring of cast-iron segments was installed under the tail of the mole; then the annular space between the outer surface of the lining and the surface of the tunnel was pressure-grouted with a thick bentonite grout. This concluded the installation of the permanent lining for the tunnel.

Use of this method permitted all operations to be carried out at atmospheric pressure and resulted in an estimated cost reduction of about one-third compared with conventional methods of driving the tunnel.

It was reported that no measurable subsidence resulted from driving the tunnel.

Figure 15-15 illustrates the assembly of equipment and the methods used to drive this tunnel.

USE OF LASER BEAMS TO GUIDE MOLES

A relatively new but effective method of guiding a mole, jumbo, or other tunnel-driving equipment is to use a laser beam for position control. These beams have been used on projects requiring accurate control of position and direction, where they have demonstrated their advantages compared with earlier or conventional methods [4, 8, 12].

The word laser stands for *l*ight *a*mplification by *s*timulated *e*mission of *r*adiation. The light from a laser tube differs from ordinary light from other sources such as the sun or an electric light bulb. Because ordinary light is a combination of many frequencies, a beam of such light will disperse in many directions. A laser produces light of only one frequency or wavelength. For this reason the rays of light travel essentially along a straight line, with little or no dispersion. If a laser generator emits a light beam of small diameter, the diameter remains small as the distances from the generator increase.

Figure 15-16 illustrates the basic method of producing a laser beam with a tube filled with helium and neon gas. A full mirror is located at one end of the tube, point A. A mirror with a small optical opening is located at the other end of the tube, point B. When an electric current is sent through the tube, the atoms of the gas are excited, which causes the gas to produce light similar to the glow of a neon light. This light in turn produces stimulated-emission light

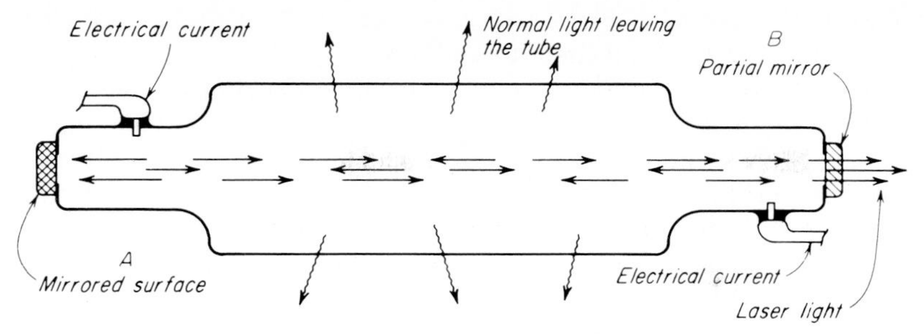

Figure 15-16 Section through a laser-beam generating tube. *(AGL Corporation.)*

from other atoms in the gas. As this second light is reflected back and forth between the mirrors, its intensity increases. When the correct level of intensity is reached, light passes out of the tube through the optical opening in mirror B as a continuous and narrow beam. It is this light that is used as a guiding beam.

Installing a laser beam system in a tunnel Figure 15-17 illustrates the basic method of installing and operating a laser beam system in a tunnel to guide a tunnel-boring machine. The same method may be used to guide a jumbo or other equipment used in driving a tunnel. Literature furnished by the manufacturers of laser beam assemblies will explain the installation procedures more fully and will assist a purchaser of such an instrument in learning how to install and operate it [13, 14].

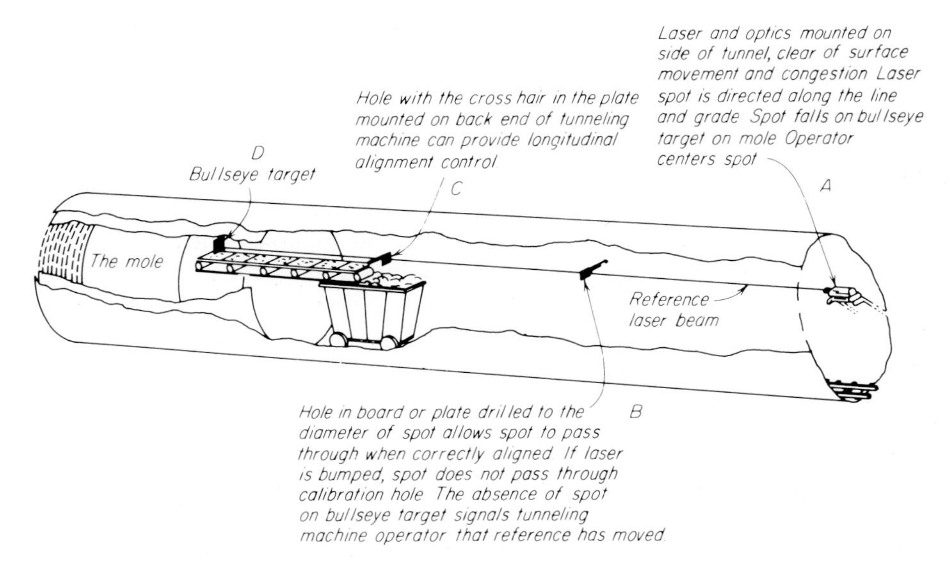

Figure 15-17 The use of a laser beam to control a mole. *(Journal of the Construction Division, American Society of Civil Engineers.)*

The installation illustrated in Fig. 15-17 is located inside a tunnel, with all items securely and accurately attached to the wall or ceiling of the tunnel lining. A surveyor's transit is used to assist in locating the items. The laser generator is located at position A. An adjustable target with a small opening is located at position B. A second target with a small opening is attached to the rear of the mole at point C, while a third target, which is illuminated by the laser beam, is attached to the forward end of the mole at position D, where it is in clear view of the operator. The position of the light beam on target D enables the operator to guide the mole.

Accuracy of a laser beam Laser beams currently used on construction projects are accurate to within $\frac{1}{8}$ in. at 100 ft and $\frac{1}{4}$ in. at 400 ft.

As the distance from the source of light to the target is increased, the vertical error in determining elevations increases in proportion to the square of the distance. This error is caused by the curvature of the earth.

ADVANTAGES OF USING MOLES

When driving tunnels through materials varying from unconsolidated earth to firm rock, moles have demonstrated that they have advantages including, but not limited to, the following:

1. Moles can drive at faster rates of advance.
2. Moles produce round, smooth, and unshattered bores.
3. Moles reduce overbreak to an average of about 5 percent, compared with 20 percent or more with drill and blast methods.
4. Tunnels driven with moles require less concrete for lining.
5. Because the ground around and adjacent to a tunnel is not disturbed and/or fractured by moles as by blasting, the formation is not weakened, which permits the use of less expensive ground support. Thus, rock bolts are more effective, and thinner concrete linings are adequate.
6. Moles reduce the danger of injury from falling rocks and from toxic gases.
7. Moles permit more continuous operations, because excavation and muck removal can be performed at the same time. Because of the reduction in nonproductive time, both labor and equipment will operate more efficiently.
8. Because the smaller sizes of the cuttings permit them to be removed from a tunnel more easily by conveyor belts or as pumped slurries, congestion in the tunnel is reduced. This leaves more space for other operations, such as installation of ground supports, linings, and ventilation ducts.
9. Because moles eliminate the need for blasting to loosen rock, there is little or no weakening of or damage to the rock adjacent to the tunnel. Also,

the danger of damaging property along the route of a tunnel by blasting is eliminated.

10. Analyses of the costs of driving tunnels with moles or by the drill, blast, and muck methods have demonstrated that reductions in costs amounting to 40 percent or more are possible.
11. Moles can be equipped with dust controls and with nozzles which permit water to be sprayed on the face of a tunnel and onto the muck as it is transported from the tunnel. The elimination of dust, in addition to the elimination of toxic gases produced by blasting, permits adequate ventilation within a tunnel at less cost.

DISADVANTAGES OF USING MOLES [4, 15]

There are several disadvantages in using moles to drive tunnels that may have limited their use. Such disadvantages include, but are not limited to, the following:

1. The high initial cost, varying with each project, but reportedly in the range of 4,000 to 6,000 times the square of the diameter in feet of a tunnel, expressed in dollars.
2. The very high cost of cutters and teeth when driving through hard rock. In driving portions of the tunnels for the Bay Area Rapid Transit system in California, the average cutter cost was $30.00 per linear foot of tunnel [5]. However, the compressive strength of the rock varied from 1,000 psi to 40,000 psi.

   In driving the 13-ft-5-in.-diameter Azotea Tunnel in New Mexico through sandstone and shale during 1964–1966, the contractor reported an initial cutter cost of $12.50 per lin ft. However, as a result of improvements that were made in the type, use, and makeup of the cutting head and cutters and in the pressures used to adapt them more nearly to the specific problems of the project, the cost was reduced to $0.66 per lin ft.
3. Because it is not possible or economically feasible to use moles in driving through hard rock, they are not now suitable for this condition. The development of cutters with longer lives may reduce this disadvantage.
4. Because most of the tunnels that have been driven by moles at this time have been completed under favorable geological conditions, there may not be enough experience to permit predetermination of the drillability of a given tunnel with a mole.
5. Because of the high initial cost of a mole, plus the relatively high cost of assembling and disassembling a unit at a project, it may not be economical to use a mole for driving a short tunnel.
6. Because moles are limited to driving circular tunnels, they cannot be used to drive tunnels having other cross sections, for example, horseshoe tunnels.

VENTILATING TUNNELS

It is necessary to ventilate a tunnel for various reasons, including the following:

1. To furnish fresh air for the workers
2. To remove obnoxious gases and the fumes produced by explosives
3. To remove the dust caused by drilling, blasting, mucking, and other operations

If a drift is driven through a tunnel from portal to portal, it may provide sufficient natural ventilation for the enlarging operations. When natural ventilation is not adequate, as is the case for most tunnels, a positive method of ventilation must be provided.

Mechanical ventilation usually is supplied by one or more electric-motor-driven fans, which may blow fresh air into a tunnel or exhaust the dust and foul air from the tunnel. If air is blown into a tunnel, it may be forced through a lightweight pipe or a fabric duct. If the air is exhausted, it is necessary to use a duct sufficiently rigid to prevent it from collapsing under partial vacuum. Many installations are designed to permit the ventilating system to operate by blowing or exhausting (see Fig. 15-18). The reversal of flow can be accomplished by a valve-and-duct arrangement, as illustrated in Fig. 15-19.

If fresh air is blown into a tunnel, it is released near the working face, and

Figure 15-18 Blower rated at 12,000 cfm at 2 psi, equipped with reversing valve. *(Ingersoll-Rand Company.)*

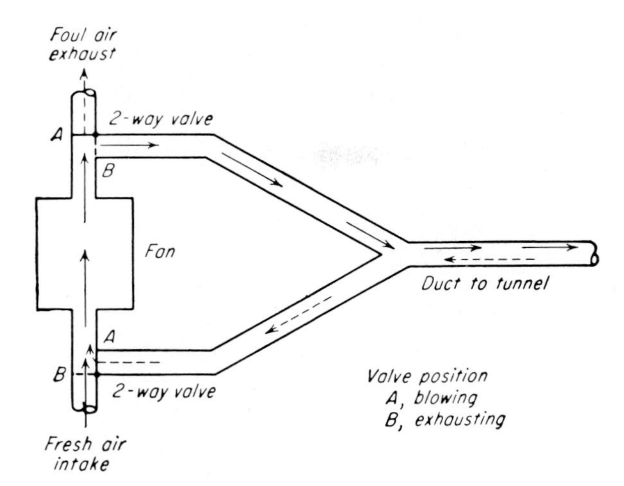

Figure 15-19 Valve and duct arrangement to reverse the direction of flow of air.

as it flows to the portal through the tunnel, it carries the dust and gases with it. If the exhaust method is used, the foul air and dust are drawn into the duct opening near the working face, thereby causing fresh air to flow into the tunnel from the portal. The latter method has the advantage of more quickly removing objectionable air from the spaces occupied by the workers.

Volume of air required for ventilation The volume of air required to ventilate a tunnel will vary with the number of workers, the frequency of blasting, the method of controlling dust, and the extent to which air-consuming equipment, if any, is used in the tunnel.

Each worker should be supplied with 200 to 500 cfm of fresh air. Compressed air, furnished to the drills, should not be included in computing the volume of air required for a project, as this air is contaminated with oil and dust before it is released by the drilling operations.

The firing of explosives to loosen rock fills the space near the face of the tunnel with gases and dust, which makes the air unfit for breathing. This foul air must be removed and replaced with fresh air before the workers can start mucking out the broken rock. The cycle of operations may be organized so that the workers retire a safe distance from the face prior to firing the explosives and eat lunch during the time required to remove the gases. If a 30-min lunch period is scheduled, the capacity of the ventilating equipment should be sufficient to clear the tunnel in that period of time.

In driving the 22- by 31-ft railroad tunnel near Aspen, Wyoming, the contractor supplied 18,000 cfm of air through a 26-in. vent pipe. The first 1,000 ft of each of the 36- by $26\frac{1}{6}$-ft twin bores of the Squirrel Hill Tunnel on the Penn-Lincoln Parkway was ventilated with a vane-axial fan capable of blowing 43,000 cfm of air through a 36-in. vent pipe, as illustrated in Fig. 15-20. After the tunnels were driven about 1,000 ft, the ventilation system was revised to

Figure 15-20 Twin radial fan used to ventilate the Squirrel Hill Tunnel. *(Joy Manufacturing Company.)*

permit air to flow from one tunnel into the other through passageways excavated between the two tunnels at 500-ft spacings. An air lock at the portal of one tunnel made it possible to control the direction of flow of air into or through either tunnel. The capacity of the ventilation system was increased to 240,000 cfm for both tunnels by installing additional fans.

Size and capacity of vent pipe After the quantity of air required to ventilate a tunnel is determined, the next step is to determine the size pipe and blower or blowers that will give the lowest total cost [16]. The total cost will include the installed costs of the blowers and pipe, with an allowance for salvage value upon completion of the project, plus the cost of electric energy required to operate the blowers. If a fixed amount of air is to be supplied at the face of a tunnel, the use of a small pipe will require the installation of a larger blower, which will result in a high installation and operating cost. If a large pipe is used, the cost of the blower and the operating cost will be less but the cost of the pipe will be higher. For every project there is a combination of sizes which will give the lowest total cost. This is the most economical installation. A typical fan used to ventilate a tunnel is shown in Fig. 15-21.

Consider a tunnel, with a 250- to 300-sq-ft bore, whose maximum length will be 16,000 ft. It is determined that 3,000 cfm of free air will be required for the tunnel. The effect which the size of pipe used has on the pressure lost in the pipe is shown in Fig. 15-22. As the tunnel progresses, the length of the pipe must be increased and the capacity of the blowers also must be increased to

Figure 15-21 Fan used for ventilating a tunnel. *(Bonanza Fans, Inc.)*

overcome the greater back pressure resulting from longer pipe. The cost of energy required to operate the blowers will increase as the length of the tunnel is increased. However, a reasonably accurate estimate of the total cost of energy can be obtained by computing the cost for each of several convenient equal lengths, such as at the quarter points.

Table 15-2 illustrates a method of determining the most economical size pipe for the tunnel under consideration. For this project the 22-in. pipe was the most economical, considering all costs.

Table 15-2 The combined cost of compressors, pipe, and energy for ventilating a tunnel*

Size pipe, in.	Cost			
	Compressor	Pipe	Energy	Total
18	$50,560	$67,480	$30,640	$148,680
20	32,920	76,900	19,580	129,400
22	19,340	85,160	11,040	115,540
24	12,260	99,500	6,960	118,720
26	10,580	116,360	4,720	131,660

*Based on 1978 prices.

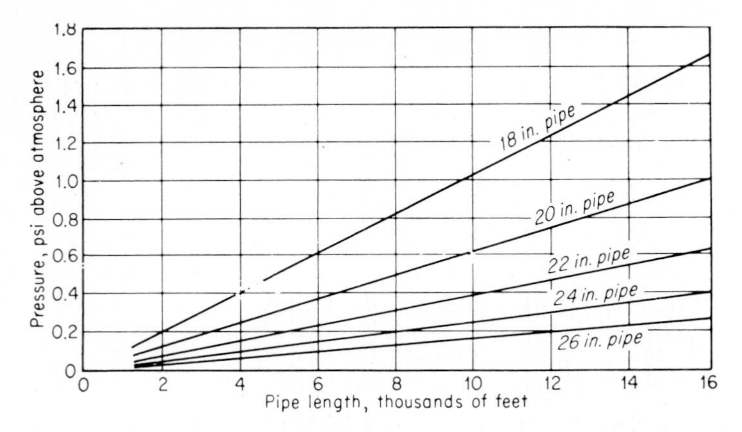

Figure 15-22 The effect of the size of the vent pipe on the loss in pressure.

Dust control The operations such as drilling, blasting, loading, and hauling muck cause dust to accumulate in the air in a tunnel. Unless precautions are taken to limit the concentration, the dust will constitute a serious health hazard to the workers. This is especially true when a tunnel is driven through rock containing a high percent of silica, as extended exposure to silica dust may cause silicosis, a lung disease for which there is no positive cure. Most states have laws governing mining and tunneling practices which are designed to protect workers against this disease by limiting the concentration of silica dust particles in the air.

Various methods are used to limit the amount of dust in the air in a tunnel, including the following:

1. The use of water instead of air to remove the cuttings from drilled holes
2. The use of a vacuum hood that fits around the drill steel at the rock face to remove the dust that comes from a hole during the drilling operation
3. Complete ventilation of the space near the face, preferably by the exhaust method, following each round of blasting
4. The use of water to keep the muck pile wet during loading operations
5. The use of a detergent dissolved in water, which is injected into the blasthole with the air when the hole is blown

A system of dust control that was used successfully on the Delaware Aqueduct was to install on the drill jumbo several suction pipes with openings near the face of the heading. These pipes drew the dust-laden air from the face and passed it through filters, located at the rear of the jumbo, which removed most of the dust.

MUCKING

The operation of loading broken rock or earth for removal from a tunnel is referred to as mucking. This operation may be performed by hand, power shovels, mucking machines, slushers, or tractor loaders.

Hand mucking is limited to small tunnels and drifts which are not large enough to justify or permit the use of mechanical muckers.

Special power shovels, with short booms and dipper sticks, have been used for mucking in large tunnels. If ventilation is not a serious problem, a diesel-engine-powered unit may be used. If the exhaust fumes are objectionable, a unit powered with an electric motor should be used.

Several types of mechanical muckers are available for use in tunnels. Figure 15-23 illustrates a model in common use. It is powered by electric motors. The machine, which operates on rails, moves forward to push the dipper into the pile of muck. When the dipper is filled, the machine backs up a short distance and tips the dipper up to discharge the muck onto a belt which conveys it back to a muck car attached temporarily to the mucking machine. As noted in the figure, the dipper can be moved from side to side to give it a wide cleanup range.

The air-powered mucking machine illustrated in Fig. 15-24, which operates on rails, is designed to discharge the muck onto a conveyor belt attached to the

Figure 15-23 A track-mounted mucking machine. *(Goodman Equipment Corporation.)*

Figure 15-24 A rail-mounted air-powered Rocker-Shovel loader. *(Eimco Mining Machinery International.)*

rear of the machine, or it can discharge directly into a muck car, attached to the rear of the machine. It is available in several dipper sizes.

The diesel-engine-powered, wheel-mounted mucking machine illustrated in Fig. 15-25 is designed to load trucks or muck cars, or to haul the muck from the tunnel. It is available with a 5-cu-yd heaped-capacity bucket.

Several types of tractor-mounted loaders are available for use in tunnels. The bucket of the loader, illustrated in Fig. 15-26, is lowered to a position in front of the tractor, filled by the forward movement of the tractor, and then lifted over the tractor to discharge the load into a truck.

Tractors or other loading equipment powered with internal-combustion

Figure 15-25 A wheel-mounted diesel-powered load-haul-dump machine. *(Eimco Mining Machinery International.)*

engines should not be used in a tunnel unless the ventilation system, natural or mechanical, is adequate to remove the exhaust fumes without injury to the workers. As carbon monoxide gas is the most dangerous fume, special care must be exercised if gasoline-engine-powered equipment is used. Because diesel engines do not produce carbon monoxide, they are safer than gasoline engines for use in tunnels.

Removing muck Muck may be removed from a tunnel by narrow-gauge muck cars pulled by locomotives, by diesel-engine-powered trucks, by conveyor belts, or as slurry pumped through pipelines.

The use of pumps and pipelines to remove muck as a slurry has been satisfactory and economical on a number of projects where tunnels were driven through earth that could be converted into a slurry. A pipeline requires less space and should not interfere with the movement of materials and equipment through a tunnel.

Tracks When muck is hauled in cars, steel rails are required. For this use relatively lightweight rails are laid to a narrow gauge, most frequently 24 or 36 in. For a long tunnel it is necessary to provide a double track in order that loaded cars may be moved out while empty cars are moved into the tunnel. As the width of a muck car is usually twice the gauge of the rails, the maximum

Figure 15-26 Overshot loader loading muck into a truck. *(Western Construction.)*

gauge is limited to slightly less than one-fourth the width of the tunnel. The weight of the rail, expressed in pounds per yard, should be heavy enough to prevent objectionable sag between the supporting ties when the locomotive and loaded cars travel over them. Also, if the same rails will be used for all haulage, including timbering, reinforcing, and concrete for the lining, considerable expense in laying the rails on a good foundation will be justified. Low rolling resistance, plus freedom from excessive maintenance cost and reduced output due to car derailments, depend largely on the use of a good track.

Muck cars Various types and sizes of cars are used to haul muck from tunnels. The capacity may be expressed in cubic feet or cubic yards. In general, the largest size that can be used in a tunnel will be the most economical, as large cars reduce the time lost in switching at the loading operation.

The cars commonly used are constructed with sides hinged at the top and fastened with latches at the bottom to permit easy dumping, as illustrated in Fig. 15-27.

Figure 15-28 illustrates a 270-hp diesel-engine-powered truck whose heaped capacity is 16.5 cu yd. This truck is designed to haul muck from tunnels.

Figure 15-27 Automatic side-dump muck cars.

Figure 15-28 Wheel-mounted diesel-powered hauling truck. *(Eimco Mining Machinery International.)*

The load is discharged at the rear end by tilting the front end up with hydraulic jacks.

LOCOMOTIVES

Three types of electric locomotives are available for tunnel hauling—the trolley, battery, and combination trolley and battery. All three are available in various weights and for operation on different track gauges, as indicated by the manufacturers.

Figure 15-29 illustrates an 8-ton electric-powered locomotive. Electric locomotives are available in sizes ranging from about 1 ton to more than 50 tons.

The trolley-type locomotive is relatively easy to operate, but it requires a bare trolley wire, which may interfere with other operations in a tunnel and which represents a source of potential danger to workers. Also, it is necessary to ground the rails which serve as a return circuit for the electricity.

The battery-type locomotive is operated from a group of storage batteries mounted directly on the locomotive. These batteries should operate a locomotive for 8 hr, after which they must be recharged, which requires about 8 hr. If a locomotive is to operate more than one 8-hr shift per day, it is necessary to provide at least two sets of batteries so that one set may be charged while the other set is in use.

The combination trolley-and-battery-type locomotive is satisfactory for use on a project which requires haulage inside and outside of a tunnel. The batteries are used in the tunnel and the trolley outside the tunnel. If the operation from the trolley is long enough, the batteries may completely recharge each round trip.

Figure 15-29 Battery-powered 8-ton locomotive. *(Goodman Equipment Corporation.)*

The size of a locomotive is indicated by its weight, expressed in tons. If it has sufficient power to slip the driving wheels when standing on dry steel rails, the maximum tractive effort will be equal to the product of the weight times the coefficient of friction between the wheels and the rails. The coefficient of friction will usually be 0.2 to 0.25. Thus, an 8-ton locomotive should provide a tractive effort of at least

$$16{,}000 \times 0.2 = 3{,}200 \text{ lb}$$

A method of determining the maximum number of cars that can be hauled is illustrated in the following examples.

Example 15-3 Determine the maximum number of cars that can be hauled on a level track by an 8-ton locomotive. The rolling resistance will be 30 lb per ton and the starting resistance 20 lb per ton of gross load, including the weight of the locomotive. The cars, which have a capacity of 80 cu ft each, weigh 2,800 lb empty and 10,800 lb loaded.

Available tractive effort, $16{,}000 \times 0.2 = 3{,}200$ lb
Total resistance, 30 lb + 20 lb = 50 lb per ton
Maximum gross load, 3,200 lb ÷ 50 lb per ton = 64 tons
Deduct weight of locomotive = 8 tons
Maximum net load = 56 tons
No. of cars, $56 \times 2{,}000 \div 10{,}800 = 10.4$, or 10
Volume of muck per trip, $10 \times 80 \div 27 = 29.6$ cu yd

Example 15-4 Determine the maximum number of loaded cars that can be hauled up a 1 percent grade with all other conditions the same as for the preceding example.

Grade resistance, 20 lb per ton
Total resistance, 70 lb per ton
Maximum gross load, 3,200 lb ÷ 70 lb per ton = 45.7 tons
Deduct weight of locomotive = 8 tons
Maximum net load = 37.7 tons
No. of cars, $37.7 \times 2{,}000 \div 10{,}800 = 7$
Volume of muck per trip, $7 \times 80 \div 27 = 20.7$ cu yd

GROUND SUPPORT

When a tunnel is driven, it may be necessary to support the ground adjacent to the tunnel until a permanent concrete lining can be installed. The temporary supports must be strong enough to resist the pressures transmitted to them by the ground. These pressures are caused by faulted, folded, or fractured rock masses or the swelling of the surrounding earth following the removal of the material from a tunnel.

The operation of placing supports in a tunnel to resist the movement of ground is referred to as timbering. The type and extent of timbering are determined to a large degree by the kind and physical condition of ground to be supported. In the early years of tunnel driving, heavy wooden timbers were commonly used for supports, but in recent years wood has been replaced by steel H beams. These beams can be fabricated from any desired section to produce ribs that fit the shape of any given tunnel bore. A rib may consist of two or more segments, which are carried into the tunnel, assembled, and bolted together as the driving progresses. Figure 15-30 illustrates the use of steel-H-beam timbering.

The advantages of steel sections instead of wood include the following:

1. Because smaller sizes are used, it is possible to reduce the size of the bore of a tunnel. This reduces the cost of excavation and permits faster driving.
2. They can be installed more quickly and economically than wood.
3. They supplement the steel reinforcing in the concrete lining.
4. Their small dimensions may permit the use of thinner concrete linings.

The safe spacing of ribs may vary from 18 in. to as much as 6 to 8 ft, depending on the physical condition of the ground. Most grounds tend to bridge over at some height above the roof of a tunnel. As the rock above the natural bridge will be self-supporting, only that rock between the roof of a

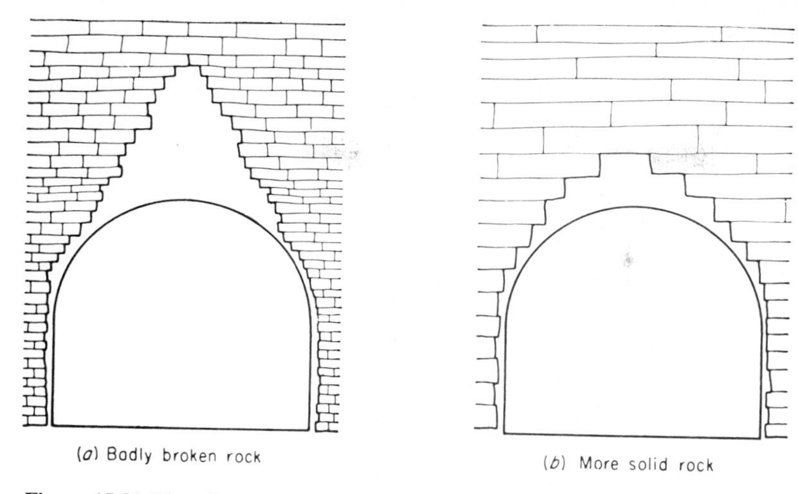

Figure 15-31 The effect of the physical condition of rock on bridging action.

tunnel and the bridge surface must be supported by the timbering. Figure 15-31 illustrates how the physical condition of the rock affects the height of the natural bridge formed over the roof of a tunnel. As shown in the figure, rock having greater solidity or larger pieces will bridge more quickly than rock which is badly broken. The timbering must support the weight of the rock lying

Figure 15-32 Hoisting sections of steel ribs into position for tunnel supports.

below the bridge surfaces. If the physical condition of the rock is known in advance of driving a tunnel, the spacing of the ribs may be determined with reasonable accuracy. However, because of the changes in the condition of rock encountered as a tunnel is driven, it is found in actual practice that the spacing of ribs should be modified to meet the conditions that exist at any given section in a tunnel.

Prior to awarding a contract for driving eight power and outlet tunnels at Garrison Dam, the United States Army Corps of Engineers drove a full-size test section 240 ft long in order to determine what forces the timbering and lining must withstand. These circular tunnels, whose excavated diameters varied from 27 to $36\frac{1}{2}$ ft, were driven through a formation of clay and shale. Figure 15-32 shows a section of a steel rib being hoisted to position for installation by a top-mounted jumbo, which rode on rails at the spring line, as shown in Fig. 15-33.

Before award of the contract for driving the twin-bore Straight Creek Tunnel, which carries Interstate Highway 70 through the Rocky Mountains in Colorado, an 8,400-ft pilot tunnel was driven through the mountain at the site of the future tunnel. Pressure cells installed in the roof and sides of this tunnel provided valuable information, which was used in designing the ground supports for both tunnels.

The pilot tunnel, which connected the two portals, was also used to haul men and materials from one portal to the other during the driving of the main tunnels.

In order that the sets of ribs may resist the forces from the ground and prevent broken pieces of rock from falling into the bore of a tunnel, it is necessary to install some type of cover, called lagging, outside of the ribs. This lagging may consist of heavy pieces of lumber, extending from rib to rib, or it may consist of a series of steel plates. The space between the lagging and the undisturbed ground or rock should be filled with timbers, large pieces of rock, or gravel packing to prevent the ground from shifting toward the timbering. The use of lagging and rock packing is illustrated in Fig. 15-33.

USING ROCK MECHANICS STUDIES TO DESIGN TUNNEL SUPPORTS AND LININGS

While driving twin tunnels in 1964 and 1965, the Wyoming Highway Department used a rock mechanics instrumentation program to determine the actual tunnel support requirements [17]. The studies were conducted by Terrametrics, Inc., Golden, Colorado [18].

Before an opening is excavated, the forces on a normal rock mass are in equilibrium. The stress field in the rock depends on the thickness of the overburden, the geological continuity and physical character of the rock, and potential extraneous factors such as tectonic stress. As the opening is increased in size, equilibrium is disturbed, and the rock adjacent to the opening under-

Figure 15-33 Forcing rib sections into place with jacks mounted on jumbos.

goes a dynamic strain readjustment. The support originally provided by the confined rock must be at least partially replaced by the timber, steel, or concrete.

If the opening is to be maintained, a major part of the induced load must ultimately be carried by the rock in the walls of the opening. The redistribution of the load to adjacent rock requires time. Peak support loads, which usually occur during the strain redistribution cycle, must be resisted by the initial steel support system. Usually these loads must be supported only for short periods of time, ranging from a few hours to a few days. Under heavy loads a support system that is too rigid may tend to prevent the redistribution process, resulting in excessive loads and possible failure of the supports. After the peak loads have diminished, the system must be strong enough to support permanently the stable rock loads.

The study conducted in the tunnels in Wyoming involved the measurements of displacement-load profiles at four representative locations, as indicated in Fig. 15-34. Each station tested consisted of four load-measuring devices, prop load cells, and three displacement-measuring devices, multi-position borehole extensometers.

Table 15-3 gives the projected rock loads that act on or might act on the tunnel supports. It is recommended that the final support system should be designed to resist the limiting rock load, which is the load that might occur if the total weight of the rock in the tension arch were transferred to the support system.

It is reported that the information gained from the studies was used to reduce the cost of the initial support system by approximately 25 percent.

Considerable information related to the use of instruments in testing soils and rocks is available. The information obtained from such tests may be used in designing the ground supports for tunnels. Such information permits the

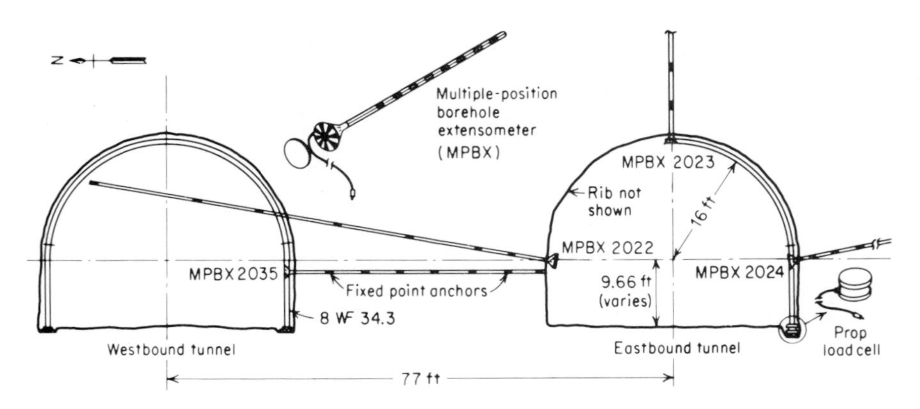

Figure 15-34 Cross section of a tunnel showing the locations of instruments for measuring rock strain and steel set loads. *(Civil Engineering.)*

Table 15-3 Projected rock loads on tunnel supports

Station		Rock loads, psf (Pa)		
From south tunnel	To south tunnel	Peak	Stable	Limiting
108 + 00	107 + 50	1,050 (50,200)	300 (14,360)	3,000 (143,600)
RMI − 1*	106 + 50	700 (33,500)	200 (9,575)	2,000 (95,750)
107 + 50	103 + 85	700 (33,500)	200 (9,575)	2,000 (95,750)
RMI + 2*	103 + 50	250 (11,950)	200 (9,575)	2,100 (100,500)
103 + 85	100 + 60	250 (11,950)	200 (9,575)	2,100 (100,500)
100 + 60	97 + 92	385 (18,400)	200 (9,575)	1,250 (59,800)
RMI − 4*	98 + 25	385 (18,400)	200 (9,575)	1,250 (59,800)
97 + 82	97 + 32	500 (23,950)	260 (12,450)	1,625 (77,750)
From north tunnel	To north tunnel			
108 + 06	107 + 56	1,050 (50,200)	300 (14,360)	3,000 (143,600)
107 + 56	105 + 50	700 (33,500)	200 (9,575)	2,000 (95,750)
RMI − 3*	105 + 15	690 (33,000)	260 (12,450)	1,700 (81,300)
105 + 50	103 + 80	690 (33,000)	260 (12,450)	1,700 (81,300)
103 + 80	100 + 75	250 (11,950)	200 (9,575)	1,250 (59,800)
100 + 75	97 + 82	385 (18,400)	200 (9,575)	1,250 (59,800)
97 + 82	97 + 32	500 (23,950)	260 (12,450)	1,625 (77,750)

* These results are based on actual tests.

design of adequate but not excessive supports. The references at the end of this chapter list publications containing relevant information.

ROCK BOLTING

Steel bolts are frequently set in holes drilled into the rock to assist in supporting the entire roof or individual rock slabs that tend to fall into a tunnel. If the characteristics of the rock are such that the bolts will suffice in supporting the roof or parts thereof, the use of bolts is both safe and economical.

The effective use of bolts requires some understanding of the natural forces that exist underground. In an underground excavation all downward-acting forces are transmitted to the walls of the excavation. Most of the rock above the excavation is supported by natural arch action that bears on the walls. The remaining rock below the arch is suspended by the arch. If this suspended rock lacks sufficient strength, it sags and tension cracks develop. As the cracks work up into the roof, weakening the suspended strata, rock begins to fall—all at once or over an extended period of time. If the rock is strong enough and free of large slips and cracks, the rock that is subject to falling usually should not exceed one-third of

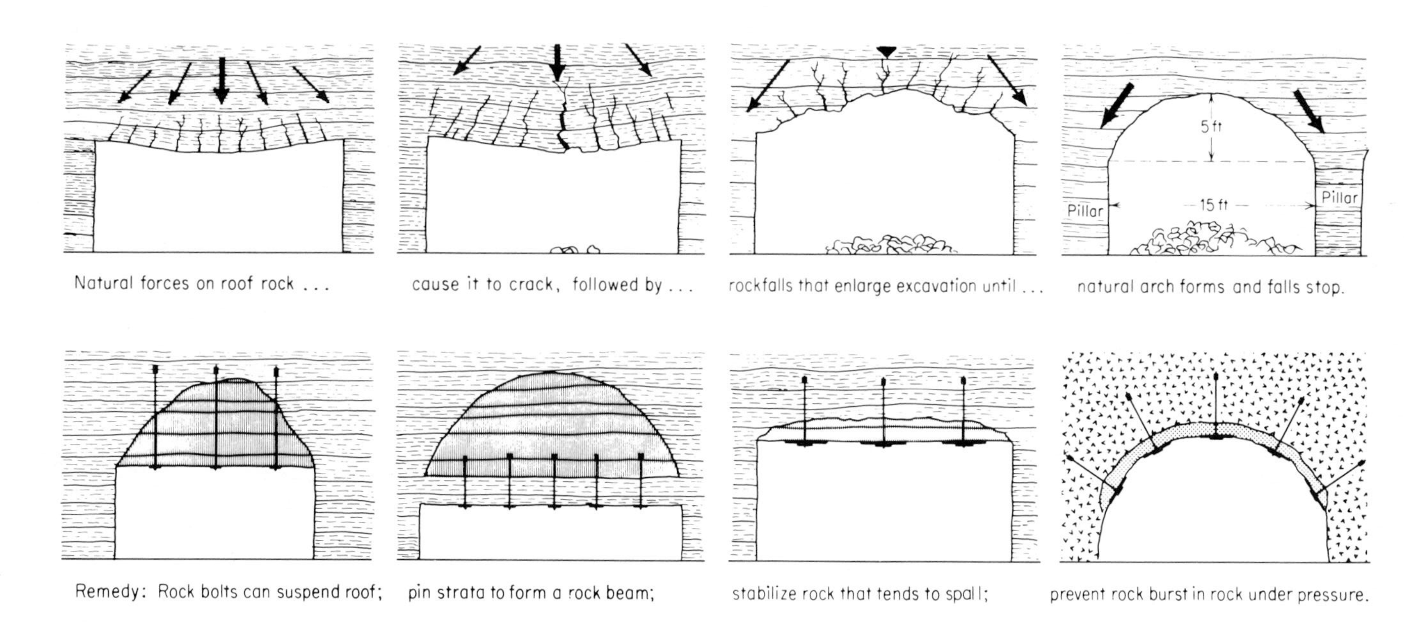

Figure 15-35 Methods of supporting tunnel roofs with rock bolts. *(Engineering News-Record.)*

the width of the roof. It is this rock that bolts can support. Figure 15-35 illustrates the natural forces that cause rock falls, as well as the remedy, which involves the use of rock bolts [19].

If bolts are to be effective, they must be long enough to be securely seated into the rock that is not subject to falling. Also, the rock into which they are seated must be strong enough to provide the necessary anchorage.

Table 15-4 gives the yield and breaking loads for frequently used rock bolts. Table 15-5 gives representative anchorage strengths of different types of rock.

Figure 15-36 illustrates one method used to anchor bolts into rock. As the bolt is tightened by applying a torque to the exposed head or nut, the tapered plug is drawn into the expansion shell, forcing it to bear against the rock adjacent to the hole. If the rock is sufficiently strong, the tensile strength of the bolt may be developed.

Another type of anchor bolt consists of a steel rod, with one end slotted and the other end threaded. Before the slotted end of the rod is inserted into the hole, a steel wedge is driven into the slot a short distance. Then the rod is inserted into the hole until the wedge presses against the bottom of the hole. A driving force is applied to the projecting end of the rod to force the wedge deeper into the slot, thus expanding the rod against the rock adjacent to the hole. A roof plate, washer, and nut are attached to the projecting end of the rod, and the required torque is applied to the nut. The torque is usually limited

Table 15-4 Safe loads on rock bolts [4]

	Load, lb (kg)	
Bolt type and diameter	Yield	Breaking
1-in.(25.4-mm) slotted rod	23,000 (10,410)	40,000 (18,150)
$\frac{3}{4}$-in.(19.1-mm) steel	15,000 (6,800)	25,000 (11,350)
$\frac{3}{4}$-in.(19.1-mm) high-strength steel	22,000 (9,950)	38,000 (17,230)
$\frac{5}{8}$-in.(15.9-mm) high-strength steel	15,000 (6,800)	24,000 (10,890)

Table 15-5 Representative anchor strength of rocks [4]

Type of rock	Approximate anchor strength, tons (kg)
Granite and basalt	20 to 30 (18,100 to 27,150)
Sandstone and quartzite	15 to 22 (13,600 to 19,950)
Limestone and marble	12 to 18 (10,850 to 16,300)
Firm sandy shale and gypsum	10 to 15 (9,072 to 13,600)
Shale	8 to 12 (7,250 to 10,850)
Wet or weak shale and coal	3 to 6 (2,720 to 5,440)
Unconsolidated weak, wet material	0 to 3 (0 to 2,720)

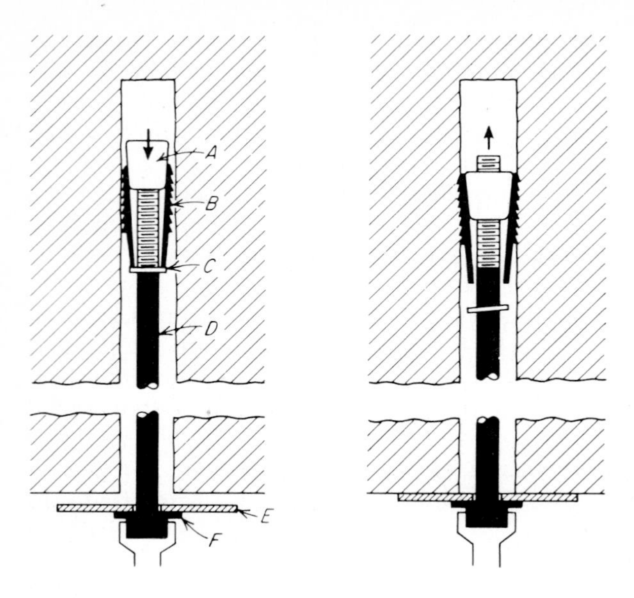

Figure 15-36 Methods of anchoring expansion bolts in rock. *(Engineering News-Record.)*

to 175 ft-lb, which should be adequate to develop the tensile strength of the bolt.

When preparing the foundation for a building at the United States Military Academy at West Point, New York, the contractor used Williams [20] hollow-core rock bolts inserted into drilled holes to strengthen the rock and hold it in position permanently. After the bolts were installed in the holes, a slurry of cement and drill cuttings was pumped through the bolt holes to fill all of the space around the bolts.

A bolt support system whose trade name is Fastloc has been developed by the Du Pont Company [21]. This system is composed of a specially designed steel bolt and a resin cartridge. When the bolt is installed in a hole, the resin covers the entire length of the bolt and anchors it to the surrounding stratum.

Celite, Inc. has developed a system of rock bolts which uses resin to anchor the entire lengths of bolts in the surrounding strata [22].

CONTROLLING GROUND WATER

In driving a tunnel, the control of water will consist of one or two operations, namely, preventing excess quantities of water from entering the tunnel and removing the water that does enter.

Most of the water in a tunnel comes from two sources, that used to wash the cuttings from the drill holes and that which flows in from the ground through which the tunnel is driven. The former may be estimated with reasonable accuracy, but the latter is subject to great variation. For example, shooting a charge of explosives may open fissures into a ground-water reservoir, thus permitting an unexpectedly large quantity of water to flow into a tunnel. It is good practice to drill exploratory holes ahead of and deeper than those drilled for explosives in order to determine whether there is badly broken rock or ground water ahead. If the exploratory holes indicate that such a condition exists, it is possible to grout off heavy flows of water and solidify the formation before the tunnel reaches the trouble zone.

Many types and sizes of pumps are available for removing water from a tunnel, as described in Chap. 18. The two types most commonly used are the air-driven centrifugal pump and the electric-motor-driven centrifugal pump. Both are compact units, with high capacities, which will operate satisfactorily under varying heads.

The water that accumulates near the face of a tunnel is collected in a sump. This water is picked up with air-driven centrifugal pumps and pumped back to another sump located nearer the portal, where most of the solid materials may settle out, after which it is pumped out of the tunnel with semipermanently installed electric-motor-driven centrifugal pumps.

An air-driven sump pump is satisfactory for use under the adverse conditions that usually exist near the face of a tunnel, but an electric-motor-driven pump usually is preferred for use in the main pumping operation. A switch control operated by a float set in the sump will make a pumping unit practically automatic.

Because of the possibility of encountering excessive flows of water into a tunnel, it is good practice to provide standby pumps, which may be placed in operation very quickly.

See Chap. 18 for more complete information on various types of pumps.

THE CROSS SECTIONS OF TUNNELS

The shape of the cross section of a concrete-lined tunnel will depend on the pressure of the ground which the lining must resist and the purpose for which the tunnel is constructed. If the ground is solid rock, any desired shape may be selected. For an aqueduct the section may be circular, while for a vehicular tunnel the section may consist of a flat invert, vertical walls, and an arched roof. If the ground is broken rock, subject to horizontal pressure, the vertical wall section of a vehicular tunnel should be replaced with horseshoe curves to resist such pressure. If the ground is highly unstable, such as soft clay or sand, it may be necessary to use a circular section, because of its greater resistance to external pressures, regardless of the purpose for which the tunnel will be used.

The most common cross sections are illustrated in Fig. 15-37. They include the circular, elliptical, horseshoe, and vertical wall with arch-roof types. The circular and elliptical sections are popular for water and sewage conduits, while the horseshoe and vertical sections are popular for vehicular tunnels where the ground conditions permit such sections to be used.

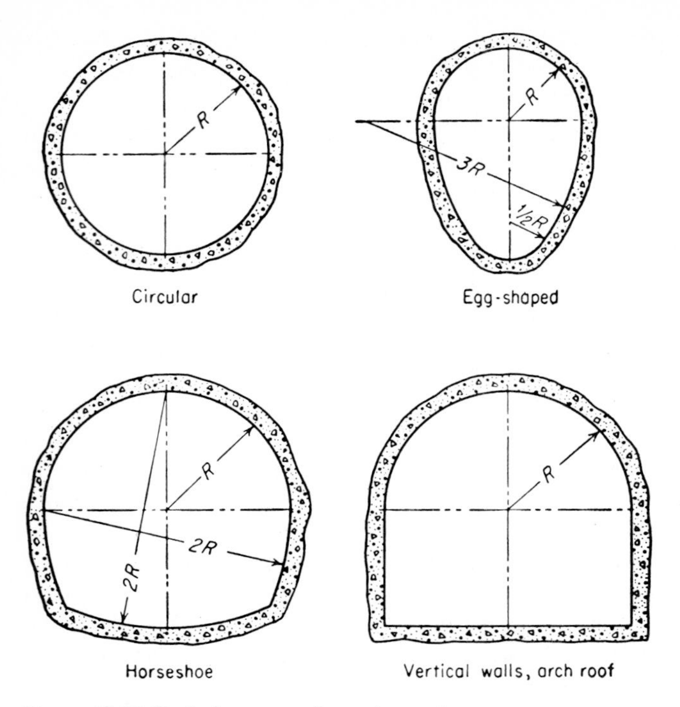

Figure 15-37 Typical cross sections of tunnels.

THICKNESS OF CONCRETE LININGS

In the interest of economy it is desirable for a concrete tunnel wall to be as thin as practical. The thickness may be determined by the condition of the ground surrounding the tunnel, the size and shape of the cross section, the requirements of construction conditions, or the internal pressure in the event it is a water conduit.

The geological survey made prior to designing a tunnel lining should indicate whether the ground is solid or broken rock or unconsolidated soil, subject to horizontal and vertical pressures. If steel ribs and reinforcing are used in a concrete wall, the thickness may be reduced, but the use of steel for the sole purpose of reducing the thickness of the lining is not economical. A rule which has been used to some extent as a guide only is to allow 1 in. of wall thickness for each foot of diameter.

As solid rock does not impose any load on a concrete lining, the thickness may be the minimum that can be placed behind the forms and cover any utility pipes, ducts, or appurtenances.

The designed thickness of a concrete lining which must resist the pressure from bad ground may include the concrete surrounding steel ribs and sections but should not include any concrete that is encroached on by wood timbers or any kind of lagging.

The 8-ft-diameter Carter Lake pressure tunnel of the Colorado-Big Thompson project, which was designed for a maximum dynamic head of 325 ft of water, has a minimum concrete thickness of $5\frac{1}{2}$ in. but is heavily reinforced with steel bars. The eight circular power and outlet tunnels at Garrison Dam, ranging from 27 to $36\frac{1}{2}$ ft in excavated diameter, have wall thicknesses varying from 30 to 42 in. The Gaviota Gorge Tunnel, a highway tunnel in California, whose inside dimensions are 35 ft 3 in. wide and 22 ft high, has three wall thicknesses, 18 in. in the cut and cover section, 24 in. in the section supported with steel ribs, and 36 in. in the section supported with timbers. All these are minimum thicknesses, which are subject to increases due to overbreak in the sizes of the bores during blasting operations.

SEQUENCE OF LINING A TUNNEL

If a tunnel is driven through solid rock, or if the supports will prevent objectionable movements of the rock until the entire bore is holed through, it is desirable to delay starting the lining until the excavation is finished. If this plan of construction is followed, the mucking and lining operations will not interfere with each other and a greater operating efficiency should be possible. However, if the ground is so unstable that it is difficult or impossible to restrain its movements, it will be necessary to install the lining as quickly as possible after blasting the mucking each round.

The sequence of installing the lining around the perimeter of a tunnel will depend on a number of factors. As illustrated in Fig. 15-38, any one of several sequences may be selected. The numbers 1, 2, and 3 indicate the sequence of placing the sections of the lining.

Plan (*a*) shows the entire wall placed in one operation. This method is limited to circular tunnels which are poured in relatively short sections. The top of the form must be rigidly blocked to prevent it from floating to the roof of the bore.

Plan (*b*) provides a rigid base on which the forms for the side walls and roof may be supported. If this plan is adopted, construction on the invert should be started at the most distant point and proceed toward the portal in order to eliminate the need of hauling materials over the concrete before it has set sufficiently.

Plan (*c*) is limited to large tunnels where it is desirable to separate the pours into the indicated sections.

Plan (*d*) offers several desirable advantages. The two curbs can be installed at the sides of a tunnel with little or no interference to the haulage tracks, located near the center of the tunnel. These tracks need not be removed or disturbed. After the curbs have cured sufficiently, they may be used to support wide-gauge rails on which the main form jumbos travel. Also, they may be

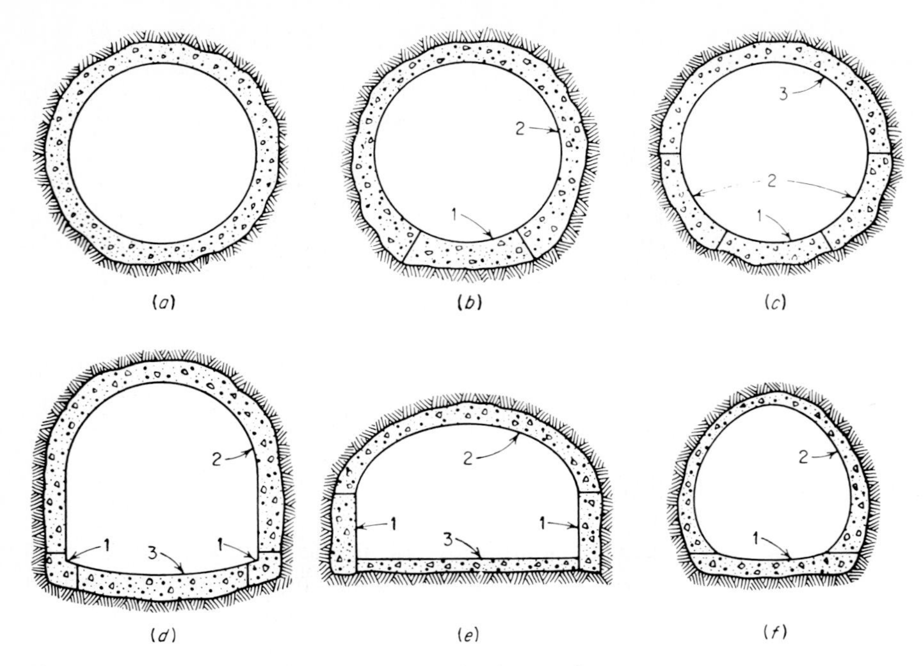

Figure 15-38 Sequence of placing concrete lining in tunnel.

used as guides, supports, and anchors for the bottoms of the wall forms. After the curbs, walls, and roof are completed, the original rails, which were used for hauling muck, supports, concrete, etc., can be removed just ahead of placing the invert. Thus, these rails need not be replaced, as would be necessary if the invert were placed first.

Plan (*e*) is used in placing the lining for large tunnels. Either the side walls or the invert may be placed first.

In plan (*f*) the invert is placed entirely across the tunnel floor; then the walls and roof are placed later in one operation. The chief objection to this method is that the haulage tracks must be removed prior to placing the invert, then relaid on top of the invert, if they are to be used to haul concrete or other materials for the walls and roof.

REINFORCING STEEL

If reinforcing steel is required in a concrete lining, it may consist of steel ribs, bars, or both. The design for a thick concrete lining may specify two layers of reinforcing bars, one near the inner and the other near the outer surface of the lining. The space between the two layers, especially near the top of the roof, must be great enough to permit the insertion of a pipe through which the concrete will be placed. Figure 15-39 shows the reinforcing steel in place for the

Figure 15-39 Placing reinforcing steel for the lining of a tunnel.

Queen Creek Tunnel in Arizona. A jumbo was constructed for use by the ironworkers in placing the reinforcing.

FORMS FOR CONCRETE LININGS

Forms used for lining tunnels are, with few exceptions, of the traveling type, constructed of steel or a combination of steel and wood. While the initial cost of steel forms will exceed the cost of wood forms, the additional uses obtained from steel compared with wood, together with the savings in time and labor required to move and set up in using steel forms, usually will make steel forms cheaper than wood for any tunnels other than short ones.

The traveling-type form is constructed of steel members which are lined with steel plate or wood to give a surface which conforms with the shape of the inside surface of the portion of the tunnel for which it will be used. Thus, a form may be used for constructing the invert, the side walls, the roof, or any desired combination thereof. Each form is mounted on a traveler or a jumbo, which in turn is mounted on wheels that permit it to be moved along rails. A traveler is equipped with adjustable jacks or screw ratchets, which permit the form to be expanded into position for a concrete pour, then collapsed slightly to pull it away from the concrete in order that it may be moved into a new location.

Figure 15-40 shows a steel-screed form, mounted on wheels that roll on wide-gauge rails, used to line the invert of a tunnel.

Figure 15-41 shows a steel form 24 ft 6 in. in diameter by 32 ft long used to

Figure 15-40 Steel screed used to line invert of a tunnel. *(Chicago Bridge & Iron Co.)*

Figure 15-41 Steel form used to line tunnels. *(Chicago Bridge & Iron Co.)*

line the walls and roof of a circular tunnel. The form, which is hinged near the upper quarter points, is supported by a prefabricated steel jumbo mounted on eight wheels. Through a series of jacks, attached between the form and the jumbo, it is possible to expand the form to full size or to retract the side walls and lower the roof to permit movement to a new location. Forms of this type may be fabricated with hinge doors along the side walls or roof to permit inspection behind the form or the placing of concrete through the doors.

In lining a highway tunnel 42 ft wide by 22 ft 10 in. high in Arizona, the contractor used two sets of steel forms, each 30 ft long. The first form was set in position, and the concrete lining for the side walls and roof was installed. The form was left in place for 72 hr after the pour, then moved ahead far enough to leave a 30-ft-long gap of unlined tunnel. The second form was used later in lining this gap. This procedure was used throughout the tunnel. Figure 15-42 shows the first set of forms in place, supported by a jumbo. Note the pipeline used in pumping concrete for the lining.

Figure 15-43 shows the three 50-ft-long sections of steel forms used to line the twin-tube Squirrel Hill Tunnel. Each form was constructed with a continuous hinge along each side just above the spring line. A set of forms could be collapsed sufficiently to permit it to pass through a form set in position for a concrete pour. After the first section was anchored in position and the concrete was poured, the second section was moved ahead and anchored and the concrete was poured. This operation was repeated for the third section. By the time the concrete was poured for the third section, the first section was ready

Figure 15-42 Two sets of steel forms used to place concrete lining in a tunnel.

Figure 15-43 Three sections of steel forms used to place concrete lining in a tunnel. *(Blaw-Knox Company.)*

to be collapsed and moved through the other two sections into a new position. This procedure was repeated until the lining was completed. One 50-ft section of lining was placed each day.

TUNNEL LINING BY THE PUMPING METHOD

A common method of placing concrete lining for a tunnel is by a concrete piston pump. The machine includes an agitator, or remixer hopper, a single- or double-cylinder piston pump, and a discharge pipe through which the concrete is pumped to the form.

Concrete that has been mixed by any convenient method is fed to the remixer hopper. This hopper, which serves as a storage reservoir ahead of the pump, is equipped with an agitator to ensure that a concrete of uniform quality will flow into the pump. The pump, which may consist of one or two horizontal cylinders, each with a single-acting piston, is located beneath the remixer hopper. When the piston is pulled back in a cylinder, concrete will flow by gravity from the remixing hopper through a valve-controlled opening into the cylinder. When the piston is pushed forward to eject the concrete from the cylinder, the inlet valve is closed and an outlet valve is opened by a system of mechanically operated levers and the concrete is forced the entire length of the discharge pipe and into the forms. Pumps are available with capacities varying from about 10 to over 100 cu yd of concrete per hour.

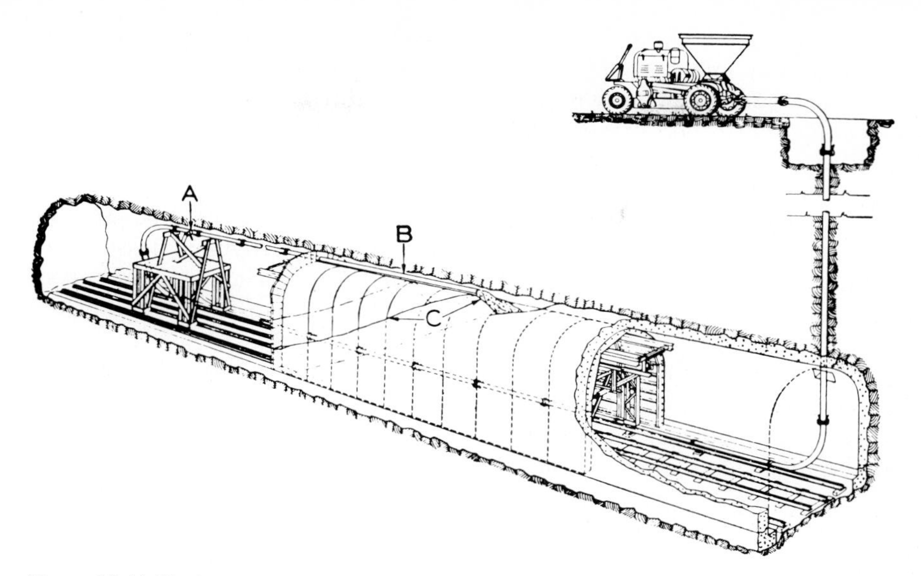

Figure 15-44 Placing concrete lining in a tunnel with a pump and a drop pipe.

A pump may be set up outside a tunnel, near a portal, or near a vertical shaft which provides passage into the tunnel. When such an installation is used, a pipe is laid and the concrete is pumped to the forms. As the maximum horizontal distance that concrete can be pumped is approximately 1,000 ft, it will be necessary, in installing lining more than 1,000 ft from a portal, either to take the pump into the tunnel or to use more than one machine, with the machine on the outside pumping concrete into the remixing hopper of a machine set up inside the tunnel.

Figure 15-44 illustrates a method of providing access to a tunnel by installing a drop pipe from a pump located at the surface of the ground above the tunnel. The drop pipe may be installed in a drilled hole or in a shaft.

PLACING THE CONCRETE LINING

The concrete walls may be placed by directing the flow of concrete from the discharge pipe through temporary openings in the form. As soon as the walls are placed to the desired height, the discharge pipe can be connected to an arch pipe, which is already installed over the top of the form. The arch pipe, frequently referred to as a slick pipe, usually extends to within a few feet of the opposite end of the form. As the concrete fills the space behind and above the form, the arch pipe is withdrawn until the entire space is filled. An alternate method is to place all the concrete from the arch pipe by letting the concrete for the side walls flow down the back sides of the form. Care should be exercised to keep the depth of concrete on each side wall nearly the same in order to eliminate unbalanced forces on the form.

The injection of a quantity of compressed air into the arch pipe through a quick-opening valve is referred to as air slugging. The slugger is operated at intervals by opening the air valve and permitting a large volume of compressed air to flow into the slick line. This has the effect of ejecting the concrete with sufficient velocity to push it away from the discharge end of the pipe into the most remote spaces. For the slugging action to be effective, at least 75 to 150 cu ft of free air at 100 psi should be injected through a $1\frac{1}{2}$- to 2-in. connection. The air usually is injected 20 to 60 ft behind the discharge end of the pipe.

As the space behind a set of forms is filled with concrete, the operation of the pump may be continued to build up considerable pressure between the form and the ground. Such a pressure is possible if the ends of the form are heavily bulkheaded to prevent the escape of concrete. This pressure forces concrete into all voids and spaces and produces a more solid lining.

The concrete should be vibrated as it is placed to eliminate voids and honeycombing. Internal vibrators may be used on the side walls if access to the concrete is possible, such as through doors in the walls of forms. If internal vibration is not practical, form vibrators may be used.

TUNNEL LINING WITH PNEUMATIC PLACERS

When a tunnel is so small that a pump cannot be set up in it and the length is so great that concrete cannot be pumped through a pipe, a pneumatic placer may be used to place the concrete lining.

Pneumatic placing of concrete for a tunnel lining involves using compressed air to force the concrete out of an airtight hopper through a discharge pipe, as illustrated in Fig. 15-45. Concrete is mixed outside the tunnel, loaded into cars, and hauled by a locomotive to the placing equipment. The concrete is transferred from the cars to the pneumatic placer or hopper, the charging door is closed tightly, and compressed air is injected into the placer to force the

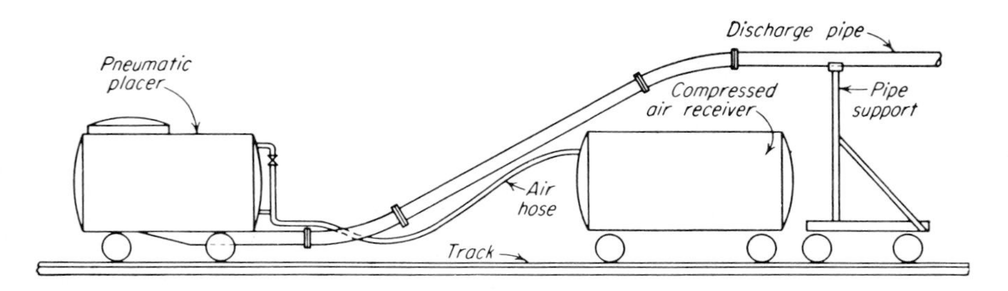

Figure 15-45 Equipment used to place concrete lining by pneumatic method.

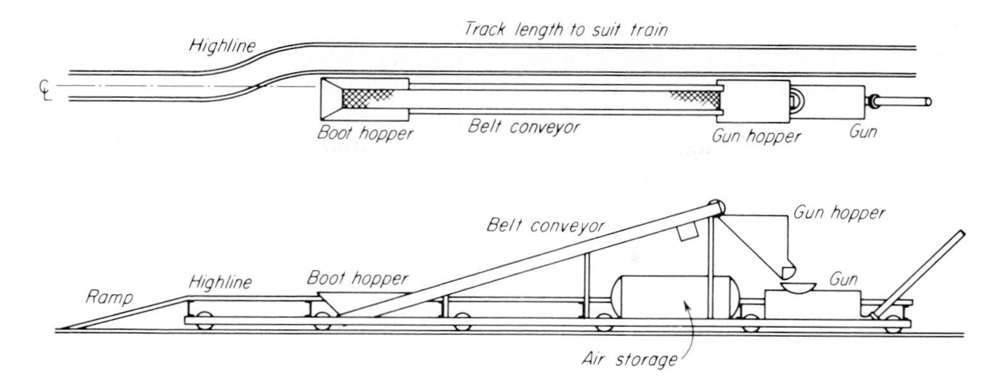

Figure 15-46 Typical concrete placing train. A train of side-dump cars is run onto the highline and dumped, one at a time, into the boot hopper, which feeds the belt conveyor. *(Elgood-Mayo Corp.)*

concrete through the discharge pipe into the forms. As the discharge pipe is generally placed above the roof of a form, it is necessary to provide a pipe support whose height can be adjusted. Air compressors, set up outside the tunnel, supply compressed air through a pipe to a portable receiver located near the placer. The purpose of the air receiver is to provide an adequate supply of compressed air during the placing operation. All the equipment used in the tunnel should be mounted on wheels to permit easy movement along the rails.

A modification of the pneumatic placer described in the previous paragraph is mounted on wheels and used as a car to haul the concrete from the mixer (see Fig. 15-46). The hopper of each car has a large door at the top, which can be sealed and made airtight. A number of these placer cars, loaded with concrete, are pushed by a small locomotive to the section of a tunnel to be lined. The front car is connected to the discharge pipe; then compressed air is injected to force the concrete into the discharge pipe. This operation is repeated until all the cars are emptied. Placers of this type were used in lining four aqueducts, 6 ft in diameter, for the city of San Diego. Each placer had a capacity of $1\frac{1}{4}$ cu yd of concrete.

When concrete is placed by the pneumatic method, care must be exercised to keep the velocity of the concrete low until the discharge end of the pipe is submerged in concrete. If this precaution is not observed, the concrete leaving the pipe will be segregated. Each pneumatic placer should be equipped with a throttling valve to regulate the flow of air into it.

THE USE OF PRECAST CONCRETE SEGMENTS TO LINE TUNNELS

In recent years a number of tunnels driven through earth by using shields or moles in the United States and in foreign countries have been lined with precast concrete segments, which, in general, were placed immediately behind the excavating operation, for example within the tailpiece of the excavating machine. The number of segments used to produce a ring has varied from two to eight or more. The lengths of the rings have been in the range of 2 to 3 ft. This method of driving a tunnel and installing the lining has been very satisfactory and economical when it was used under suitable conditions. The examples which follow describe projects where this method has been used.

Example 15-5 The 3,735-ft long, 30-ft-diameter tunnel through sand, gravel, boulders, clay, and water in Mexico City was driven with a shield. This shield had a tailpiece extension, which enabled the crew to place the two-segment precast concrete rings immediately behind the excavation [23].

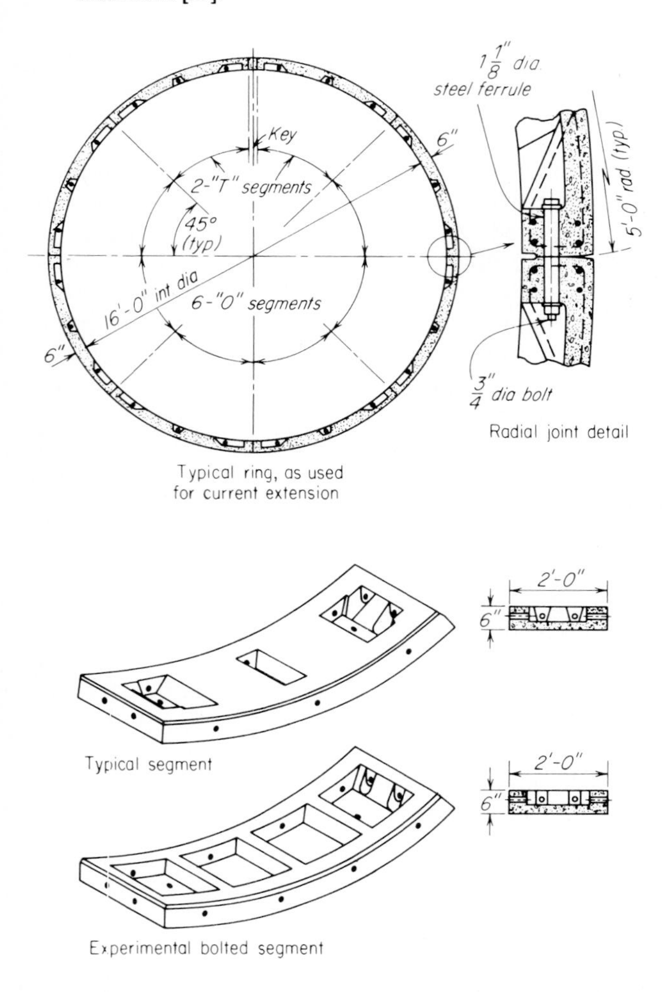

Figure 15-47 Details of prefabricated concrete linings for tunnel. *(Journal of the Construction Division, American Society of Civil Engineers.)*

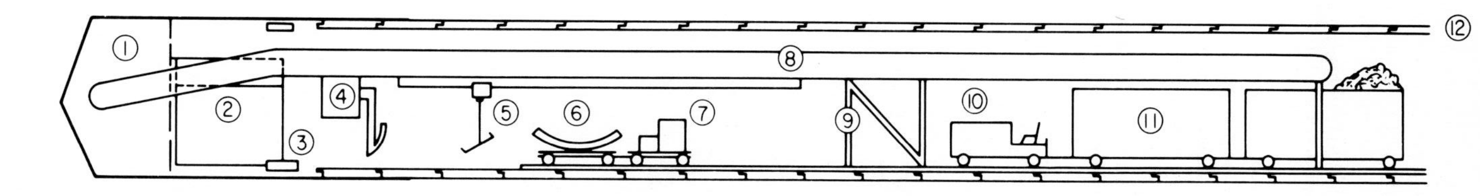

Figure 15-48 Schematic section (not to scale) of tunneling operation:
(1) Articulated head and flood doors (2) Control section
(3) Thrust jacks (4) Erector arm
(5) Monorail crane (6) Segment carriage
(7) Grout hopper and pump (8) Conveyor belt
(9) Trailing gear (10) Locomotive
(11) Muck car (12) Lining

Example 15-6 A method of driving and lining smaller tunnels through earth such as sand and clay, identified as the Mini Tunnel System, has proved effective and economical [24]. A tunnel is driven by using a cylindrical shield, usually with manual excavation. Precast concrete segments are installed immediately within the protection of the tailpiece of the shield. Cost savings varying from $50.00 to $200.00 per linear foot of tunnel have been reported, depending on ground conditions and the alternate methods of driving a tunnel.

Example 15-7 In driving a 16-ft-0-in.-inside-diameter extension for the Yonge subway tunnel in Toronto, Canada, precast reinforced concrete segments were installed to provide rings 2 ft 0 in. long. As indicated in Fig. 15-47, the segments were bolted together radially and longitudinally as they were assembled into rings under the steel tailpiece of the shield [25].

Example 15-8 A procedure using a full-face tunnel-boring machine, together with a precast concrete tunnel lining, has been used successfully in Canada in constructing a sanitary trunk

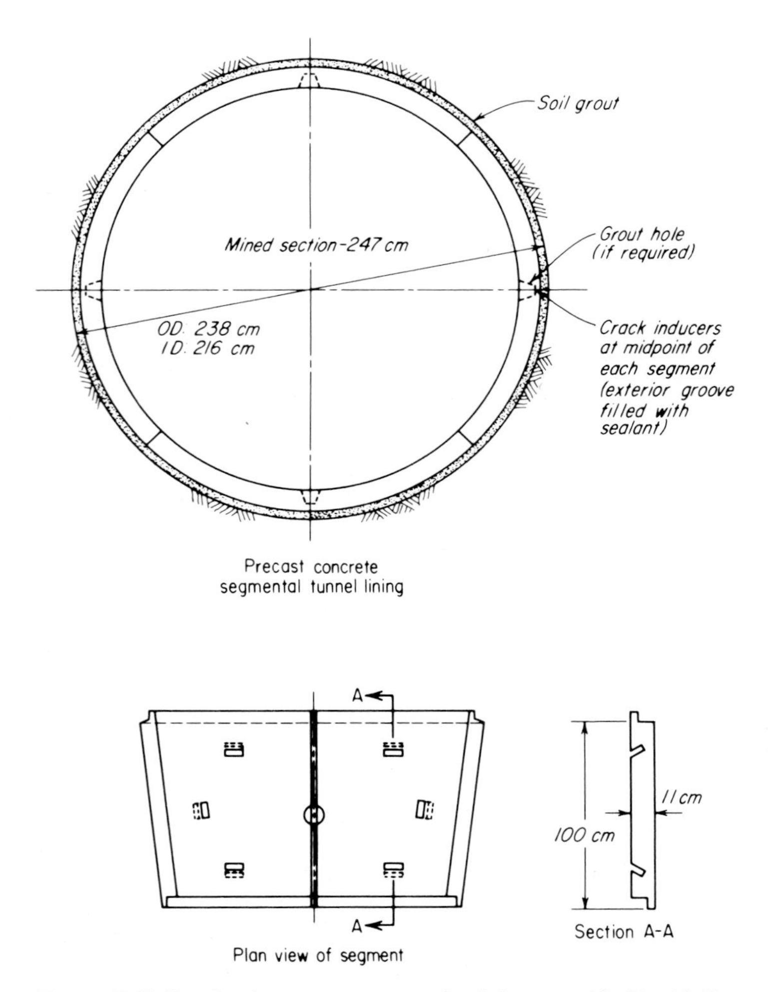

Figure 15-49 Details of concrete segments for lining tunnel in Fig. 15-48.

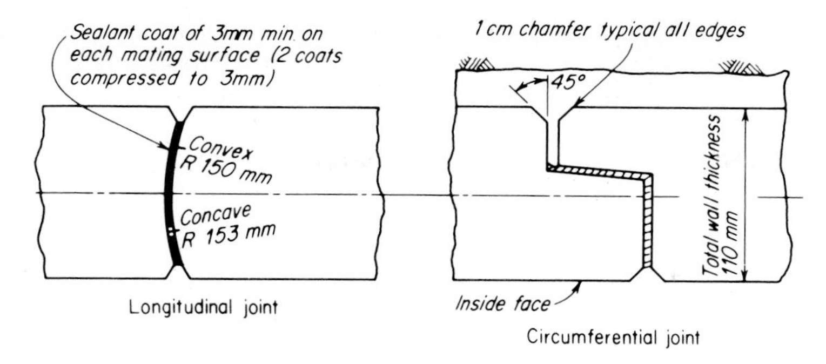

Figure 15-50 Details of joints for concrete segments for tunnel in Fig. 15-48.

sewer for Thunder Bay, Ontario. The 2.16-m-diameter tunnel was driven through soft to firm clay [26].

Each ring of the lining consisted of four trapezoidal segments of unreinforced concrete 1 m long and 11 cm thick. The rings were installed within the protection of the tailpiece of the tunnel-boring machine to serve both as the preliminary support and the final lining of the tunnel. Figures 15-48, 15-49, and 15-50 illustrate the schematic section of the tunneling operation and the details of the lining segments and of the joints for the segments.

As noted in Fig. 15-49, the segments were cast in trapezoidal shapes to enable them to be installed in rings more effectively. When the four sections were installed to form a ring, the last segment installed was at the crown or top of the tunnel. By applying sufficient force parallel to the axis of the tunnel against this segment, it was driven into position, thereby expanding the ring to the desired diameter.

After each ring was installed, a pump mounted on the boring machine forced grout, consisting of clay and water, into the annular space between the lining and the outer surface of the tunnel.

The boring machine was designed to permit the use of compressed air at the face of the tunnel, if needed.

SUMMARY [7]

Tunneling technology has received a great deal of attention in recent years. Most of it has been directed at gaining a better knowledge of geologic conditions and at the mechanization of tunnel driving. However, maintaining reasonable and planned advance rates through unpredicted and varying geologic conditions should be identified as the major development frontier.

Contractors are seeking a universal tunnel support system which will provide both immediate ground support and a final lining, or a primary lining which eliminates cleanup problems before the final lining is placed. As noted earlier in this chapter, some contractors have turned to tunnel liner segments as the answer, and this has presented a challenge to the manufacturers of tunnel-boring machines to devise methods of handling and placing these segments quickly enough to maintain pace with the rates of advance of the machines.

The major share of technological progress should continue to come from the innovative daring of tunnel builders, teamed with tunnel designers and specialists devoted to the development of equipment and systems.

REFERENCES

1. Half-mile Tunnel Driven for P.R.R. Double-track Line, *Construction Methods*, vol. 29, pp. 104–105, 170–172, May 1947.
2. Methods Spur Underground Power House, Tunnels, *Construction Methods and Equipment*, vol. 34, pp. 72–85, December 1952.
3. Hardrock Tunneler Saves 50% on Explosives, *Engineering News-Record*, vol. 170, pp. 50–55, May 23, 1963.
4. Armstrong, Ellis L.: Development of Tunneling Methods and Controls, *Journal of the Construction Division, Proceedings ASCE*, vol. 96, pp. 99–118, October 1970.
5. Tunneling Rig Takes on Drill-and-Shoot Operation, *Construction Methods and Equipment*, vol. 52, pp. 76–79, March 1970.
6. Cannon, D. E.: Record Tunnel Excavation with Boring Machines, *Civil Engineering*, vol. 37, pp. 45–48, August 1967.
7. Robbins, Richard J.: Vehicular Tunnels in Rock—Direction for Development, *Journal of the Construction Division, Proceedings ASCE*, vol. 98, pp. 235–250, September 1972.
8. Bellport, Bernard P.: Construction Innovation in Reclamation Work, *Journal of the Construction Division, Proceedings ASCE*, vol. 97, pp. 79–93, March 1971.
9. Fox, George A. and Louis E. Nicolau: Half a Century Progress in Soft Ground Tunneling, *Journal of the Construction Division, Proceedings ASCE*, vol. 102, pp. 637–667, December 1976.
10. Wakabayashi, Jiro: Japan, Tunnel Mole, *Construction Methods and Equipment*, vol. 58, pp. 44–45, July 1976.
11. Tunnel in Alluvial Soil Makes Fast Advance at Low Cost, *Roads & Streets*, vol. 116, p. 120, June 1973.
12. Sacrison, Hans: Flathead Railroad Tunnel, *Journal of the Construction Division, Proceedings ASCE*, vol. 97, pp. 127–145, March 1971.
13. AGL Corporation, P.O. Box 189, Jacksonville, Arkansas 72076.
14. Micro-Grade Laser Systems, Inc., 2352 Charleston Road, Mountain View, California 94043.
15. Norman, N. E.: "The Inevitable Marriage of Underground Mining and Big Hole Drilling," Reed Tool Company, P.O. Box 998, Sherman, TX 75090.
16. Mole Bores Tunnel No. 1, Miners No. 2, *Engineering News-Record*, vol. 175, pp. 26–33, November 11, 1965.
17. Dutro, Howard B.: Rock Mechanics Study Determines Design of Tunnel Supports and Lining, *Civil Engineering*, vol. 36, pp. 60–62, February 1966.
18. Hartman, Burt E.: "Rock Mechanics Instrumentation for Tunnel Construction," Terrametrics, Inc., Golden, CO 80401, 1967.
19. Allen, George W.: What You Should Know about Rock Bolts, *Engineering News-Record*, vol. 167, pp. 32–35, September 27, 1962.
20. Williams Form Engineering Corporation, P.O. Box 7343, Grand Rapids, MI 49510.
21. E. I. duPont deNemours & Company, Inc., 1007 Market Street, Wilmington, DE 19898.
22. Celite, Inc., 13670 York Road, Cleveland, OH 44133.
23. Custom Designed Shield Leaves No Space Behind As It Sets Tunnel Rings, *Construction Methods and Equipment*, vol. 52, pp. 64–72, August 1970.
24. Now Drive Tunnels Fast in Soft Ground without Liners, *Roads & Streets*, vol. 115, pp. 54–56, August 1972.
25. Bartlett, John V., Ted M. Noskiewicz, and James A. Ramsay: Precast Concrete Tunnel Linings for Toronto Subway, *Journal of the Construction Division, Proceedings ASCE*, vol. 97, pp. 241–256, November 1971.

26. Morton, J. D., D. D. Dunbar, and J. H. L. Palmer: Use of a Precast Segmented Concrete Lining for a Tunnel in Soft Clay, Paper prepared for submission to the International Symposium on Soft Clay, Bangkok, Thailand, July 1977. (J. D. Morton is a member of Morton, Dodds and Partners, Consulting Geotechnical and Geological Engineers, 50 Galaxy Boulevard, Rexdale, Ontario, Canada.)
27. Telescoping Forms Finish I-70 Tunnel, *Roads & Streets*, vol. 115, pp. 34–36, June 1972.
28. Automated Concrete Forms for Appalachian Tunnels, *Roads & Streets*, vol. 116, pp. 71–72, January 1973.
29. Dunbar, D. D.: Precast Concrete Liners in Tunneling, Paper presented at the 6th Annual Convention of the Ontario Sewer & Water Main Contractors' Association, February 10, 1977. R. V. Anderson Associates Limited, 194 Wilson Avenue, Toronto, Ontario, Canada.
30. Lane, K. S.: "Field Test Sections Save Cost in Tunnel Support," Underground Construction Research Council, American Society of Civil Engineers, 345 East 47th Street, New York, NY 10017, October, 1975.
31. Subsurface Exploration for Underground Excavation and Heavy Construction, Proceedings of a Specialty Conference held at New England College, Henniker, N.H., August 11–16, 1974. Available from American Society of Civil Engineers, 345 East 47th Street, New York, NY 10017.
32. Hoskins, Earl R. Jr.: "Application of Rock Mechanics," Proceedings, Fifteenth Symposium on Rock Mechanics, held at The State Game Lodge, Custer State Park, S. Dak., September 17–19, 1973. Published by American Society of Civil Engineers, 345 East 47th Street, New York, NY 10017, 1975.
33. Day, David A. and Bradford P. Boisen: Fifty-year Highlights of Tunneling Equipment, *Journal of the Construction Division, Proceedings ASCE*, vol. 101, pp. 265–280, June 1975.
34. Knight, Gail B.: Subway Tunnel Construction in New York City, *Journal of the Construction Division, Proceedings ASCE*, vol. 90, pp. 15–36, September 1964.
35. Underwood, Lloyd B.: Machine Tunneling on Missouri Dams, *Journal of the Construction Division, Proceedings ASCE*, vol. 91, pp. 1–27, May 1965.
36. Hill, George: What's Ahead for Tunneling Machines? *Journal of the Construction Division, Proceedings ASCE*, vol. 94, pp. 211–231, October 1968.
37. Pikarsky, Milton: Sixty Years of Rock Tunneling in Chicago, *Journal of the Construction Division, Proceedings ASCE*, vol. 97, pp. 189–210, November 1971.
38. Tunnel Under Alps Uses New Cost-Saving Lining Method, *Civil Engineering*, vol. 45, pp. 66–68, October 1975.
39. Rail Jumbos Handle Big Loads for Lining Tunnel, *Construction Methods and Equipment*, vol. 51, pp. 54–59, April 1969.
40. Smith, Lorraine: Auger Teams with Shield to Cut Mixed Tunnel Face, *Construction Methods and Equipment*, vol. 52, pp. 104–106, January 1970.
41. Small-bore Hydraulic Mining Machine Cuts Sewer Tunnel, *Roads & Streets*, vol. 115, pp. 66–69, August 1972.
42. Subway Freezes through Obstacles, *Engineering News-Record*, vol. 210, pp. 20–21, February 24, 1983.
43. Slurry Mole Conquers Dense Sand, The Best of Engineering-News Record Field & Office Manual, 1982.
44. Fox, George A., and Louis E. Nicolau: Half A Century Progress in Soft Ground Tunneling. *Journal of the Construction Division, Proceedings ASCE*, vol. 102, pp. 637–667, December 1976.
45. Peduzzi, Antonio: Tunnel Sealing Method with Water-proofed PVC Sheets, *Journal of the Construction Division, Proceedings ASCE*, vol. 103, pp. 1–6, March 1977.
46. Tartaglione, Louis C.: Segmental Concrete Liner for Soft Ground Tunnels, *Journal of the Construction Division, Proceedings ASCE*, vol. 103, pp. 227–243, June 1977.
47. Petrofsky, Alfred M.: Mixed Face Tunneling on Melbourne Underground, *Journal of the Construction Division, Proceedings ASCE*, vol. 106, pp. 409–425, September 1980.
48. Paulson, Boyd C.: Tokyo's Dainikoro Underwater Tube-Tunnel, *Journal of the Construction Division, Proceedings ASCE*, vol. 106, pp. 489–497, December 1980.

49. Paulson, Boyd C.: Seikan Undersea Tunnel, *Journal of the Construction Division, Proceedings ASCE*, vol. 107, pp. 508–525, September 1981.
50. Paulson, Boyd C.: Underground Transit Station Construction in Japan, *Journal of the Construction Division, Proceedings ASCE*, vol. 108, pp. 23–37, March 1982.
51. Kitamura, Akira and Yuzo Takeuchi: Seikan Tunnel, *Journal of Construction Engineering and Management, American Society of Civil Engineers*, vol. 109, pp. 25–38, March 1983.
52. Paulson, Boyd C.: Japanese Tunnel Design: Lessons for the U.S., *Civil Engineering*, vol. 51, pp. 51–53, March 1981.
53. Xin, Wang Zhen: Shanghai Tunnel Projects Spur Construction Innovations, *Civil Engineering*, vol. 52, pp. 36–38, December 1982.
54. Blaw-Knox Construction Equipment Division, Mattoon, IL 61938.
55. Caldwell Division of Smith Industries International, P.O. Box 2875, Santa Fe Springs, CA 90670.
56. Chicago Pneumatic Tool Company, 6 East 44th Street, New York, NY 10017.
57. The Eimco Corporation, 537 West Sixth South, Salt Lake City, UT 84110.
58. Ingersoll-Rand Company, Phillipsburg, NJ 08865.
59. Joy Manufacturing Company, River Road, NH 03743.
60. Reed Tool Company Mining Equipment Division, 12400 North Freeway, Houston, TX 77090.
61. The Robins Company, 650 South Orcas Street, Seattle, WA 98108.
62. Goodman Equipment Corporation, 4834 South Halsted Street, Chicago, IL 60609.

CHAPTER
SIXTEEN
FOUNDATION GROUTING

NEED FOR GROUTING

Although large deposits of rock frequently are referred to as solid rock, in many instances they are not solid. These deposits may contain fissures, cavities, slips, faults, seams, or breaks, which make the deposits unsuitable for dams, reservoirs, buildings, bridge piers, locks, tunnels, etc. When subsurface investigations disclose the existence of such structural defects, it is necessary to adopt corrective steps if a formation is to be made suitable for the intended use. If correction is impossible, or if it is unreasonably expensive, it may be necessary to abandon the site.

An operation to correct the foundation conditions is described as pressure grouting. The formation under or adjacent to a structure is grouted for several reasons, such as:

1. To solidify and strengthen the formation in order to increase its capacity to support a load
2. To reduce or eliminate the flow of water through a formation, such as under a dam or into a tunnel
3. To reduce the hydrostatic uplift under a dam

EXPLORING TO DETERMINE THE NEED FOR GROUTING

The most satisfactory method of determining whether a formation should be grouted is to obtain core samples from representative locations within the formation area. Cores may be obtained with diamond or shot drills, usually diamond for the smaller sizes and shot for the larger sizes. Several shot-drilled holes, 30 in. in diameter or larger, may be desirable in order that a man may be lowered into them for visual inspection of the formation. The size, number, depth, and spacing of the exploratory holes should be planned to provide the greatest amount of information for the lowest practical cost. Increasing the number of holes will provide more dependable information, but it will increase the cost of exploration. For each project there must be a weighted balance between the need for foundation information and the cost of obtaining this information. The decision should be made by a competent geotechnical engineer.

An accurate record should be kept for each exploratory hole. The record should show the location, size, and depth of the hole and, with the core recovered, should show the physical nature of the formation. If a core is recovered in long, continuous pieces, with little loss in length compared with the depth of the hole, this indicates a reasonably solid formation, which may require little or no grouting. However, if the core is badly broken, and if the recovered length is small in proportion to the depth of the hole, this indicates a bad foundation condition, which probably will require a large quantity of grout.

The approximate rate at which a hole will take grout may be determined by forcing water, under pressure, into the hole. For this purpose a section of pipe, $1\frac{1}{2}$ to 2 in. in diameter, and 2 to 4 ft long, is sealed into the top of a hole with a threaded end projecting from the hole. As water is forced into the hole, the rate of flow and pressure should be recorded. If the rate of flow drops quickly, with a corresponding increase in pressure, this indicates the presence of only a few thin seams or fissures, which can be closed easily with grout. If the rate of flow remains high, with little or no increase in pressure, this indicates a highly porous formation, with extensive fissures, for which a large amount of grout will be required.

MATERIAL USED FOR GROUT

The materials commonly used for grout include

1. Cement and water
2. Cement, rock flour, and water
3. Cement, clay, and water
4. Cement, clay, sand, and water
5. Cement, fly ash, and water
6. Fly ash and water
7. Asphalt
8. Clay and water
9. Chemicals

When a grout of cement and water only is to be injected into fine seams, it may be necessary to use as much as 10 parts of water to 1 part of cement, by volume, in order to obtain penetration. When the seams are large, the grout may be as dry as $\frac{3}{4}$ part water, or less, to 1 part cement. For most grouting operations the ratio will vary from 1 to 2 parts water to 1 part cement. Usually, the most satisfactory grout is the stiffest mix that can be injected effectively. This should be determined by testing the rate of injection, using varying mix ratios.

With the advent of large quantities of fly ash readily available in the United States and elsewhere, it has proved to be an economically effective grouting material when used either with portland cement or by itself. (See Chap. 20.)

Rock flour and clay may be added to cement grout in the interest of economy if the seams are small, while sand may be added if the seams are large enough to permit the sand to penetrate. Grout made of neat cement will give a higher strength than grout containing clay or sand. In grouting the foundation for the Norris Dam, the grout was mixed in the ratio of 1 part cement, 1 part rock flour, by volume, with 3 lb calcium chloride per 100 lb of cement added to speed setting. For the Chickamauga Dam the mixture was 2 parts cement, $\frac{1}{2}$ part bentonite, and 4 parts sand, by volume. As bentonite has the property of increasing up to several times its original volume when mixed with water, it is necessary to mix it thoroughly with water prior to adding cement and sand.

Table 16-1 gives the properties of certain admixtures when they are used with cement grout.

Table 16-1 Properties of admixtures used with cement grouts

Admixture	Property
Calcium chloride Sodium hydroxide Sodium silicate	Accelerates setting time
Gypsum Lime sugar Sodium tannate	Retards setting time
Finely ground bentonite	Increases plasticity Reduces grout shrinkage
Clay Ground shale Rock flour	Reduces cost of grout Reduces strength of grout

The use of asphalt, clay, and chemical grouts will be discussed later in the chapter.

DRILLING PATTERNS

After a formation has been explored and tested to determine the extent of grouting required, a drilling pattern should be adopted. The size, depth, and spacing of injection holes should give the best results at the lowest cost. It may be necessary to change the drilling pattern from time to time if the grouting operations encounter differences in formation conditions.

In general, the smallest holes which will permit the injection of grout are the most desirable, as the size of a hole seems to be secondary so long as grout can be injected through it. Because of their lower cost, small holes permit a greater number to be drilled for a limited expenditure and, thus, increase the probability of obtaining a successful grouting operation.

A simple pattern, such as a hole spacing of 20 by 20 ft, with all holes $2\frac{3}{4}$ in. in diameter and 40 ft deep, may be entirely satisfactory for one project but unsatisfactory for another. Figure 16-1 is a section through the Norris Dam showing shallow grout holes for consolidating the foundation under the dam and deeper holes for producing the impervious curtain to prevent the flow of water under the dam.

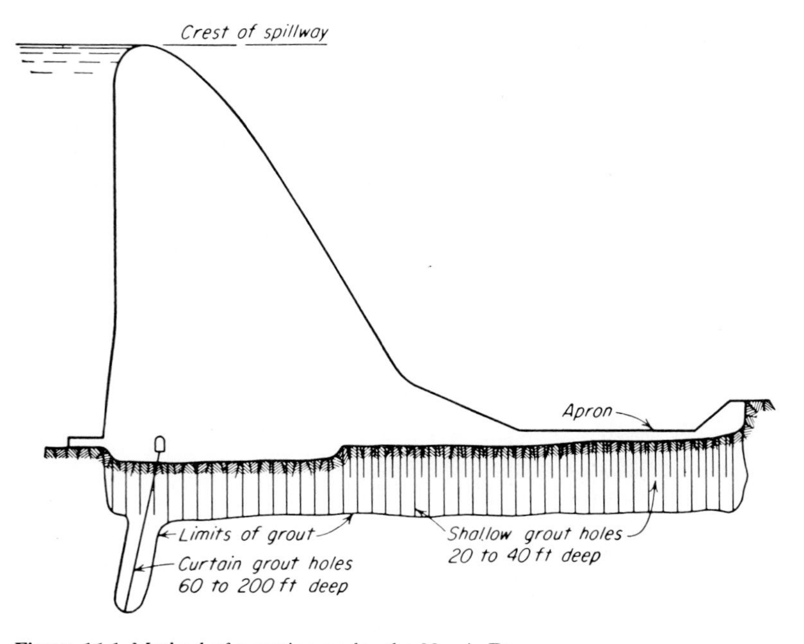

Figure 16-1 Method of grouting under the Norris Dam.

DRILLING INJECTION HOLES

Holes for the injection of grout may be drilled with jackhammers, wagon drills, diamond drills, or shot drills, depending on the terrain, class of formation material, and size and depth of holes.

Diamond drills usually give holes that are uniform in shape and size, which are more satisfactory than holes drilled by other equipment when packers must be installed for washing or grouting individual seams. Wagon drills are satisfactory for holes whose depths do not exceed 30 to 40 ft.

PREPARATIONS FOR GROUTING

The preparations for washing or grouting seams by the full-length method consist in installing a section of pipe, usually $1\frac{1}{2}$ to 2 in. in diameter, 18 to 36 in. long, in the grout hole, with the top end projecting out a short distance for connection to an air line or a pump. The space around the bottom of the pipe is closed with oakum or other suitable material; then the balance of the space is filled with cement mortar or melted sulfur, or it may be calked with lead wool.

In order to reduce the danger of weakening the formation through fractures resulting from the application of excessive pressure, uplift gauges should be installed at several locations over the area to detect any lifting of the surface during the grouting operations.

WASHING THE SEAMS

When a formation is grouted with neat cement for consolidation purposes, it is desirable to deposit the cement in clean seams from which any clay or unconsolidated materials have been removed. The most effective method of removing such materials is to force a mixture of air and water through the seams. The removal of materials may be made more effective by alternately reversing the direction of flow of the air and water.

In washing a formation, a pattern of holes is selected. Some of the holes are capped for water, some for compressed air, and others are left open to permit the outflow of the washed materials. The direction of flow may be reversed by interchanging the pipe caps. When the water flowing from the uncapped holes clears up, indicating the removal of the unconsolidated materials, the caps are moved to another pattern of holes.

If the grout holes are deep and pass through several seams of unconsolidated materials, it may be desirable to isolate each seam in order that it may be washed individually. This is done by using an injection pipe, with the lower end closed, sufficiently long to extend below the lowest seam, with a perforated section long enough to extend completely through the seam. The pipe is equipped with an expandable packer above and below the perforated section, which is set opposite the seam to be washed. When the injection pipe is lowered into a hole and the packers are expanded, any air or water delivered to the pipe will be confined to a single seam.

It is possible to determine whether a seam is open from one hole to others by injecting into the seam water containing a coloring agent, such as fluorescein dye. If the colored water appears in other holes, this indicates open passages through the seam.

GROUTING PRESSURES

The most suitable pressure for grouting operations is difficult to determine in advance. Some engineers follow a general practice of using a pressure of 1 psi for each foot of depth of hole. There is no logical proof or demonstration that this is the most satisfactory pressure.

In the interest of economy and effectiveness it is desirable to use the highest pressure that is safe. However, when grout is forced into a seam under pressure, it is possible that the total upward force on the formation above the seam may exceed the combined weight and resisting strength of the formation. If this condition is permitted to occur, the entire formation may be lifted upward, with a resulting fracture that is more serious than the original condition that grouting is supposed to correct. Thus, it is possible for grouting to do more harm than good unless it is injected under careful supervision.

If the weight of a rock formation is 150 lb per cu ft, the unit pressure on a horizontal plane, 1 ft below the surface of the rock, will be $150 \div 144$ sq in. = 1.04 psi. At a depth of 100 ft the pressure resulting from the weight only will be 104 psi. As most rocks weigh 150 lb or more per cubic foot, it is improbable that a grouting pressure equal to 1 psi for each foot of depth will endanger a foundation formation provided the pressure is confined to the intended depth.

When grout is injected by the full-length-hole method, care must be exercised to prevent the pressure of the grout near the surface of the ground from exceeding the maximum safe pressure. If the pressure at the bottom of a hole 40 ft deep is 40 psi, the pressure at a depth of 20 ft will be equal to 40 psi minus the hydrostatic pressure of 20 ft of grout. The hydrostatic pressure of cement grout may be determined from the curve in Fig. 16-2. If the grout is mixed in the ratio $1\frac{1}{2}$ parts of water to 1 part cement, the pressure change per foot of depth will be 0.66 psi. In 20 ft the reduction in pressure will be $20 \times 0.66 = 13.2$ psi. Thus, for the case cited above the pressure at a depth of 20 ft will be $40 - 13.2 = 26.8$ psi. The pressure at the surface of the ground will be $40 - (40 \times 0.66) = 13.6$ psi. If the injection of grout by the full-length-hole method produces objectionably high pressures near the top of a hole, this objection may be overcome by injecting grout in limited depth zones.

The pressures at which grout is injected frequently are varied with the depth of injection and the stage at which the grout is injected. For example, when the foundation under a concrete dam is grouted for consolidation

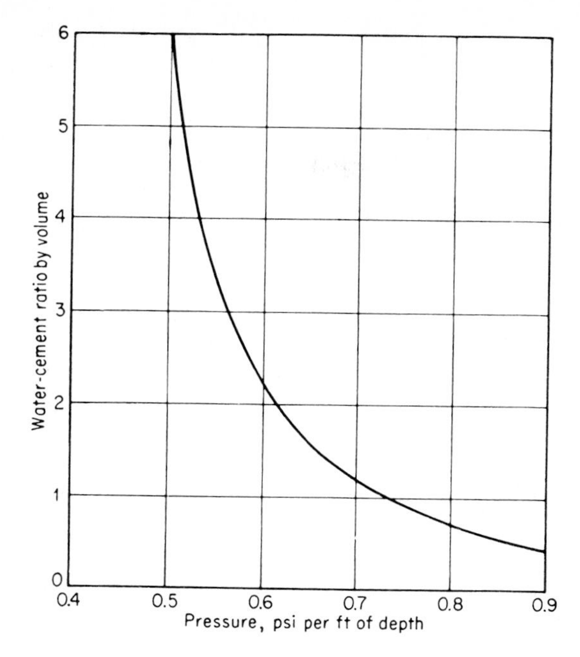

Figure 16-2 Hydrostatic pressure produced by cement grout.

purposes, the grout may be injected in two or three stages. The first stage might consist in drilling holes 20 to 40 ft deep, with a spacing of 20 ft each way. The pressure at the bottom of these holes might be limited to a maximum of 40 to 50 psi. The second stage might consist in drilling holes 40 to 60 ft deep, with a spacing of 10 ft each way. If the grout is injected after the dam is partly constructed, the pressure at the bottom of the holes might safely be increased to 80 to 100 psi. The third and last stage might consist in drilling holes 60 to 100 ft or more deep for final high-pressure grouting under the dam. These holes might be spaced as close as 5 ft apart in a single row along the axis of the dam, to permit the grouting of a solid cutoff curtain, to prevent the flow of water under the dam. If these holes are not grouted until the dam is completed, the pressure may be as high as several hundred pounds per square inch. Figure 16-1 illustrates a method of grouting the foundation for a dam.

EQUIPMENT FOR CEMENT GROUTING

The most common method of injecting cement grout is to use one or more piston-type pumps to produce the necessary pressure. The pumps usually are air-driven duplex double-acting types so constructed that the number of strokes per minute and the pressure on the grout may be varied by regulating the quantity of compressed air supplied to the pump.

The equipment will include:

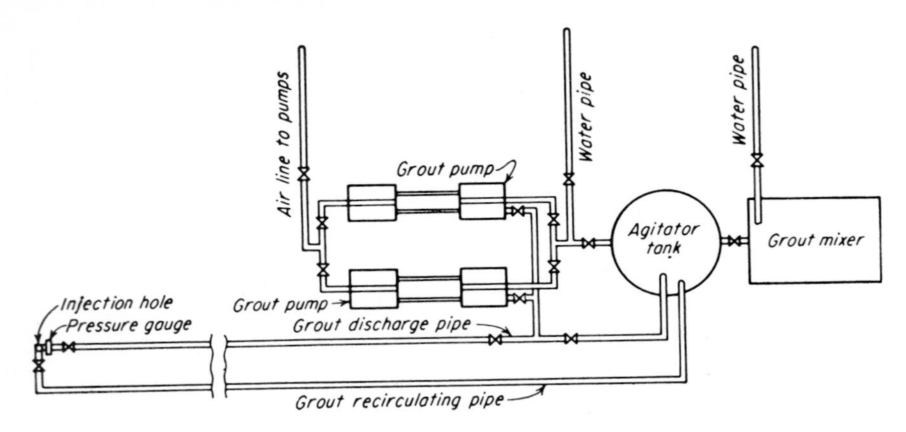

Figure 16-3 Boulder-type equipment used to inject cement grout.

1. One or more air compressors
2. One or two grout mixers
3. One agitator-type reservoir tank
4. One or more grout pumps
5. Grout discharge pipe or hose, valves, pressure gauges, etc.

The grout mixer contains a shaft with paddles, operated by a motor. After the grout is mixed, it is discharged into a tank, with an agitator to prevent separation of the solids from the water. The pumps draw their charges directly from the agitator tank.

The essential parts of the Boulder-type grout unit are illustrated in Fig. 16-3. The location of the component parts may be modified to fit any particular injection conditions. The grout discharge line may be a pipe, a rubber hose, or a combination thereof. The use of a hose will facilitate moving from one grout hole to another.

Figure 16-3 shows a grout recirculating pipe or line whose primary function is to permit grout to flow through the pumps and discharge pipe at a uniform rate even though the rate of injection into a hole is reduced as the cavities are filled.

It is good practice to install two grout pumps, even though one pump can supply all the grout that will be needed. In the event of a pump failure, the auxiliary pump can be placed in operation immediately, thereby reducing the danger of losing a partly filled hole or group of holes.

INJECTING CEMENT AND/OR FLY ASH GROUT

The records that were kept at the time the grout holes were drilled, together with the records obtained from washing the holes, if such an operation was performed, should serve as a guide in estimating the grout mix to use. The best results are obtained by using the thickest grout that can be injected without plugging the hole. It may be necessary to start with a batch of thin grout, and then thicken each succeeding batch by reducing the water-cement ratio until the maximum practical thickness is determined.

The specifications covering the grouting of a project may require the injection of grout at a given pressure until the rate of injection for a given hole small diminish to a specified amount or until a hole will not take any more grout at a specified pressure.

Grout may be injected into the full length of a hole at one time, or it may be injected into a portion of the length only. If the latter method is used, it is possible to apply a high pressure in injecting at the bottom of a hole. As the depth of injection is reduced, the pressure may be reduced accordingly. This is referred to as the zone method of grouting. In order to inject grout by the zone method, it is necessary to use an injection pipe that is long enough to reach the lowest zone of injection. The zone to be grouted at a given time is isolated

Figure 16-4 Self-contained gasoline-engine-powered grout mixer and pump. *(Acker Drill Company.)*

from the rest of the hole by means of a packer, which is set near the bottom of the injection pipe and just above the top of the zone. Most packers are of removable types, which permit them to be reused many times.

PRESSURE GROUTING WITH ASPHALT

If a formation contains fissures with water flowing through them, it will be difficult or even impossible to consolidate the formation with cement grout. The velocity of the water tends to sweep the grout through the openings without giving it an opportunity to solidify.

In numerous instances the injection of asphalt grout into fissures containing flowing water has sealed the fissures and stopped or reduced the flow of water. After the flow of water is stopped, it is possible to inject cement grout to complete the consolidation operation. Thus, the primary function of asphalt grout is to seal off the flow of water in order that cement grout may be retained in the fissures.

The heated asphalt is injected through a perforated pipe, which may have a stem line running through it, as a means of keeping the asphalt at the desired temperature until it flows into the formation. An alternate method of heating the asphalt in the injection pipe is to install an electric wire inside the pipe, with the pipe completing the electric circuit. An electric current through the wire will heat the asphalt. When hot asphalt flows from an injection pipe into the openings in a formation, the outside surface of the asphalt tends to solidify, but, because of the low heat conductivity, the inside tends to remain a liquid for some time. The pressure from the injection pipe will keep the interior part of the asphalt flowing for a considerable distance, several hundred feet in some instances. As the grout solidifies under pressure, it conforms with the shape of the fissures and seals them against the flow of water.

The equipment required to inject asphalt grout consists of a heating kettle, a piston-type pump, an air compressor or an electric motor to operate the pump, a source of electric current, plus a supply of pipe hose, valves, and pressure gauges.

An interesting example of the use of asphalt grout to stop the flow of water through a fissurized formation was developed in correcting the leakage from the reservoir at Great Falls Dam [1]. The limestone abutments to the dam had developed extensive fissures below the level of the water in the reservoir, through which a large quantity of water flowed from the reservoir into the river channel below the dam. Injection holes, which were drilled through the fissurized formations, were grouted first with asphalt to stop the flow of water, after which the consolidation was completed using cement grout. Asphalt grout, at temperatures varying from 300 to 350°F, was injected at rates varying from 40 to 60 cu ft per hr.

CLAY GROUTING

Grout mixtures of clay and water or cement, clay, and water have been used successfully to fill large seams and cavities subjected to low hydrostatic heads. While clay adds little, if any, strength to a foundation, its high resistance to the flow of water makes it an excellent barrier to water seepage. It can be mixed with water to any desired consistency and injected with equipment similar to that used for cement grouting. Pressures in excess of 100 psi have been used to inject clay grout into formations.

Clay is not a satisfactory grout for use in fissures which contain flowing water. The primary advantage of clay as a grouting material is its low cost compared with cement and asphalt, especially when a deposit of clay is available near the site to be grouted.

In sealing large seams and solution channels in the limestone along the ridges of the reservoir at Madden Dam in the Canal Zone, 70,000 cu yd of clay grout was used. The water content of the mixture varied from 43 to 55 percent by weight, depending on the back pressure of the fissures.

CHEMICAL GROUTING

During recent years considerable success has been experienced with chemical grouting. The chemical method has several advantages when compared with other methods, including the following:

1. Because it is a liquid, it can be pumped into and through very small openings, which other grouts cannot penetrate.
2. The chemical forms a barrier to the flow of water through a formation by changing from a liquid to a gel at a predetermined time after it enters the formation. The gel time may be varied from 3 sec to several hours.
3. Fewer grout holes are required.
4. The time required to inject the chemical is usually less than for other grouting materials.

If a formation to be grouted is highly porous, with relatively large voids, preliminary grouting should be done using such materials as cement, a mixture of cement and clay, or a mixture of cement and bentonite, individually or in combination, to reduce the rate of flow of water to approximately 10 percent of the initial rate. Then the final grouting can be done with chemicals.

Chemicals that have been used with success include sodium silicate and calcium chloride and a chemical designated as AM-9, which is manufactured by the American Cyanamid Company. The latter is an aqueous solution of two acrylic monomers and a catalyst, dimethylaminopropionitrile (DMAPN), which

Figure 16-5 Two-tank grout plant with mixers and pump for slurries and chemicals. *(ChemGrout, Inc.)*

is mixed with an aqueous solution of catalyst ammonium persulfate (AP) just before it is pumped into the ground [2].

During the construction of a tunnel for a storm sewer in Houston, Texas, the contractor injected a solution of 2 lb of calcium chloride dissolved in 1 gal of sodium silicate into the formation of wet sandy clay, jointed clay, and extended areas of water-bearing sand, ahead of excavation to stabilize and seal the formation [3].

Prior to constructing the Round Butte Dam in Oregon, the contractor injected cement grout into the curtain under the dam to stop 95 percent of the flow of water through the seamy rock formation; then he used AM-9 chemicals to complete the grouting [4].

RECENT TECHNIQUES IN GROUTING

These techniques apply to methods of drilling, methods of grouting, and the materials used for grouting. The new materials used for grouting consist primarily of fluids that have been developed by the field of organic chemistry. They include such materials as hard gels, silicates with organic hardeners, aqueous resins, acrylamides, phenoplasts and aminoplasts, and others. The use of these materials, which are liquids when injected, permit the grouting of almost any type of formation.

A comprehensive discussion of the uses and methods of these techniques is beyond the scope and space in this book. A more extensive study of this subject is available in a publication titled *Proceedings of the Conference on Grouting in Geotechnical Engineering* sponsored by the Geotechnical Engineering Division of the American Society of Civil Engineers, 1982, 345 East 47th Street, New York, NY 10017. The major divisions of this publication are as follows:

1. Materials for cement and mortar grouts
2. Dam grouting technology
3. Application of dam grouting technology
4. Design and control for dam grouting
5. Chemical grouts materials
6. Behavior of chemically grouted soils
7. Chemical grouting technology and applications
8. Grouting for tunnels, shafts, and mines
9. Alternate grouting technology
10. Testing and control for grouting
11. Alternative grouting technology
12. Application of grouting technology

EXAMPLES DESCRIBING GROUTING OPERATIONS

The following examples describe and discuss projects where grouting has been used to correct or control soil conditions to permit construction of desirable structures.

Example 16-1 Logan Martin Dam [5] This is a multipurpose project on the Coosa River in Alabama. The foundation for the dam is a cavernous limestone, which required considerable pressure grouting to fill the cavities under the dam. Curtain grouting was used to consolidate and strengthen the foundation.

Tests which were conducted to assist in selecting the drilling equipment demonstrated that diamond core drills were most suitable for drilling the injection holes.

A maximum initial or primary hole spacing of 40 ft was chosen as one workable for exploratory purposes because it prescribed a recognizable area of influence confined to reasonable dimensions. Cores were required from this set of holes to give geologic information on bedrock structure and its influence on the occurrence of cavitation. The election of any practical drilling method capable of detecting cavities was permitted for all holes located between cored primary holes. The depths of the holes were varied as information gained from the drilling indicated such variations to be desirable and effective.

Figure 16-6 illustrates the layout for a grout mixing and pumping plant. After clays, sand, rock dust, and fly ash were tested, it was determined that a grout consisting of cement and fly ash was most suitable. When fly ash was in short supply, rock dust was used as a substitute.

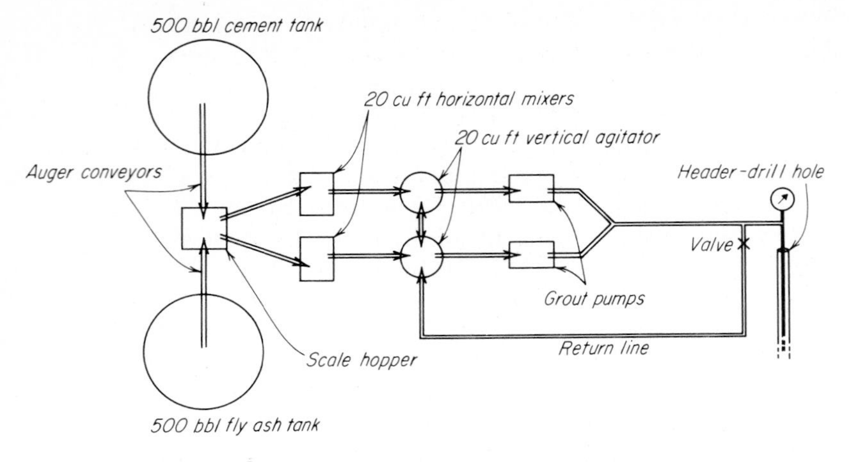

Figure 16-6 Layout of grout mixing plant for the Logan Martin Dam.

Example 16-2 Paris rapid transit tunnels [6] During the driving of vehicular tunnels in Paris, France, chemical grout was applied under pressure to consolidate water-bearing sand ahead of the excavation, with considerable success.

Two methods were used to accomplish the objectives. One method consisted of impregnating the sands with a diluted silica gel, giving a low strength, then partitioning the impregnated sand by cement grouting under high pressure. The other method consisted of grouting the sand through two closely drilled holes, using a pure silicate and a reagent such as calcium chloride. The main inconvenience with these methods is the required high pressure, which is not always permitted with the soil cover or structures.

For grouting medium to fine sands, viscous materials, generally colloidal, are used. These materials are designated as gels. They have a viscosity at the time of placing of about 25 centipoises. Their main characteristic, as far as grouting is concerned, is that their viscosity increases as a function of time. For equal strength, these silica gels are cheaper than other gels having the same viscosity-increase characteristics, such as lignosulfonates or derived products, vulcanizable oils, and alginates. In Europe, it has been determined that sodium silicate-based gels are the best.

When grouting clayey sands, gels are not suitable, and resins are used. A resin is a grouting material having a viscosity similar to that of water, which changes quickly at a given time by setting.

The essential difference between the two types of materials, gels and resins, is only a matter of viscosity characteristics, and it does not matter whether the materials are of organic or mineral origin. This variation in viscosity is considered the main criterion for selection of grout material. In each type, the materials are then classified according to the final strength they produce. In practice the two types of grout are generally used in combination on the same project. They must be compatible and have the same pH value. It is possible to use the more costly materials, resins, only when the more permeable areas have been treated with gels. However, gel grouting may follow, in some cases, an inexpensive clay cement pregrouting. This latter procedure was used on the project described here. This use was made possible by development of a new gel, sometimes referred to as Carongel. This reagent, of organic origin, is an ester, which has reactive properties on the silicate only after saponification. The reaction time and final strength are adjustable.

The use of these elaborate materials would not have been possible without the development of injection methods that permitted grouting to be competitive with other methods of stabilizing the soil. For this project, high-productivity drills and automatic grout mixing plants were used, ensuring regularity in the proportioning of component materials.

Because of variations in the characteristics and properties of the soil that was grouted, it was highly desirable to inject grouts that were most suitable for given conditions. In order to accomplish this objective, grout holes were drilled to their full depths and cased where necessary. Then there was introduced into each hole a polyvinylchloride pipe perforated with rings of small holes at intervals of 1 ft. Each ring of holes was covered with a short rubber sleeve, which fitted tightly around the tube and acted as a one-way valve, to permit grout to flow out from but nothing to flow into the tube. Injection of the grout into the formation was accomplished by lowering into this tube a grout pipe fitted with a double packer. The spacing between the two packers could be preset to limit the flow of grout to the desired part of the soil formation. The space in the grout pipe between the packers contained perforations to permit the grout to flow into the formation at controlled locations. Before injecting the grout, the casing, if used during the drilling of the hole, was removed from the hole.

Thus use of this technique provided several advantages, including:

1. It permitted the injection of grout in short stages and a better spread of the grout material.
2. It permitted the injection of grout that was best adapted to a given soil condition.
3. It permitted the injection of grout in several steps, using different mixes as desired, and it provided a method of repeating the procedure at a given level if desired.
4. It permitted separate drilling and grouting operations, thereby reducing possible interference between the two operations.

Because the polyvinylchloride tubes were easily broken, they did not impede the excavation that followed.

DETERMINING THE EFFECTIVENESS OF GROUTING

A question that usually arises in connection with a grouting operation is how to determine whether the operation has been successful. Several methods have been used with varying degrees of success.

To determine the extent of flow of grout from the injection holes through a formation, several holes are left open to see whether grout will appear in them. The appearance of grout in these holes serves as a guide to indicate the extent of flow.

Prior to concluding a grouting operation, additional exploratory holes may be drilled at various locations within the area that has been grouted in order to obtain cores from the formation. If these cores show the existence of sufficient grout to produce good consolidation where voids originally existed, this indicates that the grouting operation has been successful. The effectiveness of the grouting operation also may be tested by attempting to inject water or grout into the holes from which the cores were obtained. If these holes refuse to take grout, the test indicates that the formation has been consolidated adequately by previous injections.

FREEZING SOIL FOR TEMPORARY GROUND SUPPORT

Introduction Controlled ground freezing for mining and construction applications has been practiced for over a century. Despite the technological developments that have occurred during this period, ground freezing is still used on projects today with considerable success when project and soil conditions warrant the use of this method of stabilization.

General advantages and disadvantages of freezing soil Ground freezing may be used in any soil or rock formation, regardless of structure, grain size, or permeability. However, it is best suited to soft ground, rather than rock conditions. Freezing may be used for any reasonable size, shape, or depth of excavation, and the same physical plant can be used from job to job despite wide variation in these factors.

Freezing is normally used to provide structural underpinning or temporary support for an excavation or to prevent ground water from flowing into an excavation area. Because the impervious frozen earth barrier is constructed prior to excavation, it generally eliminates the need for compressed air or dewatering and the concern for adjacent ground subsidence during dewatering or excavation. However, later ground-water flows may result in failure of the freezing program if not properly considered during the planning stage. Further, although subsidence may not be of concern, ground movements resulting from frost expansion of the soil during freezing may occur under certain conditions, and this condition must be considered in planning.

The freezing of soil may be accomplished rapidly if necessary or desirable. However, rapid freezing will usually be more expensive than slower freezing.

Frozen ground behaves as a viscoplastic material with strength properties which are primarily dependent on the ice content, duration of the applied load, and the temperature of the ground. The type and texture of the ground are relatively less important. Within limits, it is relatively insensitive to advance

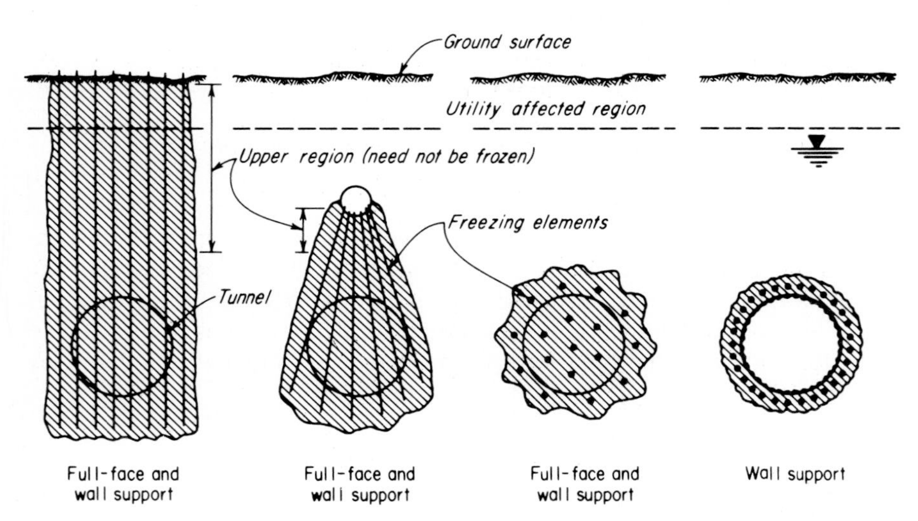

Figure 16-7 Alternate techniques for temporary support of a tunnel heading by freezing. *(Geofreeze.)*

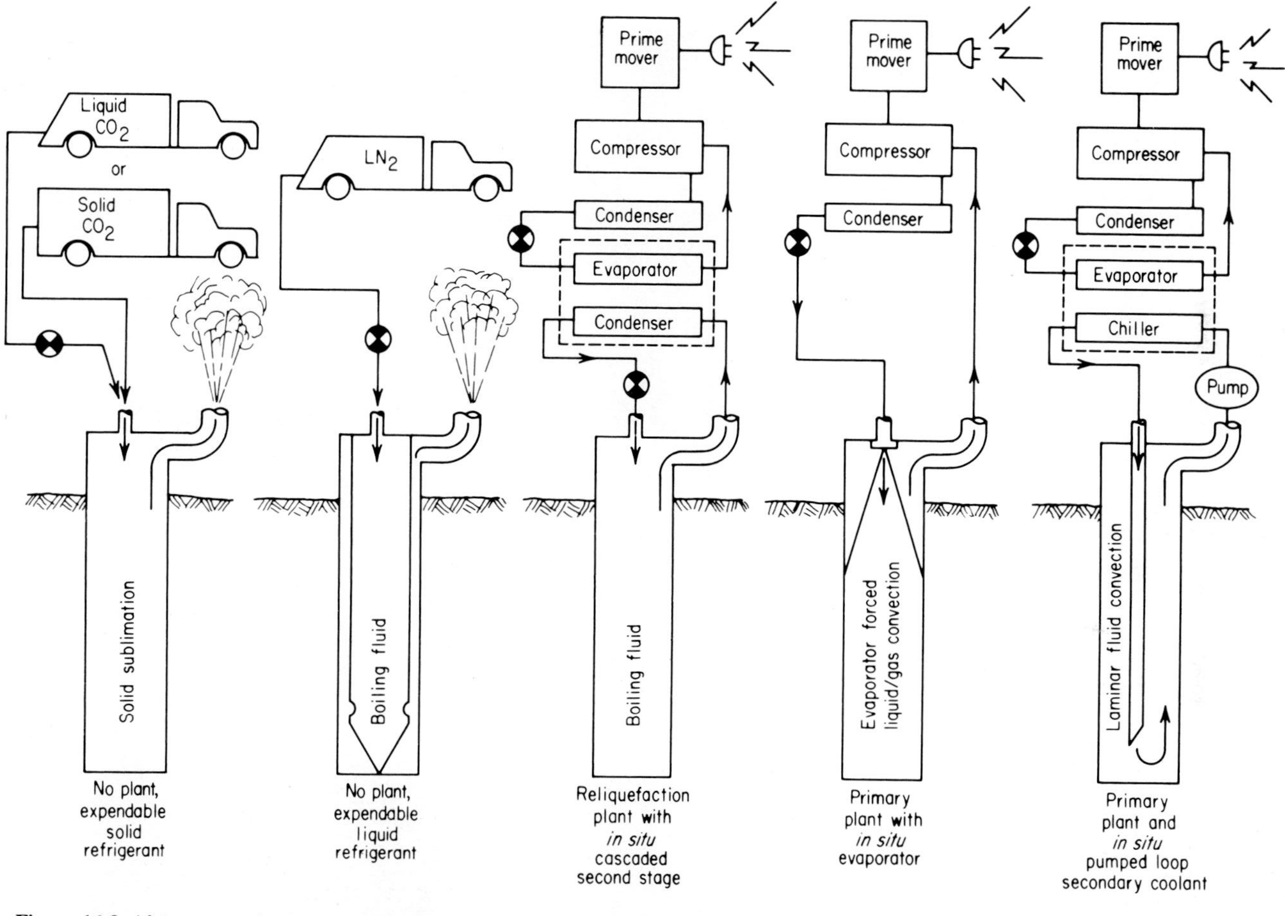

Figure 16-8 Alternate methods of freezing ground (CO_2 is carbon dioxide; LN_2 is liquid nitrogen). *(Geofreeze.)*

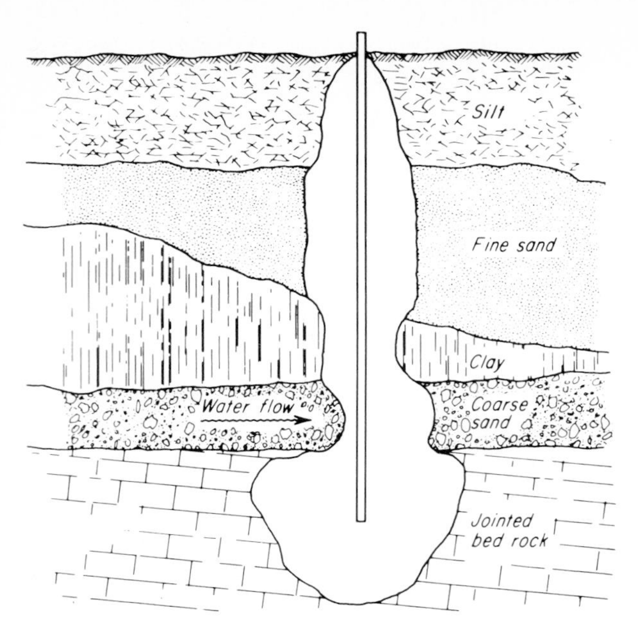

Figure 16-9 Irregular shape of frozen zone in heterogeneous ground. *(Geofreeze.)*

Figure 16-10 Excavation for 115- by 157- by 75-ft-deep pit for nuclear power plant stabilized by freezing. *(Geofreeze.)*

geologic prediction, and by changing the temperature or duration of loading, it will usually be possible to accommodate all types of ground conditions with one type of freezing method. As an example of the flexibility of support provided by freezing the ground, Fig. 16-7 illustrates some of the basic configurations that have been used in tunnel work.

Cost of freezing ground The direct costs of freezing for a specific project will depend largely on the ground conditions, the spacing of the freezing elements, the time available, and the type of refrigeration system used.

The relative economics of ground freezing are dependent on the specific conditions and requirements of a project. To evaluate the relative economics of various alternative methods of temporary ground support, it is necessary to consider their effect on the total project costs rather than on the direct costs of the specific alternative as a single item. In many instances, the direct costs of freezing alone may appear somewhat higher than the direct costs of an alternate method. However, when the added items of work required by the alternate method are considered, the total costs may favor the freezing method.

Alternate methods of freezing the ground As illustrated in Fig. 16-8, there are various methods or systems used for freezing ground. Each method has

Figure 16-11 View of tunnel excavated through frozen ground. *(Geofreeze.)*

advantages and disadvantages compared with other methods, depending on job conditions, the location of the project, and other factors. For a more comprehensive treatment of this subject, several publications are available.

The effect of the properties of the ground on the shape of the frozen zone As illustrated in Fig. 16-9, when heterogeneous ground is frozen, the shape of the frozen zone may be quite irregular. This possibility should be considered in selecting the spacings of the injection pipes.

REFERENCES

1. Weber, A. H.: Correction of Reservoir Leakage at Great Falls Dam, *Proceedings ASCE*, p. 101, January 1950.
2. Chemical Grout Seals Shaft through Wet Sand, *Construction Methods and Equipment*, vol. 44, pp. 140–147, May 1962.
3. Murphy, W. D.: Machine Tunneling under Houston, *Civil Engineering*, vol. 34, pp. 44–45, September 1964.
4. Under Dams or over Subways Chemical Grout Plugs Leaks, *Engineering News-Record*, vol. 171, p. 62, August 8, 1963.
5. Grant, Leland F., and John S. Winefordener: Grouting a Dam Cutoff in Cavernous Limestone, *Journal of the Construction Division, Proceedings ASCE*, vol. 92, pp. 1–15, September 1966.
6. Janin, Jean J., and Guy F. LeSchiellour: Chemical Grouting for Paris Rapid Transit Tunnels, *Journal of the Construction Division, Proceedings ASCE*, vol. 96, pp. 61–74, June 1970.
7. Jaques, W. B.: Stopping Water with Chemical Grout, *Civil Engineering*, vol. 51, pp. 59–62, December 1981.
8. Seltz-Petrash, Ann: Baltimore's Got the Subway Everyone Loves, *Civil Engineering*, vol. 51, pp. 42–45, January 1981.
9. Acker Drill Company, Inc., P. O. Box 830, Scranton, PA 18501.
10. ChemGrout, Inc., P. O. Box 1140, La Grange Park, IL 60625.
11. Celtite, Inc., 13670 York Road, Cleveland, OH 44133.
12. Parish, W. C. Pete, W. H. Baker, and R. M. Rubright: Underpinning with Chemical Grout, *Civil Engineering*, vol. 53, pp. 42–45, August 1983.
13. Geofreeze, P. O. Box 277, Lorton, VA 22079.

CHAPTER

SEVENTEEN

PILES AND PILE-DRIVING EQUIPMENT

INTRODUCTION

This chapter deals with the selection of load-bearing piles and the equipment required to drive the piles.

Load-bearing piles, as the name implies, are used primarily to transmit loads through soil formations with poor supporting properties into or onto formations that are capable of supporting the loads. If the load is transmitted to the soil through skin friction between the surface of the pile and the soil, the pile is called a *friction pile*. If the load is transmitted to the soil through the lower tip, the pile is called an *end-bearing pile*. Many piles depend on a combination of friction and end bearing for their supporting strengths.

TYPES OF PILES

Piles may be classified on the basis of their use or the materials from which they are made. On the basis of use there are two major classifications, *sheet* and *load-bearing*.

Sheet piling is used primarily to resist the flow of water and loose soil. Typical uses include cutoff walls under dams, cofferdams, bulkheads, trench sheeting, etc. On the basis of the materials from which they are made sheet piling may be classified as *steel*, *wood*, and *concrete*.

On the basis of the material from which they are made and the method of constructing and driving them, load-bearing piles may be classified as follows:

1. Timber
   a. Untreated
   b. Treated with a preservative
2. Concrete
   a. Precast
   b. Cast in place
3. Steel
   a. H section
   b. Steel pipe
4. Composite

Each type of load-bearing pile has a place in the field of construction, and for some projects more than one type may seem satisfactory. It is the duty of the engineer to select that type of pile which is most satisfactory for a given project, considering all the factors that affect the selection. Among the factors that will influence his decision are the following:

1. Type, size, and weight of the structure to be supported
2. Physical properties of the soil at the site
3. Depth to a stratum capable of supporting the piles
4. Possibility of variations in the depth to a supporting stratum
5. Availability of materials for piles
6. Number of piles required
7. Facilities for driving piles
8. Comparative costs in place
9. Durability required
10. Types of structures adjacent to the project
11. Depth and kind of water, if any, above the ground into which the piles will be driven

To illustrate the effect which these factors have on the selection of types of piles, consider factor 4. If soil borings at the site of a project indicate that the depth to a stratum capable of supporting piles varies considerably, precast concrete piles should not be selected. Regardless of other desirable factors the difficulty and expense of increasing or decreasing the length of such piles should eliminate them from consideration. If concrete piles are desired, one of the cast-in-place types should be selected.

TIMBER PILES

Timber piles are made from the trunks of trees. While such piles are available in most sections of the nation and the world, it is becoming more difficult to obtain long, straight timber piles. Pine piles are reasonably available in lengths up to 60 ft, while Douglas fir piles are available in lengths in excess of 100 ft from the Pacific Northwest.

Among the advantages of timber piles are the following:

1. The more popular lengths and sizes are available on short notice.
2. They are economical in cost.
3. They are handled easily, with little danger of breakage.
4. They can be cut off to any desired length after they are driven.
5. They can be pulled easily in the event removal is necessary.

Among the disadvantages of timber piles are the following:

1. It may be difficult to obtain piles sufficiently long and straight for some projects.
2. It may be difficult or impossible to drive them into hard formations.
3. It is difficult to splice them to increase their lengths.
4. While they are satisfactory when used as friction piles, they are not suitable for use as end-bearing piles under heavy loads.
5. The length of life may be short unless the piles are treated with a preservative.

PRECAST CONCRETE PILES

Square and octagonal piles are cast in horizontal forms, while round piles are cast in vertical forms. After the piles are cast, they should be cured under damp sand, straw, or mats for the period required by the specifications, frequently 21 days, if cured at ambient temperatures.

With the exception of short lengths, precast concrete piles must be reinforced with sufficient steel to prevent damage or breakage while they are being handled from the casting beds to the driving positions. The Foundation Code of the City of New York, adopted in 1948, specifies that precast concrete piles shall contain longitudinal reinforcing steel in an amount not less than 2 percent of the volume of a pile. Lateral steel shall be at least $\frac{1}{4}$-in.-diameter round bars, spaced not more than 12 in. apart, except at the top and bottom 3 ft of a pile, where the spacing shall not exceed 3 in. The concrete cover over the reinforcing steel shall be at least 2 in. Figure 17-1 shows typical details of precast concrete piles.

Concrete piles should be cast as near the site of use as possible in order to reduce the cost of handling them from the casting beds to the pile driver. In the event it is necessary to transport them to a driver, this may be done by a truck. For handling concrete piles, care must be exercised to prevent breakage or damage due to flexural stresses. Long piles should be picked up at several points to reduce the unsupported lengths.

Concrete piles may be cast in any desired sizes and lengths. Those used in

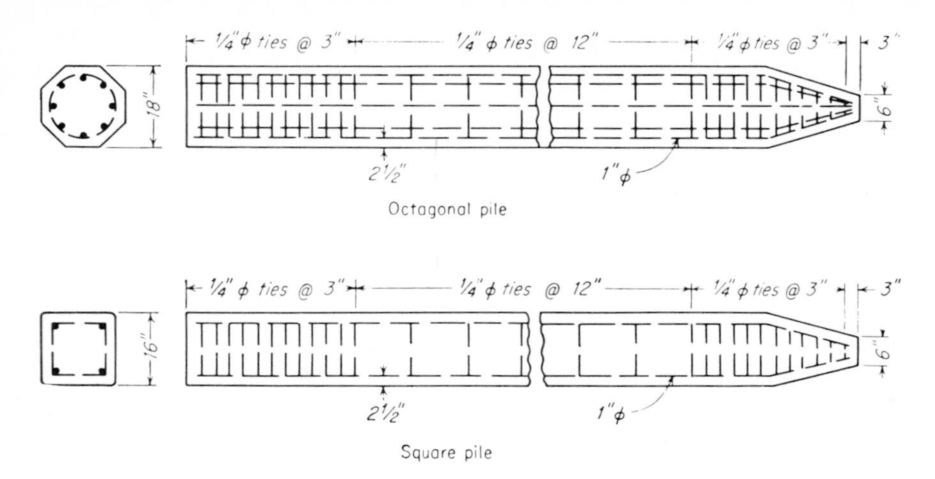

Figure 17-1 Typical details of precast concrete piles.

constructing the Morganza Floodway on the Mississippi River were square and octagonal in cross section, 20 in. wide, and varied in length to more than 100 ft. Almost 360,000 lin ft of piles was driven on this project. The piles were cast in prefabricated steel forms, loaded on railroad cars by a gantry crane with a 135-ft span, and hauled to the driving site, where they were driven by special

Figure 17-2 Preparing forms and reinforcing for precast concrete piles.

Figure 17-3 Crawler-mounted pile-driving rigs. *(Raymond International Builders, Inc.)*

rigs. For a project as big as this the large investment in special equipment is justified, but for a small project the required investment in equipment probably would make the cost prohibitive. Thus, the maximum-size concrete piles that can be used on a project may be determined by economy.

One of the disadvantages of using precast concrete piles, especially for a project where different lengths are required, is the difficulty of reducing or increasing the lengths of the piles.

If a pile proves to be too long, it is necessary to cut off the excess length. This is done, after a pile is driven to its maximum penetration, by chipping the concrete away from the reinforcing steel, cutting the reinforcing with a gas torch, then removing the surplus length of the concrete core. This operation represents a waste of material and time, which can be very expensive.

When a precast concrete pile does not develop sufficient driving resistance to support the design load, it may be necessary to increase the length and drive the pile to a greater depth. Unless the reinforcing bars extend above the top of a pile, it will be necessary to chip the concrete back far enough to permit additional reinforcing to be welded to the original longitudinal bars. Then the concrete is placed for the added length.

Among the advantages of precast concrete piles are the following:

1. They have high resistance to chemical and biological attacks.
2. They have great strength.
3. A pipe may be installed along the center of a pile to facilitate jetting.

Among the disadvantages of precast concrete piles are the following:

1. It is difficult to reduce or increase the length.
2. Large sizes require heavy and expensive handling and driving equipment.
3. Inability to obtain piles by purchase may delay the starting of a project.
4. Possible breakage of piles during handling or driving produces a delay hazard.

CAST-IN-PLACE CONCRETE PILES

As the name implies, cast-in-place concrete piles are constructed by depositing the freshly mixed concrete in place in the ground and letting it cure there. The two principal methods of constructing such piles are:

1. Driving a metallic shell, leaving it in the ground, and filling it with concrete
2. Driving a metallic shell and filling it with concrete as the shell is pulled from the ground

There are several modifications for each of the two methods.

The more commonly used piles constructed by these two methods are described in the following sections.

RAYMOND STEP-TAPER CONCRETE PILES

The step-taper pile is installed by driving a spirally corrugated steel shell, made up of sections 4, 8, 12, and 16 ft long, with successive increases in diameter for each section. A corrugated sleeve at the bottom of each section is screwed into the top of the section immediately below it. Piles of the necessary length, up to a maximum of 80 ft, are obtained by joining the proper number of sections at

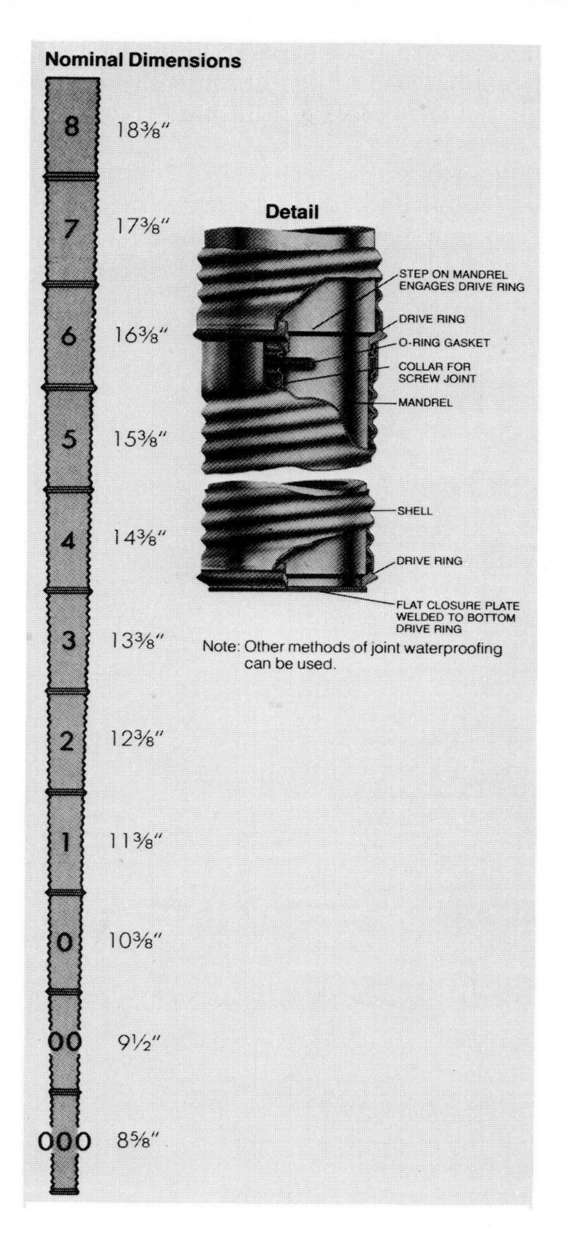

STEP-TAPER PILES

Figure 17-4 Dimensions and detail for Raymond Step-taper piles.

the job. The shells are available in various gauges of metal to fit different job conditions. The bottom of the shell, whose diameter can be varied from $8\frac{5}{8}$ to $13\frac{3}{8}$ in., is closed prior to driving by a flat steel plate or a hemispherical steel boot.

After a shell is assembled in the desired length, a step-tapered rigid-steel core is inserted and the shell is driven to the desired penetration. The core is removed, and the shell is filled with concrete. Figure 17-4 gives the dimensions

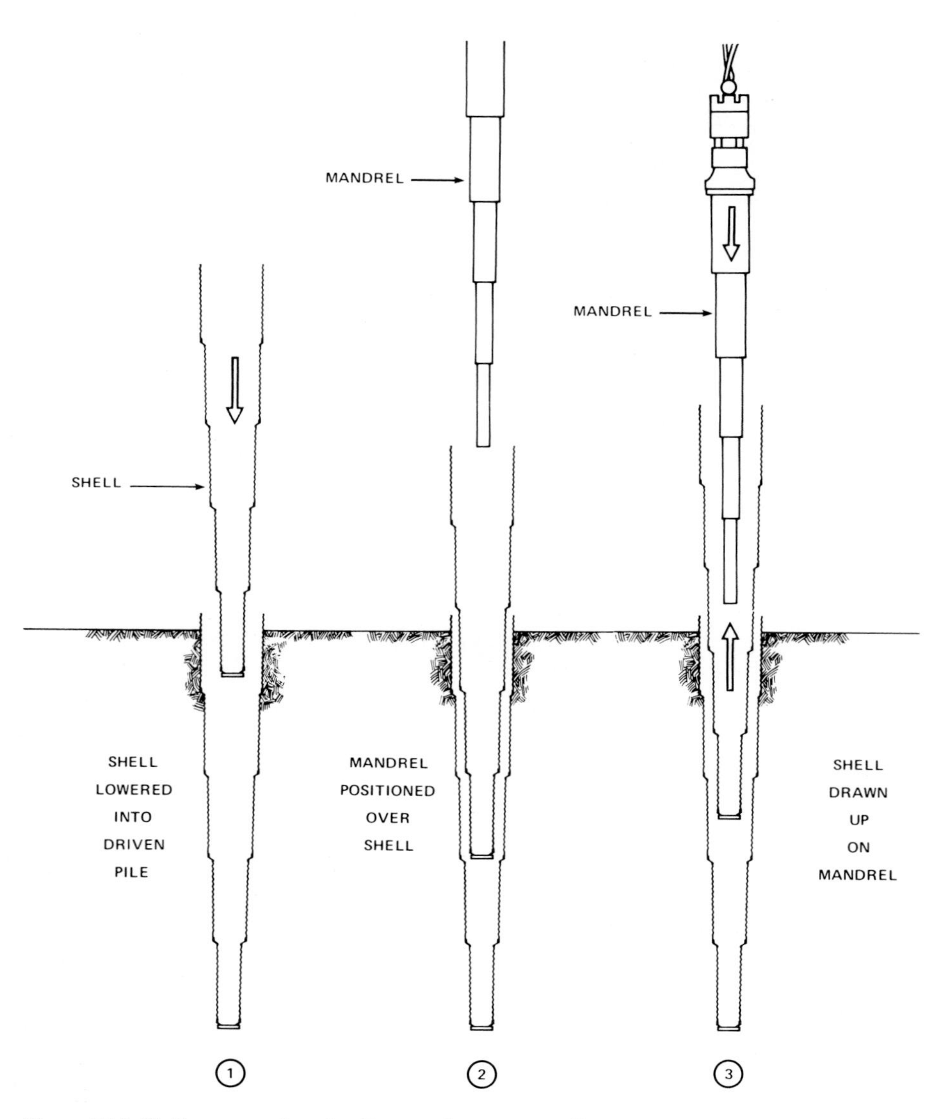

Figure 17-5 Shell-up procedure for Raymond step-taper piles.

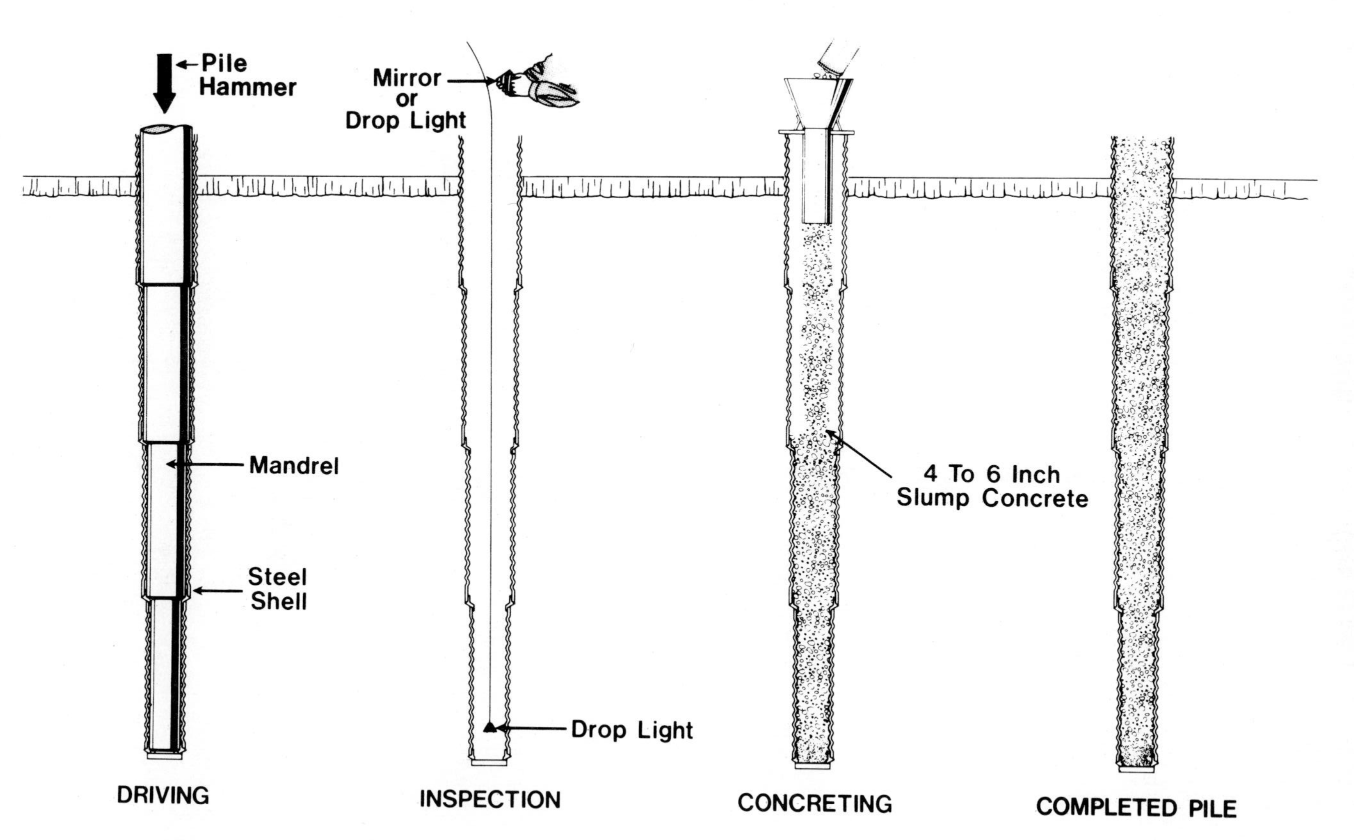

Figure 17-6 Installation sequence for Raymond step-taper piles.

of a step-taper pile shell. Figure 17-5 illustrates steps in driving these piles. Figure 17-6 shows the installation sequence for these piles.

MONOTUBE PILES

The monotube pile is obtained by driving a fluted, tapered steel shell, closed at the tip with an 8-in.-diameter driving point, to the desired penetration. The

Figure 17-7 Driving No. 5 gauge 12 in. Monotube piles up to 200 ft long with a 37,000 ft-lb steam hammer. *(The Union Metal Manufacturing Company.)*

Table 17-1 Data on Monotube piles

Type	Size, point diameter × butt diameter × length	Weight, lb per ft				Volume of concrete, cu yd
		9 ga	7 ga	5 ga	3 ga	
F	$8\frac{1}{2}'' \times 12'' \times 25'$	17	20	24	28	0.43
Taper	$8'' \times 12'' \times 30'$	16	20	23	27	0.55
0.14 in.	$8\frac{1}{2}'' \times 14'' \times 40'$	19	22	26	31	0.95
per ft	$8'' \times 16'' \times 60'$	20	24	28	33	1.68
	$8'' \times 18'' \times 75'$		26	31	35	2.59
J	$8'' \times 12'' \times 17'$	17	20	23	27	0.32
Taper	$8'' \times 14'' \times 25'$	18	22	26	30	0.58
0.25 in.	$8'' \times 16'' \times 33'$	20	24	28	32	0.95
per ft	$8'' \times 18'' \times 40'$		26	30	35	1.37
Y	$8'' \times 12'' \times 10'$	17	20	24	28	0.18
Taper	$8'' \times 14'' \times 15'$	19	22	26	30	0.34
0.40 in.	$8'' \times 16'' \times 20'$	20	24	28	33	0.56
per ft	$8'' \times 18'' \times 25'$		26	31	35	0.86

Extensions (Overall lengths 1 ft greater than indicated)

Type	Diameter × length	9 ga	7 ga	5 ga	3 ga	cu yd per ft
N 12	$12'' \times 12'' \times 20'/40'$	20	24	28	33	0.026
N 14	$14'' \times 14'' \times 20'/40'$	24	29	34	41	0.035
N 16	$16'' \times 16'' \times 20'/40'$	28	33	39	46	0.045
N 18	$18'' \times 18'' \times 20'/40'$		38	44	52	0.058

shell is driven without a mandrel, inspected, and filled with concrete. Any desired length of shell, up to approximately 125 ft, may be obtained by welding extensions to a standard-length shell.

Table 17-1 gives dimensions and other information on one type of tube. Figure 17-7 illustrates a rig driving one of these piles; Fig. 17-7(*a*) illustrates these piles in position to support a bridge.

FRANKI PRESSURE-INJECTED FOOTINGS

When a structure is located at a site whose upper soil formations are unable to provide the required load-supporting capacities, it is necessary to transfer the loads to a deeper formation with adequate capacity. The Franki pressure-injected footing technique provides one method of obtaining the required supporting capacity. The steps in providing this capacity are illustrated in Fig. 17-8.

As illustrated in Fig. 17-8(*a*), a steel-drive tube of the desired diameter and length is fitted with an expendable steel boot to close the bottom end. Using a

Figure 17-7a Monotube piles used to support a bridge structure. *(Union Metal Manufacturing Company.)*

pile-driving hammer of adequate size, usually with 50,000 to 100,000 ft-lb per blow, the tube is driven to the desired depth. As illustrated in Fig. 17-8(*b*), after the tube is driven to the desired depth, the hammer is removed and a charge of dry concrete is dropped into the tube and compacted with a drop hammer to form a compact watertight plug. Following step (*b*) the tube is raised slightly and held in that position while repeated blows of the drop hammer expel the plug of concrete from the lower end of the tube, leaving in the tube sufficient concrete to prevent the intrusion of water or soil. Additional concrete is dropped into the tube and forced by the drop hammer to flow out of the bottom of the tube to form an enlarged bulb or pedestal, as illustrated by step (*c*).

The final step in this operation consists of raising the tube in increments and dropping additional dry concrete into the tube, as illustrated in Fig. 17-8(*d*). This concrete is hammered with sufficient energy to force it to flow out of the tube and fill the exposed hole. This operation is repeated until the hole is filled to the desired elevation.

The shaft may be reinforced by inserting a spirally wound steel cage, inbedded in the bulb and extended upward for the full height of the shaft. The placing of the concrete is accomplished as described heretofore.

Other methods of providing these footings are available.

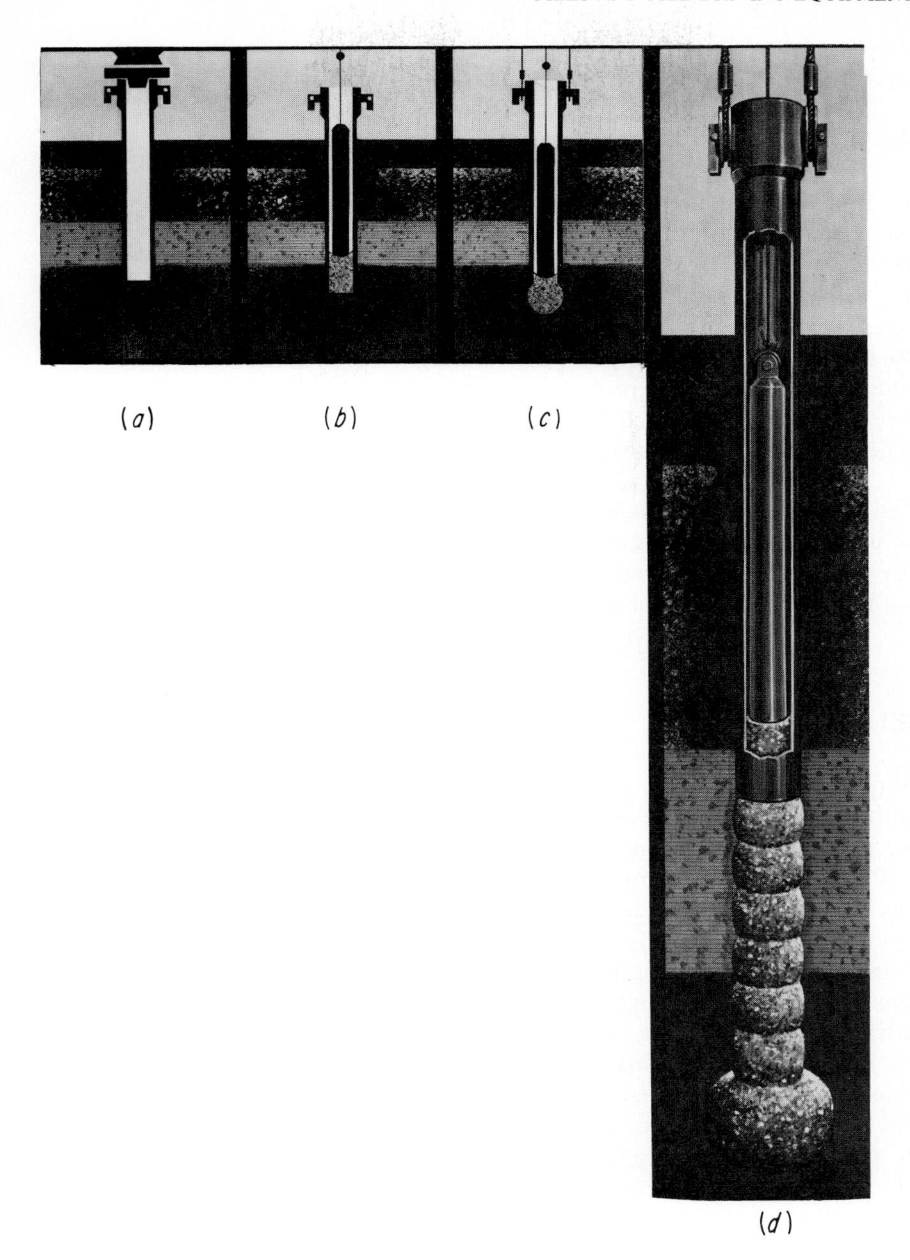

Figure 17-8 Steps in providing Franki pressure injected footings: (*a*) The drive tube driven to the desired depth. (*b*) Concrete dropped into the drive tube and compacted, (*c*) Drop hammer expels concrete from tube. (*d*) Additional steps complete the footing.

ADVANTAGES AND DISADVANTAGES OF CAST-IN-PLACE CONCRETE PILES

Among the advantages of cast-in-place concrete piles are the following:

1. The lightweight shells may be handled and driven easily.
2. Variations in length do not present a serious problem. The length of a shell may be increased or decreased easily.
3. The shells may be shipped in short lengths and assembled at the job.
4. Excess reinforcing, to resist stresses caused by handling only, is eliminated.
5. The danger of breaking a pile while driving is eliminated.
6. Additional piles may be provided quickly if they are needed.

Among the disadvantages of cast-in-place concrete piles are the following:

1. A slight movement of the earth around an unreinforced pile may break it.
2. An uplifting force, acting on the shaft of an uncased and unreinforced pile, may cause it to fail in tension.
3. The bottom of a pedestal pile may not be symmetrical.

STEEL-PIPE PILES

These piles are installed by driving pipes to the desired depth and, if desired, filling them with concrete. A pipe may be driven with the lower end closed with a plate or a steel driving point, or the pipe may be driven with the lower end open. Pipes varying in diameter from 6 to 30 in. or more have been driven in lengths varying from a few feet to several hundred feet.

A closed-end pipe pile is driven in any conventional manner, usually with a pile hammer. If it is necessary to increase the length of a pile, two or more sections may be welded together or sections may be connected by using an inside sleeve for each joint. This type of pile is particularly advantageous for use on jobs when the headroom for driving is limited and short sections must be added to obtain the desired total length.

An open-end pipe pile is installed by driving the pipe to the required depth, removing the material from inside, and filling the space with concrete. Because open-end pipe piles offer less driving resistance than closed-end piles, a smaller pile hammer may be used. The use of a light hammer is desirable when piles are driven near a structure whose foundation might be damaged by the impact of the blows from a large hammer. Open-end piles may be driven to depths which could never be reached with closed-end piles.

After a pile is driven to the desired depth, the material inside is removed by bursts of compressed air, a mixture of water and compressed air, an earth auger, or a small orange-peel bucket; then the pipe is filled with concrete.

STEEL PILES

In constructing foundations that require piles driven to great depths, steel H piles probably are more suitable than any other type. Steel piles may be driven through hard materials to a specified depth to eliminate the danger of failure due to scouring, such as under a pier in a river. Also, steel piles may be driven to great depths through poor soils to bear on a solid rock stratum. The great strength of steel combined with the small displacement of soil permits a large portion of the energy from a pile hammer to be transmitted to the bottom of a pile. As a result, it is possible to drive steel piles into soils which could not be penetrated by any other type of pile. However, in spite of the great strength of these piles the author has seen jobs where it was necessary to drill pilot holes into compacted sand ahead of steel H piles in order to obtain the specified penetration. By weld-splicing sections together, lengths in excess of 200 ft have been driven.

THE RESISTANCE OF PILES TO PENETRATION

In general, the forces which enable a pile to support a load also cause the pile to resist the efforts made to drive it. The total resistance of a pile to penetration will equal the sum of the forces produced by skin friction and end bearing. The portion of the resistance supplied by either skin friction or end bearing may vary from almost 0 to 100 percent, depending on the soil more than on the type of pile. A steel H pile driven to refusal in stiff clay should be classified as a skin-friction pile, while the same pile driven through a mud deposit to rest on solid rock should be classified as an end-bearing pile.

Numerous tests have been conducted to determine values for skin friction for various types of piles and soils. A representative value for skin friction can be obtained by determining the total force required to pull a pile up slightly, using hydraulic jacks with calibrated pressure gauges.

The value of the skin friction is a function of the coefficient of friction between the pile and the soil and the pressure of the soil normal to the surface of the pile, or for a soil such as some types of clay the value of the skin friction may be limited to the shearing strength of the soil immediately adjacent to the pile. Consider a concrete pile driven into a soil that produces a normal pressure of 100 psi on the vertical surface of the pile. This is not an unusually high pressure for certain soils such as compacted sand. If the coefficient of friction is 0.25, the value of the skin friction will be $0.25 \times 100 \times 144 = 3{,}600$ psf. Table 17-2 gives representative values of skin friction on piles. The author of the table states that the information is intended as a qualitative guide, not as correct information to be used in any and all cases.

The magnitude of end-bearing pressure can be determined by driving a button-bottom-type pile and leaving the driving casing in place. A second steel

Figure 17-9 Diesel hammer driving steel H pile on 1:1 batter. *(Pileco, Inc.)*

Table 17-2 Approximate allowable value of skin friction on piles*

Material	Skin friction, psf (kg/sq m) — Approximate depth: 20 ft (6.1 m)	60 ft (18.3 m)	100 ft (30.5 m)
Soft silt and dense muck	50–100 (244–488)	50–120 (244–586)	60–150 (273–738)
Silt (wet but confined)	100–200 (488–976)	125–250 (610–1,220)	150–300 (738–1,476)
Soft clay	200–300 (976–1,464)	250–350 (1,220–1,710)	300–400 (1,476–1,952)
Stiff clay	300–500 (1,464–2,440)	350–550 (1,710–2,685)	400–600 (1,952–2,928)
Clay and sand mixed	300–500 (1,464–2,440)	400–600 (1,952–2,928)	500–700 (2,440–3,416)
Fine sand (wet but confined)	300–400 (1,464–1,952)	350–500 (1,710–2,440)	400–600 (1,952–2,928)
Medium sand and small gravel	500–700 (2,440–3,416)	600–800 (2,928–3,904)	600–800 (2,928–3,904)

* Some allowance is made for the effect of using piles in small groups.

pipe, slightly smaller than the driving casing, is lowered onto the concrete button. The force, applied through the second pipe, required to drive the button into the soil is a direct measure of the supporting strength of the soil. This is true because there is no skin friction on the inside pipe.

PILE HAMMERS

The function of a pile hammer is to furnish the energy required to drive a pile. Pile-driving hammers are designated by type and size.

The types commonly used include the following:

1. Drop
2. Single-acting steam
3. Double-acting steam
4. Differential-acting steam
5. Diesel
6. Vibratory
7. Hydraulic

The size of a drop hammer is designated by its weight, while the size of each of the other hammers is designated by the theoretical energy per blow, expressed in foot-pounds.

For each type of hammer listed the driving energy is supplied by a falling mass, which strikes the top of a pile. The various types are described in the following sections.

Drop hammers A drop hammer is a heavy metal weight that is lifted by a rope, then released and allowed to fall on top of the pile. The hammer may be released by a trip and fall freely, or it may be released by loosening the friction band on the hoisting drum and permitting the weight of the hammer to unwind the rope from the drum. The latter type of release reduces the effective energy of a hammer because of the friction loss in the drum and rope. Leads are used to hold the pile in position and to guide the movement of the hammer so that it will strike the pile with a solid blow.

Standard drop hammers are made in sizes which vary from about 500 to 3,000 lb. The height of drop or fall most frequently used varies from 5 to 20 ft. When a large energy per blow is required to drive a pile, it is better to use a heavy hammer with a small drop than a light hammer with a large drop.

Drop hammers are suitable for driving piles on remote projects which require only a few piles and for which the time of completion is not an important factor.

Among the advantages of drop hammers are the following:

1. Small investment in equipment

2. Simplicity of operation
3. Ability to vary the energy per blow by varying the height of fall

Among the disadvantages of drop hammers are the following:

1. Slow rate of driving piles
2. Danger of damaging piles by lifting a hammer too high
3. Danger of damaging adjacent buildings as a result of the heavy vibration caused by a hammer
4. Cannot be used directly for underwater driving

Single-acting steam hammers A single-acting steam hammer is a freely falling

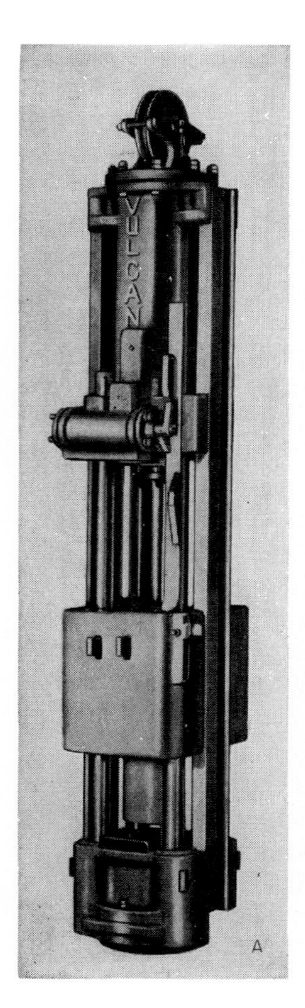

Figure 17-10 Single-acting steam hammers. (*a*) Open type *(Vulcan Iron Works)*. (*b*) Enclosed type *(MKT Geotechnical Systems)*.

weight, called a ram, which is lifted by steam or compressed air, whose pressure is applied to the underside of a piston that is connected to the ram through a piston rod. When the piston reaches the top of the stroke, the steam pressure is released and the ram falls freely to strike the top of a pile. The energy supplied by this type hammer is delivered by a heavy weight striking with a lower velocity, due to the relatively low fall. Whereas a drop hammer may strike four to eight blows per minute, a single-acting steam hammer will strike 50 or more blows per minute when delivering the same energy per blow.

Single-acting steam hammers may be open or enclosed. Figure 17-10 shows one of each type. This hammer is available in sizes varying from a few hundred to more than 30,000 ft-lb of energy per blow. Table 17-3 gives data on several of the more popular sizes of single-acting steam hammers. The length of the stroke and the energy per blow for this type of hammer may be decreased slightly by reducing the steam pressure below that recommended by the

Figure 17-11 Single-acting steam hammer driving steel pile shell. *(The Union Metal Manufacturing Company.)*

Table 17-3 Specifications for impact-type pile drivers

Make and model	Type[a]	Rated energy, ft·lb	Speed, blows per min	Weight of ram, lb	Maximum stroke, in	Boiler hp (ASME)	Air or steam, psi	Net weight, lb
BOLT CPD-15	Air-cushion	30,000	55	14,200	36		250	30,000
	Single action		85		23.5		120	
BSP								
BSP B15	Dbl. Act. Diesel	26,200	80–100	3,300	100[d]			9,000
BSP B25	Dbl. Act. Diesel	45,700	800–100	5,500	100[d]			15,200
BSP B35	Dbl. Act. Diesel	63,900	80–100	7,700	100[d]			21,200
BSP B45	Dbl. Act. Diesel	80,000	80–100	10,000	100[d]			27,500
BSP 500N	Dbl. Act. Air	1,200	330	200	9		90	2,000
BSP 600N	Dbl. Act. Air	3,000	250	500	12		90	3,800
BSP 700N	Dbl. Act. Air	4,700	225	850	13		90	6,500
BSP 900	Dbl. Act. Air/Steam	8,750	145	1,600	17	85	90	7,100
BSP 1000	Dbl. Act. Air/Steam	13,100	105	3,000	19	104	90	10,850
BSP 1100	Dbl. Act. Air/Steam	19,150	95	5,000	19	126	90	14,000
CONMACO								
300	Sgl. Act. Air/Steam	90,000	55	30,000	36	247	150	55,390
200	Sgl. Act. Air/Steam	60,000	60	20,000	36	217	120	44,560
160	Sgl. Act. Air/Steam	48,750	60	16,250	36	198	120	33,200
140	Sgl. Act. Air/Steam	42,000	60	14,000	36	179	110	30,750
160D	Differ. Air/Steam	41,280	103	16,000	$15\frac{1}{2}$	237	160	35,400
125	Sgl. Act. Air/Steam	40,625	50	12,500	39	120	125	21,940
115	Sgl. Act. Air/Steam	37,375	50	11,500	39	99	120	20,780
140D	Differ. Air/Steam	36,000	103	14,000	$15\frac{1}{2}$	211	140	31,200
100	Sgl. Act. Air/Steam	32,500	50	10,000	39	85	100	19,280
80	Sgl. Act. Air/Steam	26,000	50	8,000	39	75	85	17,280
65	Sgl. Act. Air/Steam	19,500	60	6,500	36	67	100	11,200
50	Sgl. Act. Air/Steam	15,000	60	5,000	36	56	80	9,700
DELMAG[e,f]								
D-55		62,500 to 117,175	36–47	11,860	137			26,300
D-46-02[e,f]		48,400 to 105,000	37–53	10,100	128			19,900
D-44		43,500 to 87,000	37–56	9,460	137			22,440
D-36-02		38,000 to 83,100	37–53	7,940	128			17,700
D-30-05[e,f]		31,800 to 62,900	38–52	6,600	127			13,150
D-30		23,800 to 54,250	39–60	6,600	126			12,350
D-22-02[e,f]		24,600 to 48,400	38–52	4,850	127			11,400
D-22		39,700	40–60	4,850	127			11,100
D-15		27,100	40–60	3,300	132			6,600
D-12		22,500	40–60	2,750	130			6,050
D- 5		9,100	42–60	1,100	116			2,730
D- 4		3,630	50–60	836	52			1,360
D- 2		1,815	60–70	484	49			792
FOSTER/KOBE								
K150	Sgl. Act. Diesel	298,000	42–60	33,100	108			87,000
K60	Sgl. Act. Diesel	105,600	42–60	13,200	112			37,500
K45	Sgl. Act. Diesel	91,100	39–60	9,920	112			25,600
K42	Sgl. Act. Diesel	79,000	40–60	9,260	102			24,000
K35	Sgl. Act. Diesel	70,800	39–60	7,720	112			18,700
K32	Sgl. Act. Diesel	60,100	40–60	7,050	102			17,750
K25	Sgl. Act. Diesel	50,700	39–60	5,510	112			13,100
K22	Sgl. Act. Diesel	41,300	40–60	4,850	102			12,350
K13	Sgl. Act. Diesel	25,200	34–60	2,870	106			8,000
LINK-BELT								
660	Dbl. Act. Diesel	45,000	80–84	7,564	71[d]			23,500
520	Dbl. Act. Diesel	26,300	80–84	5,070	63[d]			12,545
440	Dbl. Act. Diesel	18,200	86–90	4,000	56[d]			9,839
180	Dbl. Act. Diesel	8,100	90–95	1,725	57[d]			4,546

NOTE: A diesel pile hammer has a variable stroke and hence a variable energy rating based upon the amount of resistance built up in the pile and the amount of compression in the cylinder.

[a] Type "U" indicates hammer is suitable for underwater driving. Type "E" indicates hammer can be converted to extractor.

[b] MKT air/steam hammers are listed with flat anvils. Vulcan hammers are listed with standard base, no drive cap. All diesels are listed without drive caps.

[c] MKT DA 35 diesel is convertible from single to double acting. Rated energy is 21,000 ft-lb double-acting, and 35,000 ft-lb single-acting.

[d] Equivalent ram stroke.

[e] Models showing two energy figures are equipped with adjustable fuel pumps—the numbers indicate the minimum and maximum settings. (Intermediate steps not shown.)

[f] Require revised "U" cylinder to operate in box leads.

Table 17-3 Specifications for impact-type pile drivers (*Continued*)

	Type[a]	Rated energy, ft·lb	Speed, blows per min	Weight of ram, lb	Maximum stroke, in	Boiler hp (ASME)	Air steam, psi	Net weight, lb
MENCK								
MRBS 8000	Sgl. Act. Air Steam	867,960	32	176,365	59	1,500	156	330,695
MRBS 4600	Sgl. Act. Air Steam	499,070	36	101,410	59	850	156	176,371
MRBS 3000	Sgl. Act. Air Steam	325,480	40	66,135	59	520	156	108,025
MRBS 1800	Sgl. Act. Air Steam	189,850	40	38,580	59	320	156	64,596
MRBS 850	Sgl. Act. Air Steam	93,340	40	18,960	59	160	156	27,800
MITSUBISHI								
MB-70	Sgl. Act. Diesel	141,000	38–60	15,840	102			46,000
MH-45	Sgl. Act. Diesel	84,300	42–60	9,920	102			24,500
M-43	Sgl. Act. Diesel	84,000	40–60	9,460	102			22,660
MH-35	Sgl. Act. Diesel	65,900	42–60	7,720	102			18,500
M-33	Sgl. Act. Diesel	64,000	40–60	7,260	102			16,940
MH-25	Sgl. Act. Diesel	46,900	42–60	5,510	102			13,200
M-23	Sgl. Act. Diesel	45,000	42–60	5,060	102			11,220
MH-15	Sgl. Act. Diesel	28,100	42–60	3,310	102			8,400
M-14S	Sgl. Act. Diesel	26,000	42–60	2,970	102			7,260
MKT[b]								
DE70B	Sgl. Act. Diesel	63,000	40–50	7,000	126			15,400
S20	U Sgl. Act. Air Steam	60,000	60	20,000	36	190	150	38,650
MRBS-50	Sgl. Act. Air Steam	46,350	40	11,300			115	15,550
DE-50B	Sgl. Act. Diesel	45,000	40–50	5,000	126			12,250
DA55B	Sgl. Act. Diesel	45,000	40–50	5,000	126			17,000
DA55B	Dbl. Act. Diesel	38,200	78–82	5,000				17,000
S14-B	U Sgl. Act. Air Steam	37,500	60	14,000	32	155	100	31,700
S10	U Sgl. Act. Air Steam	32,500	55	10,000	39	130	80	22,380
DA35[c]	Sgl. Act. Diesel	25,200	40–50	2,800	150			10,000
DE30B	Sgl. Act. Diesel	25,200	45–50	2,800	129			7,500
DA35[c]	Dbl. Act. Diesel	21,000	78–82	2,800				10,000
11B3	U Dbl. Act. Air Steam	19,150	95	5,000	19	128	100	14,000
DE20	Sgl. Act. Diesel	16,000	40–50	2,000	111			5,375
10B3	U Sgl. Act. Air Steam	13,100	105	3,000	19	104	100	10,850
9B3	U Dbl. Act. Air Steam	8,750	145	1,600	17	85	100	7,000
7	E Dbl. Act. Air Steam	4,150	225	800	$9\frac{1}{2}$	65	100	5,000
6	E Dbl. Act. Air Steam	2,500	275	400	$8\frac{1}{4}$	45	100	2,900
5	E Dbl. Act. Air Steam	1,000	300	200	7	35	100	1,500
VULCAN[b]								
3100	Sgl. Act. Air/Steam	300,000	58	100,000	36	1,021	130	174,500
560	Sgl. Act. Air/Steam	300,000	45	62,500	60	875	150	134,060
060	Sgl. Act. Air/Steam	180,000	62	60,000	36	750	130	125,000
540	Sgl. Act. Air/Steam	200,000	48	40,900	60	635	130	102,980
040	Sgl. Act. Air/Steam	120,000	60	40,000	36	600	120	98,000
400C	Sgl. Act. Air/Steam	113,488	100	40,000	$16\frac{1}{2}$	700	150	91,180
030	Sgl. Act. Air/Steam	90,000	55	30,000	36	247	150	55,410
020	Sgl. Act. Air/Steam	60,000	60	20,000	36	217	120	43,785
200C	Dbl. Act. Air/Steam	50,200	98	20,000	$15\frac{1}{2}$	260	142	39,000
016	Sgl. Act. Air/Steam	48,750	60	16,250	36	210	120	33,340
014	Sgl. Act. Air/Steam	42,000	60	14,000	36	200	110	29,590
140C	Dbl. Act. Air/Steam	36,000	103	14,000	$15\frac{1}{2}$	211	140	27,984
010	Sgl. Act. Air/Steam	32,500	50	10,000	39	157	105	19,500
0R	Sgl. Act. Air/Steam	30,225	80	9,300	39	140	100	18,050
08	Sgl. Act. Air/Steam	26,000	50	8,000	39	127	83	16,750
80C	Dbl. Act. Air/Steam	24,450	111	8,000	$16\frac{1}{2}$	180	120	17,885
0	Sgl. Act. Air/Steam	24,375	50	7,500	39	128	80	16,250
06(106)	Sgl. Act. Air/Steam	19,500	60	6,500	36	94	100	11,200
65C	Dbl. Act. Air/Steam	19,200	117	6,500	$15\frac{1}{2}$	152	150	14,886
50C	Dbl. Act. Air/Steam	15,100	120	5,000	$15\frac{1}{2}$	125	120	11,782
1(106)	Sgl. Act. Air/Steam	15,000	60	5,000	36	81	80	9,700
30C	Dbl. Act. Air/Steam	7,260	133	3,000	$10\frac{1}{2}$	40	120	7,036
2	Sgl. Act. Air/Steam	7,260	70	3,000	29	49	80	6,700
DGH-900	Dbl. Act. Air/Steam	5,750	360	900	10	115	178	5,000
18C	Dbl. Act. Air/Steam	3,600	150	1,800	$10\frac{1}{2}$	45	120	4,139
DGH-100D	Dbl. Act. Air/Steam	643	505	100	6		100	786
5100	Sgl. Act. Air/Steam	500,000	48	100,000	60	1,400	150	197,000
4250	Sgl. Act. Air/Steam	1,000,000	54	250,000	48	3,000	150	400,000

manufacturer. The reduced pressure has the effect of decreasing the height to which the piston will rise before it begins its free fall.

Among the advantages of single-acting steam compared with drop hammers are the following:

1. Greater number of blows per minute permits faster driving.
2. Greater frequency of blows reduces the increase in skin friction between blows.
3. Heavier ram falling at lower velocity transmits a greater portion of the energy to driving piles.
4. Reduction in the velocity of the ram decreases the danger of damage to piles during driving.
5. The enclosed types may be used for underwater driving.

Among the disadvantages of single-acting steam hammers compared with drop hammers are the following:

1. They require more investment in equipment such as a steam boiler or an air compressor.
2. They are more complicated, with higher maintenance cost.
3. They require more time to set up and take down.
4. They require a larger operating crew.

Double-acting steam hammers In the double-acting steam hammer steam pressure is applied to the underside of the piston to raise the ram; then during the downward stroke steam is applied to the top side of the piston to increase the energy per blow. Thus, with a given weight ram, it is possible to attain a desired amount of energy per blow with a shorter stroke than with a single-acting hammer. The number of blows per minute will be approximately twice as great as for a single-acting hammer with the same energy rating.

The lighter ram and higher striking velocity of the double-acting hammer may be advantageous when driving light- to medium-weight piles into soils having normal frictional resistance. It is claimed that the high frequency of blows will keep a pile moving downward continuously, thus preventing static skin friction from developing between blows. However, when heavy piles are driven, especially into soils having high frictional resistance, the heavier weight and slower velocity of a single-acting hammer will transmit a greater portion of the rated energy into driving the piles. Figure 17-12 shows the essential parts of a double-acting hammer. The hammer is fully enclosed by a steel case.

Table 17-3 gives dimensions and other data for several of the more popular sizes of double-acting hammers. As indicated in this table, the energy per blow as well as the number of blows per minute can be modified by varying the steam pressure.

Among the advantages of double-acting compared with single-acting hammers are the following:

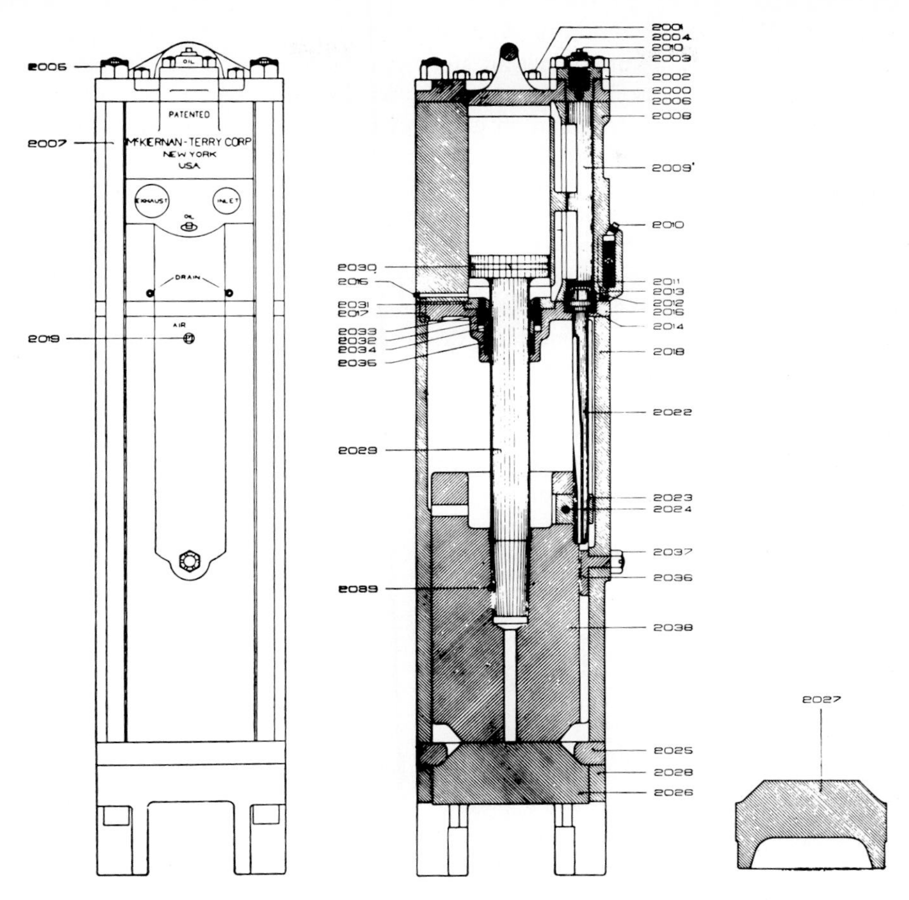

Figure 17-12 Section through a double-acting pile hammer. *(MKT Geotechnical Systems.)*

1. The greater number of blows per minute reduces the time required to drive piles.
2. The greater number of blows per minute reduces the development of static skin friction between blows.
3. Piles can be driven more easily without leads.

Among the disadvantages of double-acting compared with single-acting hammers are the following:

1. The relatively light weight and high velocity of the ram make this type of hammer less suitable for use in driving heavy piles into soils having high frictional resistance.
2. The hammer is more complicated.

Differential-acting steam hammers A differential-acting steam hammer is a modified double-acting hammer in that steam pressure is used to lift the ram and to accelerate the ram on the downstroke. As shown in Fig. 17-13, the ram has a large piston which operates in an upper cylinder and a small piston which operates in a lower cylinder. The lifting of the ram is effected by the difference in the pressure forces acting on the two pistons. The number of blows per minute is comparable with that for a double-acting hammer, while the weight and the equivalent free fall of the ram are comparable with those of a single-acting hammer. Thus, it is claimed that this type of hammer has the advantages of the single- and double-acting hammers. It is reported that this hammer will drive a pile in one-half the time required by the same-size single-acting hammer and in doing so will use 25 to 35 percent less steam. This hammer is available in open or closed types. Table 17-3 gives dimensions and data for these hammers. The values given in the table for rated energy per blow are correct provided the steam pressure is sufficient to produce the indicated normal blows per minute.

Hydraulic hammers These hammers operate on the differential principle of

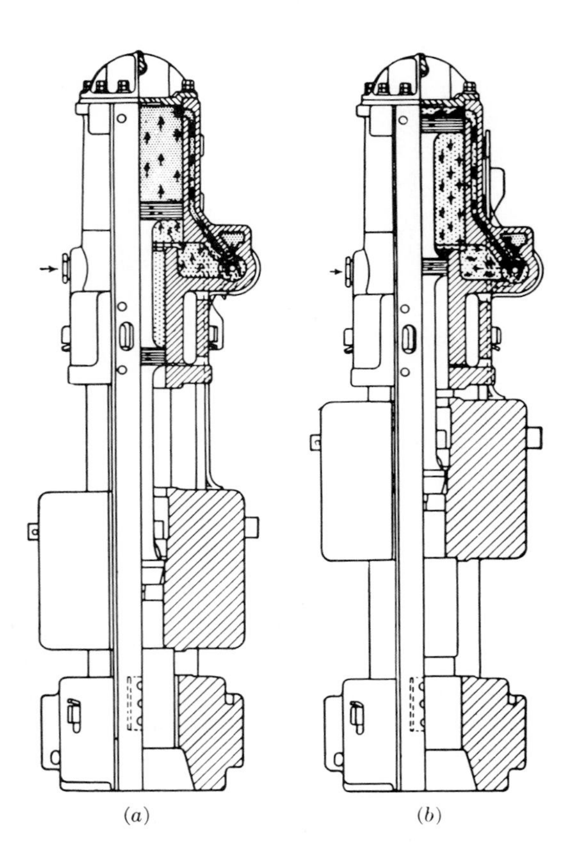

Figure 17-13 Section through a differential-acting steam hammer. (*a*) Piston in lower position. (*b*) Piston in upper position. *(Vulcan Iron Works.)*

hydraulic fluid instead of steam or compressed air used by conventional hammers. See Fig. 17-14. Table 17-3 gives the specifications for the sizes manufactured by MKT Geotechnical Systems.

Dynamic pile-driving equations in current use are applicable to these hammers.

Diesel hammers A diesel pile-driving hammer is a self-contained driving unit which does not require an external source of energy such as a steam boiler or an air compressor. In this respect it is simpler and more easily moved from one

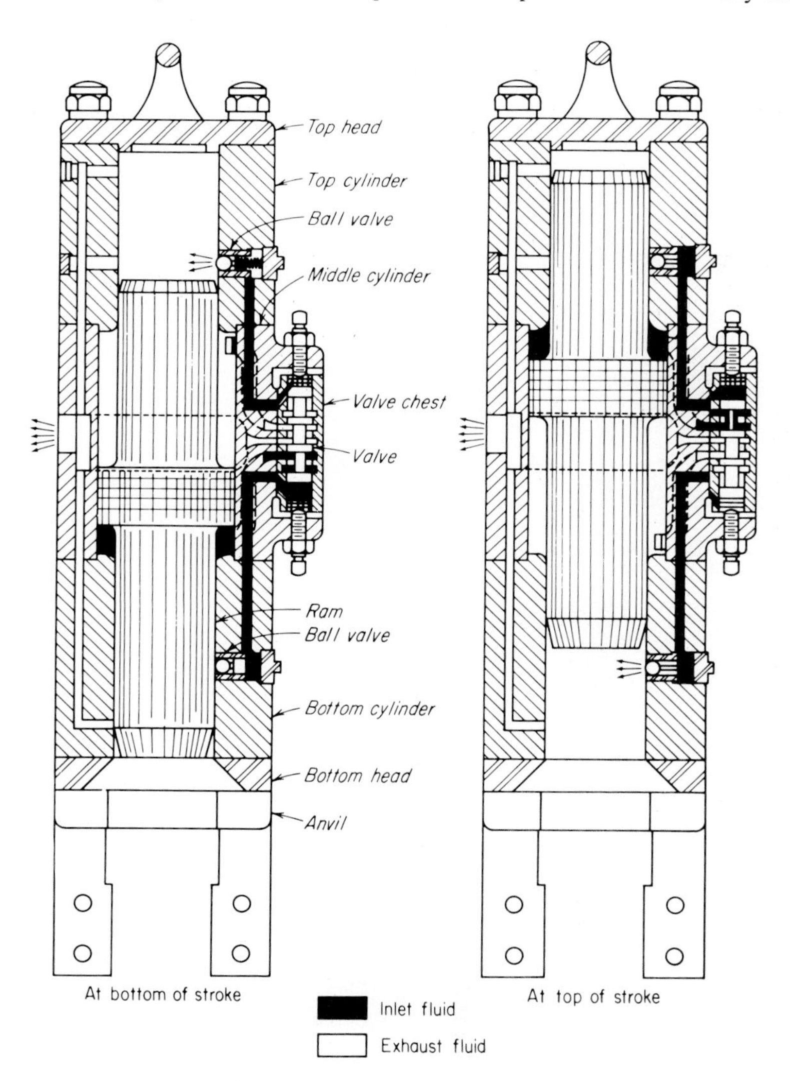

Figure 17-14 Operation of a fluid-valve double-acting hammer. *(MKT Geotechnical Systems.)*

location to another than a steam hammer. A complete unit consists of a vertical cylinder, a piston or ram, an anvil, fuel- and lubricating-oil tanks, a fuel pump, injectors, and a mechanical lubricator.

After a hammer is placed on top of a pile, the combined piston and ram are lifted to the upper end of the stroke and released to start the unit operating. As the ram nears the end of the downstroke, it activates a fuel pump that injects the fuel into the combustion chamber between the ram and the anvil. The continued downstroke of the ram compresses the air and the fuel to ignition heat. The resulting explosion drives the pile downward and the ram upward to repeat its stroke. The energy per blow, which can be controlled by the operator, may be varied over a wide range. Table 17-3 lists the specifications for several makes and models of diesel hammers. Figure 17-9

Figure 17-15 Diesel hammer driving sheet piling. *(Pileco, Inc.)*

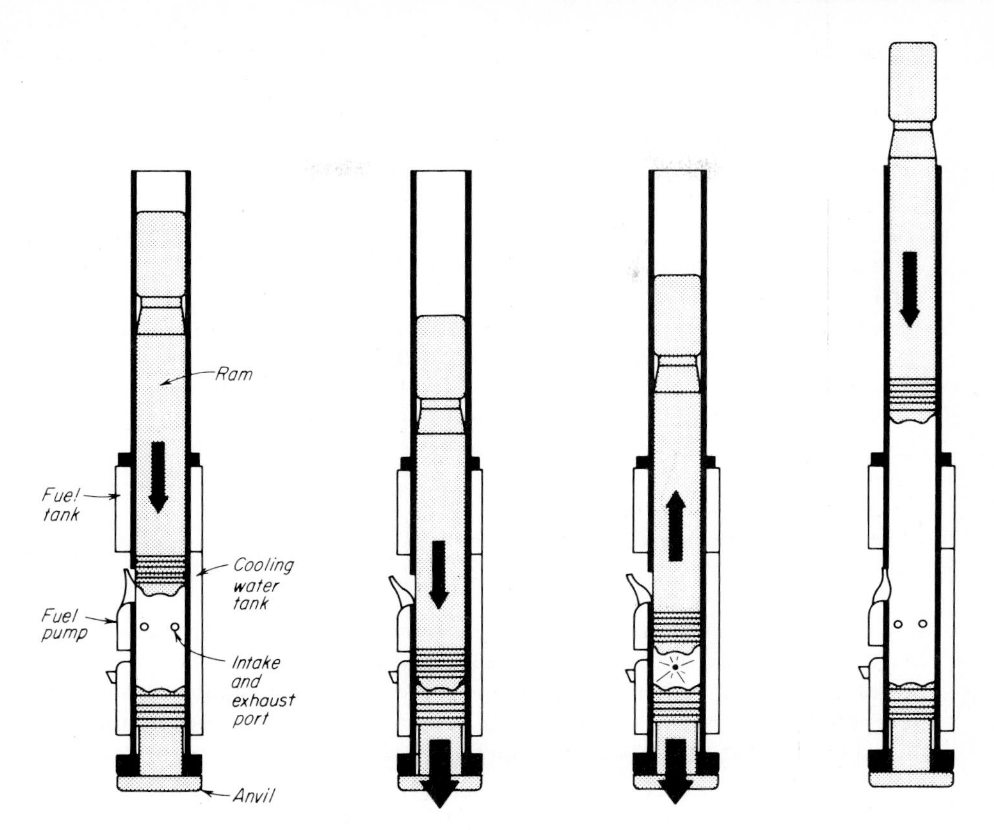

Figure 17-16 The operation of a diesel hammer. *(L. B. Foster Company.)*

illustrates a diesel hammer driving a steel H pile, while Fig. 17-15 illustrates a diesel hammer driving steel-sheet piling. Figure 17-16 illustrates the operation of a diesel hammer.

Advantages of the diesel hammer When compared with the steam hammer, the diesel hammer has several potential advantages, including the following:

1. The hammer needs no external source of energy. Thus, it is more mobile and it requires less time to set up and start operating.
2. The hammer is economical to operate. The rated fuel consumption for a 24,000-ft-lb hammer is 3 gal per hr when it is operating. Because a hammer does not operate continuously, the actual consumption is less.
3. It is convenient to operate in remote areas. Because it uses diesel oil as a source of energy, it is not necessary to provide a boiler, water for steam, and fuel oil.
4. It operates well in cold weather. Diesel hammers have been used at

temperatures well below 0°F, where it would be difficult or impossible to provide steam.

5. The hammer is light in weight when compared with the weight of a steam hammer of equal rating.
6. Maintenance and servicing are simple and fast.
7. The energy per blow increases as the driving resistance of a pile increases.
8. Because the resistance of a pile to driving is necessary for continuing operation of a diesel hammer, this hammer will not operate if a pile breaks or falls out from under a hammer.
9. Because of the low velocity in easy driving, and also because the piston reacts to the impact needed for each blow by rebounding up its cylinder, a diesel hammer is less likely to batter the piles when driving them.
10. The energy per blow and the number of blows per minute can be varied easily to permit a diesel hammer to operate most effectively for an existing condition [1].

Disadvantages of the diesel hammer Among the disadvantages of a diesel hammer are the following:

1. It is difficult to determine the energy per blow for this hammer. Because the height to which the piston ram will rise following the explosion of the fuel in the combustion chamber is a function of the driving resistance, the energy per blow will vary with the driving resistance. For this reason there is uncertainty about the accuracy of applying dynamic pile-driving formulas to diesel hammers.
2. The hammer may not operate well when driving piles into soft ground. Unless a pile offers sufficient driving resistance to activate the ram, the hammer will not operate.
3. The number of strokes per minute is less than for a steam hammer. This is especially true for a diesel hammer with an open end or top.
4. The length of a diesel hammer is slightly greater than the length of a steam hammer of comparable energy rating.

VIBRATORY PILE DRIVERS

Vibratory pile drivers have demonstrated their effectiveness in speed and economy in driving piles into certain types of soils. These drivers are especially effective when the piles are driven into water-saturated noncohesive soils. The drivers may experience difficulty in driving piles into dry sand, or similar materials, or into tight cohesive soils that do not respond to the vibrations.

The drivers are equipped with horizontal shafts, to which eccentric weights are attached. As the shafts rotate in pairs, in opposing directions, at speeds that can be varied to well in excess of 1,000 rpm, the forces produced by the rotating weights produce vibrations that are transmitted to a pile, and thence into the soil adjacent to the pile. The agitation of the soil, especially when it is saturated with water, reduces the skin friction between the soil and the pile materially. The combined weight of the pile and the driver resting on the pile will drive the pile quite rapidly.

Figure 17-17 illustrates the basic principle of the rotating weights, using six shafts. As noted in the figure, the two inner shafts, with lighter weights, rotate at twice the speeds of the two top and bottom shafts. During each revolution of the two top and bottom shafts the forces contributed by all weights will act downward at 0° and 360°, whereas at 180° the forces tend to counteract each other, as indicated in the exciting force curve [4]. Figure 17-18 illustrates a vibratory driver driving sheet piling.

Performance factors for vibratory drivers There are five performance factors that determine the effectiveness of a vibratory driver.

1. *Amplitude* This is the magnitude of the vertical movement of the pile produced by the vibratory unit. It may be expressed in inches or millimeters.
2. *Eccentric moment* The eccentric moment of a vibratory unit is a basic measure or indication of the size of a driver. It is the product of the weight of the eccentrics multiplied by the distance from the center of rotation of the shafts to the center of gravity of the eccentrics. The heavier the eccentric

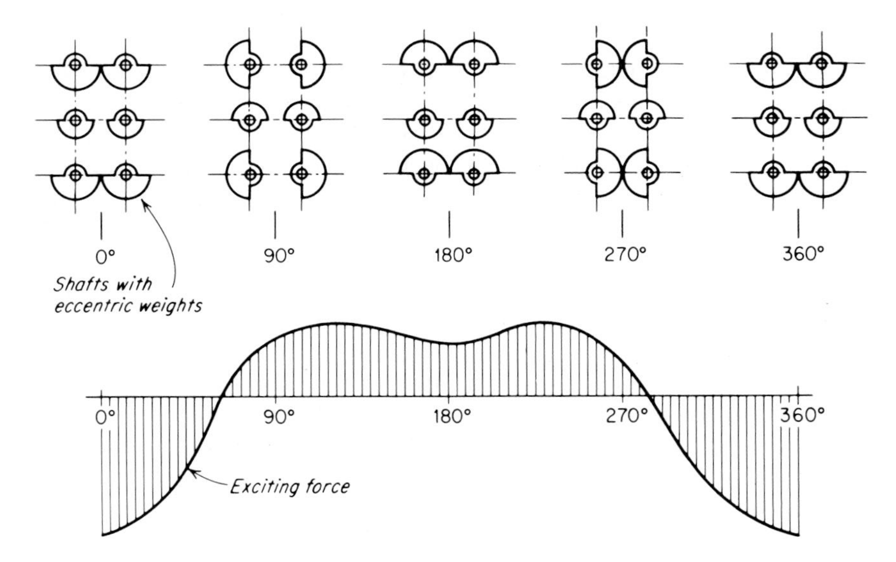

Figure 17-17 Operation of the eccentric weights for a vibratory pile driver. Diagram shows how exciting force of a six-shaft vibrator varies with the position of the eccentric weights attached to the shafts.

Figure 17-18 A vibratory driver driving sheet piling. *(L. B. Foster Company.)*

weights and the farther they are from the center of rotation of the shaft, the greater the eccentric moment of the unit.

3. *Frequency* This is expressed as the number of vertical movements of the vibrator per minute, which is also the number of revolutions of the rotating shafts per minute. Tests conducted on piles driven by vibratory drivers have indicated that the frictional forces between the piles being driven and the soil into which they are driven are at minimum values when frequencies are maintained in the range of 700 to 1,200 vibrations per minute. In general, the frequencies for piles driven into clay soils should be lower than for piles driven into sandy soils.
4. *Vibrating weight* The vibrating weight includes the vibrating case and the vibrating head of the vibrator unit, plus the pile being driven.
5. *Nonvibrating weight* This is the weight part of the system which does not vibrate, including the suspension mechanism and the motors. Nonvibrating weights push down on and aid in driving piles.

The Foster vibro driver/extractor This driver uses two electric motors, powered by a portable generator or a commercial source of electrical energy, to drive eccentrically loaded shafts, whose speeds of rotation may be varied to produce the most effective frequency for a given soil (see Fig. 17-19).

Figure 17-19 A vibratory hammer driving steel sheet piling. *(L. B. Foster Company.)*

JETTING PILES

The use of a water jet to assist in driving piles into sand or fine gravel frequently will speed the driving operation. The water, which is discharged through a nozzle at the lower end of a jet pipe, keeps the soil around a pile in agitation, thereby reducing the resistance due to skin friction. Successful jetting requires a plentiful supply of water at a pressure high enough to loosen the soil and remove it from the hole ahead of the penetration of the pile. Jet pipes commonly used vary in size from 2 to 4 in. in diameter, with nozzles varying from $\frac{1}{2}$ to $1\frac{1}{2}$ in. in diameter. The water pressure at the nozzle may vary from approximately 100 to more than 300 psi, with the quantity commonly varying from 300 to 500 gpm, but as high as 1,000 gpm in some instances.

Although some piles have been jetted to final penetration, this is not considered good practice, primarily because it is impossible to determine the safe supporting capacity of a pile so driven. Most specifications require that piles shall be driven the last few feet without the benefit of jetting. The Foundation Code of the City of New York requires a contractor to obtain special permission prior to jetting piles and specifies that piles shall be driven the last 3 ft with a pile hammer.

DRIVING PILES BELOW WATER

If it is necessary to drive piles below water, either of two methods may be used. When the driving unit is a drop hammer, an open-type steam hammer, or a diesel hammer, the pile is driven until the top is just above the surface of the water. Then a follower is placed on top of the pile, and the driving is continued through the follower. The follower may be made of wood or steel and must be strong enough to transmit the energy from the hammer to the pile.

When the driving unit is an enclosed steam hammer, the driving may be continued below the surface of the water, without a follower. It is necessary to install an exhaust hose to the surface of the water for the steam. Also, it is necessary to supply about 60 cfm of compressed air to the lower part of the hammer housing to prevent water from flowing into the casing and around the ram. An air pressure of $\frac{1}{2}$ psi for each foot of depth below the surface of water will be satisfactory.

PILE-DRIVING EQUATIONS

There are many pile-driving equations, each of which is intended to give the supporting strength of a pile. The equations are empirical, with coefficients that have been determined for certain existing or assumed conditions. While each formula may give dependable values for the conditions under which it was developed, there is no equation that will give dependable values for the supporting strength of piles for all the varying conditions that exist on foundation jobs.

It is not within the scope of this book to analyze the various pile-driving equations or the theory related to them. For a more comprehensive study of this subject it is suggested that the reader consult the books listed at the end of this chapter [3, 4]. Perhaps the most popular equation in the United States is the *Engineering News* equation. Its popularity seems due primarily to its simplicity rather than its accuracy. For the three types of pile-driving hammers in current use it has the following forms:

For a drop hammer

$$R = \frac{2WH}{S + 1.0} \tag{17-1}$$

For a single-acting steam hammer

$$R = \frac{2WH}{S + 0.1} \tag{17-2}$$

For a double- and differential-acting steam hammer

$$R = \frac{2E}{S + 0.1} \tag{17-3}$$

where R = safe load on a pile, lb
W = weight of falling mass, lb
H = height of free fall for mass W, ft
E = total energy of ram at bottom of its downward stroke, ft-lb
S = average penetration per blow for last five or ten blows, in.

MICHIGAN PILE-DRIVING TESTS

In 1961 the Michigan Highway Department and the United States Bureau of Public Roads began the most comprehensive tests ever conducted on pile driving [2, 3, 4]. The eight main objectives of the tests were

1. To develop a method for determining the driving energy output of various types of pile-driving hammers, including air, single- or double-acting steam, and diesel hammers
2. To determine by load tests the load-bearing capacities of piles driven under test conditions
3. To determine what factors, if any, relate the measured pile-driving energy to the load-bearing capacity of a pile
4. To determine the proper wall thickness of pipe piles under certain driving conditions
5. To determine the correlations between the tested load-bearing capacity and estimates of load-bearing capacity as obtained by nine of the best-known pile-driving formulas
6. To determine the best methods or procedures for jetting piles through intermediate soil layers when the driving resistance is large but the bearing capacity of the pile in these layers is not satisfactory
7. To determine the effect of the pile cross section or the surface configuration on the energy required for driving piles
8. To determine the effect of pile cross section or surface configuration on the load-bearing capacity of pile

A total of 88 piles were driven for test and analysis purposes. They included 12-in.-diameter pipe, open-end and closed-end, and 12-in. WF sections. The results of the tests appear in Table 17-4. Enthru, as used in the table,

Table 17-4 Results of Michigan State Highway Department pile tests

Site	Hammer model	Maximum rated energy, ft-lb	Type pile	Accepted enthru determinations	Peak force, kips	Peak acceleration, G's	Average enthru, ft-lb	Ratio enthru/enthru		
								Minimum	Maximum	Average
1	Vulcan No. 1	15,000	H	11	230	90	5,283	0.27	0.45	0.35
			Pipe	5	180	75	5,094	0.26	0.41	0.34
	Link-belt 312	18,000	H	27	380	170	7,726	0.28	0.56	0.43
			Pipe	12	440	210	8,226	0.32	0.62	0.46
	McKiernan-Terry DE-30	22,400	H	22	390	270	5,769	0.19	0.36	0.26
			Pipe	9	490	240	8,682	0.31	0.59	0.39
	Delmag D-12	22,500	H	30	800	360	9,123	0.19	0.53	0.41
			Pipe	13	690	350	11,870	0.39	0.64	0.53
2	Vulcan No. 1	15,000	H, Pipe	12	200	110	5,339	0.30	0.43	0.36
	Vulcan 50C	15,100	Pipe	3	490	270	9,822	0.55	0.76	0.65
	Link-belt 312	18,000	H, Pipe	3	200	100	7,554	0.37	0.50	0.42
	McKiernan-Terry DE-30	22,400	H, Pipe	14	370	190	8,417	0.31	0.46	0.38
	Delmag D-12	22,500	H, Pipe	9	600	320	10,033	0.24	0.60	0.45
3	Vulcan No. 1	15,000	Pipe	10	220	80	6,359	0.32	0.48	0.42
	Vulcan 80C	24,500	Pipe	6	650	310	13,872	0.52	0.64	0.57
	Link-belt 520	30,000	Pipe	8	560	150	16,637	0.44	0.66	0.55
	McKiernan-Terry DE-40	32,000	Pipe	4	700	310	18,088	0.52	0.65	0.57
	Delmag D-22	39,700	Pipe	9	1,050	470	24,660	0.53	0.78	0.62

is the actual energy delivered to a pile by a hammer, which was determined by instruments. As noted in the table, there was considerable variation in the net energy delivered to the piles by any given hammer. Also, the tests revealed that no dynamic formula currently in use can consistently provide an accurate estimate of the bearing capacity of a pile [7].

After observing the tests and analyzing the results, the highway department modified the *Engineering-News* equation as follows:

$$R = \frac{2.5E}{S + 0.1} \frac{W_r + e^2 W_p}{W_r + W_p} \tag{17-4}$$

where R = computed design pile load capacity, lb
E = manufacturer's maximum rated energy per blow, ft-lb
S = final average penetration of pile per blow, in.
W_r = weight of ram, lb
W_p = weight of pile, including driving appurtenances, lb
e = coefficient of restitution, whose values are:
0.55 for steel hammer on steel pile, with no cushion
0.50 for well-compacted cushion in driving pipe piles
0.50 for double-acting steel hammer striking on steel anvil and driving steel piles or precast concrete piles
0.40 for ram of double-acting steam hammer striking steel anvil and driving wood piles
0.40 for medium-compacted wood cushion in driving steel or pipe piles
0.40 for ram of single-acting steam hammer or drop hammer striking directly on head of precast concrete pile
0.25 for ram of single-acting steam hammer or drop hammer striking on well-conditioned wood cap in driving precast concrete piles or directly on wood-pile heads
0.00 for badly broomed wood piles

Dynamic formulas During the past 20 years considerable progress has been made in conducting tests and studies, analyzing the results, and developing formulas for determining the load-bearing and other properties of piles.

In 1964 the Ohio Department of Transportation sponsored a research program at the Case Western Reserve University, Cleveland, Ohio, whose objective was to develop reliable techniques to predict static capacities for load-bearing piles. One of the systems developed by this program was a pile-driving analyzer, a programmed field computer used by engineers and others to analyze a pile during the driving operation. A reusable transducer was developed that can be attached to a pile prior to driving it. During driving, the transducer produces signals which are processed by an analyzer to give printouts indicating load-bearing capacities and other desired information. Reported uses of this equipment indicate that it is very reliable.

As a result of tests and analyses performed on piles and pile driving, numerous formulas have been developed to provide dependable load-bearing capacities and other information for piles, and also for the pile drivers.

Limited space in this book does not permit the presentation or discussions of these equations. Persons who are interested in further studies of this subject will find a list of some of the publications related to these equations, driving procedures, and results. See the references appearing at the end of this chapter.

SELECTING A PILE-DRIVING HAMMER

Selecting the most suitable pile hammer for a given project involves a study of several factors, such as the size and type of piles, the number of piles, the character of the soil, the location of the project, the topography of the site, the type of rig available, whether driving will be done on land or in water, etc. A pile-driving contractor usually is concerned with selecting the hammer that will drive the piles for a project at the lowest practical cost. As most contractors must limit their ownership to a few representative sizes and types of hammers, a selection should be made from those hammers already owned unless conditions are such that it is economical or necessary to secure an additional size or type. Naturally, more consideration should be given to the selection of a hammer for a project that requires several hundred piles than for a project that requires only a few piles.

As previously stated, the function of a pile hammer is to furnish the energy required to drive a pile. This energy is supplied by a weight which is raised and permitted to drop on top of a pile, under the effect of gravity alone or with steam acting during the downward stroke. The theoretical energy per blow will equal the product of the weight times the equivalent free fall. Since some of this energy is lost in friction as the weight travels downward, the net energy per blow will be less than the theoretical energy, the actual amount depending on the efficiency of the particular hammer. The efficiencies of pile hammers vary from 50 to 100 percent.

Table 17-5 gives recommended sizes of hammers for different types and sizes of piles and driving resistances. The sizes are indicated by the theoretical foot-pounds of energy delivered per blow. Table 17-3 gives the theoretical energy per blow for several types and sizes of hammers. For each hammer listed the specified theoretical energy per blow is correct provided the hammer is operated at the designated number of strokes per minute.

In general, it is good practice to select the largest hammer that can be used without overstressing or damaging a pile. As previously shown, when a large hammer is used, a greater portion of the energy is effective in driving a pile, which produces a higher operating efficiency. Therefore, the hammer sizes given in Table 17-5 should be considered as the minimum sizes. In some instances hammers as much as 50 percent larger may be used advantageously.

Table 17-5 Recommended sizes of hammers for driving various types of piles
Size expressed in foot-pounds of energy per blow

Length of piles, ft	Depth of penetration	Weight of various types of piles, lb per lin ft						
		Steel sheet*			Timber		Concrete	
		20	30	40	30	60	150	400
Driving through ordinary earth, moist clay, and loose gravel, normal frictional resistance								
25	$\frac{1}{2}$	2,000	2,000	3,600	3,600	7,000	7,500	15,000
	Full	3,600	3,600	6,000	3,600	7,000	7,500	15,000
50	$\frac{1}{2}$	6,000	6,000	7,000	7,000	7,500	15,000	20,000
	Full	7,000	7,000	7,500	7,500	12,000	15,000	20,000
75	$\frac{1}{2}$	.....	7,000	7,500	......	15,000	......	30,000
	Full	.....	.....	12,000	......	15,000	......	30,000
Driving through stiff clay, compacted sand, and gravel, high frictional resistance								
25	$\frac{1}{2}$	3,600	3,600	3,600	7,500	7,500	7,500	15,000
	Full	3,600	7,000	7,000	7,500	7,500	12,000	15,000
50	$\frac{1}{2}$	7,000	7,500	7,500	12,000	12,000	15,000	25,000
	Full	.....	7,500	7,500	......	15,000	......	30,000
75	$\frac{1}{2}$	.....	7,500	12,000	......	15,000	......	36,000
	Full	.....	.....	15,000	......	20,000	......	50,000

* The indicated energy is based on driving two steel-sheet piles simultaneously. In driving single piles, use approximately two-thirds of the indicated energy.

REFERENCES

1. Rausche, Frank, and George G. Goble: Performance of Pile-driving Hammers, *Journal of the Construction Division, Proceedings ASCE*, vol. 98, pp. 201–218, September 1972.
2. Michigan Pile Test Program Results are Released, *Engineering News-Record*, vol. 164, pp. 26–34, May 20, 1965.
3. "A Performance Investigation of Pile-driving Hammers and Piles, Final Report," Michigan State Highway Commission, Lansing, Michigan, March 1965.
4. Housel, W. S.: Michigan Study of Pile-driving Hammers, *Journal of the Soil Mechanics and Foundation Division, Proceedings ASCE*, vol. 91, pp. 37–64, September 1965.
5. Sullivan, Richard A., and Charles J. Ehlers: Planning for Driving Offshore Piles, *Journal of the Construction Division, Proceedings ASCE*, vol. 49, pp. 59–79, July 1973.
6. McClelland, B., J. A. Focht, Jr., and W. J. Emrich: Problems in Design and Installation of Offshore Piles, *Journal of the Soil Mechanics and Foundation Division, Proceedings ASCE*, vol. 95, pp. 1491–1514, November 1969.
7. Samson, C. H., T. J. Hirsch, and L. L. Lowery: Computer Study of Dynamic Behavior of Piling, *Journal of the Structural Division, Proceedings ASCE*, vol. 89, pp. 413–440, August 1963.

8. Goble, G. G., G. E. Likens, and F. Rausche: "Bearing Capacity of Piles from Dynamic Measurements," Final Report, Department of Civil Engineering, Case Western University, Cleveland, OH 44106, March 1975.
9. Rausche, F., Fred Moses, and G. G. Goble: Soil Resistance Predictions from Pile Dynamics, *Journal of the Soil Mechanics and Foundation Division, Proceedings ASCE*, vol. 98, pp. 917–937, September 1972.
10. Goble, G. G., K. Fricke, and G. E. Likens, Jr.: Driving Stresses in Concrete Piles, *Journal of Prestressed Concrete Institute*, vol. 21, pp. 70–88, January–February 1976.
11. Sandhu, Balbir S.: Predicting Driving Stresses in Piles, *Journal of the Construction Division, Proceedings ASCE*, vol. 108, pp. 485–503, December 1982.
12. L. B. Foster Company, Seven Parkway Center, Pittsburgh, PA 15220.
13. MKT Geotechnical Systems, P. O. Box 793, Dover, NJ 07801.
14. Pileco, Inc., P. O. Box 16099, Houston, TX 77027.
15. Raymond International Builders, Inc., 2801 Post Oak Boulevard, Houston, TX 77027.
16. The Union Metal Manufacturing Company, P. O. Box 8530, Canton, OH 44711.
17. Vulcan Iron Works, Inc., 2725 North Australian Avenue, West Palm Beach, FL 33407.
18. Pile Dynamics, Inc., 4423 Emery Industrial Parkway, Warrensville Heights, OH 44128.
19. Franki Foundation Company, 920 Statler Office Building, Boston, MA 02116.

CHAPTER
EIGHTEEN
PUMPING EQUIPMENT

INTRODUCTION

Pumps are used extensively on construction projects for such operations as

1. Removing water from pits, tunnels, etc.
2. Unwatering cofferdams
3. Furnishing water for jetting and sluicing
4. Furnishing water for many types of utility services
5. Lowering the water table for excavations
6. Foundation grouting

Most projects require the use of one or more water pumps at various stages during the period of construction. Construction pumps frequently are required to perform under severe conditions, such as those resulting from variations in the pumping head or from handling water that is muddy, sandy and trashy, or highly corrosive. The rate of pumping may vary several hundred percent during the period of construction. The most satisfactory solution to the pumping problem may be a single all-purpose pump, or it may be to use several types and sizes of pumps, to permit flexibility in the operations. The proper solution is to select the equipment which will take care of the pumping needs adequately at the lowest total cost, considering the investment in pumping equipment, the cost of operating the pumps, and any losses that will result from possible failure of the pumps to operate satisfactorily.

For some projects a pump may be the most critical item of construction equipment. In constructing a multimillion-dollar concrete and earth-fill dam, a contractor used a single centrifugal pump to supply water from a nearby stream. The water was used to wash all concreted aggregate, for mixing and curing the concrete, and for moisture in the earth-filled dam. When the pump developed mechanical trouble and the rate of pumping dropped below the job requirements for several days, progress on the project suffered a loss of approximately 25 percent. With the fixed costs exceeding $4,000 per day, the loss due to the partial failure of the pump exceeded $1,000 per day.

Among the factors that should be considered in selecting construction pumps are the following:

1. Dependability
2. Availability of parts for making repairs
3. Simplicity to permit easy repairs
4. Economical installation and operation

CLASSIFICATION OF PUMPS

The pumps most commonly used on construction projects may be classified as:

1. Displacement
   a. Reciprocating
   b. Diaphragm
2. Centrifugal
   a. Conventional
   b. Self-priming
   c. Air-operated

Reciprocating pumps A reciprocating pump operates as the result of the movement of a piston inside a cylinder. When the piston is moved in one direction, the water ahead of the piston is forced out of the cylinder. At the same time additional water is drawn into the cylinder behind the piston. Regardless of the direction of movement of the piston, water is forced out of one end and drawn into the other end of the cylinder. This is classified as a double-acting pump. If water is pumped during a piston movement in one direction only, the pump is classified as single-acting. If a pump contains more than one cylinder, mounted side by side, it is classified as a duplex for two cylinders, triplex for three cylinders, etc. Thus a pump might be classified as duplex double-acting, duplex single-acting, etc.

The volume of water pumped in one stroke will equal the area of the cylinder times the length of the stroke, less a small deduction for slippage through the valves or past the piston, usually about 3 to 5 percent. If this volume is expressed in cubic inches, it may be converted to gallons by dividing by 231, the number of cubic inches in a gallon. The volume pumped in gpm by a simplex double-acting pump will be

$$Q = c\,\frac{\pi d^2 ln}{4 \times 231} \tag{18-1}$$

where Q = capacity of a pump, gpm
$c = 1 -$ slip allowance; varies from 0.95 to 0.97
d = diameter of cylinder, in.
l = length of stroke, in.
n = number of strokes per min (*Note*: The movement of the piston in either direction is a stroke.)

The volume pumped per minute by a multiplex double-acting pump will be

$$Q = Nc\,\frac{\pi d^2 ln}{4 \times 231} \tag{18-2}$$

where N = number of cylinders in pump.

The energy required to operate a pump will be

$$W = \frac{wQh}{e}$$

where W = energy, ft-lb per min
w = weight of 1 gal of water, lb
h = total pumping head, ft, including friction loss in pipe
e = efficiency of pump, expressed decimally

The horsepower required by the pump will be

$$P = \frac{W}{33{,}000} = \frac{wQh}{33{,}000e} \tag{18-3}$$

where P = power, hp
33,000 = ft-lb of energy per min for 1 hp

Example 18-1 How many gallons of fresh water will be pumped per minute by a duplex double-acting pump, size 6 by 12 in., driven by a crankshaft making 90 rpm? If the total head is 160 ft and the efficiency of the pump is 60 percent, what is the minimum horsepower required to operate the pump? The weight of water is 8.34 lb per gal.

SOLUTION Assume a water slippage of 4 percent. If we apply Eq. (18-2), the rate of pumping will be

$$Q = Nc\,\frac{\pi d^2 ln}{924}$$

$$= \frac{2 \times 0.96 = \pi \times 36 \times 12 \times 180}{924} = 518 \text{ gpm}$$

Applying formula (18-3), the power required by the pump will be

$$P = \frac{wQh}{33{,}000e}$$

$$= \frac{8.34 \times 518 \times 160}{33{,}000 \times 0.60} = 34.9 \text{ hp}$$

The capacity of a reciprocating pump depends essentially on the speed at which the pump is operated and is independent of the head. The maximum head against which a reciprocating pump will deliver water depends on the strength of the component parts of the pump and the power available to operate the pump. The capacity of this type of pump may be varied considerably by varying the speed of the pump.

Because the flow of water from each cylinder of a reciprocating pump stops and starts every time the direction of piston travel is reversed, a characteristic of this type of pump is to deliver water with pulsations. The amplitude of the pulsations may be reduced by using more cylinders and by installing an air chamber on the discharge side of a pump.

Among the advantages of reciprocating pumps are the following:

1. They are able to pump at a uniform rate against varying heads.
2. Their capacity can be increased by increasing the speed.
3. They have reasonably high efficiency regardless of the head and speed.
4. They are usually self-priming.

Among the disadvantages of reciprocating pumps are the following:

1. Heavy weight and large size for given capacity
2. Possibility of valve trouble, especially in pumping water containing trash
3. Pulsating flow of water
4. Danger of damaging a pump in operating against a high head

Diaphragm pumps The principle under which a diaphragm pump operates is illustrated in Fig. 18-1. The central portion of the flexible diaphragm is

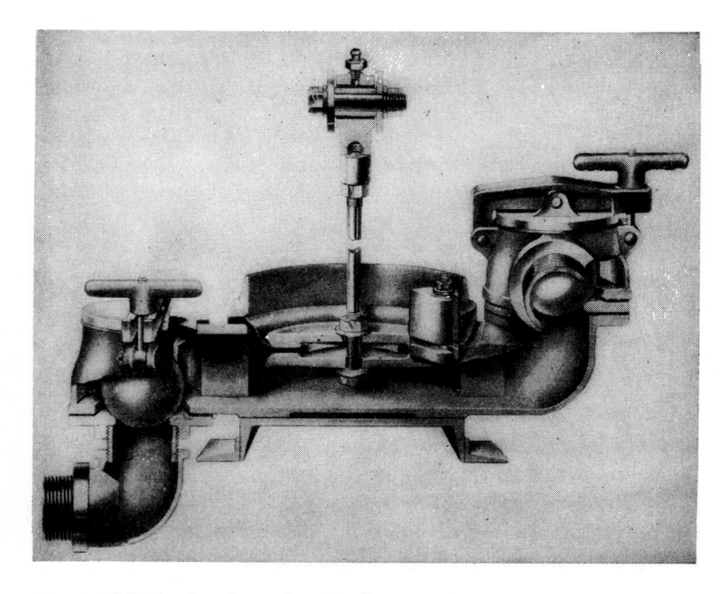

Figure 18-1 Section through a diaphragm pump.

Table 18-1 Minimum capacities for diaphragm pumps at 10-ft suction lifts*†

Size	Capacity, gal per hr
2-in. single	2,000
3-in. single	3,000
4-in. single	6,000
4-in. double	9,000

* Contractors Pump Bureau.
† Diaphragm pumps shall be tested with standard contractor's type suction hose 5 ft longer than the suction lift shown.

alternately raised and lowered by the pump rod, which is connected to a walking beam. This action draws water into and discharges it from the pump. Because this type of pump will handle clear water or water containing large quantities of mud, sand, sludge, and trash, it is popular as a construction pump. It is suitable for use on jobs where the quantity of water varies considerably, as the loss of prime during low flow does not prevent it from automatically repriming when the quantity of water increases. The accessible diaphragm may be replaced easily.

The Contractors Pump Bureau specifies that diaphragm pumps shall be manufactured in the size and capacity ratings given in Table 18-1.

Centrifugal pumps A centrifugal pump contains a rotation element, called an impeller, which imparts to water passing through the pump a velocity sufficiently great to cause it to flow from the pump even against considerable pressure. A mass of water may possess energy due to its height above a given datum or due to its velocity. The former is potential, while the latter is kinetic, energy. One type of energy can be converted into the other under favorable conditions. The kinetic energy imparted to a particle of water as it passes through the impeller is sufficient to cause the particles to rise to some determinable height.

The principle of the centrifugal pump may be illustrated by considering a drop of water at rest at a height h above a surface. If the drop of water is permitted to fall freely, it will strike the surface with a velocity given by the equation

$$V = \sqrt{2gh} \tag{18-4}$$

where V = velocity, fps
g = acceleration of gravity, equal to 32.2 ft per sec at sea level
h = height of fall, ft

If the drop falls 100 ft, the velocity will be 80.4 fps. If the same drop is given an upward velocity of 80.4 fps, it will rise 100 ft. These values assume no loss in energy due to friction through air. It is the function of the centrifugal pump to give the water the necessary velocity as it leaves the impeller. If the speed of the pump is doubled, the velocity of the water will be increased from 80.4 to 160.8 fps, neglecting any increase in friction losses. With this velocity the water can be pumped to a height given by the equation

$$h = \frac{V^2}{2g} = \frac{160.8^2}{64.4} = 400 \text{ ft}$$

This indicates that if a centrifugal pump is pumping water against a total head of 100 ft, the same quantity of water can be pumped against a total head of 400 ft by doubling the speed of the impeller. In actual practice the maximum possible head for the increased speed will be less than 400 ft because of increases in losses in the pump due to friction. These results illustrate the effect which increasing the speed or the diameter of an impeller has on the performance of a centrifugal pump.

A centrifugal pump may be equipped with an open or an enclosed impeller. Although an enclosed impeller usually has higher efficiency, it will not handle water containing trash as well as an open impeller.

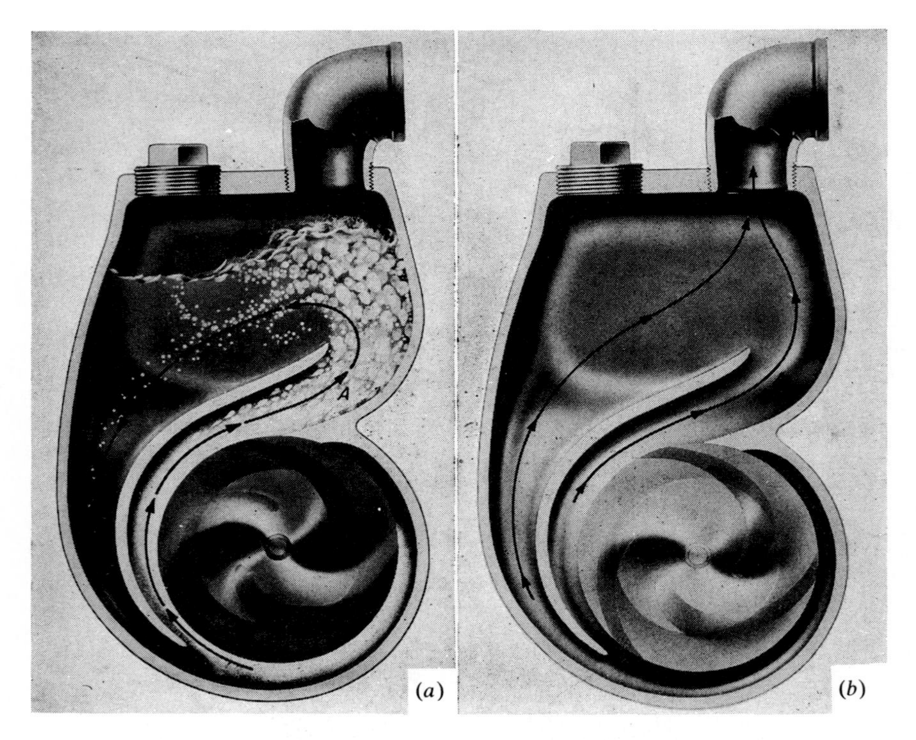

Figure 18-2 Section through a self-priming centrifugal pump. (*a*) Priming action. (*b*) Pumping action.

The power required to operate a centrifugal pump is given by Eq. (18-3). The efficiencies of these pumps may be as high as 75 percent.

Self-priming centrifugal pumps The centrifugal pumps most commonly installed in water and sewage plants are set below the level of water on the suction side because they are not self-priming. However, on construction projects, pumps frequently must be set up above the surface of the water to be pumped. Consequently, self-priming centrifugal pumps are more suitable than the conventional types for use on construction projects. The operation of a self-priming pump is illustrated in Fig. 18-2. A check valve on the suction side of the pump permits the chamber to be filled with water prior to starting the pump. When the pump is started, the water in the chamber produces a seal which enables the pump to draw air from the suction pipe. The air and water flow through channel A into the chamber, where the air escapes through the discharge, and the water flows down through channel B to the impeller. This action continues until all the air is exhausted from the suction line and water enters the pump. When a pump is stopped, it will retain its charge of priming water indefinitely. Such a pump is self-priming to heights in excess of 25 ft when in good mechanical condition.

Submersible pumps Figure 18-3 illustrates an electric-motor-operated sub-

Figure 18-3 Electric-motor-operated submersible pump. (*The Gorman-Rupp Co.*)

mersible pump which may be very useful in dewatering tunnels, foundation pits, trenches, and similar places. Figure 18-4 is a performance curve for this pump when operated against varying heads of water. The figure also includes pertinent information related to the pump. Other types and models have different performance characteristics.

Multistage centrifugal pumps If a centrifugal pump has a single impeller, it is described as a single-stage pump, whereas if there are two or more impellers and the water discharged from one impeller flows into the suction of another, it is described as a multistage pump. Multistage pumps are especially suitable for pumping against high heads or pressures, as each stage imparts an additional pressure to the water. Pumps of this type are used frequently to supply water for jetting, where the pressure may run as high as several hundred psi.

Model S3B1:..........Centrifugal,single stage
Discharge:..........3 inches
Solids handled:..........3/8"
Horsepower:..........6
Hertz:..........60
RPM:..........3450
Voltage:..........230 volt, 1 phase, 7.2 kw
230/460 dual voltage, 3 phase, 6.8 kw or
575 volt, 3 phase, 6.8 kw
Cable:..........# 10 gauge, 50' length
Weight: (pump and cable)..........125 lbs. (approx.)

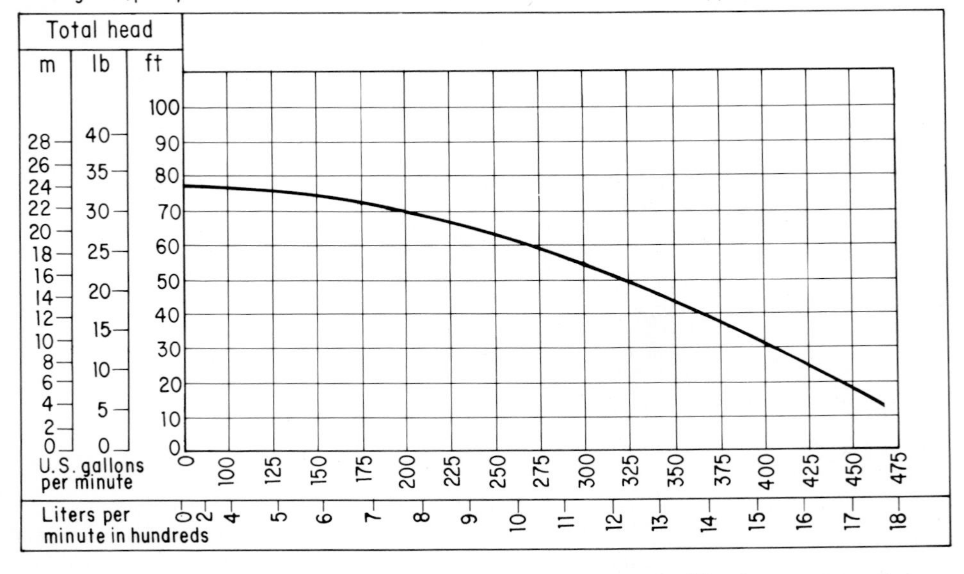

Figure 18-4 Performance curve for Gorman-Rupp submersible pump *(The Gorman-Rupp Co.)*

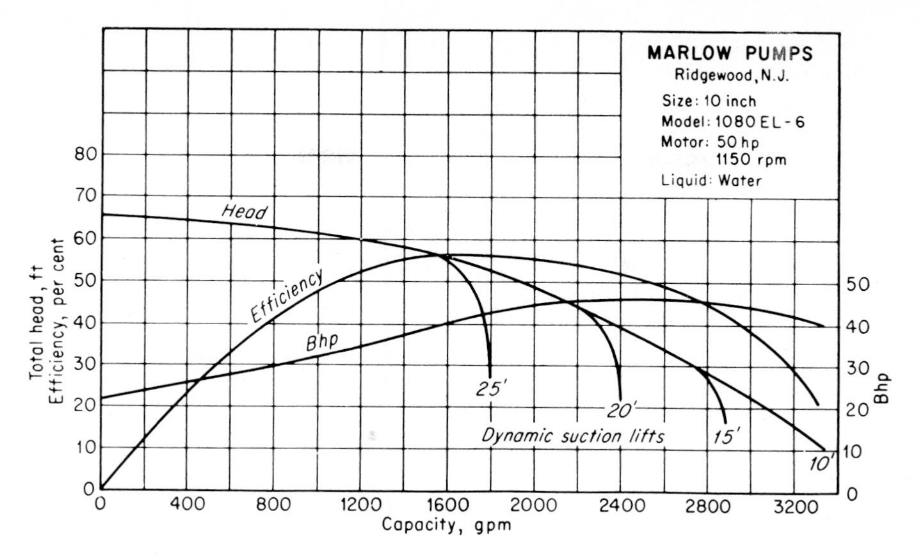

Figure 18-5 Performance curves for centrifugal pump.

Performance of centrifugal pumps The pump manufacturers will furnish sets of curves showing the performance of their pumps under different operating conditions. A set of curves for a given pump will show the variations in capacity, efficiency, and horsepower for different pumping heads. These curves can be very helpful in selecting the pump that is most suitable for a given pumping condition. Figure 18-5 illustrates a set of performance curves for a

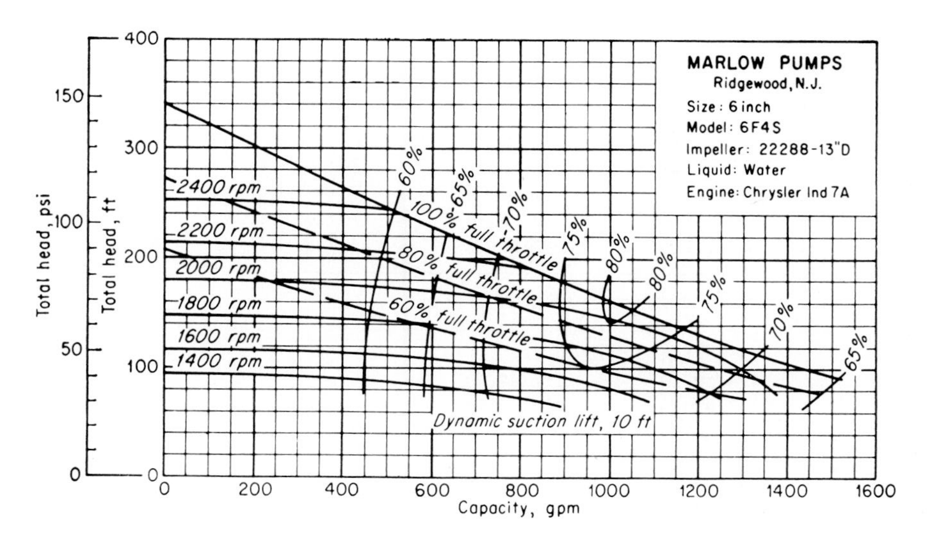

Figure 18-6 The effect of varying the speed on the performance of a centrifugal pump.

10-in. centrifugal pump. For a total head of 60 ft the capacity will be 1,200 gpm, the efficiency 52 percent, and the required power 35 bhp. If the total head is reduced to 50 ft and the dynamic suction lift does not exceed 23 ft, the capacity will be 1,930 gpm, the efficiency 55 percent, and the required power 44 bhp. This pump will not deliver any water against a total head in excess of 66 ft, which is called the shutoff head.

Since a construction pump frequently is operated under varying heads, it is desirable to select a pump with relatively flat head-capacity and horsepower

Table 18-2 Minimum capacities for M-rated self-priming centrifugal pumps manufactured in accordance with standards of the Contractors Pump Bureau

Total head including friction, ft (meters)	Capacity, gpm (liters per min) — Height of pump above water, ft (meters)			
	10 (3.0)	15 (4.6)	20 (6.1)	25 (7.6)
Model 5-M(1 ½ in.)				
15 (4.6)	85 (321.7)	—	—	—
20 (6.1)	84 (317.9)	68 (257.4)	—	—
25 (7.6)	82 (310.4)	67 (253.6)	—	—
30 (9.1)	79 (299.0)	66 (249.8)	49 (185.5)	35 (132.5)
40 (12.2)	71 (268.7)	60 (227.1)	46 (174.1)	33 (124.9)
50 (15.2)	59 (223.3)	52 (196.8)	41 (155.2)	28 (106.0)
60 (18.3)	42 (159.0)	40 (151.4)	32 (121.1)	22 (83.3)
70 (21.3)	22 (83.3)	22 (83.3)	20 (75.0)	12 (45.4)
Model 7-M (2 in.)				
20 (6.1)	117 (442.8)	—	—	—
30 (9.1)	116 (439.1)	102 (386.1)	82 (310.4)	—
40 (12.2)	105 (397.4)	100 (378.5)	80 (302.8)	58 (219.5)
50 (15.2)	92 (348.2)	90 (340.7)	76 (287.7)	55 (208.2)
60 (18.3)	70 (265.0)	70 (265.0)	70 (265.0)	55 (208.2)
70 (21.3)	40 (151.4)	40 (151.4)	40 (151.4)	40 (151.4)
Model 8-M (2 in.)				
20 (6.1)	135 (511.0)	—	—	—
25 (7.6)	134 (507.2)	117 (442.8)	—	—
30 (9.1)	132 (499.6)	115 (435.3)	93 (352.0)	65 (246.0)
40 (12.2)	123 (465.6)	109 (412.6)	88 (333.1)	63 (238.5)
50 (15.2)	109 (412.6)	99 (373.7)	81 (306.6)	59 (223.3)
60 (18.3)	90 (340.7)	84 (317.9)	70 (265.0)	51 (193.0)
70 (21.3)	66 (249.8)	65 (246.0)	57 (215.7)	41 (155.2)
80 (24.4)	40 (151.4)	40 (151.4)	40 (151.4)	28 (106.0)

Table 18-2

Total head including friction, ft (meters)	Capacity, gpm (liters per min) — Height of pump above water, ft (meters)			
	10 (3.0)	15 (4.6)	20 (6.1)	25 (7.6)
Model 10-M (2 in.)				
25 (7.6)	166 (628.3)	—	—	—
30 (9.1)	165 (624.5)	140 (529.9)	110 (416.4)	—
40 (12.2)	158 (598.0)	140 (529.9)	110 (416.4)	75 (283.9)
50 (15.2)	145 (548.8)	130 (492.1)	106 (401.2)	70 (265.0)
60 (18.3)	126 (476.9)	117 (442.8)	97 (367.1)	68 (257.4)
70 (21.3)	102 (386.1)	100 (378.5)	85 (321.7)	60 (227.1)
80 (24.4)	74 (280.1)	74 (280.1)	68 (257.4)	48 (181.7)
90 (27.4)	40 (151.4)	40 (151.4)	40 (151.4)	32 (121.1)
Model 15-M (3 in.)				
20 (6.1)	259 (980.3)	—	—	—
30 (9.1)	250 (946.3)	210 (794.9)	200 (757.0)	—
40 (12.2)	241 (912.2)	207 (783.5)	177 (669.9)	160 (605.6)
50 (15.2)	225 (851.6)	202 (764.6)	172 (651.0)	140 (529.9)
60 (18.3)	197 (745.6)	197 (745.6)	169 (639.7)	140 (529.9)
70 (21.3)	160 (605.6)	160 (605.6)	160 (605.6)	138 (522.3)
80 (24.4)	125 (473.1)	125 (473.1)	125 (473.1)	125 (473.1)
90 (27.4)	96 (363.4)	96 (363.4)	96 (363.4)	96 (363.4)
Model 18-M (3 in.)				
25 (7.6)	301 (1,139.3)	—	—	—
30 (9.1)	295 (1,116.6)	255 (965.2)	200 (757.0)	—
40 (12.2)	276 (1,044.7)	250 (946.3)	200 (757.0)	162 (613.2)
50 (15.2)	250 (946.3)	237 (897.0)	198 (749.4)	159 (601.8)
60 (18.3)	216 (817.6)	212 (8–2.4)	182 (688.9)	146 (552.6)
70 (21.3)	174 (658.6)	174 (658.6)	158 (598.0)	127 (480.7)
80 (24.4)	129 (488.3)	129 (488.3)	125 (473.1)	104 (393.6)
90 (27.4)	82 (310.4)	82 (310.4)	82 (310.4)	74 (280.1)
95 (29.0)	57 (215.7)	57 (215.7)	57 (215.7)	57 (215.7)
Model 20-M (3 in.)				
30 (9.1)	333 (1,260.0)	280 (1,059.8)	235 (889.5)	165 (624.5)
40 (12.2)	315 (1,192.3)	270 (1,022.0)	230 (870.6)	162 (613.2)
50 (15.2)	290 (1,097.7)	255 (965.2)	220 (832.7)	154 (582.9)
60 (18.3)	255 (965.2)	235 (889.5)	205 (775.9)	143 (541.3)
70 (21.3)	212 (802.4)	209 (791.1)	184 (696.4)	130 (492.1)
80 (24.4)	165 (624.5)	165 (624.5)	157 (594.2)	114 (431.5)
90 (27.4)	116 (439.1)	116 (439.1)	116 (439.1)	94 (355.8)
100 (30.5)	60 (227.1)	60 (227.1)	60 (227.1)	60 (227.1)

Model 40-M (4 in.)				
25 (7.6)	665 (2,517.0)	—	—	—
30 (9.1)	660 (2,498.1)	575 (2,176.4)	475 (1,797.9)	355 (1,343.7)
40 (12.2)	645 (2,441.3)	565 (2,138.5)	465 (1,760.0)	350 (1,324.8)
50 (15.2)	620 (2,346.7)	545 (2,062.8)	455 (1,722.2)	345 (1,305.8)
60 (18.3)	585 (2,214.2)	510 (1,930.3)	435 (1,646.5)	335 (1,268.0)
70 (21.3)	535 (2,025.0)	475 (1,797.9)	410 (1,551.9)	315 (1,192.3)
80 (24.4)	465 (1,760.0)	410 (1,551.9)	365 (1,381.5)	280 (975.8)
90 (27.4)	375 (1,419.4)	325 (1,230.1)	300 (1,135.5)	220 (832.7)
100 (30.5)	250 (946.3)	215 (813.8)	195 (738.1)	145 (548.8)
110 (33.5)	65 (246.0)	60 (227.1)	50 (189.2)	40 (151.4)
Model 90-M (6 in.)				
25 (7.6)	1,500 (5,677.5)	—	—	—
30 (9.1)	1,480 (5,601.8)	1,280 (4,844.8)	1,050 (3,974.3)	790 (2,990.1)
40 (12.2)	1,430 (5,412.6)	1,230 (4,655.6)	1,020 (3,860.7)	780 (2,952.3)
50 (15.2)	1,350 (5,109.8)	1,160 (4,390.6)	970 (3,671.5)	735 (2,782.0)
60 (18.3)	1,225 (4,636.6)	1,050 (3,974.2)	900 (3,406.5)	690 (2,611.7)
70 (21.3)	1,050 (3,974.2)	900 (3,406.5)	775 (2,933.4)	610 (2,308.9)
80 (24.4)	800 (3,028.0)	680 (2,573.8)	600 (2,271.0)	490 (1,854.7)
90 (27.4)	450 (1,703.3)	400 (1,514.0)	365 (1,381.5)	300 (1,135.5)
100 (30.5)	100 (378.5)	100 (378.5)	100 (378.5)	100 (378.5)
Model 125-M (8 in.)				
25 (7.6)	2,100 (7,948.5)	1,850 (7,002.3)	1,570 (5,942.5)	—
30 (9.1)	2,060 (7,797.1)	1,820 (6,888.7)	1,560 (5,904.6)	1,200 (4,542.0)
40 (12.2)	1,960 (7,418.6)	1,740 (6,585.9)	1,520 (5,753.2)	1,170 (4,428.5)
50 (15.2)	1,800 (6,813.0)	1,620 (6,131.7)	1,450 (5,488.3)	1,140 (4,314.9)
60 (18.3)	1,640 (6,207.4)	1,500 (5,677.5)	1,360 (5,147.6)	1,090 (4,125.7)
70 (21.3)	1,460 (5,526.1)	1,340 (5,071.9)	1,250 (4,731.3)	1,015 (3,840.8)
80 (24.4)	1,250 (4,731.1)	1,170 (4,428.5)	1,110 (4,201.4)	950 (3,595.8)
90 (27.4)	1,020 (3,860.7)	980 (3,709.3)	940 (3,557.9)	840 (3,179.4)
100 (30.5)	800 (3,028.0)	760 (2,876.6)	710 (2,687.4)	680 (2,573.8)
110 (33.5)	570 (2,157.5)	540 (2,043.9)	500 (1,892.5)	470 (1,779.0)
120 (36.6)	275 (1,040.9)	245 (927.3)	240 (908.4)	240 (908.4)
Model 200-M (10 in.)				
20 (6.1)	3,350 (12,679.8)	3,000 (11,355.0)	—	—
30 (9.1)	3,000 (11,355.0)	2,800 (10,598.0)	2,500 (9,462.5)	1,550 (5,866.8)
40 (12.2)	2,500 (9,462.5)	2,500 (9,462.5)	2,250 (8,516.3)	1,500 (5,677.5)
50 (15.2)	2,000 (7,570.0)	2,000 (7,570.0)	2,000 (7,570.0)	1,350 (5,109.8)
60 (18.3)	1,300 (4,920.5)	1,300 (4,920.5)	1,300 (4,920.5)	1,150 (4,352.8)
70 (21.3)	500 (1,892.5)	500 (1,892.5)	500 (1,892.5)	500 (1,892.5)

Table 18-3 Minimum capacities for MT-rated, solids-handling, self-priming centrifugal pumps manufactured in accordance with the standards of the Contractors Pump Bureau

Total head including friction, ft	(meters)	Capacity, gpm (liters per min) — Height of pump above water, ft (meters): 10	(3.0)	15	(4.6)	20	(6.1)	25	(7.6)
Model 5MT (1 1/2 in.)									
20	(6.1)	82	(310.4)	64	(242.2)				
30	(9.1)	70	(265.0)	64	(242.2)	50	(189.3)	33	(124.9)
40	(12.2)	50	(189.3)	50	(189.3)	50	(189.3)	33	(124.9)
50	(15.2)	25	(94.6)	22	(83.3)	22	(83.3)	22	(83.3)
Model 10MT (2 in.)									
20	(6.1)	164	(620.7)	132	(499.6)				
30	(9.1)	164	(620.7)	132	(499.6)	105	(397.4)	75	(283.9)
40	(12.2)	164	(620.7)	132	(499.6)	105	(397.4)	75	(283.9)
50	(15.2)	164	(620.7)	132	(499.6)	105	(397.4)	75	(283.9)
60	(18.3)	135	(511.0)	132	(499.6)	105	(397.4)	75	(283.9)
70	(21.3)	88	(333.1)	88	(333.1)	88	(333.1)	68	(257.4)
80	(24.4)	40	(151.4)	40	(151.4)	40	(151.4)	40	(151.4)
Model 18MT (3 in.)									
20	(6.1)	310	(1,173.4)	265	(1,003.0)				
30	(9.1)	305	(1,154.4)	265	(1,003.0)	200	(757.0)	115	(435.3)
40	(12.2)	300	(1,135.5)	265	(1,003.0)	200	(757.0)	110	(416.4)
50	(15.2)	275	(1,040.9)	260	(984.1)	200	(757.0)	105	(397.4)
60	(18.3)	215	(813.8)	215	(813.8)	200	(757.0)	100	(378.5)
70	(21.3)	170	(643.5)	170	(643.5)	170	(643.5)	100	(378.5)
80	(24.4)	87	(329.3)	87	(329.3)	87	(329.3)	87	(329.3)
90	(27.4)	25	(94.6)	25	(94.6)	25	(94.6)	25	(94.6)
Model 33MT (4 in.)									
30	(9.1)	550	(2,081.8)	460	(1,741.1)	350	(1,324.8)	240	(908.4)
40	(12.2)	540	(2,043.9)	455	(1,722.2)	350	(1,324.8)	240	(908.4)
50	(15.2)	500	(1,892.5)	430	(1,627.6)	340	(1,286.9)	230	(870.6)
60	(18.3)	450	(1,703.3)	395	(1,495.1)	320	(1,211.2)	220	(832.7)
70	(21.3)	370	(1,400.5)	360	(1,362.6)	300	(1,135.5)	210	(794.9)
80	(24.4)	275	(1,040.9)	275	(1,040.9)	260	(984.1)	180	(681.3)
90	(27.4)	190	(719.2)	190	(719.2)	190	(719.2)	150	(567.8)
100	(30.5)	100	(378.5)	100	(378.5)	100	(378.5)	100	(378.5)

Table 18-3 (Continued)

				Model 35MT (4 in.)					
25	(7.6)	585	(2,214.2)	500	(1,892.5)	350	(1,324.8)		
30	(9.1)	585	(2,214.2)	500	(1,892.5)	350	(1,324.8)	240	(908.4)
40	(12.2)	585	(2,214.2)	500	(1,892.5)	350	(1,324.8)	240	(908.4)
50	(15.2)	585	(2,214.2)	500	(1,892.5)	350	(1,324.8)	240	(908.4)
60	(18.3)	545	(2,062.8)	500	(1,892.5)	350	(1,324.8)	240	(908.4)
70	(21.3)	495	(1,873.6)	480	(1,816.8)	350	(1,324.8)	240	(908.4)
80	(24.4)	430	(1,627.6)	420	(1,589.7)	340	(1,286.9)	240	(908.4)
90	(27.4)	320	(1,211.2)	320	(1,211.2)	260	(984.1)	220	(832.7)
100	(30.5)	100	(378.5)	100	(378.5)	100	(378.5)	100	(378.5)
				Model 70MT (6 in.)					
20	(6.1)	1,195	(4,523.1)	975	(3,690.4)				
30	(9.1)	1,180	(4,466.3)	975	(3,690.4)	715	(2,706.3)	350	(1,324.8)
40	(12.2)	1,175	(4,447.4)	950	(3,595.8)	715	(2,706.3)	350	(1,324.8)
50	(15.2)	1,160	(4,390.6)	935	(3,539.0)	715	(2,706.3)	350	(1,324.8)
60	(18.3)	1,150	(4,352.8)	925	(3,501.1)	715	(2,706.3)	350	(1,324.8)
70	(21.3)	1,120	(4,239.2)	900	(3,406.5)	715	(2,706.3)	350	(1,324.8)
80	(24.4)	950	(3,595.8)	875	(3,311.9)	700	(2,649.5)	350	(1,324.8)
90	(27.4)	700	(2,649.5)	700	(2,649.5)	600	(2,271.0)	350	(1,324.8)
100	(30.5)	450	(1,703.3)	450	(1,703.3)	450	(1,703.3)	300	(1,135.5)
110	(33.5)	200	(757.0)	200	(757.0)	200	(757.0)	200	(757.0)

curves, even though efficiency must be sacrificed in order to obtain these conditions. A pump with a flat horsepower demand permits the use of an engine or an electric motor that will provide adequate power over a wide pumping range, without a substantial surplus or deficiency, regardless of the head.

The effect of varying the speed of a centrifugal pump is illustrated by the curves in Fig. 18-6.

Capacity tables for self-priming centrifugal pumps The Contractors Pump Bureau of the AGC publishes pump standards for several types of pumps, including self-priming centrifugal pumps. The standard capacities, which were approved in 1976, are given in Tables 18-2 and 18-3 [1].

LOSS OF HEAD DUE TO FRICTION IN PIPE

Table 18-4 gives the nominal loss of head due to water flowing through new steel pipe. The actual losses may differ from the values given in the table because of variations in the diameter of a pipe and in the condition of the inside surface.

The relationship between the head of fresh water in feet and pressure in psi

Table 18-4 Friction loss for water, in feet per 100 ft of clean wrought-iron or steel pipe*

Flow, gpm	Nominal diameter of pipe, in.											
	1	$1\frac{1}{4}$	$1\frac{1}{2}$	2	$2\frac{1}{2}$	3	4	5	6	8	10	12
5	1.93	0.51										
10	6.86	1.77	0.83	0.25	0.11							
14	12.8	3.28	1.53	0.45	0.19							
20	25.1	6.34	2.94	0.87	0.36	0.13						
24	35.6	8.92	4.14	1.20	0.50	0.17						
30	54.6	13.6	6.26	1.82	0.75	0.26	0.07					
40	.....	23.5	10.79	3.10	1.28	0.44	0.12					
50	.....	36.0	16.4	4.67	1.94	0.66	0.18	0.06				
75	.....	.....	35.8	10.1	4.13	1.39	0.28	0.12				
100	.....	.....	62.2	17.4	8.51	2.39	0.62	0.20	0.08			
120	.....	.....	.....	24.7	10.0	3.37	0.88	0.29	0.12			
150	.....	.....	.....	38.0	15.4	5.14	1.32	0.33	0.17			
170	.....	.....	.....	48.4	19.6	6.53	1.67	0.54	0.22			
200	.....	.....	.....	66.3	26.7	8.90	2.27	0.74	0.30	0.08		
220	.....	.....	.....	.....	32.2	10.7	2.72	0.88	0.36	0.09		
260	.....	.....	.....	.....	44.5	14.7	3.24	1.20	0.49	0.13		
280	.....	.....	.....	.....	51.3	16.9	4.30	1.38	0.56	0.14		
300	.....	.....	.....	.....	.....	19.2	4.89	1.58	0.64	0.16		
340	.....	.....	.....	.....	.....	24.8	6.19	2.00	0.81	0.21		
400	.....	.....	.....	.....	.....	33.9	8.47	2.72	1.09	0.28	0.09	
500	.....	.....	.....	.....	.....	52.5	13.0	4.16	1.66	0.42	0.14	0.06
600	.....	.....	.....	.....	.....	.....	18.6	5.88	2.34	0.60	0.19	0.08
700	.....	.....	.....	.....	.....	.....	25.0	7.93	3.13	0.80	0.26	0.11
800	.....	.....	.....	.....	.....	.....	32.4	10.22	4.03	1.02	0.33	0.14
900	.....	.....	.....	.....	.....	.....	40.8	12.9	5.05	1.27	0.41	0.17
1,000	.....	.....	.....	.....	.....	.....	50.2	15.8	6.17	1.56	0.50	0.21
1,100	.....	.....	.....	.....	.....	.....	.....	19.0	7.41	1.87	0.59	0.25
1,200	.....	.....	.....	.....	.....	.....	.....	22.5	8.76	2.20	0.70	0.30
1,300	.....	.....	.....	.....	.....	.....	.....	.....	10.2	2.56	0.82	0.34
1,400	.....	.....	.....	.....	.....	.....	.....	.....	11.8	2.95	0.94	0.40
1,500	.....	.....	.....	.....	.....	.....	.....	.....	13.5	3.37	1.07	0.45
2,000	.....	.....	.....	.....	.....	.....	.....	.....	23.8	5.86	1.84	0.78
3,000	.....	.....	.....	.....	.....	.....	.....	.....	.....	12.8	4.00	1.68
4,000	.....	.....	.....	.....	.....	.....	.....	.....	.....	22.6	6.99	2.92
5,000	.....	.....	.....	.....	.....	.....	.....	.....	.....	.....	10.80	4.47

* Reprinted from "Tentative Standards of Hydraulic Institute, Pipe Friction," Copyright 1948 by the Hydraulic Institute, 122 E. 42d St., New York, New York, 10017.

is given by the equation

$$h = 2.31\,p$$

or

$$p = 0.434h$$

Table 18-5 Length of steel pipe, in feet, equivalent to fittings and valves*

Item	Nominal size, in.											
	1	$1\frac{1}{4}$	$1\frac{1}{2}$	2	$2\frac{1}{2}$	3	4	5	6	8	10	12
90° elbow	2.8	3.7	4.3	5.5	6.4	8.2	11.0	13.5	16.0	21.0	26.0	32.0
45° elbow	1.3	1.7	2.0	2.6	3.0	3.8	5.0	6.2	7.5	10.0	13.0	15.0
Tee, side outlet	5.6	7.5	9.1	12.0	13.5	17.0	22.0	27.5	33.0	43.5	55.0	66.0
Close return bend	6.3	8.4	10.2	13.0	15.0	18.5	24.0	31.0	37.0	49.0	62.0	73.0
Gate valve	0.6	0.8	0.9	1.2	1.4	1.7	2.5	3.0	3.5	4.5	5.7	6.8
Globe valve	27.0	37.0	43.0	55.0	66.0	82.0	115.0	135.0	165.0	215.0	280.0	335.0
Check valve	10.5	13.2	15.8	21.1	26.4	31.7	42.3	52.8	63.0	81.0	105.0	125.0
Foot valve	24.0	33.0	38.0	46.0	55.0	64.0	75.0	76.0	76.0	76.0	76.0	76.0

* Courtesy the Gormon-Rupp Company.

Table 18-6 Friction loss for water, in feet per 100 ft of rubber or rubber-substitute hose

Flow gpm	ID of hose, in.								
	$\frac{3}{4}$	1	$1\frac{1}{4}$	$1\frac{1}{2}$	2	$2\frac{1}{2}$	3	4	5
5	9.25	2.54	0.93	0.46					
10	32.3	9.25	2.78	1.15	0.46	0.23			
15	69.3	20.6	5.78	2.54	0.93	0.46			
20	125.0	32.3	9.93	4.16	1.62	0.69			
25	......	50.8	15.0	6.70	2.31	0.93			
30	......	71.7	21.2	9.25	3.23	1.15	0.23		
35	......	94.8	27.7	12.5	4.15	1.38	0.46		
40	......	125.0	34.6	15.5	5.55	1.85	0.69		
50	......	......	55.5	23.1	8.32	2.31	1.15		
60	......	......	80.9	32.3	11.8	3.23	1.38		
70	......	......	103.8	43.9	15.2	4.15	1.85		
80	......	......	134.0	55.5	19.9	5.31	2.54		
90	......	......	......	69.3	25.4	6.93	3.23	0.69	
100	......	......	......	85.5	28.8	8.1	3.93	1.15	0.23
125	......	......	......	......	46.2	12.2	5.78	1.38	0.46
150	......	......	......	......	62.4	17.3	8.1	1.62	0.69
175	......	......	......	......	85.5	23.1	10.6	2.54	0.93
200	......	......	......	......	106.1	30.0	13.6	3.23	1.15
250	......	......	......	......	......	37.0	16.1	4.16	1.38
300	......	......	......	......	......	43.8	21.0	4.85	1.62
350	......	......	......	......	......	62.3	27.7	6.70	2.31
400	......	......	......	......	......	106.0	48.5	9.25	3.93
450	......	......	......	......	......	......	60.0	14.5	4.85
500	......	......	......	......	......	......	74.0	17.1	6.00
1,000	......	......	......	......	......	......	......	62.3	22.2

where h = depth of water or head, ft
p = pressure at depth h, psi

Table 18-5 gives the equivalent length of straight steel pipe having the same loss in head due to water friction as fittings and valves.

LOSS OF HEAD DUE TO FRICTION IN RUBBER HOSE

The flexibility of rubber hose makes it a desirable substitute for pipe for use with pumps on many jobs. Such hose may be used on the suction side of a pump if it is constructed with a wire insert to prevent collapse under partial vacuum. Rubber hose is available with end fittings corresponding with those for iron or steel pipe.

Table 18-6 gives the loss in head in feet per 100 ft due to friction caused by water flowing through the hose. The values in the table apply for rubber substitutes.

SELECTING A PUMP

Before a pump for a given job is selected, it is necessary to analyze all information and conditions that will affect the selection. The most satisfactory pumping equipment will be the combination of pump and pipe that will provide the required service for the least total cost. The total cost includes the installed and operating cost of the pump and pipe for the period that it will be used, with an appropriate allowance for salvage value at the completion of the project. In order to analyze the cost of pumping water, it is necessary to have certain information, such as the following:

1. Rate at which the water is to be pumped
2. Height of lift from the existing water surface to the point of discharge
3. Pressure head at discharge, if any
4. Variations in water level at suction or discharge
5. Altitude of the project
6. Height of the pump above the surface of water to be pumped
7. Size of pipe to be used, if already determined
8. Number, sizes, and types of fittings and valves in the pipeline

The examples which follow are intended to illustrate methods of selecting pumps and pumping systems:

Example 18-2 Select a self-priming centrifugal pump, with a capacity of 600 gpm, for the project illustrated in Fig. 18-7. All pipe, fittings, and valves will be 6 in. with threaded connections.

Use the information given in Table 18-5 to convert the fittings and valves into equivalent lengths of pipe.

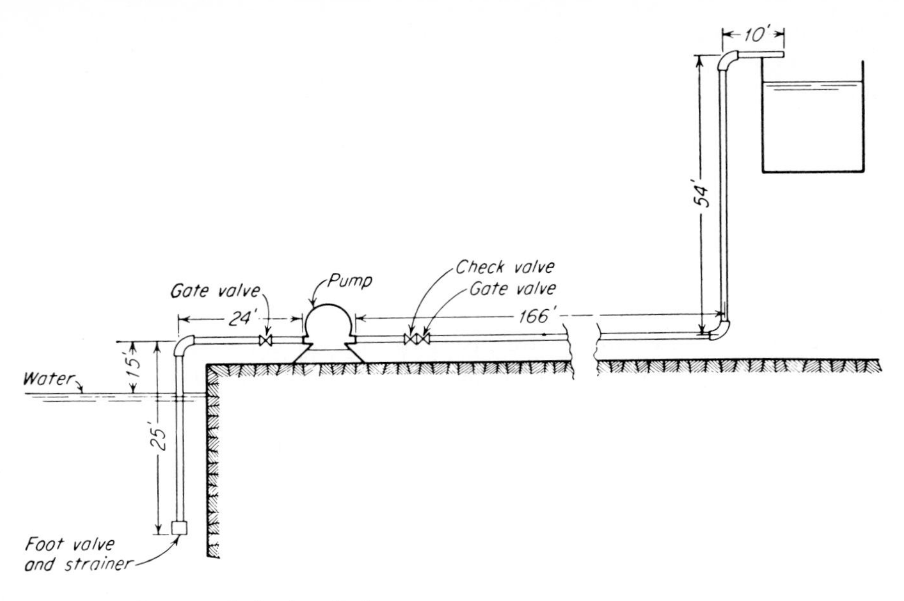

Figure 18-7 Pump and pipe installation.

Item	Equivalent length of pipe, ft
1 foot valve and strainer	= 76
3 elbows @ 16 ft	= 48
2 gate valves @ 3.5 ft	= 7
1 check valve	= 63
Total	= 194
Add length of pipe	= 279
Total equivalent length of 6-in. pipe	= 473

From Table 18-4 the friction loss per 100 ft of 6-in. pipe will be 2.34 ft. The total head, including lift plus head lost in friction, will be

Lift, 15 + 54	= 69.0 ft
Head lost in friction, 473 ft @ 2.34 ft per 100 ft	= 11.1 ft
Total head	= 80.1 ft

Table 18-2 indicates that a model 90-M pump will deliver the required quantity of water.

Sometimes the problem is to select the pump and pipeline that will permit water to be pumped at the lowest total cost. The following example illustrates a method that may be used to select the most economical pumping system:

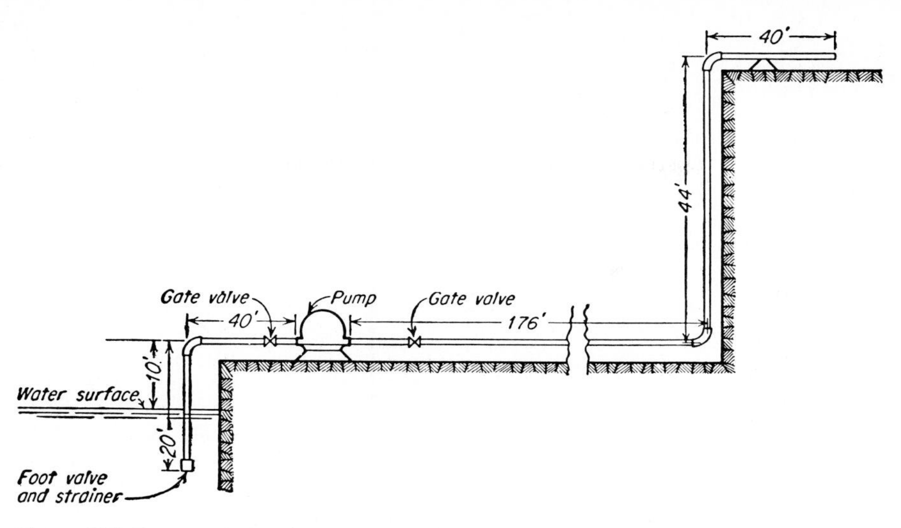

Figure 18-8 Pump and pipe installation.

Example 18-3 In operating a rock quarry it is necessary to pump 400 gpm of clear water. The pump and pipeline selected will be installed as illustrated in Fig. 18-8. It is estimated that the pump will be operated a total of 1,200 hr per year. Compare the economy of using 4-in. and 6-in. steel pipe for the water line. Assume that the pump will have an economic life of 5 years and that the pipeline and fittings will have a life of 10 years. Also, assume that the cost of installing the pipeline will be the same regardless of the size, so that this cost may be disregarded.

Consider the use of 4-in. pipe. The total equivalent length of pipe will be:

Item	Equivalent length of pipe, ft
1 foot valve and strainer	= 75
3 elbows @ 11 ft	= 33
2 gate valves @ 2.5 ft	= 5
Pipe	= 320
Total equivalent length	= 433

The total head, including lift and head lost in friction, will be

Lift, 10 + 44	= 54.0 ft
Head lost in friction, 433 ft @ 8.47 ft per 100 ft	= 36.7 ft
Total head	= 90.7 ft

A model 90-M self-priming pump, with a capacity of approximately 450 gpm, will be required for this installation.

Consider the use of 6-in. pipe. The total equivalent length of pipe will be:

Item	Equivalent length of pipe, ft
1 foot valve and strainer	= 76
3 elbows @ 16 ft	= 48
2 gate valves @ 3.5 ft	= 7
Pipe	= 320
Total equivalent length	= 451

The total head, including lift and head lost in friction, will be

Lift, 10 + 44	= 54.0 ft
Head lost in friction, 451 ft @ 1.09 ft per 100 ft	= 4.9 ft
Total head	= 58.9 ft

A model 30-M self-priming pump, with a capacity of approximately 450 gpm, will be satisfactory for this installation. The excess capacity of this pumping system is an advantage in favor of using 6-in. pipe.

The cost of each size pipeline, fittings, and valves will be

Item	Size pipe	
	4 in.	6 in.
320-ft pipeline	$776.00	$1,340.00
3 elbows	12.00	24.00
1 foot valve	24.00	36.00
2 gate valves	168.00	216.00
Total cost	$980.00	$1,616.00
Depreciation cost per year, based of 10-yr life	98.00	161.60
Depreciation cost per hr, based on 1,200 hr per yr	0.08	0.14

The combined cost per hour for each size pump and pipeline system will be:

Item	Cost per hr	
	4-in. pipe	6-in. pipe
Pump	$2.62	$1.74
Pipe, fittings, and valves	0.08	0.14
Total cost per hr	$2.70	$1.88

This analysis shows that the additional cost of the 6-in. pipe is more than offset by the reduction in the cost of the smaller pump.

WELLPOINT SYSTEMS

In excavating below the surface of the ground, it is not uncommon practice to encounter ground water before reaching the bottom of a pit. For pits excavated into sand and gravel, the flow of water will be large if some method is not adopted to remove the water before it enters the pit. While the water may be permitted to flow into sumps located in the pit, then removed by pumps, the presence of such water usually creates a nuisance and interferes with the construction operations. The installation of a wellpoint system along or around the pit may lower the water table below the bottom of the excavation, thus permitting the work to be done under relatively dry conditions.

A wellpoint is a perforated tube enclosed in a screen, which is installed below the surface of the ground to collect water in order that the water may be removed from the ground. The essential parts of a wellpoint are illustrated in Fig. 18-9. The top of a wellpoint is attached to a riser pipe, which extends a short distance above the surface of the ground, where it is connected to a large pipe called a header. The header pipe is connected to the suction of a centrifugal pump. A wellpoint system may include a few or several hundred wellpoints, all connected to one or more headers and pumps.

The principle by which a wellpoint system operates is illustrated in Fig. 18-10. Figure 18-10(a) shows how a single point will lower the surface of the

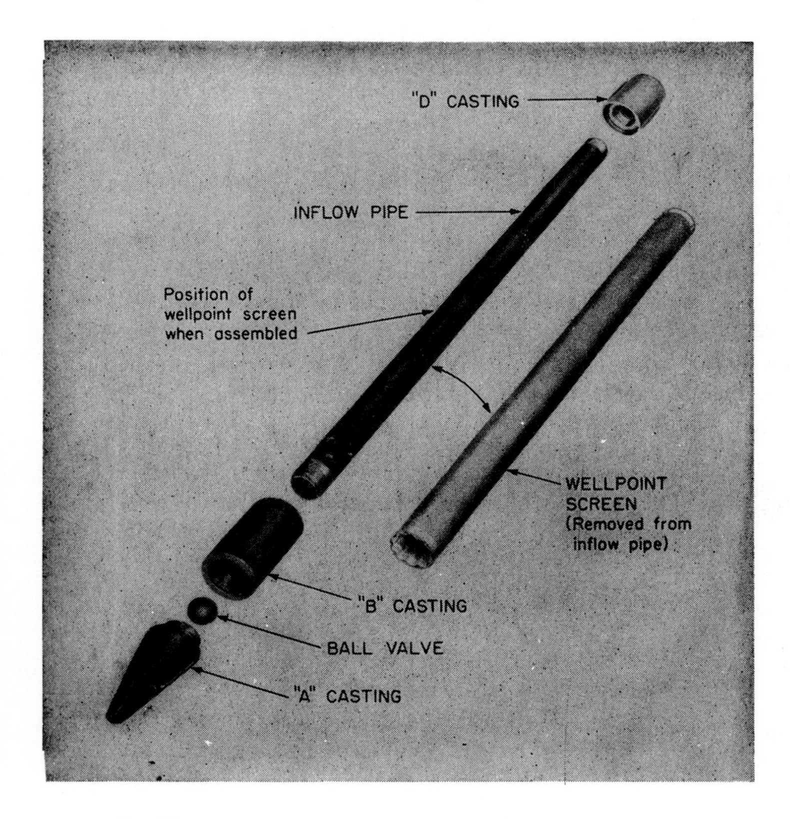

Figure 18-9 The essential parts of a wellpoint system.

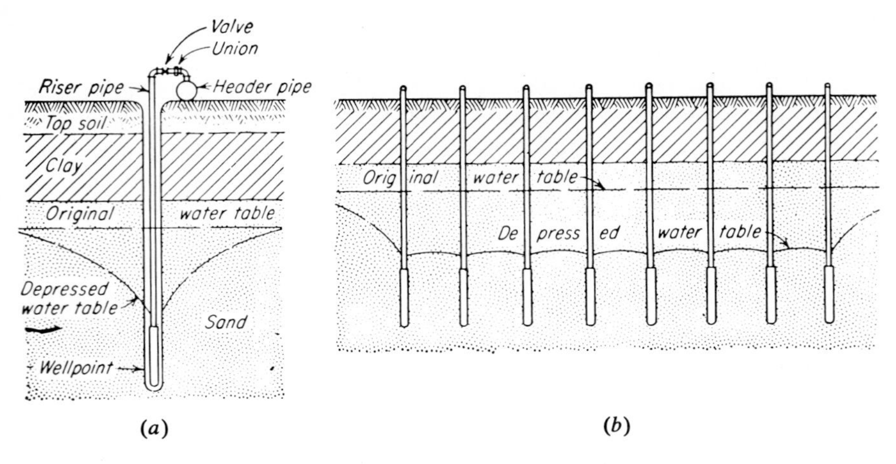

Figure 18-10 Lowering the water table adjacent to wellpoints.

water table in the soil adjacent to the point. Figure 18-10(*b*) shows how several points, installed reasonably close together, lower the water table over an extended area. A group of wellpoints properly installed along a trench or around a foundation pit will lower the water table below the depth of excavation.

Wellpoints will operate satisfactorily if they are installed in a permeable soil such as sand or gravel. If they are installed in a less permeable soil, such as silt, it may be necessary first to sink a large pipe, say 6 to 10 in. in diameter, for each point, remove the soil from inside the pipe, install a wellpoint, fill the space inside the pipe with sand or fine gravel, then withdraw the pipe. This leaves a volume of sand around each wellpoint to act as a water collector and a filter to increase the rate of flow for each point.

Wellpoints may be installed at any desired spacing, usually varying from 2 to 5 ft, along the header. The maximum height that water can be lifted is about 18 to 20 ft. If it is necessary to lower the water table to a greater depth, one or more additional stages should be installed, each stage at a lower depth within the excavation. Figure 18-11 shows a project on which two stages were used. Figure 18-12 shows a typical wellpoint installation prior to starting excavation.

Installing a wellpoint system If the soil conditions are suitable, a wellpoint is jetted into position by forcing water through an opening at the bottom of the point. After each point is jetted into position, it is connected through a pipe or a rubber hose to a header pipe, usually 6, 8, or 10 in. in diameter. A valve is installed between each wellpoint and the header to regulate the flow of water. The header is connected to a self-priming centrifugal pump, which is equipped with an auxiliary air pump to remove any air from the water before it enters the pump proper.

Figure 18-11 Two-stage wellpoint installation. *(L. B. Foster Company.)*

Capacity of a wellpoint system The capacity of a wellpoint system depends on the number of points installed, the permeability of the soil, and the amount of water present. An engineer who is experienced in this kind of work can make tests which will enable him to estimate with reasonable accuracy the capacity necessary to lower the water to the desired depth. The flow per wellpoint may vary from 3 or 4 gpm to as much as 30 or more on some installations.

When excavating 45 to 50 ft below the surface of the water in the Colorado River for the cutoff wall for the Morelos Dam, the contractor installed three main stages, with a supplemental fourth stage of wellpoints to enclose an area of 15 acres. A total of 2,750 wellpoints were serviced by 49 pumps. The maximum pumping rate was 17,400 gpm, with 2,150 wellpoints in operation. This gave an average yield of 8.1 gpm per point and 528 gpm per pump.

Prior to designing the wellpoint system for the Davis Dam on the Colorado River, a 45- by 58-ft test area was enclosed with 66 wellpoints, spaced $3\frac{1}{2}$ ft apart, each 21 ft long. The test, which was run for 172 hr, using two 8-in. pumps, gave an average yield of 13 gpm per wellpoint.

Figure 18-12 Single-stage wellpoint installation. *(Moretrench-American Corporation.)*

Figure 18-13 Three-stage wellpoint installation. *(Moretrench-American Corporation.)*

PROBLEMS

18-1 A two-cylinder duplex double-acting pump, size 6 by 12 in., is driven by a crankshaft which makes 120 rpm. If the water slippage is 7 percent, how many gallons of water will the pump deliver per minute? If the total head is 100 feet and the efficiency of the pump is 60 percent, what is the minimum horsepower required to operate the pump?

18-2 The centrifugal pump whose performance curves are given in Fig. 18-5 will be used to pump water against a total head of 50 ft. The dynamic suction lift will be 10 ft. Determine the capacity and efficiency of the pump and the horsepower required to operate the pump.

18-3 A centrifugal pump is to be used to pump all the water from a cofferdam whose dimensions are 70 ft long, 50 ft wide, and 12 ft deep. The water must be pumped against an average total head of 45 ft. The average height of the pump above the water will be 12 ft. If the cofferdam must be emptied in 15 hr, determine the minimum model self-priming pump Class M to be used based on the ratings of the Contractors Pump Bureau.

18-4 Use Table 18-2 to select a centrifugal pump to handle 500 gpm of water. The water will be pumped from a pond through 460 ft of 6-in. pipe to a point 30 ft above the level of the pond, where it will be discharged into the air. The pump will be set 10 ft above the surface of the water in the pond. What is the designation of the pump selected?

18-5 Select a self-priming centrifugal pump to handle 600 gpm of water for the project illustrated in Fig. 18-7. Increase the height of the vertical pipe from 54 ft to 60 ft. All other conditions will be as shown in the figure.

18-6 Select a self-priming centrifugal pump to handle 600 gpm of water for the project illustrated in Fig. 18-7. Change the size of the pipe, fittings, and valves to 6 in.

REFERENCE

1. "Contractors Pump Manual," Contractors Pump Bureau, 13975 Connecticut Avenue, Silver Spring, MD 20906.

CHAPTER
NINETEEN
THE PRODUCTION OF CRUSHED-STONE AGGREGATE

INTRODUCTION

The production of crushed-stone aggregate involves drilling, blasting, loading, transporting, crushing, screening, handling, and storing the aggregate. As the first four operations have already been discussed, this chapter will be devoted to studies of the last four operations.

In operating a quarry and a crushing plant, the drilling pattern, the amount of explosives, the size power shovel to load the stone, and the size of the primary crusher should be coordinated to assure that all stone from the quarry can pass through the opening to the crusher. It is desirable for the loading capacity of the shovel and the capacity of the crushing plant to be approximately equal. Table 19-1 gives the recommended minimum sizes of jaw and gyratory crushers required to handle the stone passing through power-shovel dippers of the specified capacities.

TYPES OF CRUSHERS

Crushers may be classified according to the stage of crushing which they accomplish, such as primary, secondary, tertiary, etc. A primary crusher receives the stone directly from a quarry and produces the first reduction in size. The output of the primary crusher is fed to a secondary crusher, which further reduces the size. Some of the stone may pass through four or more crushers before it is reduced to the necessary fineness.

Table 19-1 Recommended minimum sizes of primary crushers for use with shovel dippers of the indicated capacities

Capacity of dipper, cu yd (cu m)		Jaw crusher, in. (mm)*		Gyratory crusher, size of openings,† in. (mm)	
$\frac{3}{4}$	(0.575)	28 × 36	(712 × 913)	16	(406)
1	(0.765)	28 × 36	(712 × 913)	16	(406)
$1\frac{1}{2}$	(1.145)	36 × 42	(913 × 1,065)	20	(508)
$1\frac{3}{4}$	(1.340)	42 × 48	(1,065 × 1,200)	26	(660)
2	(1.530)	42 × 48	(1,065 × 1,200)	30	(760)
$2\frac{1}{2}$	(1.910)	48 × 60	(1,260 × 1,525)	36	(915)
3	(2.295)	48 × 60	(1,260 × 1,525)	42	(1,066)
$3\frac{1}{2}$	(2.668)	48 × 60	(1,260 × 1,525)	42	(1,066)
4	(3.060)	56 × 72	(1,420 × 1,830)	48	(1,220)
5	(3.820)	66 × 86	(1,675 × 2,182)	60	(1,520)

* The first two digits are the width of the opening at the top of the crusher, measured perpendicular to the jaw plates. The second two digits are the width of the opening, measured across the jaw plates.

† The recommended sizes are for gyratory crushers equipped with straight concaves.

While there is no rigid classification of crushers, the following is representative of common crusher uses.

1. Primary crushers
   a. Jaw
   b. Gyratory
   c. Hammer mill
2. Secondary crushers
   a. Cone
   b. Roll
   c. Hammer mill
3. Tertiary crushers
   a. Roll
   b. Rod mill
   c. Ball mill

As stone passes through a crusher, it undergoes a reduction in size, which may be expressed as a ratio of reduction. The ratio of reduction is the ratio of the distance between the fixed and moving faces at the top divided by the distance at the bottom of a crusher. Thus, if the distance between the two faces of a jaw crusher at the top is 16 in. and at the bottom is 4 in., the ratio of reduction is 4.

The ratio of reduction for a roller crusher is the ratio of the dimension of the largest stone that can be nipped by the rolls divided by the setting of the rolls, which is the smallest distance between the faces of the rolls.

Jaw crushers This machine is very popular as a primary crusher. It operates by allowing stone to flow into the space between two jaws, one of which is stationary, while the other is movable. The distance between the jaws diminishes as the stone travels downward under the effect of gravity and the movable jaw, until it ultimately passes through the lower opening. The movable jaw is capable of exerting a pressure sufficiently high to crush the hardest rock.

The Blake type, illustrated in Fig. 19-1, is a double-toggle crusher. The movable jaw is suspended from a shaft mounted on bearings on the crusher frame. The crushing operation is effected by rotating an eccentric shaft, which raises and lowers the pitman, which actuates the two toggles. As the two toggles are raised by the pitman, a high pressure is exerted near the bottom of the swing jaw, which partially closes the opening between the bottoms of the two jaws. This operation is repeated as the eccentric shaft is rotated.

The jaw plates, which are made of manganese steel, may be removed, replaced, or, in some cases, reversed. The jaws may be smooth, or, in the event the stone tends to break into slabs, corrugated jaws may be used to reduce the slabbing. The swing jaw may be straight, or it may be curved to reduce the danger of choking.

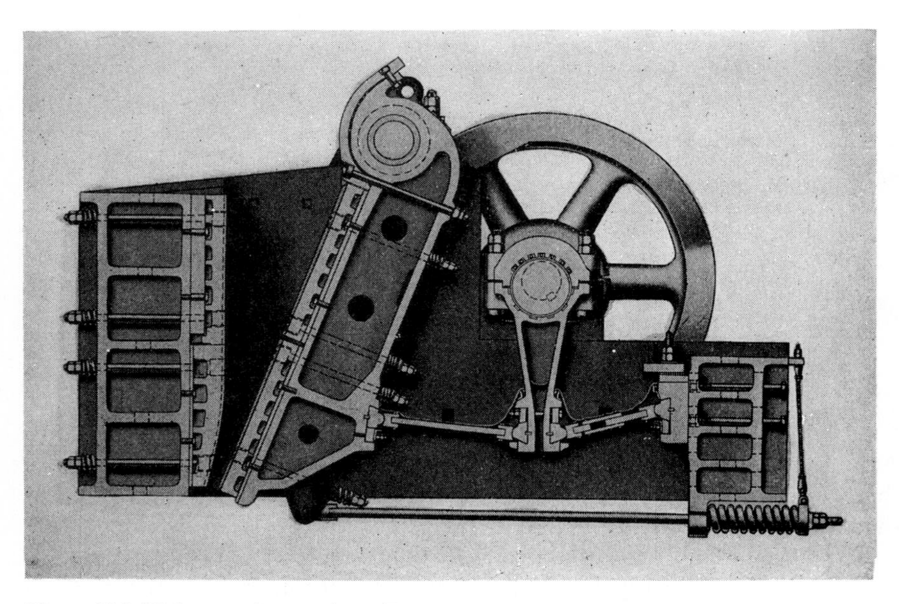

Figure 19-1 Blake-type jaw crusher. *(Fiat-Allis Construction Machinery, Inc.)*

Figure 19-2 Toggle-type jaw crusher. *(Portec, Inc., Pioneer Division.)*

When the eccentric shaft of the single-toggle crusher, illustrated in Fig. 19-2, is rotated, it gives the movable jaw a vertical and horizontal motion. This type crusher is used quite frequently in portable rock-crushing plants because of its compact size, light weight, and reasonably sturdy construction. The capacity of the single-toggle crusher is usually less than that of the Blake-type unit.

When a jaw crusher is used as a primary crusher, the size may be determined by the capacity of the shovel dipper, as indicated in Table 19-1, in which case the capacity of the crusher may be secondary. A jaw crusher should have a top opening at least 2 in. wider than the largest stones that will be fed to it.

Table 19-2 gives representative capacities for various sizes of jaw crushers. As the setting may be based on the open or closed position of the bottom of the swing jaw, a capacity table should specify which setting applies. The closed

Table 19-2 Representative capacities of Blake-type jaw crushers, in tons per hour (metric tons per hour) of stone

Size crusher, in. (mm)*	Maximum rpm	Maximum hp (kW)	Closed setting of discharge opening, in. (mm)										
			1 (25.4)	$1\frac{1}{2}$ (38.1)	2 (50.8)	$2\frac{1}{2}$ (63.5)	3 (76.2)	4 (102)	5 (137)	6 (152)	7 (178)	8 (203)	9 (229)
10 × 6	300	15	11	16	20								
(254 × 406)		(11.2)	(10)	(14)	(18)								
10 × 20	300	20	14	20	25	34							
(254 × 508)		(14.9)	(13)	(18)	(23)	(31)							
15 × 24	275	30		27	34	42	50						
(381 × 610)		(22.4)		(24)	(31)	(38)	(45)						
15 × 30	275	40		33	43	53	62						
(381 × 762)		(29.8)		(30)	(39)	(48)	(56)						
18 × 36	250	60		46	61	77	93	125					
(458 × 916)		(44.8)		(42)	(55)	(69)	(84)	(113)					
24 × 36	250	75			77	95	114	150					
(610 × 916)		(56.0)			(69)	(86)	(103)	(136)					
30 × 42	200	100				125	150	200	250	300			
(762 × 1,068)		(74.6)				(113)	(136)	(181)	(226)	(272)			
36 × 42	175	115				140	160	200	250	300			
(916 × 1,068)		(85.5)				(127)	(145)	(181)	(226)	(272)			
36 × 48	160	125				150	175	225	275	325	375		
(916 × 1,220)		(93.2)				(136)	(158)	(202)	(249)	(294)	(339)		
42 × 48	150	150				165	190	250	300	350	400	450	
(1,068 × 1,220)		(111.9)				(149)	(172)	(226)	(272)	(318)	(364)	(408)	
48 × 60	120	180					220	280	340	400	450	500	550
(1,220 × 1,542)		(134.7)					(200)	(254)	(309)	(364)	(408)	(454)	(500)
56 × 72	95	250						315	380	450	515	580	640
(1,422 × 1,832)		(186.3)						(286)	(345)	(408)	(468)	(527)	(580)

* The first number indicates the width of the feed opening, while the second number indicates the width of the jaw plates.

position is most commonly used and is the basis for the values given in Table 19-2. The capacity is given in tons per hour for stone weighing 100 lb per cu ft when crushed.

Gyratory crushers A section through a gyratory crusher is illustrated in Fig. 19-3. The crusher unit consists of a heavy cast-iron or steel frame, with an eccentric shaft setting and driving gears in the lower part of the unit. In the upper part there is a cone-shaped crushing chamber, lined with hard-steel or manganese-steel plates called the concaves. The crushing member includes a hard-steel crushing head mounted on a vertical steel shaft. This shaft and head are suspended from the spider at the top of the frame, which is so constructed that some vertical adjustment of the shaft is possible. The eccentric support at the bottom causes the shaft and the crushing head to gyrate as the shaft rotates, thereby varying the width of the space between the concaves and the head. As the rock which is fed in at the top of the crushing chamber moves downward, it undergoes a reduction in size until it finally passes through the opening at the bottom of the chamber.

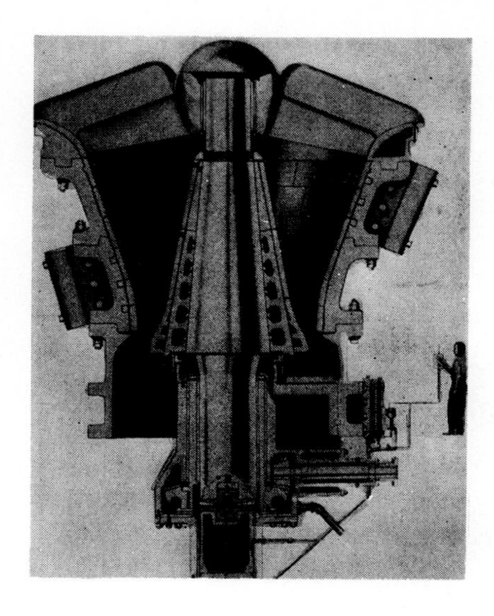

Figure 19-3 Hydroset gyratory crusher. *(Fiat-Allis Construction Machinery, Inc.)*

The size of a gyratory crusher is the width of the receiving opening, measured between the concaves and the crusher head. The setting is the width of the bottom opening and may be the open or closed dimension. When a setting is given, it should be specified if it is the open or closed dimension.

The ratio of reduction for gyratory crushers usually varies from about 5.5 to 7.5, with an average value around 6.5 for the sizes up to 42 in.

If a gyratory crusher is used as a primary crusher, the size selected may be dictated by the size of the rock from the quarry or it may be dictated by a desired capacity. When this machine is used as a secondary crusher, the capacity of a gyratory crusher may be increased by increasing the speed of the machine within reasonable limits.

Table 19-3 gives representative capacities of gyratory crushers, expressed in tons per hour, based on a continuous feed of stone weighing 100 lb per cu ft when crushed. The crushers with straight concaves are commonly used as primary crushers, while those with nonchoking concaves are commonly used as secondary crushers.

Cone crushers Cone, or reduction, crushers are used as secondary or tertiary crushers. They are capable of producing large quantities of uniformly fine crushed stone. A cone crusher differs from a gyratory crusher in the following respects:

1. It has a shorter cone.
2. It has a smaller receiving opening.
3. It rotates at a higher speed, from 430 to 580 rpm.
4. It produces a more uniformly sized stone with the maximum size equal to the width of the closed-side setting.

Figure 19-4 Gyrosphere crusher. *(Telsmith Division, Barber-Greene Company.)*

Figure 19-5 shows a section through a Symons standard cone crusher. The conical head, made usually of manganese steel and mounted on the vertical shaft, serves as one of the crushing surfaces. The other surface is the concave, which is attached to the upper part of the crusher frame. The bottom of the shaft is set in an eccentric bushing to produce the gyratory effect as the shaft rotates.

While the maximum diameter of the crusher head may be used to designate the size of a cone crusher, the size of the feed opening, which limits the size of rocks that may be fed to the crusher, is the width of the opening at the entrance to the crushing chamber. The magnitude of the eccentric throw and the setting of the discharge opening may be varied within reasonable limits. Because of the high speed of rotation all particles passing through a crusher will be reduced to sizes no larger than the closed-size setting, which should be used to designate the size of the discharge opening.

Table 19-3 Representative capacities of gyratory crushers, in tons per hour (metric tons per hour) of stone

Size of crusher, in. (mm)	Approximate power required, hp (kW)	Open-side setting of crusher, in. (mm)											
		$1\frac{1}{2}$ (38)	$1\frac{3}{4}$ (44)	2 (51)	$2\frac{1}{4}$ (57)	$2\frac{1}{2}$ (63)	3 (76)	$3\frac{1}{2}$ (89)	4 (102)	$4\frac{1}{2}$ (114)	5 (127)	$5\frac{1}{2}$ (140)	6 (152)
Straight concaves													
8 (20.0)	15–25 (11–19)	30 (27)	36 (33)	41 (37)	47 (42)								
10 (25.4)	25–40 (19–30)		40 (36)	50 (45)	60 (54)								
13 (33.1)	50–75 (37–56)				85 (77)	100 (90)	133 (120)						
16 (40.7)	60–100 (45–75)						160 (145)	185 (167)	210 (190)				
20 (50.8)	75–125 (56–93)							200 (180)	230 (208)	255 (271)			
30 (76.2)	125–175 (93–130)								310 (281)	350 (317)	390 (353)		
42 (106.7)	200–275 (150–205)										500 (452)	570 (515)	630 (569)
Modified straight concaves													
8 (20.0)	15–25 (11–19)	35 (32)	40 (36)	45 (41)									
10 (25.4)	25–40 (19–30)		54 (49)	60 (54)	65 (59)								
13 (33.1)	50–75 (37–56)					95 (86)	130 (117)						
16 (40.7)	60–100 (45–75)						150 (135)	172 (155)	195 (176)				
20 (50.8)	75–125 (56–93)							182 (165)	200 (180)	220 (199)			
30 (76.2)	125–175 (93–130)								340 (308)	370 (335)	400 (362)		
42 (106.7)	200–275 (150–205)										607 (550)	650 (589)	690 (625)
Nonchoking concaves													
8 (20.0)	15–25 (11–19)	42 (38)	46 (42)										
10 (25.4)	25–40 (19–30)	51 (46)	57 (52)	63 (57)	69 (62)								
13 (33.1)	50–75 (37–56)	79 (71)	87 (79)	95 (86)	103 (93)	111 (100)							
16 (40.7)	60–100 (45–75)			107 (96)	118 (106)	128 (115)	150 (135)						
20 (50.8)	75–125 (56–93)				155 (140)	169 (152)	198 (178)	220 (198)	258 (233)	285 (257)	310 (279)		

Figure 19-5 Symons standard cone crusher. *(Nordberg Manufacturing Company.)*

Table 19-4 gives representative capacities for the Symons standard cone crusher, expressed in tons of stone per hour for material weighing 100 lb per cu ft when crushed.

Hammer mills The hammer mill, which is the most widely used impact crusher, may be used for primary or secondary crushing. The basic parts of a unit include a housing frame, a horizontal shaft extending through the housing, a number of arms and hammers attached to a spool which is mounted on the shaft, one or more manganese-steel or other hard-steel breaker plates, and a series of grate bars whose spacings may be adjusted to regulate the width of openings through which the crushed stone flows. These parts are illustrated in the section through the crusher shown in Fig. 19-6.

As the stone to be crushed is fed to the mill, the hammers, which travel at a high speed, strike the particles, breaking them and driving them against the breaker plates, which further reduces their sizes.

The size of a hammer mill may be designated by the size of the feed opening. The capacity will vary with the size of the unit, the kind of stone crushed, the size of the material fed to the mill, and the speed of the shaft. Table 19-5 gives representative capacities of hammer mills expressed in tons of stone per hour for material weighing 100 lb per cu ft when crushed.

Roll crushers Roll crushers are used for producing additional reductions in the sizes of stone after the output of a quarry has been subjected to one or more stages of prior crushing. A roll crusher consists of a heavy cast-iron frame equipped with two hard-steel rolls, each mounted on a separate horizontal shaft. Most crushers are so constructed that each roll is driven independently by a flat-belt pulley or a V-belt sheave. One of the rolls is mounted on a slide frame, to permit an adjustment in the width of the discharge opening between the two rolls. The movable roll is spring-loaded to provide safety against

Table 19-4 Representative capacities of Symons standard cone crushers, in tons per hour (metric tons per hour) of stone*

Size of crusher, ft (m)	Size of feed opening, in. (mm)	Minimum discharge settings, in. (mm)	Discharge setting, in. (mm)										
			$\frac{1}{4}$ (6.3)	$\frac{3}{8}$ (9.5)	$\frac{1}{2}$ (12.7)	$\frac{5}{8}$ (15.9)	$\frac{3}{4}$ (19.1)	$\frac{7}{8}$ (22.3)	1 (25.4)	$1\frac{1}{4}$ (31.8)	$1\frac{1}{2}$ (38.0)	2 (50.8)	$2\frac{1}{2}$ (63.5)
2 (0.61)	$2\frac{1}{4}$ (57)	$\frac{1}{4}$ (5.6)	15 (14)	20 (18)	25 (23)	30 (27)	35 (32)						
2 (0.61)	$3\frac{1}{4}$ (82)	$\frac{3}{8}$ (9.5)		20 (18)	25 (23)	30 (27)	35 (32)	40 (36)	45 (41)	50 (45)	60 (54)		
3 (0.91)	$3\frac{7}{8}$ (96)	$\frac{3}{8}$ (9.5)		35 (32)	40 (36)	55 (50)	70 (63)	75 (68)					
3 (0.91)	$5\frac{1}{8}$ (130)	$\frac{1}{2}$ (12.7)			40 (36)	55 (50)	70 (63)	75 (68)	80 (72)	85 (77)	90 (81)	95 (86)	
4 (1.22)	5 (127)	$\frac{3}{8}$ (9.5)		60 (54)	80 (72)	100 (90)	120 (109)	135 (122)	150 (136)				
4 (1.22)	$7\frac{3}{8}$ (187)	$\frac{3}{4}$ (19.0)					120 (109)	135 (122)	150 (136)	170 (154)	177 (160)	185 167)	

Table 19-4 (*continued*)

$4\frac{1}{4}$	$4\frac{1}{2}$	$\frac{1}{2}$			100	125	140	150					
(1.29)	(114)	(12.7)			(90)	(113)	(126)	(136)					
$4\frac{1}{4}$	$7\frac{3}{8}$	$\frac{5}{8}$				125	140	150	160	175			
(1.29)	(187)	(15.8)				(113)	(126)	(136)	(145)	(158)			
$4\frac{1}{4}$	$9\frac{1}{2}$	$\frac{3}{4}$					140	150	160	175	185	190	
(1.29)	(241)	(19.0)					(126)	(136)	(145)	(158)	(167)	(172)	
$5\frac{1}{2}$	$7\frac{1}{8}$	$\frac{5}{8}$				160	200	235	275				
(1.67)	(181)	(15.8)				(145)	(181)	(213)	(249)				
$5\frac{1}{2}$	$8\frac{5}{8}$	$\frac{7}{8}$						235	275	300	340	375	450
(1.67)	(219)	(22.2)						(213)	(249)	(272)	(304)	(340)	(407)
$5\frac{1}{2}$	$9\frac{7}{8}$	1							275	300	340	375	450
(1.67)	(248)	(25.4)							(249)	(272)	(304)	(340)	(407)
7	10	$\frac{3}{4}$					330	390	450	560	600		
(2.30)	(254)	(19.0)					(300)	(353)	(407)	(507)	(543)		
7	$11\frac{1}{2}$	1							450	560	600	800	
(2.30)	(292)	(25.4)							(407)	(507)	(543)	(725)	
7	$13\frac{1}{2}$	$1\frac{1}{4}$								560	600	800	900
(2.30)	(343)	(31.7)								(507)	(543)	(725)	(815)

* Courtesy Nordberg Manufacturing Company.

Figure 19-6 Cutaway of hammer mill rock crusher showing breaking action. *(Iowa Manufacturing Company.)*

Table 19-5 Representative capacities for hammer mills, in tons per hour (metric tons per hour) of stone*

Size feed opening, in. (mm)	Size feed, in. (mm)	Power required, hp (kW)	Width of openings between grate bars, in. (mm)						
			$\frac{1}{8}$ (3.2)	$\frac{3}{16}$ (4.7)	$\frac{1}{4}$ (6.4)	$\frac{3}{8}$ (9.5)	$\frac{1}{2}$ (12.7)	1 (25.4)	$1\frac{1}{4}$ (31.8)
$6\frac{1}{4} \times 9$ (159 × 229)	3 (76.2)	15–20 (11–15)	2.5 (2.3)	3.5 (3.2)	5.0 (4.5)	8.0 (7.2)	10.0 (9.1)		
12 × 15 (304 × 380)	3 (76.2)	50–60 (37–45)	9.0 (8.2)	13.0 (11.8)	17.0 (15.4)	23.0 (20.8)	29.0 (26.2)	36.0 (32.6)	39.0 (35.2)
15 × 25 (380 × 635)	6 (152.4)	100–125 (75–93)	18.0 (16.4)	25.0 (22.6)	31.0 (28.1)	40.0 (36.3)	47.0 (42.6)	65.0 (59.0)	70.0 (63.5)
15 × 37 (380 × 940)	6 (152.4)	150–200 (112–149)	27.0 (24.5)	37.0 (33.6)	47.0 (42.6)	60.0 (54.5)	71.0 (64.4)	97.0 (88.0)	105.0 (95.4)
15 × 49 (380 × 1,245)	6 (152.4)	200–250 (149–187)	36.0 (32.6)	50.0 (45.4)	63.0 (57.0)	80.0 (72.5)	95.0 (85.9)	130.0 (117.5)	140.0 (126.8)

* Courtesy Fiat-Allis Construction Machinery, Inc.

Figure 19-7 Cutaway showing single-impeller impact breaker. *(Iowa Manufacturing Company.)*

damage to the rolls when trap iron or other noncrushable material passes through the machine. Figures 19-9, 19-10, and 19-11 illustrate roll crushers.

The maximum size of material that may be fed to a crusher is directly proportional to the diameter of the rolls. If the feed contains stones that are too large, the rolls will not grip them and pull them through the crusher. The angle of nip in Fig. 19-12, which is constant for smooth rolls, has been found to be 16°54′. The maximum-size particles that can be crushed is determined as follows. Let

R = radius of rolls
B = angle of nip
$D = R \cos B = 0.9575R$
A = maximum-size feed
C = roll setting = size of finished product

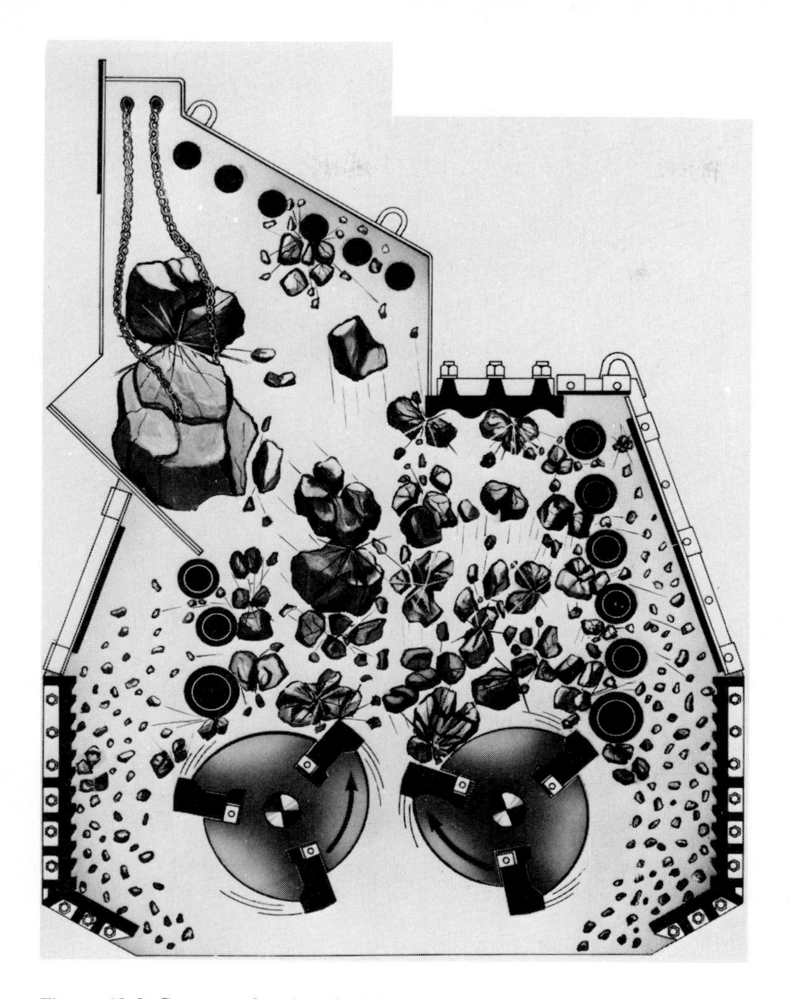

Figure 19-8 Cutaway showing double-impeller impact breaker. *(Iowa Manufacturing Company.)*

Then

$$\begin{aligned} X &= R - D \\ &= R - 0.9575R = 0.0425R \\ A &= 2X + C \\ &= 0.085R + C \end{aligned} \tag{19-1}$$

Example 19-1 Determine the maximum-size stone that may be fed to a smooth-roll crusher whose rolls are 40 in. in diameter, when the roll settings is 1 in.

$$\begin{aligned} A &= 0.085 \times 20 + 1 \\ &= 2.7 \text{ in.} \end{aligned}$$

Figure 19-9 Cutaway showing triple-roll crusher. *(Portec, Inc., Pioneer Division.)*

The capacity of a roll crusher will vary with the kind of stone, size of feed, size of the finished product, width of rolls, speed at which the rolls rotate, and extent to which the stone is fed uniformly into the crusher. Referring to Fig. 19-12, the theoretical volume of a solid ribbon of material passing between the two rolls in 1 min would be the product of the width of the opening times the width of the rolls times the speed of the surface of the rolls. The volume may be expressed in cubic inches per minute or cfm. In actual practice the ribbon of crushed stone will never be solid. A more realistic volume should approximate one-fourth to one-third of the theoretical volume. An equation which may be used as a guide in estimating the capacity is derived as follows. Let

C = distance between rolls, in.
W = width of rolls, in.
S = peripheral speed of rolls, in. per min
N = speed of rolls, rpm
R = radius of rolls, in.

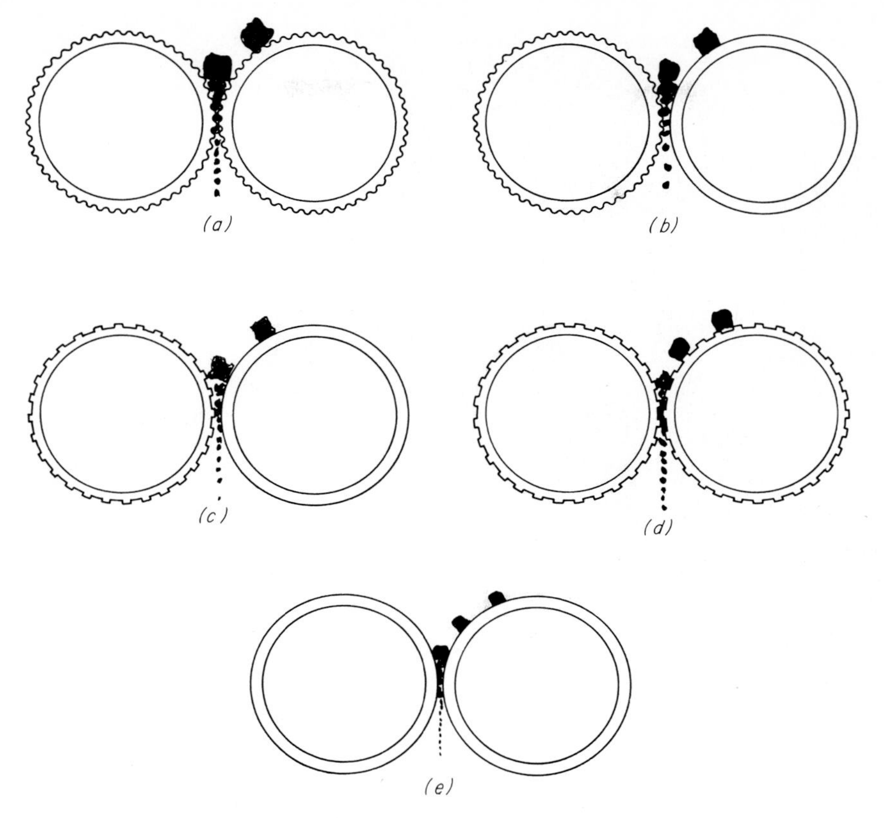

Figure 19-10 Different types of shells for roll crushers. (*a*) Two corrugated shells. (*b*) One corrugated and one smooth shell. (*c*) One step-tooth and one smooth shell. (*d*) Two step-tooth shells synchronized. (*e*) Two smooth shells. *(Grundler Crusher & Pulverizer Company.)*

V_1 = theoretical volume, cu in. or cfm
V_2 = actual volume, cu in. or cfm
Q = probable capacity, tons per hr

Then

$$V_1 = CWS$$

Assume $V_2 = V_1/3$.

$$V_2 = \frac{CWS}{3} \text{ cu in. per min}$$

Divide by 1,728 cu in. per cu ft.

$$V_2 = \frac{CWS}{5{,}184} \text{ cfm}$$

Figure 19-11 Cutaway of single-roll crusher showing deep-ribbed pittman and segment-type rotor Breaker plate with ribs provides concentrated crushing action between teeth and ribs and acts as a sizing anvil for close control of product size. *(Grundler Crusher & Pulverizer Company.)*

Assume the crushed stone weighs 100 lb per cu ft.

$$Q = \frac{100 \times 60 V_2}{2{,}000} = 3 V_2$$

$$= \frac{CWS}{1{,}728} \text{ tons per hr} \tag{19-2}$$

S may be expressed in terms of the diameter of the roll and the speed in rpm.

$$S = 2\pi RN$$

Substituting this value of S in Eq. (19-2) gives

$$Q = \frac{CW\pi RN}{864} \tag{19-3}$$

Table 19-6 gives representative capacities for smooth-roll crushers, expressed in tons of stone per hour for material weighing 100 lb per cu ft when crushed. These capacities should be used as a guide only in estimating the probable output of a crusher. The actual capacity may be more or less than the given values.

If a roll crusher is producing a finished aggregate, the reduction ratio

Table 19-6 Representative capacities of smooth-roll crushers, in tons per hour (metric tons per hour) of stone*

Size of crusher,† in. (mm)	Speed, rpm	Power required, hp (kW)	Width of opening between rolls, in. (mm)						
			$\frac{1}{4}$ (6.3)	$\frac{1}{2}$ (12.7)	$\frac{3}{4}$ (19.1)	1 (25.4)	$1\frac{1}{2}$ (38.1)	2 (50.8)	$2\frac{1}{2}$ (63.5)
16 × 16 (414 × 416)	120	15–30 (11–22)	15.0 (13.6)	30.0 (27.2)	40.0 (36.2)	55.0 (49.7)	85.0 (77.0)	115.0 (104.0)	140.0 (127.0)
24 × 16 (610 × 416)	80	20–35 (15–26)	15.0 (13.6)	30.0 (27.2)	40.0 (36.2)	55.0 (49.7)	85.0 (77.0)	115.0 (104.0)	140.0 (127.0)
30 × 18 (763 × 456)	60	50–70 (37–52)	15.0 (13.6)	30.0 (27.2)	45.0 (40.7)	65.0 (59.0)	95.0 (86.0)	125.0 (113.1)	155.0 (140.0)
30 × 22 (763 × 558)	60	60–100 (45–75)	20.0 (18.1)	40.0 (36.2)	55.0 (49.7)	75.0 (67.9)	115.0 (104.0)	155.0 (140.0)	190.0 (172.0)
40 × 20 (1,016 × 508)	50	60–100 (45–75)	20.0 (18.1)	35.00 (31.7)	50.0 (45.2)	70.0 (63.4)	105.0 (95.0)	135.0 (122.0)	175.0 (158.5)
40 × 24 (1,016 × 610)	50	60–100 (45–75)	20.0 (18.1)	40.0 (36.2)	60.0 (54.3)	85.0 (77.0)	125.0 (113.1)	165.0 (149.5)	210.0 (190.0)
54 × 24 (1,374 × 610)	41	125–150 (93–112)	24.0 (21.7)	48.0 (43.5)	71.0 (64.3)	95.0 (86.0)	144.0 (130.0)	192.0 (173.8)	240.0 (217.5)

* Courtesy Iowa Manufacturing Company.
† The first number indicates the diameters of the rolls, and the second indicates the widths of the rolls.

should not be greater than 4:1. However, if a roll crusher is used to prepare feed for a fine grinder, the reduction may be as high as 7:1.

Rod and ball mills These mills are used to produce fine aggregate, such as sand, from stone that has been crushed to suitable sizes by other crushing equipment. It is not uncommon for specifications for concrete to require the use of a homogeneous aggregate, regardless of size. If crushed stone is used for the coarse aggregate, sand manufactured from the same stone will satisfy the specifications.

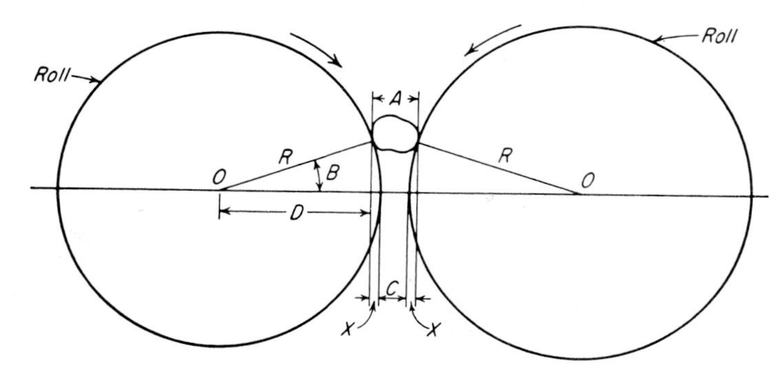

Figure 19-12 Crushing rock between two rolls.

Figure 19-13 Section through a rod mill.

A rod mill is a circular steel shell, lined on the inside with a hard mineral wearing surface, equipped with a suitable support or trunnion arrangement at each end, with a driving gear at one end. It is operated with its axis in a horizontal position. It is charged with steel rods, whose lengths are slightly less than the length of the mill. Crushed stone, which is fed through the trunnion at one end of the mill, flows to the discharge at the other end. As the mill rotates slowly, the stone is constantly subjected to the impact of the tumbling rods, which produce the desired grinding. A mill may be operated wet or dry, with or without water added. The size of a rod mill is specified by the diameter and the length of the shell, such as 8 by 12 ft, respectively. Figure 19-13 shows a section through a rod mill.

A ball mill, which uses steel balls instead of rods to supply the impact necessary to grind the stone, will produce fines with smaller grain sizes than those produced by a rod mill. Figure 19-14 shows a section through a ball mill.

Sizes of stone produced by jaw and roll crushers While the setting of the discharge opening of a crusher will determine the maximum-size stone produced, the aggregate sizes will range from slightly greater than the crusher

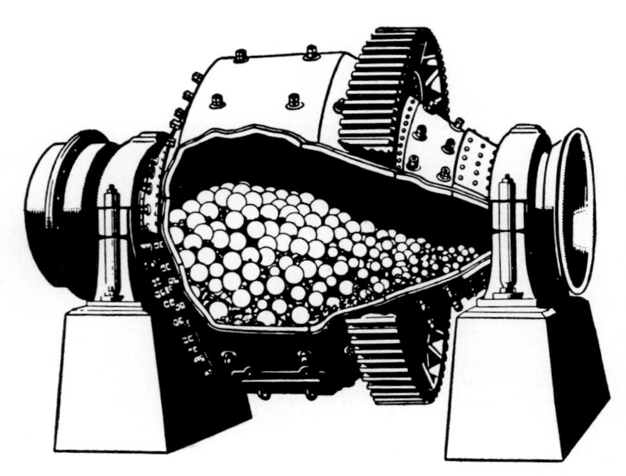

Figure 19-14 Section through a conical ball mill.

setting to fine dust. Experience gained in the crushing industry indicates that for any given setting for a jaw or roll crusher approximately 15 percent of the total amount of stone passing through the crusher will be larger than the setting. If the openings of a screen which receives the output from such a crusher are the same size as the crusher setting, 15 percent of the output will not pass through the screen. Figure 19-15 gives representative values for the percent of crushed stone passing through or retained on screens having various sizes of openings for different crusher settings. The information in this figure will apply to jaw and roll crushers only.

In operating a crusher, it generally is desirable to know how the product varies in sizes from the maximum to the minimum sizes. The chart in Fig. 19-15 gives the percent of material passing or retained on screens having the size openings indicated. The chart can be applied to both jaw- and roll-type crushers. To read the chart, select the vertical line corresponding to the crusher setting. Then go down this line to the number which indicates the size of the screen opening. From the size of the screen opening proceed horizontally to the left to determine the percent of material passing through the screen or to the right to determine the percent of material retained on the screen.

Example 19-2 A jaw crusher, with a closed setting of 3 in., produces 50 tph of crushed stone. Determine the amount of stone produced in tons per hour within the following size ranges: in excess of 2 in.; between 2 and 1 in.; between 1 and $\frac{1}{4}$ in.; less than $\frac{1}{4}$ in.

From Fig. 19-15 the amount retained on a 2-in. screen is 42 percent of 50 = 21 tph. The amount in each of the size ranges is determined as follows:

Size range, in.	% passing screens	% in size range	Total output of crusher, tons per hr	Amount produced in size range, tons per hr
Over 2	100–58	42	50	21.0
2–1	58–33	25	50	12.5
1–$\frac{1}{4}$	33–11	22	50	11.0
$\frac{1}{4}$–0	11–0	11	50	5.5
Total		100		50.0

LOG WASHERS

When natural deposits of aggregate, such as sand and gravel, or individual pieces of crushed stone contain deleterious material as a part of the matrix or as deposits on the surface of the aggregate, it will be necessary to remove these materials before using the aggregate. One method of removing the material is to pass the aggregate through a machine called a log washer, which is illustrated in Fig. 19-16. This unit consists of a steel tank with two electric-motor-driven shafts, to which numerous replaceable paddles are attached.

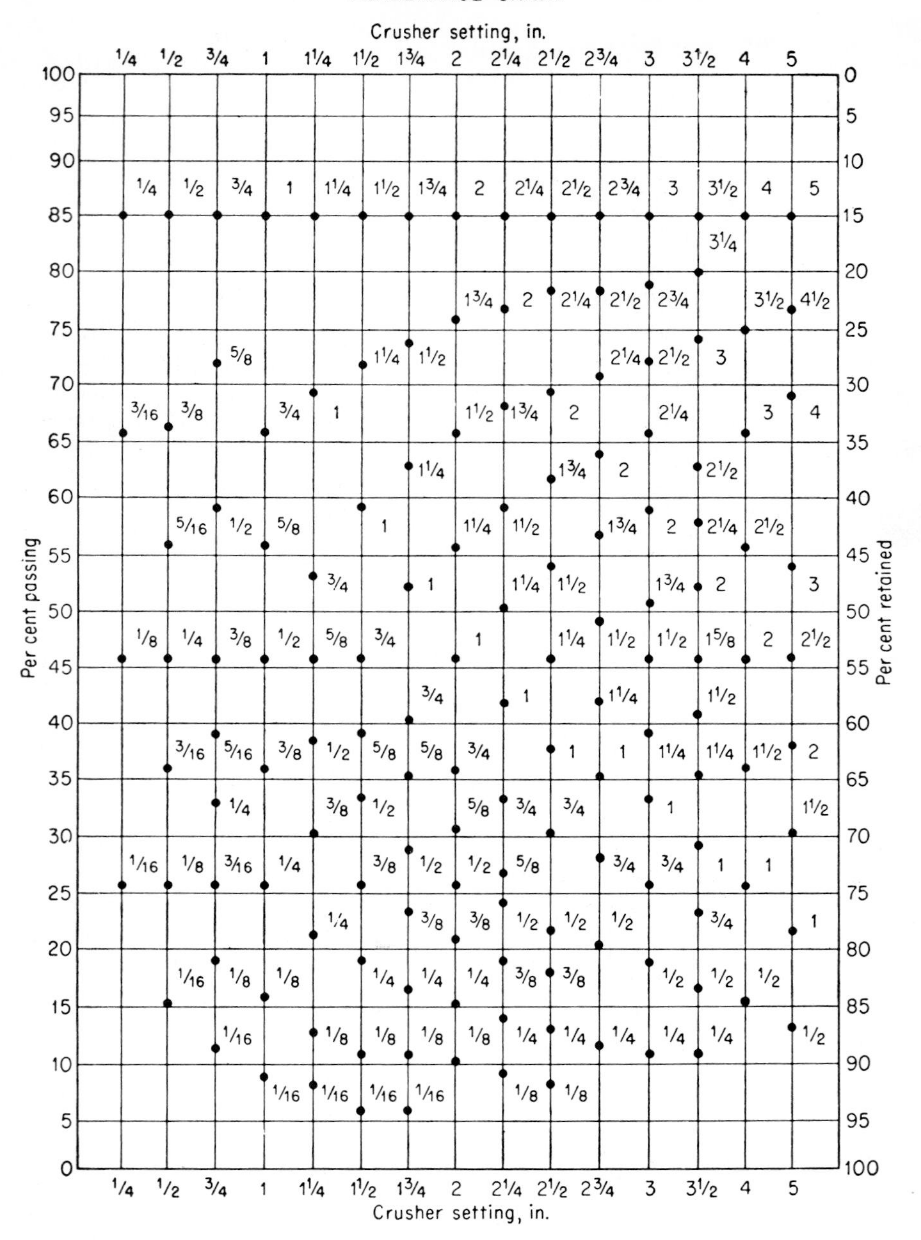

Figure 19-15 Analysis of the size of aggregate produced by jaw and roll crushers. *(Universal Engineering Company.)*

Figure 19-16 Log washer for scrubbing coarse aggregate. *(Kolberg Manufacturing Corporation.)*

When the washer is placed in operation, the end of the tank on which the motor is mounted is raised above the opposite end. The aggregate to be processed is fed into the unit at the lower end, while a constant supply of water flows into the upper end. As the shafts are rotated in opposite directions, the paddles move the aggregate toward the upper end of the tank, while producing a continuing scrubbing action between the particles. The stream of water will remove the undesirable material and discharge it from the tank at the lower end, while the processed aggregate will be discharged at the upper end.

When a contractor submitted samples of a local aggregate which he proposed to use in producing concrete for an airport runway in New Mexico, he was advised that the particles were coated with a deposit that made the aggregate unfit for use. After the aggregate was processed with a log washer, it was found to be entirely acceptable.

SAND PREPARATION AND CLASSIFICATION MACHINES

When the specifications for sand and other fine aggregates require the materials to meet size gradations, it is frequently necessary to produce the gradations by mechanical equipment. Several types of equipment are available for this purpose. Figure 19-17 shows a plan and two sections of a machine that is used to classify sand into eight sizes. Sand and water are fed to the classifier as indicated at the left end of the unit. As the water flows to the outlet end of

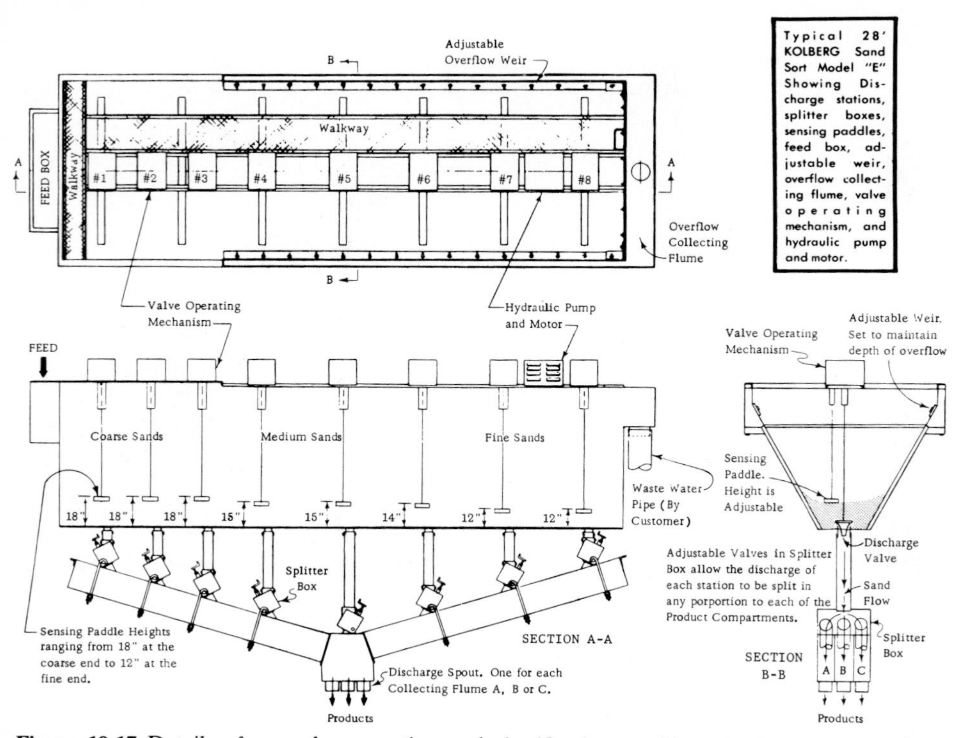

Figure 19-17 Details of a sand preparation and classification machine. *(Kolberg Manufacturing Corporation.)*

the tank, the sand particles settle to the bottom of the tank, the coarse ones first and the fine ones last. When the depth of a given size reaches a predetermined level, a sensing paddle will actuate a discharge valve at the bottom of the compartment to permit that material to flow into the splitter box, from which it can be removed and used.

Figure 19-18 illustrates the classification machine and four piles of sand from the machine, arranged by size of the grains.

Figure 19-19 illustrates a screw-type classifier which may be used to produce specification sand. When the machine is placed in operation, the discharge end, where the electric motor is mounted, is elevated above the opposite end. Sand and water are fed into the hopper. As the spiral screws rotate, the sand is moved up the tank to the discharge outlet under the motor. Undesirable material is flushed out of the tank by the overflowing water.

SELECTING CRUSHING EQUIPMENT

In selecting crushing and screening equipment it is essential that certain information be known prior to making the selection. The information needed

Figure 19-18 Various sizes of sand produced by a preparation and classification machine. *(Kolberg Manufacturing Corporation.)*

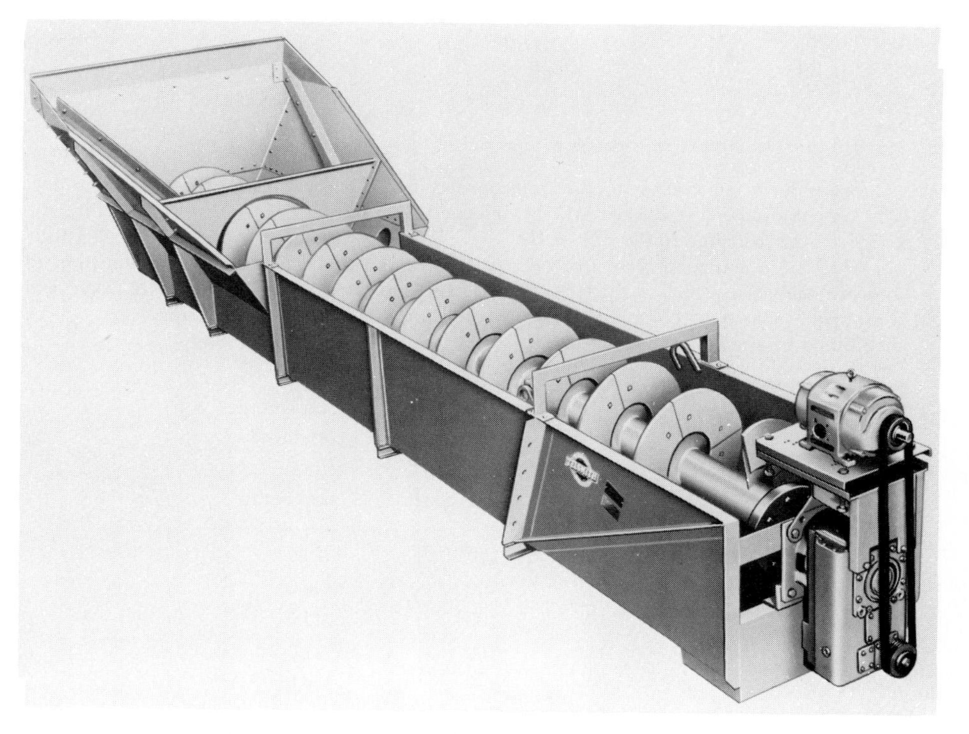

Figure 19-19 Screw classifier for producing specification sand. *(Telsmith Division, Barber-Greene Company.)*

should include, but will not necessarily be limited to, the following items:

1. The kind of stone to be crushed
2. The maximum size and perhaps the size ranges of the feed to the plant
3. The method of feeding the crushers
4. The required capacity of the plant
5. The percent of material falling within specified size ranges

The following example will illustrate a method which may be used to select crushing equipment:

Example 19-3 Select a primary and a secondary crusher to produce 100 tph of crushed limestone. The maximum-size stones from the quarry will be 16 in. The quarry stone will be hauled by truck, dumped into a surge bin, and fed to the primary crusher by an apron feeder, which will maintain a reasonably uniform rate of feed. The aggregate will be used on a project whose specifications require the following size distributions:

Size screen opening, in.		
Passing	Retained on	Percent
$1\frac{1}{2}$		100
$1\frac{1}{2}$	$\frac{3}{4}$	42–48
$\frac{3}{4}$	$\frac{1}{4}$	30–36
$\frac{1}{4}$	0	20–26

Consider a jaw crusher for the primary and a roll crusher for the secondary crushing. The output of the jaw crusher will be screened to remove specification sizes before the oversize material is fed to the roll crusher.

Assume a setting of 3 in. for the jaw crusher. This will give a ratio of reduction of approximately 5:1, which is satisfactory. Table 19-2 indicates a size 24- by 36-in. crusher with a probable capacity of 114 tph. Figure 19-15 indicates that the product of the crusher will be distributed by sizes as follows:

Size range, in.	% passing screens	% in size range	Total output of crusher, tons per hr	Amount produced in size range, tons per hr
Over $1\frac{1}{2}$	100–46	54	100	54.0
$1\frac{1}{2}-\frac{3}{4}$	46–26	20	100	20.0
$\frac{3}{4}-\frac{1}{4}$	26–11	15	100	15.0
$\frac{1}{4}$–0	11–0	11	100	11.0
Total		100		100.0

As the roll crusher will receive the output from the jaw crusher, the rolls must be large enough to handle 3-in. stone. Assume a setting of $1\frac{1}{2}$ in. From Eq. (19-1) the minimum radius will be 17.7 in. Try a 40- by 20-in. crusher with a capacity of approximately 105 tph for a $1\frac{1}{2}$-in. setting.

For any given setting the crusher will produce about 15 percent stone having at least one dimension larger than the setting. Thus, for a given setting 15 percent of the stone that passes through the roll crusher will be returned for recrushing. The total amount of stone passing through the crusher, including the returned stone, is determined as follows. Let

$$Q = \text{total amount of stone through the crusher}$$

Then

$$0.15Q = \text{amount of returned stone}$$

$$0.85Q = \text{amount of new stone}$$

$$Q = \frac{\text{amount of new stone}}{0.85}$$

$$= \frac{54}{0.85} = 63.5 \text{ tph}$$

The 40- by 20-in. roll crusher will handle this amount of stone easily. The distribution of the output of this crusher by size range will be as follows:

Size range, in.	% passing screens	% in size range	Total amount through crusher, tph	Amount produced in size range, tph
$1\frac{1}{2}-\frac{3}{4}$	85–46	39	63.5	24.8
$\frac{3}{4}-\frac{1}{4}$	46–18	28	63.5	17.8
$\frac{1}{4}-0$	18–0	18	63.5	11.4
Total		85		54.0

Now combine the output of each crusher by specified sizes.

Size range, in.	From jaw crusher, tph	From roll crusher, tph	Total amount, tph	% in size range
$1\frac{1}{2}-\frac{3}{4}$	20.0	24.8	44.8	44.8
$\frac{3}{4}-\frac{1}{4}$	15.0	17.8	32.8	32.8
$\frac{1}{4}-0$	11.0	11.4	22.4	22.4
Total	46.0	54.0	100.0	100.0

SCALPING CRUSHED STONE

The term scalping, as used in this chapter, refers to a screening operation which is performed to remove from the main mass of stone to be processed that stone which is too large for the crusher opening or is small enough to be used without further crushing. Scalping may be performed ahead of a primary crusher, and it represents good crushing practice to scalp all crushed stone following each successive stage of reduction.

Scalping ahead of a primary crusher serves two purposes. The use of a grizzly, which consists of a number of widely spaced parallel bars, will prevent oversize stones from entering the crusher and blocking the opening. If the product of the quarry contains such stones, it is desirable to remove them ahead of the crusher.

The product of the quarry may contain dirt, mud, or other debris which is not acceptable in the finished product and therefore must be removed from the stone. Scalping should accomplish this removal. Also, the product of the quarry may contain an appreciable amount of stone which was reduced by the blasting operation to specification sizes. In this event it may be good economy to remove such stone ahead of the primary crusher, thereby reducing the total load on the crusher and increasing the overall capacity of the plant. Figure 19-20 illustrates a commercial bar grizzly.

It usually is good practice and economical to install a scalper after each stage of reduction to remove specification sizes. This stone may be transported to grading screens, where it can be sized and placed in appropriate storage.

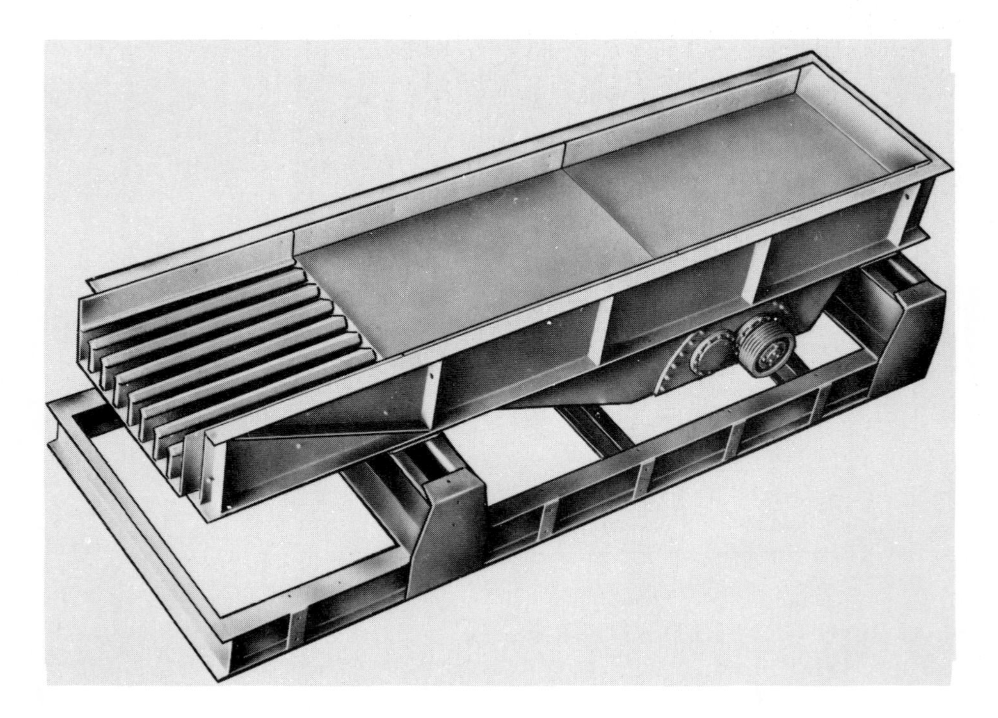

Figure 19-20 Vibrating grizzly feeder. *(Iowa Manufacturing Company.)*

Any stone removed ahead of a crusher will reduce the total load on the crusher, which will permit the use of a smaller crusher or an increase in the output of the plant.

FEEDERS

The capacity of a crusher will be increased if the stone is fed to it at a uniform rate. Surge feeding tends to overload a crusher, and then the surge is followed by an insufficient supply of stone. This type of feeding, which reduces the capacity of a crusher, may be eliminated by using a mechanical feeder ahead of a crusher. The installation of such a feeder may increase the capacity of a jaw crusher as much as 15 percent. An apron-type feeder, as illustrated in Fig. 19-21, is suitable for use ahead of a primary crusher.

SURGE PILES

A stationary stone-crushing plant may include several types and sizes of crushers, each probably followed with a set of screens and a belt conveyor to transport the stone to the next crushing operation or to storage. A plant may be designed to provide temporary storage for stone between the successive stages of crushing. This plan has the advantage of eliminating or reducing the surge effect that frequently exists when the crushing, screening, and handling

Figure 19-21 Apron-type feeder. *(Universal Engineering Corp.)*

operations are conducted on a straight-line basis. The stone in temporary storage, which is referred to as a surge pile, ahead of a crusher may be used to keep at least a portion of a plant in operation. Within reasonable limits the use of a surge pile ahead of a crusher permits the crusher to be fed uniformly at the most satisfactory rate, regardless of variations in the output of other equipment ahead of the crusher. The use of surge piles has enabled some plants to increase the production by as much as 10 to 20 percent.

Among the arguments against the use of surge piles are the following:

1. They require additional area for storage.
2. They require the construction of storage bins or reclaiming tunnels.
3. They increase the amount of handling of stone.

The decision to use or not use surge piles should be based on an analysis of the advantages and disadvantages for each plant.

SCREENING AGGREGATE

Screening of crushed stone is necessary in order that the aggregate may be separated by size ranges. Most specifications covering the use of aggregate stipulate that the different sizes shall be combined to produce a blend having a given size distribution. Persons who are responsible for preparing the specifications for the use of aggregate realize that crushing and screening cannot be done with complete precision, and, accordingly, they allow some tolerance in the size distribution. The extent of tolerance may be indicated by a statement such as: The quantity of aggregate passing a 1-in. screen and retained on a $\frac{1}{4}$-in. screen shall be not less than 30 or more than 40 percent of the total quantity of aggregate.

Revolving screens Revolving screens have several advantages over other types of screens, especially when they are used to wash and screen sand and gravel. The operating action is slow and simple, and the maintenance and repair costs are low. If the aggregate to be washed contains silt and clay, a scrubber can be installed near the entrance end of a screen in order that the material may be agitated in water. At the same time streams of water may be sprayed on the aggregate as it moves through the screen. Figure 19-22 shows a revolving screen with a scrubber in operation. The aggregate, which is separated by sizes, is stored temporarily in the bins below the screen.

Vibrating screens The vibrating screen is the most widely used screen for aggregate production. Figures 19-23 and 19-24 show multiple-deck screens units of this type. The steel frame may be designed to permit the installation of one or more screens, one above the other. Each screen is referred to as a deck. The vibration is obtained by means of an eccentric shaft, a counterweighted shaft, or electromagnets attached to the frame or to the screens.

Figure 19-22 Revolving screen with scrubber in operation.

Figure 19-23 Double-deck vibrating screen. *(Portec, Inc., Pioneer Division.)*

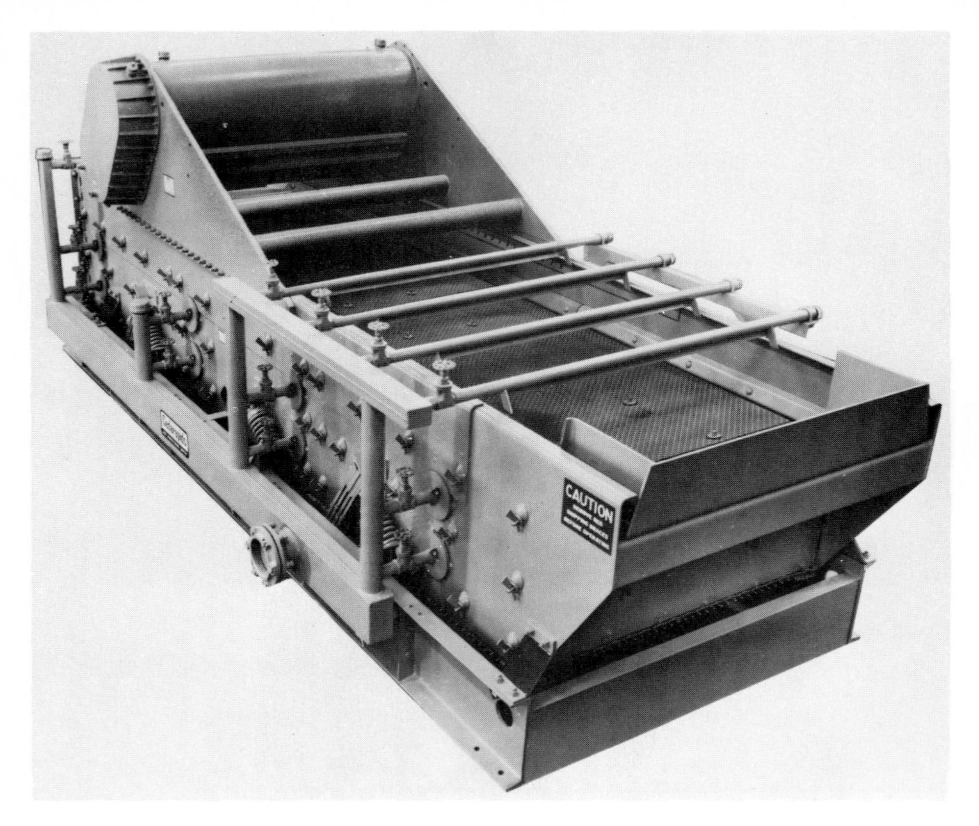

Figure 19-24 Triple-deck screen with spray bars for washing aggregate. *(Iowa Manufacturing Company.)*

A unit is installed with a slight slope from the receiving to the discharge end, which, combined with the vibrations, causes the aggregate to flow over the surface of the screen. Most of the particles that are smaller than the openings in a screen will drop through the screen, while the oversize particles will flow off the screen at the discharge end. For a multiple-deck unit the sizes of the openings will be progressively smaller for each lower deck.

A screen will not pass all material whose sizes are equal to or less than the dimensions of the openings in the screen. Some of this material may be retained on and carried over the discharge end of a screen. The efficiency of a screen may be defined as the ratio of the amount of material passing through a screen divided by the total amount that is small enough to pass through, with the ratio expressed as a percent. The highest efficiency is obtained with a single-deck screen, usually amounting to 90 to 95 percent. As additional decks are installed, the efficiencies of these decks will decrease, being above 85 percent for the second deck and 75 percent for the third deck.

The capacity of a screen is the number of tons of material that 1 sq ft will pass per hour. The capacity will vary with the size of the openings, kind of material screened, moisture content, and other factors. Because of the factors that affect the capacity of a screen, it will seldom if ever be possible to calculate in advance the exact capacity of a screen. If a given number of tons of material must be passed per hour, it is good practice to select a screen whose total calculated capacity is 10 to 25 percent greater than the quantity to be screened.

The chart in Fig. 19-25 gives capacities for dry screening which may be used as a guide in selecting the correct size screen for a given flow of material. The capacities given in the chart should be modified by the application of appropriate correction factors. Representative values of these factors are given hereafter.

Efficiency factors If a low screening efficiency is permissible, the capacity of a screen may be higher than the values given in Fig. 19-25. Table 19-7 gives factors by which the chart values of capacities may be multiplied to obtain corrected capacities for given efficiencies.

Deck factors This is a factor whose value will vary with the particular deck position for multiple-deck screens. The values are given in Table 19-8.

Aggregate-size factors The capacities of screens given in Fig. 19-25 are based on screening dry material which contains particle sizes such as would be found in the output of a representative crusher. If the material to be screened contains a surplus of small sizes, the capacity of the screen will be increased,

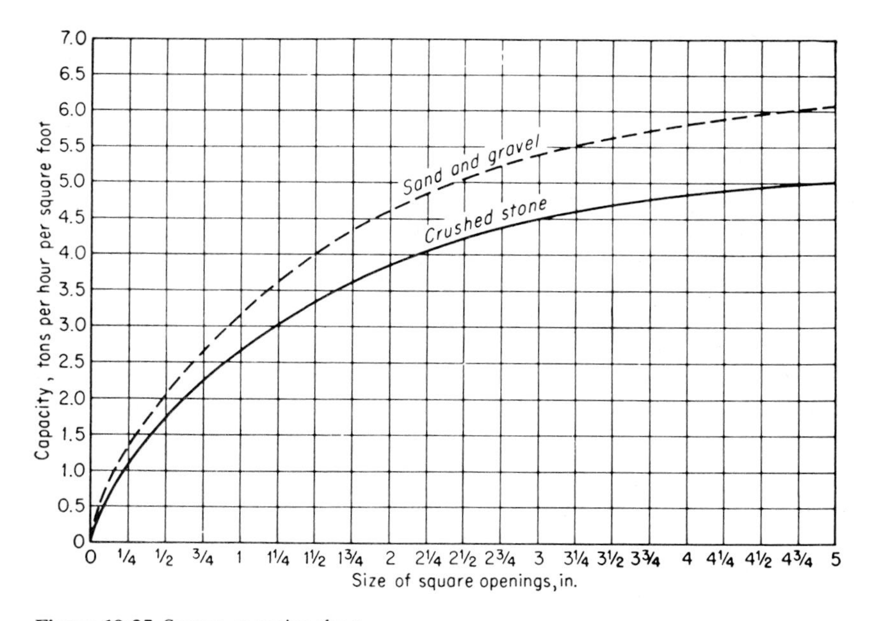

Figure 19-25 Screen-capacity chart.

Table 19-7 Efficiency factors

Permissible screen efficiency, %	Efficiency factor
95	1.00
90	1.25
85	1.50
80	1.75
75	2.00

Table 19-8 Deck factors

For deck no.	Deck factor
1	1.00
2	0.90
3	0.75
4	0.60

Table 19-9 Aggregate-size factors

% of aggregate less than $\frac{1}{2}$ the size of screen opening	Aggregate-size factor
10	0.55
20	0.70
30	0.80
40	1.00
50	1.20
60	1.40
70	1.80
80	2.20
90	3.00

whereas if the material contains a surplus of large sizes, the capacity of the screen will be reduced. Table 19-9 gives representative factors which may be applied to the capacity of a screen to correct for the effect of fine or coarse particles.

Determining the screen required Figure 19-25 gives the theoretical capacity of a screen in tons per hour per square foot based on material weighing 100 lb per cu ft when crushed. The corrected capacity of a screen is given by the equation

$$Q = ACEDG \tag{19-4}$$

where Q = capacity of screen, tph
A = area of screen, sq ft
C = theoretical capacity of screen, tph per sq ft
E = efficiency factor
D = deck factor
G = aggregate-size factor

The minimum area of a screen to provide a given capacity is determined from the equation

$$A = \frac{Q}{CEDG} \tag{19-5}$$

Example 19-4 Determine the minimum-size single-deck screen, having $1\frac{1}{2}$-in.-square openings, for screening 120 tph of dry crushed stone, weighing 100 lb per cu ft when crushed. A screening efficiency of 90 percent is satisfactory. An analysis of the aggregate indicates that approximately 30 percent of it will be less than $\frac{3}{4}$ in. in size. The values of the factors to be used in equation (19-5) are

$Q = 120$ tph

$C = 3.32$ tph per sq ft (Fig. 19-25)

$E = 1.25$ (Table 19-7)

$D = 1.0$ (Table 19-8)

$G = 0.8$ (Table 19-9)

Substituting these values in equation (19-5), we get

$$A = \frac{120}{3.32 \times 1.25 \times 1.0 \times 0.8} = 36.1 \text{ sq ft}$$

In view of the possibility of variations in the factors used, and to provide a margin of safety, it is recommended that a 4- by 10-ft screen be selected.

Portable crushing and screening plants Many types and sizes of portable crushing and screening plants are used in the construction industry. When there is a satisfactory deposit of stone near a project that requires stone aggregate, it frequently will be more economical to set up a portable plant and produce the crushed stone instead of purchasing it from a commercial source.

Typical portable crushing and screening plants are illustrated in Figs. 19-26 and 19-27. The stone from the quarry is fed to the plant by a belt conveyor at the right. This particular machine is designed to permit the quarry product to pass over a bottom-deck screen, which removes the material smaller than the screen, thereby reducing the load on the primary crusher. The oversize stone from this screen is fed to a jaw crusher, thence to a belt conveyor, which returns it to a top-deck screen, where the specification sizes are removed. The oversize stone is fed to a roll crusher. Portable plants commonly use a jaw

Figure 19-26 Portable aggregate plant in operation. *(Iowa Manufacturing Company.)*

crusher for primary crushing and a roll crusher for secondary crushing. Changes in the specification sizes may be met, over a reasonably wide range, by adjusting the crusher settings and changing the sizes of the screens.

FLOW DIAGRAMS OF AGGREGATE-PROCESSING PLANTS

Figures 19-28 and 19-29 illustrate flow diagrams for two portable aggregate-processing plants. By passing the stone for the quarry over a screen before it

Figure 19-27 Cutaway of portable rock-crushing plant. *(Portec, Inc., Pioneer Division.)*

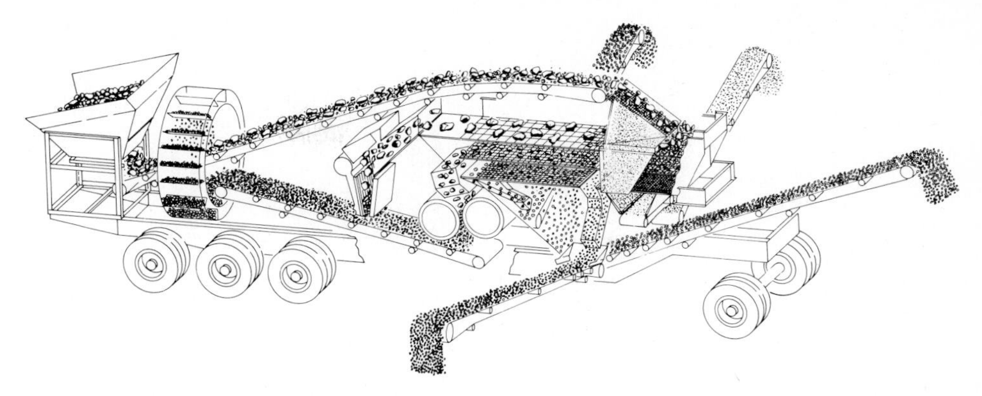

Figure 19-28 Flow diagram of a four-product portable aggregate plant. *(Iowa Manufacturing Company.)*

goes to the primary crusher, any stone within the specification sizes will be removed prior to crushing. This arrangement should increase the output of the plant.

Representative sizes of crushers and other equipment for this plant are as follows:

Jaw crusher, 10×36 in.
Roll crusher, 40×22 in.
Vibrator screen, 4×12 ft, $3\frac{1}{2}$ decks
Feeder, 4-ft hopper
Feeder conveyor, 30 in. wide, 50 ft long
Return conveyor, 24 in. wide

Figure 19-30 illustrates the flow diagram of the aggregate-processing plant

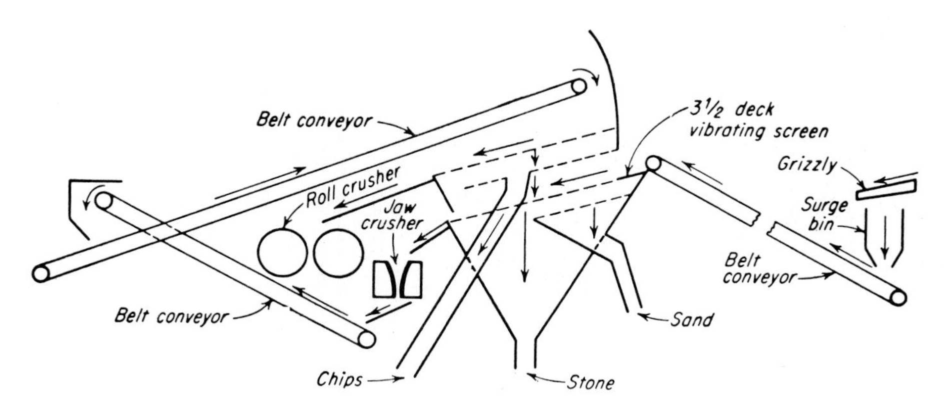

Figure 19-29 Flow diagram of a portable aggregate-processing plant.

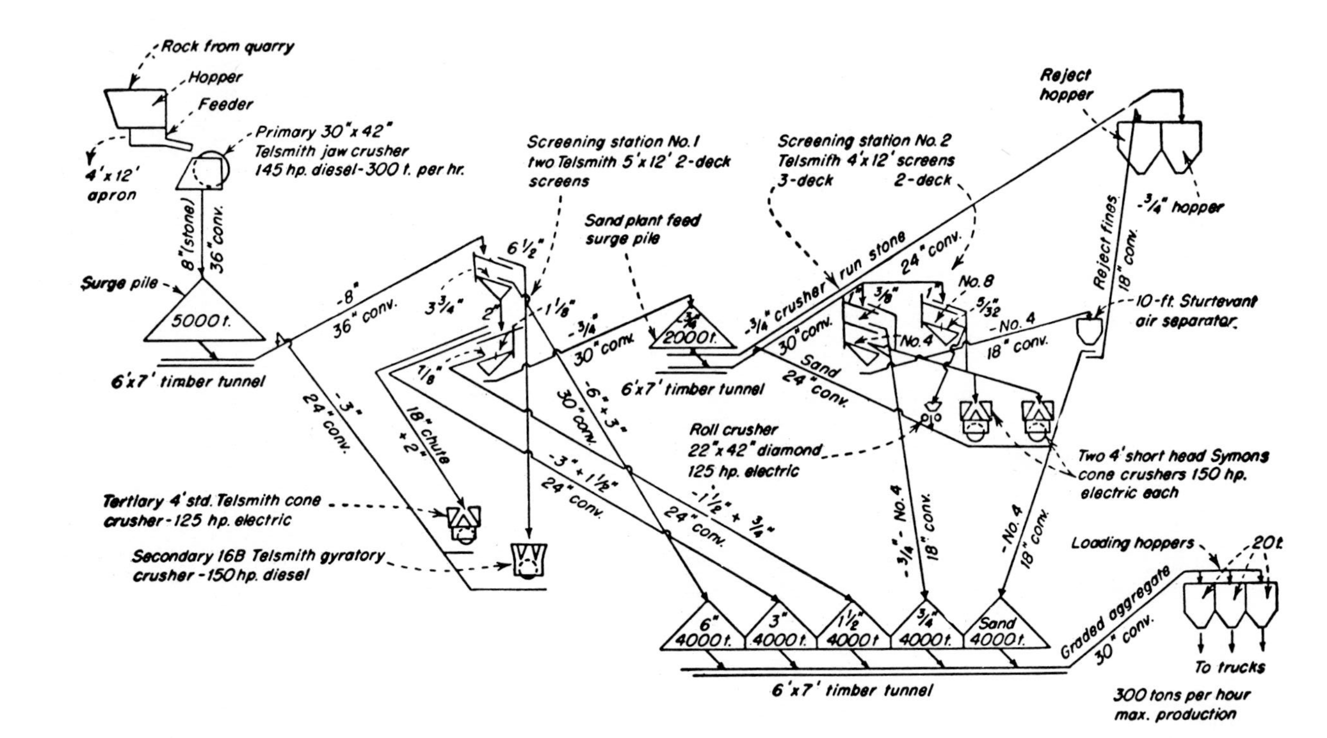

Figure 19-30 Flow diagram for the aggregate-processing plant at the Philpott Dam.

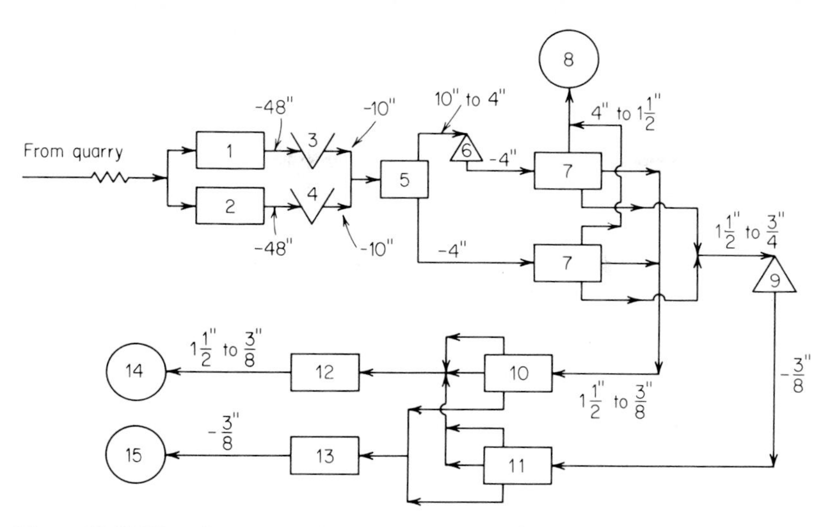

Figure 19-31 Flow diagram for the aggregate production plant at the Perris Dam: (1) 5 × 20′ grizzly feeder; (2) 5 × 16′ vibrating feeder; (3) 48 × 48″ jaw crusher; (4) 42 × 48″ jaw crusher; (5) 6 × 16′ scalping screen; (6) 7′ standard cone crusher; (7) 6 × 16′ doubledeck screens; (8) stock pile for 4 to $1\frac{1}{2}$″ aggregate; (9) $5\frac{1}{2}$′ shorthead cone crusher; (10) 8 × 20′ doubledeck screen; (11) 8 × 20′ doubledeck screen; (12) double screw classifier; (13) double screw classifier; (14) stock pile for $1\frac{1}{2}$ to 3/8″ aggregate; (15) stock pile for 3/8″ minus aggregate.

Figure 19-32 Belt-conveyor system for handling aggregate at production plant. *(Kolberg Manufacturing Corporation.)*

for the Philpott Dam. This plant was located near the quarry, and trucks were used to haul the finished aggregate to the concrete mixing plant at the dam.

Example 19-5 When constructing the Perris Dam in California, the contractor opened a rock quarry to drill, blast, crush, and deliver 50,000 tons per week of aggregate to the project [1].

He drilled 4-in. diameter holes to depths of 40 ft, loaded them with Du Pont Pourvex slurry, then blasted 300 to 400 holes at a time to produce 40,000 to 50,000 tons of rock. All rock in excess of 48 in. in size was removed for riprap before sending the smaller-size material to a crusher plant.

Because the project required more than 2,500,000 cu yd of crushed rock, the crushing plant was designed to provide the quantities and sizes of material needed for the job. As illustrated in Fig. 19-31, the plant produced aggregate in three size ranges.

HANDLING CRUSHED-STONE AGGREGATE

After stone is crushed and screened to provide the desired size ranges, it is necessary to handle it carefully or the large and small particles may be separated, thereby destroying the blend in sizes, which frequently is essential to a satisfactory aggregate. If aggregate is permitted to flow freely off the end of a belt conveyor, especially at some height above the storage bin, the material will be segregated by sizes, as illustrated in Fig. 19-33. A strong cross wind tends to separate the smaller sizes from the larger sizes. Specifications covering the production of aggregate frequently stipulate that the aggregate transported by a belt conveyor shall not be permitted to fall freely from the discharge end of a

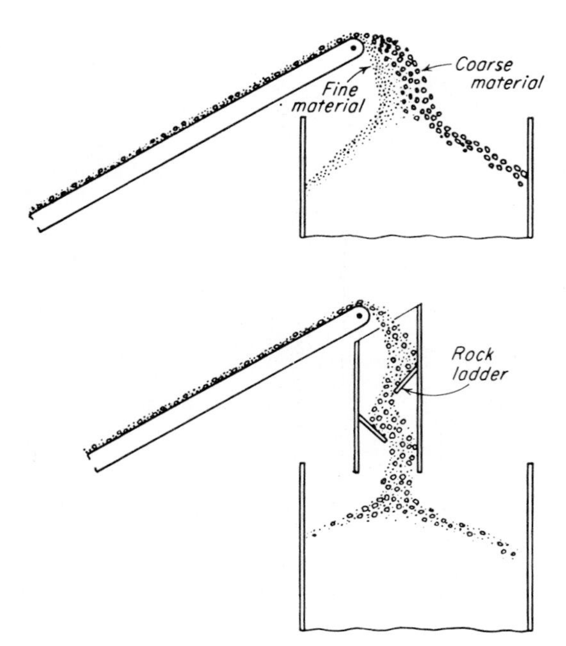

Figure 19-33 Method of preventing the segregation of aggregate discharge from a conveyor belt.

Figure 19-34 Pile of aggregate showing segregation.

Figure 19-35 A rock ladder used to reduce the segregation of aggregate.

belt. The end of the belt should be kept as low as possible, and the aggregate should be discharged through a rock ladder, containing baffles, to prevent segregation.

Figure 19-34 shows a pile of badly segregated aggregate, while Fig. 19-35 shows a rock ladder used to reduce segregation.

The references appearing at the end of this chapter list other publications containing information related to the production of aggregate.

PROBLEMS

19-1 A jaw crusher, with a closed setting of 3 in., produces 200 tph of crushed stone. Determine the number of tons per hour produced in each of the following size ranges: in excess of $2\frac{1}{2}$ in.; between $2\frac{1}{2}$ and $1\frac{1}{2}$ in.; between $1\frac{1}{2}$ in. and $\frac{1}{4}$ in.; less than $\frac{1}{4}$ in.

19-2 A roll crusher, set at 2 in., produces 120 tph of crushed stone. Determine the number of tons per hour produced in each of the following size ranges: in excess of $1\frac{1}{2}$ in.; between $1\frac{1}{2}$ in. and $\frac{3}{4}$ in.; between $\frac{3}{4}$ in. and $\frac{1}{4}$ in.; less than $\frac{1}{4}$ in.

19-3 Select a jaw crusher for primary crushing and a roll crusher for secondary crushing to produce 200 tph of limestone rock. The maximum-size stone from the quarry will be 12 in. The stone is to be crushed to the following specifications:

Size screen opening, in.		
Passing	Retained on	Percent
$2\frac{1}{2}$		100
$2\frac{1}{2}$	$1\frac{1}{2}$	30–40
$1\frac{1}{2}$	$\frac{3}{4}$	22–30
$\frac{3}{4}$	$\frac{1}{4}$	20–30
$\frac{1}{4}$	0	15–25

Specify the size and setting for each crusher selected.

19-4 A jaw crusher and a roll crusher are used in an attempt to crush 140 tph of stone to the following specifications:

Size screen openings, in.		
Passing	Retained on	Percent
$2\frac{1}{2}$		100
$2\frac{1}{2}$	$1\frac{1}{2}$	40–48
$1\frac{1}{2}$	$\frac{3}{4}$	30–36
$\frac{3}{4}$	$\frac{1}{4}$	20–25
$\frac{1}{4}$	0	10–20

Select crushers to produce this aggregate. Is it possible to produce aggregate to meet these specifications with the indicated crushers without surpluses in any of the size ranges? Can the product of these crushers be processed to provide the desired sizes in the specified percentages? If so, tell how.

19-5 A 30- by 42-in. jaw crusher is set to operate with a $2\frac{1}{2}$-in. opening. The output from the crusher is discharged onto a screen with $1\frac{1}{2}$-in. openings, whose efficiency is 90 percent. The aggregate that does not pass through the screen goes to a 40- by 20-in. roll crusher, set at $1\frac{1}{4}$ in. The output from the roll crusher is fed back over the screen.

Determine the maximum output of the plant in tons per hour.

Determine the output of the plant in tons per hour in each of the two sizes 1 in. to $\frac{1}{2}$ in., and less than $\frac{1}{2}$ in.

19-6 A portable crushing plant is equipped with the following units:

1 jaw crusher, size 15 by 30 in.
1 roll crusher, size 30 by 22 in.
1 set of horizontal vibrating screens, two decks, with $1\frac{1}{2}$- and $\frac{3}{4}$-in. openings

The specifications require that 100 percent of the aggregate shall pass a $1\frac{1}{2}$-in. screen and 50 percent shall pass a $\frac{3}{4}$-in. screen.

Assume that 10 percent of the stone from the quarry will be smaller than $1\frac{1}{2}$ in. and that this aggregate will be removed by passing the quarry product over the screen before sending it to the jaw crusher. The aggregate will weigh 110 lb per cu ft.

Determine the maximum output of the plant, expressed in tons per hour. Include the aggregate removed by the screens prior to sending it to the crushers.

19-7 The output from a 36- by 42-in. crusher, with a closed opening of $2\frac{1}{2}$ in. is passed over a single horizontal vibrating screen with $1\frac{1}{4}$ in. openings. If the permissible screen efficiency is 90 percent, use the information in this book to determine the minimum-size screen, expressed in square feet, required to handle the output of the crusher.

19-8 The output from a 36- by 42-in. jaw crusher, with a closed setting of 4 in., is to be screened into the following sizes: $2\frac{1}{2}$ to $1\frac{1}{2}$ in.; $1\frac{1}{2}$ to $\frac{3}{4}$ in.; less than $\frac{3}{4}$ in. A three-deck horizontal vibrating screen will be used to separate the three sizes. The stone weighs 115 lb per cu ft. If the permissible screen efficiency is 90 percent, determine the minimum-size screen for each deck, expressed in square feet, required to handle the output of the crusher.

REFERENCES

1. Contractor-planned Crushing Plant Keeps Dam on Schedule, *Roads & Streets*, vol. 115, pp. 46–48, August 1972.
2. Higgins, Lindley R.: Aggregates Today; Breakthrough in Methodology, Advances in Automation and Movement, *Construction Methods & Equipment*, vol. 54, pp. 72–75, January 1972.
3. Newman, Donald: Mathematical Method for Blending Aggregates, *Journal of the Construction Division, Proceedings ASCE*, vol. 90, pp. 1–13, September 1964.
4. Hancher, Donn E., and John A. Havers: "Mathematical Model of Aggregate Plant Production," American Society of Civil Engineers, 345 East 47th Street, New York, NY 10017, no date.
5. "Telsmith Mineral Processing Handbook," 1976, Telsmith Division, Barber-Greene Company, 532 East Capitol Drive, Milwaukee, WI 53212.
6. "Cedar Rapids Reference Book," 4th Pocket Edition, 1968, Iowa Manufacturing Company, 916 16th Street NE, Cedar Rapids, IA 52402.
7. "Facts and Figures Booklet," 16th ed., 1969, Portec Inc., Pioneer Division, 3200 Como Avenue SE, Minneapolis, MN 55414.
8. Aggregate Reference Guide, 1976, Kolberg Manufacturing Corporation, 20 West 21st Street, Yankton, SD 57078.

9. Fiat-Allis Construction Machinery, Inc., Box 1213, Milwaukee, WI 53201.
10. Grundler Crusher & Pulverizer Company, 2915 North Market Street, St. Louis MO 63106.
11. Iowa Manufacturing Company, 916 16th Street NE, Cedar Rapids, IA 52402.
12. Kolberg Manufacturing Corporation, 20 West 21st Street, Yankton, SD 57078.
13. Nordberg Division, Rex Chainbelt, Inc., 4710 West Greenfield Avenue, Milwaukee, WI 53201.
14. Portec, Inc., Pioneer Division, 3200 Como Avenue SE, Minneapolis, MN 55414.
15. Telsmith Division, Barber-Greene Company, 532 East Capitol Drive, Milwaukee, WI 53212.
16. Universal Engineering Corporation, Subsidiary of Pettibone Corporation, 625 C Avenue NW, Cedar Rapids, IA.

CHAPTER

TWENTY

CONCRETE

INTRODUCTION

Portland cement concrete is unquestionably the most widely used structural material in the world for civil engineering projects. Its versatility, economy, adaptability, and worldwide availability, and especially its low maintenance requirements, make it very useful. The term concrete is applicable for many products, but is most generally used with portland cement concrete. It consists of portland cement, water, and aggregate which have been mixed together, placed, consolidated, and allowed to solidify and harden. The portland cement and water form a paste, which acts as the glue, or binder. When fine aggregate is added (aggregate whose size range lies between the No. 200 mesh sieve and the No. 4 sieve), the resulting mixture is termed mortar. Then when coarse aggregate is included (aggregate sizes larger than the No. 4 sieve), concrete is produced. Normal concrete consists of about three-fourths aggregate and one-fourth paste, by volume. The paste usually consists of water-cement ratios between 0.4 and 0.7 by weight. Admixtures are sometimes added for specific purposes, such as to entrain numerous microscopic air bubbles, impart color, retard the initial set of the concrete, waterproof the concrete, etc.

The operations involved in the production of concrete will vary with the type of end use for the concrete, but, in general, the operations include the following (see Fig. 20-1):

1. Batching the materials
2. Mixing
3. Transporting

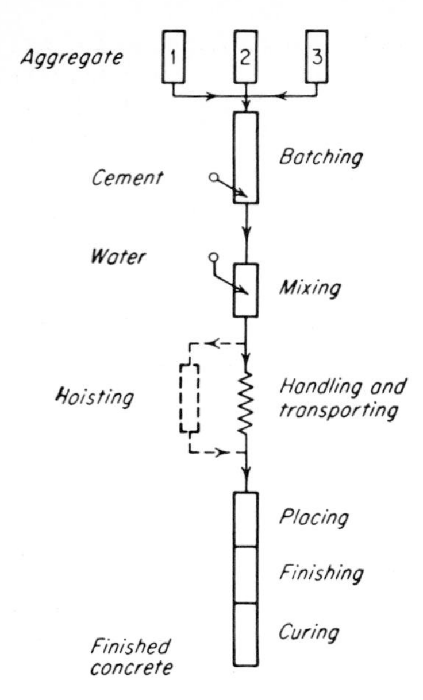

Figure 20-1 Flow diagram showing the operations performed in constructing a concrete project.

4. Placing
5. Consolidating
6. Finishing
7. Curing

HISTORY OF CONCRETE

Concrete, as we know it today, had its beginning in 1824 when Joseph Aspdin took out a patent in England on "Portland" cement. This product, which consists of limestone and clay burned at temperatures in excess of 2,700°F, is termed hydraulic in that it will react with water and harden under water. Concrete became widely used in Europe during the late 1800s and was brought to the United States late in that century. Its use continued to spread rapidly as knowledge about it and experience with it grew. Today, in the United States alone, in excess of 250 million cu yd of concrete are produced each year [1].

PROPORTIONING CONCRETE MIXTURES

For successful concrete utilization, the mixture must be properly proportioned. The American Concrete Institute (ACI) has a number of excellent recommended practices, including one on proportioning concrete mixtures [2]. While it is beyond the scope of this text to cover the specific details of proportioning, a few observations are in order. First, although it takes water to initiate the hydraulic reaction, the higher the water-cement ratio, the lower the resulting strength and durability. Second, the more water that is used (which is not to be confused with water-cement ratio), the higher will be the slump. Third, the more aggregate that is used, the lower the cost of the concrete. Fourth, the larger the maximum size of coarse aggregate, the less the amount of cement paste that will be needed to coat all the particles and provide necessary workability. Fifth, the more the concrete is consolidated, the better it becomes. Sixth, The use of properly entrained air enhances almost all concrete properties with little, or no, decrease in strength if the mix proportions are adjusted for the air. And seventh, the surface abrasion resistance of the concrete is almost entirely a function of the properties of the fine aggregate.

FRESH CONCRETE

To the designer, fresh concrete is of little importance. To the constructor, fresh concrete is *all-important*, because it is the fresh concrete that must be mixed, transported, placed, consolidated, finished, and cured. To satisfy both the designer and the constructor, the concrete should [2]:

1. Be easily mixed and transported
2. Be uniform throughout, both within a given batch and between batches
3. Be of proper workability so that it can be consolidated, will completely fill the forms, will not segregate, and will finish properly

The major property of importance to the constructor is the workability, which is difficult to define in precise terms, Like the terms *warm* and *cold*, workability depends upon the situation. One measure of workability is slump, which is a measurable value based on an American Society for Testing and Materials standard test (ASTM C143) [3]. The test is very simple to perform. Fresh concrete is placed into a hollow frustrum of a cone, 4 in. in diameter at the top, 8 in. in diameter at the bottom, and 12 in. high. After filling, the cone is raised from the concrete, allowing the fresh concrete to "slump" down. The amount of slump is measured in inches (or millimeters) from its original height of 12 in., with the stiffest concrete having zero slump and the most fluid concrete having slumps in excess of 8 in. While the slump only measures one attribute of workability, the flowability of fresh concrete, it is the most widely used measure. Table 20-1 gives the recommended slumps for various types of concrete construction [4].

HANDLING AND BATCHING CONCRETE MATERIALS

Most concrete batches, although designed on the basis of absolute volumes of the ingredients, are ultimately controlled in the batching process on the basis of

Table 20-1 Recommended slumps for various types of construction (ACI 211.1-81) [4]

Types of Construction	Slump, in.	
	Maximum	Minimum
Reinforced foundation walls and footings	3	1
Plain footings, caissons, and substructure walls	3	1
Beams and reinforced walls	4	1
Building columns	4	1
Pavements and slabs	3	1
Mass concrete	2	1

weight. Therefore it is necessary to know the weight-volume relationships of all the ingredients. Then each ingredient must be accurately weighed if the resulting mixture is to have the desired properties. It is the function of the batching equipment to perform this weighing measurement.

Handling cement Cement may be supplied to the project in paper bags, each containing 1 cu ft loose measure and weighing 94 lb net. However, for most large projects, the cement is supplied in bulk quantities from cement transport trucks, each holding 25 tons or more, or from railroad cars.

Bag cement must be stored in a dry place on pallets and should be left in the original bags until used for concrete. If the batching of concrete requires one or more whole bags of cement, the use of bag cement simplifies the batching operation.

Bulk cement usually is unloaded for cars or trucks and stored in overhead silos or bins. Figure 20-2 shows an overhead silo capable of holding cement received from railroad cars or cement trucks. The cement is transported by screw conveyor to the storage bin. A weighing hopper beneath the bin is used to measure the correct amount of cement. Weighing may be done with either a beam scale or a springless dial scale, the latter being more expensive but more dependable.

Batching and concrete Usually, concrete specifications require the concrete to be batched with aggregate having at least two size ranges (coarse and fine) and up to six ranges. Figure 20-3 illustrates the proper and improper methods of batching. Aggregate from each size range must be accurately measured. The aggregate, water, cement, and admixtures (if used) are introduced into a concrete mixer and mixed for a suitable period of time until all the ingredients are adequately blended together. ASTM C94 [5] recommends, where no mixer performance tests are available, a minimum mixing time of 1 min for mixers of 1 cu yd or less, with an increase in mixing time of $\frac{1}{4}$ min for each additional cu yd of concrete. Thus, an 8-cu-yd mixer should mix the concrete a minimum of $1 + 7 \times \frac{1}{4} = 3\frac{3}{4}$ min.

Figure 20-2 Silo for the storage of bulk cement.

Interestingly, most modern plants have performance data on their mixers to show that they can adequately mix the concrete in less time than that specified by ASTM [3]. Quite often, modern plants completely mix up to 8 cu yd of concrete in 1 min.

Batch plants and mixers There are two types of concrete-mixing operations in use, job-batched concrete and central-batched concrete. Today, unless the project is in a remote location or is relatively large, more and more of the concrete is batched in a central batch plant and transported to the job site in ready-mixed concrete trucks. Figure 20-4 shows a portable concrete batch plant, and Fig. 20-5 shows a large central batch plant of the type in general use today throughout the United States. This type of concrete is controlled by ASTM specification C94 [5], and there is a national organization promoting its use (National Ready Mixed Concrete Association [6]). Figure 20-6 shows the layout of the central batch plant for the Phillpot Dam. Note that a concrete batch required four different sizes of coarse aggregate, plus sand, two types of cement, flaked ice, and water. The water and liquid admixtures are normally measured by volume, while the cement and aggregates are measured by weight. To control the batching, close tolerances are maintained. Table 20-2 gives the permissible tolerances in accordance with ACI 304 [2]. Batch plants are available in three categories: manual, semiautomatic, and fully automatic. Manual batching is generally used for small jobs or low output values (less than

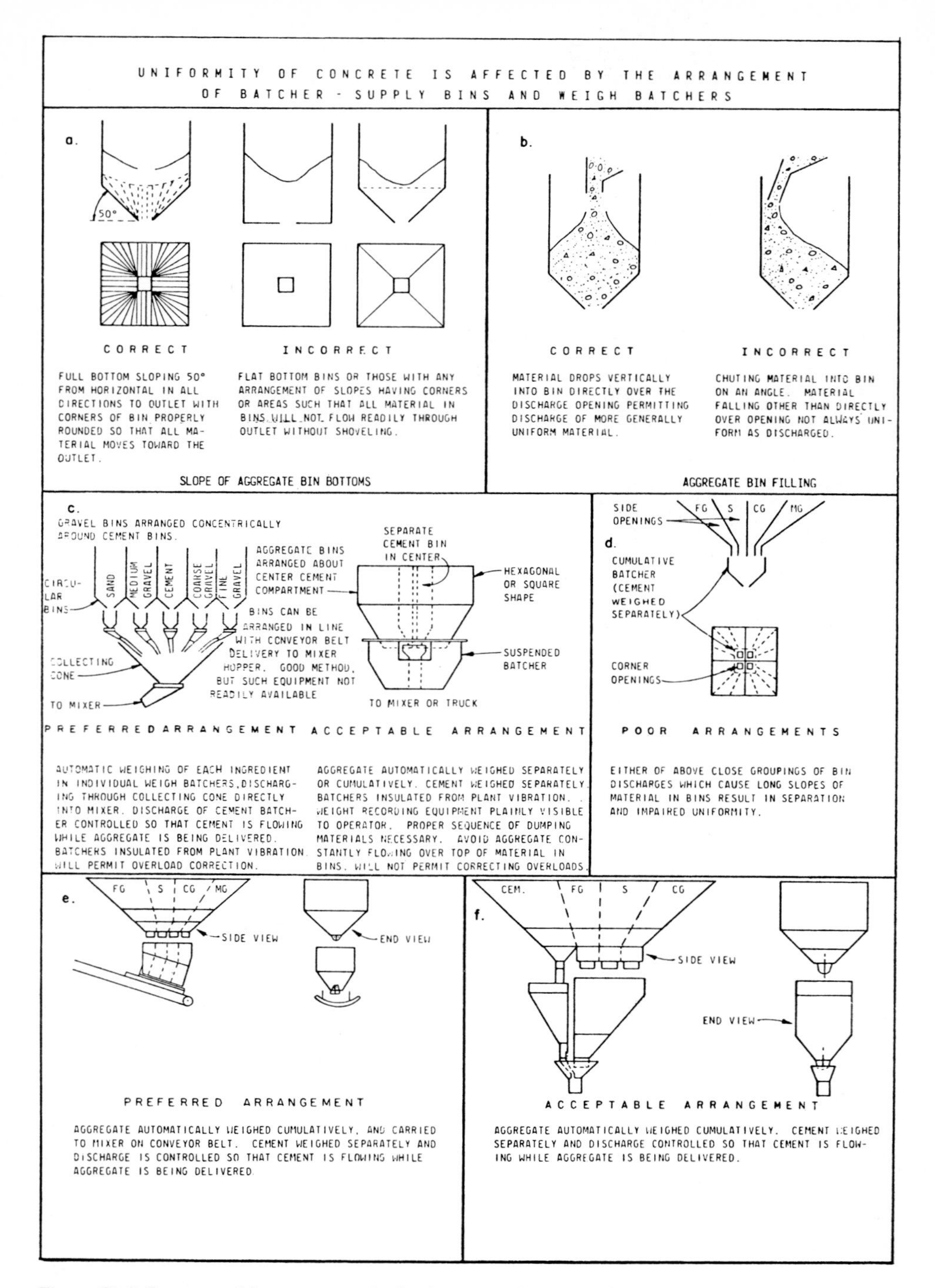

Figure 20-3 Proper and improper methods of concrete batching, (from ACI 304) [1].

Figure 20-4 Portable concrete batch plant. *(Rexworks.)*

about 500 cu yd total or around 20 cu yd per hr). In semiautomatic plants the charging and discharging of the batchers are activated manually but are automatically terminated. In a fully automatic batch plant, a single starter switch activitates the batching sequence, the weights and volumes of which have been previously programmed into the system. Figure 20-7 illustrates the control console for a fully automated plant.

Present-day plants usually have mixers capable of mixing up to 8 cu yd of concrete in each batch (although plants have been built with mixers capable of mixing 12 cu yd of concrete in each batch), and can produce up to about 200 cu yd of concrete per hour. The mixer either tilts to discharge the concrete into a truck or a chute is inserted into the mixer to catch and discharge the concrete. To increase efficiency, many large plants contain two mixers connected in series. The back mixer premixes the aggregates and cement, which reduces the time necessary for the front mixer to completely mix the batch.

Figure 20-5 Large central concrete batch plant. *(Rexworks.)*

Although the figures and discussion presented herein cover drum mixing of concrete, there are two other types of mixers in use—the pan mixer and the continuous mixer. These latter two will not be covered, as their use is limited to special situations.

In determining the quantities needed and the output for a given plant, any delays in productivity resulting from reduced operating factors should be included. The following example illustrates the method of calculating quantities and output.

Example 20-1 Determine the quantities of materials required per batch and the probable output for an 8-cu-yd central-mix plant. The quantities of materials per cubic yard are:

Cement, 5.6 bags
Sand, 1,438 lb
Gravel, 1,699 lb
Water, 39 gal

If the batch is 8.0 cu yd, the required amount of (1) cement would be $8 \times 5.6 \times 94 =$ 4,211 lb, (2) sand would be $8 \times 1{,}438 = 11{,}504$ lb, (3) gravel would be $8 \times 1{,}699 = 13{,}592$ lb, and (4) water would be $8 \times 39 = 312$ gal. The calculated fresh unit weight of the concrete would be

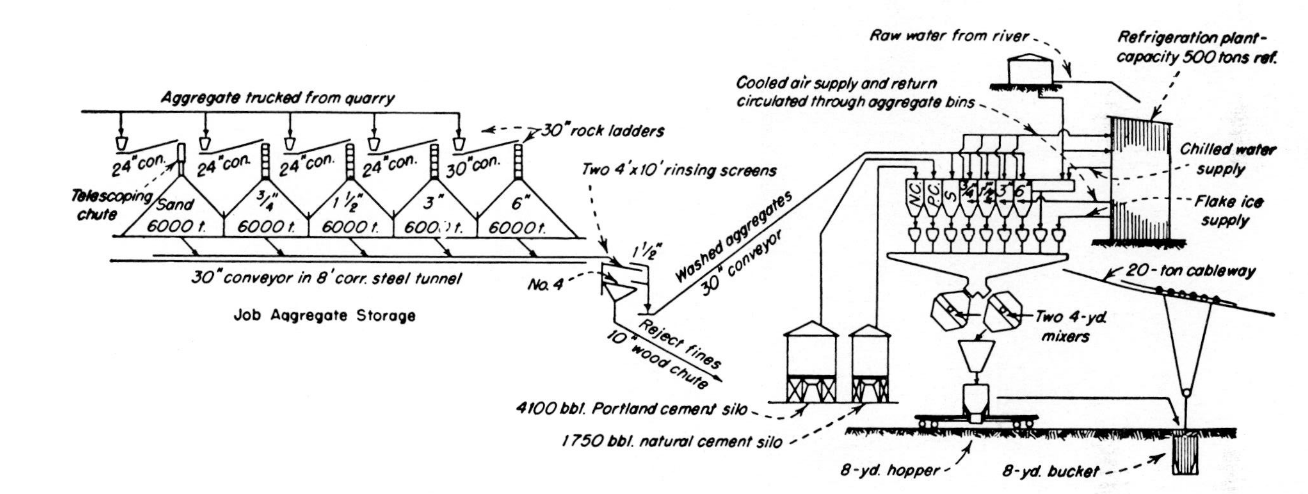

Figure 20-6 Flow diagram for the concrete-mixing plant at Philpott Dam.

Figure 20-7 Transit mixer (8 cu yd).

Table 20-2 Recommended tolerances for batching concrete materials (ACI 304) [2]

Ingredient	Batch weights greater than 30% of scale capacity: Individual batching	Batch weights greater than 30% of scale capacity: Cumulative batching	Batch weights less than 30% of scale capacity: Individual batching	Batch weights less than 30% of scale capacity: Cumulative batching
Cement and other cementitious materials	±1% or ±0.3% of scale capacity, whichever is greater		Not less than required weight or 4% more than required weight	
Water (by volume or weight), %	±1	Not recommended	±1	Not recommended
Aggregates, %	±2	±1	±2	±0.3% of scale capacity or ±3% of required cumulative weight, whichever is less
Admixtures (by volume or weight), %	±3	Not recommended	±3	Not recommended

(in lb per cu yd):

Cement, $5.6 \times 94 =$	526.4 lb
Sand	1,438 lb
Gravel	1,699 lb
Water, $39 \times 7.48 =$	291.7 lb
Total	3,955.1 lb

The fresh unit wt = 3,955.1/27 = 146.5 lb per cu ft.

If the mixer discharges the entire load into one truck, the time per cycle could be about:

Charging mixer	= 0.50 min
Mixing concrete	= 2.25 min
Discharging mixer	= 0.50 min
Lost time	= 0.25 min
Total time	= 3.50 min

No. batches per hr = 60/3.5 = 17.1
Output = $17.1 \times 8 =$ 136.8 cu yd per hr

If the lost time could be eliminated, then the production could be raised to 147.7 cu yd per hr. Furthermore, if the mixing time could be reduced to 1 min, the production could be raised to 240 cu yd per hr. However, most of the time the plant will not operate the full 60 min out of each hour at maximum capacity.

READY-MIXED CONCRETE

Increasingly, concrete is mixed in a central location and transported to the purchaser in a fresh state, mixed at the plant or en route. This type of concrete is termed ready-mixed concrete and is governed by ASTM C94 [5]. Because of its economy and quality, and through the efforts of the National Ready Mixed Concrete Association [6], concrete purchased in this manner enjoys wide acceptance. Obviously, to be useful, ready-mixed concrete must be available within a reasonable distance from the project. On remote sites and sites requiring large quantities of concrete, generally field concrete batch plants are used.

Concrete purchased from a ready-mixed concrete plant can be provided in several ways. These include:

1. *Central-mixed concrete.* This is concrete which is mixed completely in a stationary mixer and transported to the project either in a truck agitator, a truck mixer operating at agitating speed, or in a nonagitating truck.
2. *Shrink-mixed concrete.* This is concrete which is partially mixed in a stationary mixer and then mixed completely in a truck mixer (usually en route to the project).
3. *Truck-mixed concrete.* This is concrete that is completely mixed in a truck mixer, with 70 to 100 revolutions to be at a speed sufficient to completely mix the concrete. This type of concrete is usually termed transit-mixed concrete because it is generally mixed en route.

The specifications for the batch plant and the transport trucks, to include the transit mixers, are covered in detail in ASTM C94 [5]. Of particular importance is the elapsed time from the introduction of water to placement of the concrete into the forms. ASTM C94 allows a maximum of $1\frac{1}{2}$ hr, or before the drum has revolved 300 revolutions, whichever comes first.

Transit mixers are available in several sizes up to about 14 cu yd, but the most popular size is 8 cu yd (Fig. 20-7). They are capable of thoroughly mixing the concrete within about 100 revolutions of the mixing drum (see Fig. 20-8) at mixing speed (generally 8 to 12 rpm). This mixing during transit usually results in stiffening the mixture, and ASTM C94 allows the addition of water at the job site to restore the slump, followed by remixing. This has caused problems and raised questions concerning the uniformity of ready-mixed concrete. ACI 304 [1] recommends that some of the water be withheld until the mixer arrives at the project site (especially in hot weather), then the remaining water be added and an additional 30 revolutions of mixing be required. To offset any stiffening, small amounts of additional water are permitted, *provided the design water-cement ratio is not exceeded.* The uniformity requirements of ready-mixed concrete are given in Table 20-3.

Concrete may be ordered in several ways. They are:

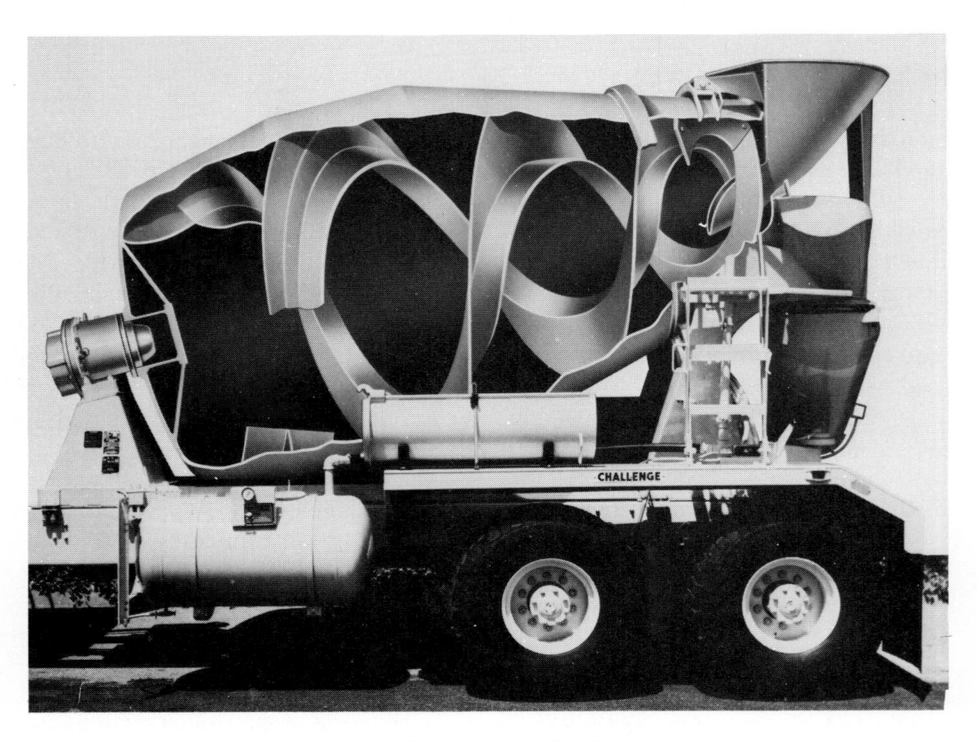

Figure 20-8 Sectional view through the drum of a transit mixer.

Table 20-3 Uniformity requirements for ready-mixed concrete (ASTM C94) [5]

Test	Requirement, expressed as maximum permissible difference in results of tests of samples taken from two locations in the concrete batch
Weight per cu ft calculated to an air-free basis	1.0 lb/cu ft
Air content, volume percent of concrete	1.0%
Slump:	
If average slump is 4 in. or less	1.0 in.
If average slump is 4 to 6 in.	1.5 in.
Coarse aggregate content, portion by weight retained on No. 4 sieve	6.0%
Unit weight of air-free mortar based on average for all comparative samples tested	1.6%
Average compressive strength at 7 days for each sample, based on average strength of all comparative test specimens	7.5%

1. *Recipe batch.* The purchaser assumes responsibility for proportioning the concrete mixture, to include specifying the cement content, the maximum allowable water content, and the admixtures required. The purchaser may also specify the amounts and type of coarse and fine aggregate. Under this approach, the purchaser assumes full responsibility for the resulting strength and durability of the mixture, providing the stipulated amounts are furnished as specified.
2. *Performance batch.* The purchaser specifies the requirements for the strength of the concrete, and the manufacturer assumes full responsibility for the proportions of the various ingredients that go into the batch.
3. *Part performance and part recipe.* The purchaser generally specifies a minimum cement content, the required admixtures, and the strength requirements, allowing the manufacturer to proportion the concrete mixture within the constraints imposed.

Today, most purchasers of concrete use the third approach, part performance and part recipe, as it ensures a minimum durability while still allowing the ready-mixed concrete supplier some flexibility to supply the most economical mixture.

MOVING AND PLACING CONCRETE

Once the concrete arrives at the project site, it must be moved to its final position without segregation and before it has achieved an initial set. This movement may be accomplished in several ways, depending upon the distance, elevation, and other constraints imposed. These methods include buckets or hoppers, chutes and drop pipes, belt conveyors, and concrete pumps.

Buckets or hoppers Normally properly designed bottom dump buckets permit concrete placement at the lowest practical slump (Fig. 20-9). Care should be exercised to prevent the concrete from segregating as a result of discharging from too high above the surface or allowing the fresh concrete to fall past obstructions. Gates should be designed so that they can be opened and closed at any time during the discharge of the concrete.

Manual or motor propelled buggies Hand buggies and wheelbarrows are usually capable of carrying from 4 to 9 cu ft of concrete, and thus are suitable on many projects, provided there are smooth and rigid runways upon which to operate. Hand buggies are safer than wheelbarrows because they have two wheels rather than one. Hand buggies and wheelbarrows are recommended for distances less than 200 ft, while power-driven or motor-driven buggies—with capacities up to around 14 cu ft—can traverse up to 1,000 ft economically (see Fig. 20-10).

Chutes and drop pipes Chutes are often used to transfer concrete from a higher elevation to a lower elevation. They should have a round bottom, and

Figure 20-9 Discharging concrete directly from bucket onto forms.

Figure 20-10 Motor-driven concrete buggies.

the slope should be steep enough for the concrete to flow continuously without segregation. Drop pipes are circular pipes used to transfer the concrete vertically. The pipe should have a diameter at least eight times the maximum aggregate size at the top 6 to 8 ft, and may be tapered to approximately six times the maximum aggregate size below [1]. Drop pipes are usually used when concrete is placed in a wall or column to avoid segregation from allowing the concrete to free-fall through the reinforcement. In such areas, pipes should always be used.

Belt conveyors Conveyors are classified into three types: (1) portable or self-contained conveyors; (2) feeders or series conveyors; and (3) side-discharge or spreader conveyors. All types must have the proper belt size and speed to achieve the desired rate of placement. Figure 20-11 shows the use of several portable conveyors to place concrete for a floor slab. This type of conveyor is capable of moving large quantities of concrete quite rapidly. Particular attention must be given to points where the concrete leaves one conveyor and either continues on another conveyor or is discharged, as segregation can easily occur. Conveyors lend themselves to moving concrete over long distances (Fig. 20-12) or up slopes (Fig. 20-13). The major disadvantage is the time necessary to set them up and to change them. The optimum concrete slump for conveyors is from $2\frac{1}{2}$ to 3 in. [4].

Concrete pumps The placement of concrete through rigid or flexible lines is not new. In fact, a patent for this method of moving concrete was issued in 1913 [7]. Pumping was not used extensively until the 1930s when German pumping equipment was introduced in this country. The pump is an extremely simple machine. By applying pressure to a column of fresh concrete in a pipe, it can be moved through the pipe if a lubricating outer layer is provided and if the mixture is properly proportioned for pumping. In order to work properly,

Figure 20-11 Use of conveyor belts to place concrete for floor slab. *(Morgen Manufacturing Company.)*

Figure 20-12 Use of conveyor belts to place concrete for a bridge. *(Morgen Manufacturing Company.)*

Figure 20-13 Use of conveyor belts to place concrete in column forms. *(Morgen Manufacturing Company.)*

the pump must be fed concrete of uniform workability and consistency. Today, concrete pumping is one of the fastest growing specialty contracting fields in the United States, as perhaps one-fourth of all concrete is placed by pumping. Pumps are available in a variety of sizes, capable of delivering concrete at sustained rates of 10 to 150 cu yd per hr. Effective pumping range varies from 300 to 1,000 ft horizontally, or 100 to 300 ft vertically [8], although occasionally pumps have moved concrete more than 5,000 ft horizontally and 1,000 ft vertically [9].

Pumps require a steady supply of *pumpable* concrete to be effective. Today there are three types of pumps being manufactured: piston pumps, pneumatic pumps, and squeeze pressure pumps. They are shown diagrammatically in Fig. 20-14(*a*), (*b*), and (*c*), respectively. Most piston pumps today contain two pistons, with one retracting during the forward stroke of the other to give a more continuous flow of concrete. The pneumatic pumps normally use a reblending discharge box at the discharge end to bleed off the air and to prevent segregation and spraying. In squeeze pressure pumps, hydraulically powered rollers rotate on the flexible hose within the drum and squeeze the concrete out at the top. The vacuum keeps a steady supply of concrete in the tube from the receiving hopper.

Pumps may be mounted on trucks, trailers, or skids. The truck-mounted pump and boom combination is particularly efficient and cost-effective in saving labor and eliminating the need for pipelines to carry the concrete. Hydraulically operated and articulated, booms come in lengths up to 100 ft and more (Fig. 20-15).

Successful pumping of concrete is no accident. A common fallacy is to assume that any good placeable concrete will pump successfully. The basic principle of pumping is that the concrete moves as a cylinder through a lubricated line, with the lubrication continually being replenished by the cylinder of concrete. To pump concrete successfully, a number of rules should be carefully followed. They are:

1. Use a minimum cement factor of 517 lb of cement per cubic yard of concrete ($5\frac{1}{2}$ sacks per cu yd).
2. Use a combined gradation of coarse and fine aggregate that ensures *no* gaps in sizes that will allow paste to be squeezed through the coarser particles under the pressures induced in the line. In particular, it is important for the fine aggregate to have at least 5 percent passing the No. 100 sieve and about 3 percent passing the No. 200 sieve (see gradations given in reference [8]). Line pressures of 300 psi are common, and they can reach as high as 1,000 psi. This is the most often overlooked aspect of good pumping!
3. Use a minimum pipe diameter of 5 in.
4. Always lubricate the line with cement paste or mortar before beginning the pumping operation.
5. Ensure a steady, uniform supply of concrete, with a slump of between 2 and 5 in. as it enters the pump.

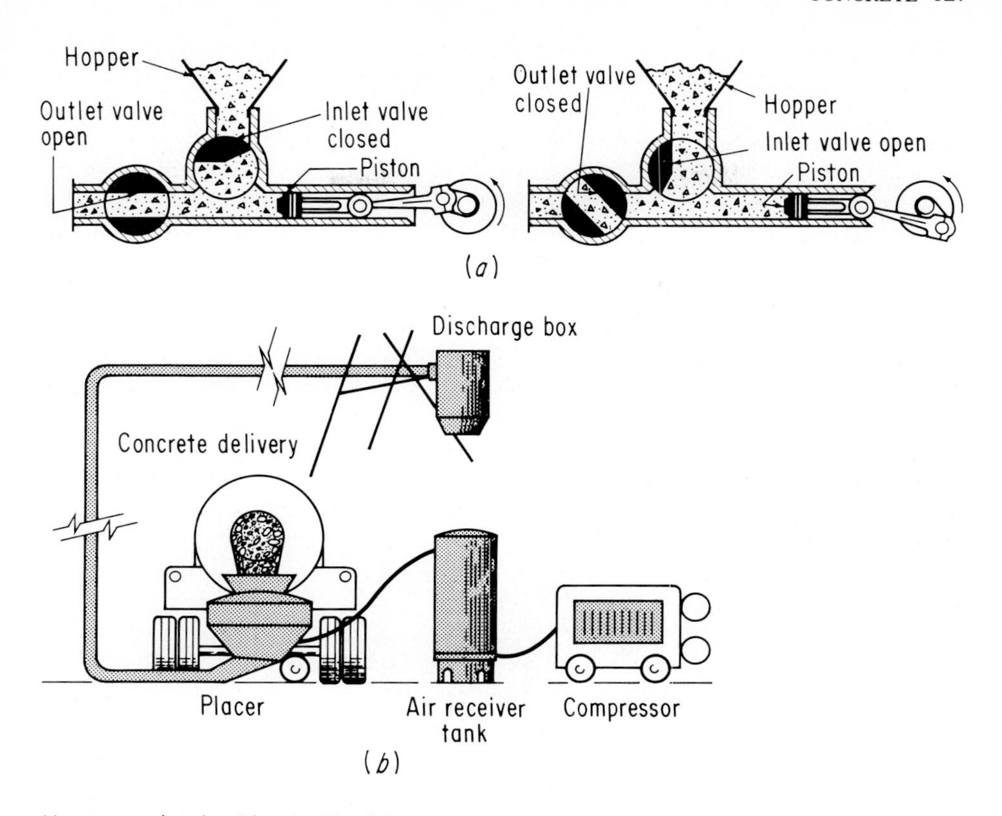

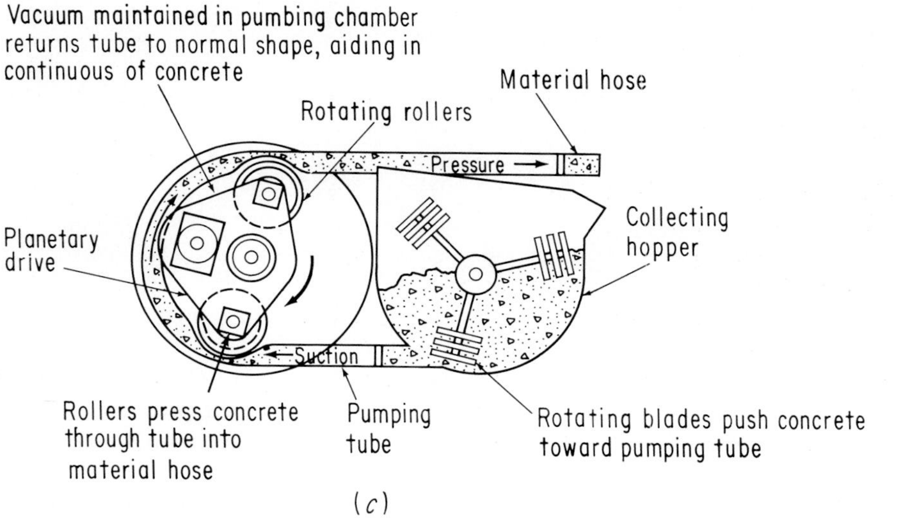

Figure 20-14 (*a*) Schematic drawing of piston-type concrete pump (from ACI 304.2R) [8]. (*b*) Schematic drawing of pneumatic-type concrete pump (from ACI 304.2R) [8]. (*c*) Schematic drawing of squeeze pressure–type concrete pump (from ACI 304.2R) [8].

Figure 20-15 Squeeze pressure–type concrete pump delivering concrete. *(Challenge-Cook Bros., Inc.)*

6. Always presoak the aggregates before mixing them in the concrete to prevent their soaking up mix water under the imposed pressure. This is especially important when aggregates are used which have a high absorption (such as structural lightweight aggregate).
7. Avoid the use of reducers in the conduit line. One common problem is the use of a 5-in. to 4-in. reducer at the discharge end so that workers will have only a 4-in. flexible hose to move around. This creates a constriction and significantly raises the pressure necessary to pump the concrete.

8. Never use aluminum lines. Aluminum particles will be scraped from the inside of the pipe as the concrete moves through and will become part of the concrete. Aluminum and portland cement react, liberating hydrogen gas, which can rupture the concrete—with disastrous results.

CONSOLIDATING CONCRETE

Concrete, being a heterogeneous mixture of water and solid particles in a stiff condition, will normally contain a large quantity of voids when placed into the forms. It is the purpose of consolidation to remove these entrapped air voids. The importance of proper consolidation cannot be overemphasized, as entrapped air can render the concrete totally unstable. Entrapped air can be reduced two ways—use more water or consolidate the concrete. Figure 20-16 shows qualitatively the benefits of consolidation, especially on low-water-content concrete.

Consolidation is normally achieved through the use of mechanical vibrators. There are three general types [10]: internal, surface, and form vibrators. Internal, or spud vibrators as they are often called, have a vibrating casing or head which is immersed into the concrete and vibrates at a high frequency (often as high as 10,000 to 15,000 vibrations per min) against the concrete.

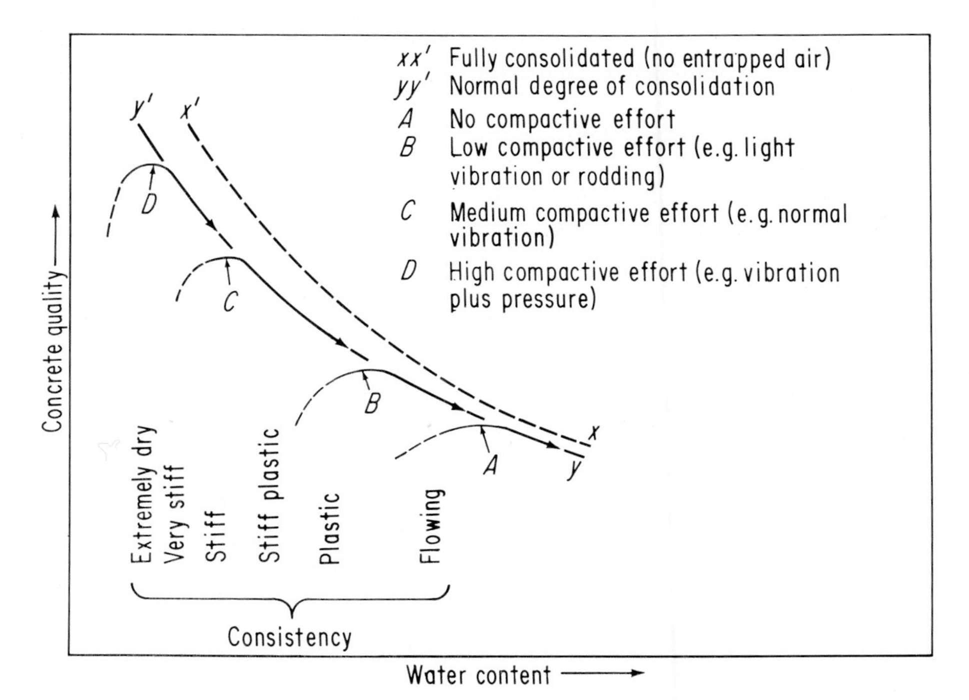

Figure 20-16 Effect of compactive effort on concrete quality (from ACI 309) [10].

Currently these vibrators are the rotary type and come in sizes from $\frac{3}{4}$ in. to 7 in. (Fig. 20-17), each with an effective radius of action [10]. They are powered by electric motors or compressed air. Manufacturers have extensive data on their vibrators.

Surface vibrators exert their effects at the top surface of the concrete and consolidate the concrete from the top down. They are used mainly in slab construction, and there are four general types: the vibrating screed, the pan-type vibrator, the plate or grid vibratory tamper, and the vibratory rolling screed. These surface vibrators operate in the range of 3,000 to 6,000 vibrations per min.

Form vibrators are external vibrators attached to the outside of the form or mold. They vibrate the form, which in turn vibrates the concrete. These types of vibrators are generally used in large precast concrete plants.

Recommended vibration practices Internal vibration is generally best suited for ordinary construction provided the section is large enough for the vibrator to be manipulated. As each vibrator has an effective radius of action, vibrator insertions should be vertical at about $1\frac{1}{2}$ times the radius of action. The vibrator should never be used to move concrete laterally, as segregation can easily occur. The vibrator should be rapidly inserted to the bottom of the layer (usually 12 to 18 in. maximum lift thickness) and at least 6 in. into the previous layer. It should then be held stationary for about 5 to 15 sec until the consolidation is considered adequate. The vibrator should then be withdrawn slowly. Where several layers are being placed, each layer should be placed while the preceding layer is still plastic.

Vibration accomplishes two actions. First, it "slumps" the concrete, removing a large portion of air that is entrapped when the concrete is deposited. Then, continued vibration consolidates the concrete, removing most of the remaining entrapped air. Generally, it will not remove entrained air. The question concerning overvibration is often raised: When does it occur and

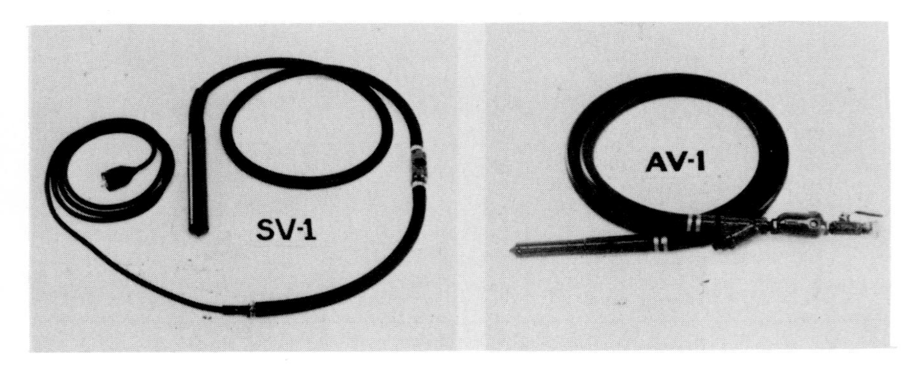

Figure 20-17 Internal vibrator. *(Allen Engineering Corp.)*

Figure 20-18 Using the shotcrete method to place concrete lining for a canal. *(Challenge-Cook Bros. Inc.)*

how harmful is it? The fact is that on low-slump concrete (concrete with less than 3 in. slump) it is almost impossible to overvibrate it with internal vibrators! When in doubt as to how much vibration to impart to low-slump concrete, vibrate it some more. The same cannot be said of concrete whose slump is 3 in. or more. This concrete can be overvibrated, which results in segregation as a result of coarse aggregate moving away from the vibrating head. Here the operator should note the presence of air bubbles escaping to the concrete surface as the vibrator is inserted. When these bubbles cease, vibration is generally complete and the vibrator should be withdrawn. Another point of caution concerns surface vibrators. They too can overvibrate the concrete at the surface, significantly weakening it if they remain in one place too long.

Another concern is the vibration of reinforcing steel. Such vibration *improves* the bond between the reinforcing steel and the concrete and thus is desirable. The undesirable side effects include damage to the vibrator and possible movement of the steel from its intended position.

Finally, revibration is the process whereby the concrete is vibrated again after it has been allowed to remain undisturbed for some time. Such revibration can be accomplished at any time the running vibrator will sink of its own weight into the concrete and liquefy it momentarily [10]. Such revibration will improve the concrete through increased consolidation.

FINISHING AND CURING CONCRETE

It cannot be stated too strongly that *any* work you do to a concrete surface after it has been consolidated will weaken the surface. All too often, concrete technicians overlook this fact and manipulate the surface of the concrete to produce a smooth, attractive surface. On walls and columns, an attractive surface may be desirable and the surface strength may not be too important, but on a floor slab, sidewalk, or pavement, the surface strength is very important. On the latter types of surfaces only the absolute minimum finishing necessary to impart the desired texture should be permitted, and the use of "jitterbugging" (the forcing of coarse aggregate down into the concrete with a steel grate tool) should not be permitted, as the surface can be weakened significantly. Furthermore, each step in the finishing operation, from first floating to the final floating or troweling, should be delayed as long as possible and still permit the desired grade and surface smoothness to be obtained. In no case should finishing commence if any free bleed water has not been blotted up, nor should neat cement or mixtures of sand and cement be worked into such surfaces to dry them up.

Along with placement and consolidation, proper curing of the concrete is extremely important. Curing may be considered as the method whereby the concrete is assured of adequate time, temperature, and supply of water for the cement to continue to hydrate. The time normally required is 3 days, and optimum temperatures are between 40 and 80°F. As most concrete is batched with sufficient water for hydration, the only problem is to ensure that the concrete does not become dried out. This may be accomplished by ponding with water (for slabs), covering with burlap or polyethylene sheets, or spraying with an approved curing compound. Curing is one of the least costly operations in the production of quality concrete, and one that is all too frequently overlooked. Concrete, if allowed to dry out during the curing stage, will attempt to shrink. The developing bonds from the cementitious reaction will attempt to restrain the shrinkage from taking place. But the end result is *always* the same: the shrinkage wins out and a crack forms as the shrinkage stresses are always higher than the tensile strength of the concrete. Proper curing does reduce the detrimental effects of cracking and develops the intended strength of the concrete.

SHOTCRETING

Shotcreting is mortar or concrete conveyed through a hose and pneumatically projected at high velocity onto a surface [11]. The force of the jet impacting on the surface compacts the mixture. Usually a relatively dry mixture is used, and thus it is able to support itself without sagging or sloughing, even for vertical and overhead applications. Shotcrete, or gunite, as it is more commonly known, is used most often in special applications involving repair work, thin

Figure 20-19 Equipment used to produce shotcrete concrete. *(Challenge-Cook Bros. Inc.)*

layers, or fiber-reinforced layers. Figure 20-18 illustrates the use of shotcrete to line a canal. Figure 20-19 shows the equipment used to produce shotcrete. There are two methods of producing shotcrete: the dry-mix process, and the wet-mix process. The dry-mix process, in which the cement and damp sand are thoroughly mixed and carried to the nozzle, where water is introduced under pressure and intimately mixed with the cement and sand before being jetted onto the surface, has been used successfully for more than 50 years. The wet-mix process is newer (around 20 years old) and involves mixing all the ingredients including water before being delivered through a hose under pressure to the desired surface. Using the wet-mix process, aggregates up to $\frac{3}{4}$ in. in size have been shotcreted. With either process there is some rebound of the mortar or concrete, resulting in some loss (usually held to 5 to 10 percent).

FLY ASH

Fly ash is produced as a by-product in the production of electricity from the burning of coal. It is the fine residue which would "fly" out the stack of a modern power plant if it were not captured. With the ever-increasing drive to wring every possible Btu out of the coal, modern power plants pulverize the coal to almost 100 percent, passing the No. 200 sieve before burning it. In the furnace, the coal is heated to in excess of 2,700°F, melting all the incombustibles.

Because of the strong induced air currents in the furnace, the molten residue from the coal becomes spherical in shape and the majority is carried out at the top of the furnace. Captured in massive electrostatic precipitators or bag houses, the fly ash is extremely fine (often finer than portland cement). Fly ash has been found to be an excellent mineral admixture in portland cement concrete, improving almost all properties of the concrete [12]. Recently, the Environmental Protection Agency ruled that fly ash must be allowed in all concrete construction involving federal funds [13], so the use of fly ash will become the rule rather than the exception.

There are basically two types of fly ash: Class F from bituminous coal and Class C from subbituminous and lignitic coal. Their quality is governed by ASTM C618 [14]. Portland cement, when it combines with water, releases calcium hydroxide (the white streaks noticeable adjacent to exposed concrete around cracks). This calcium hydroxide does not contribute to strength or durability. By introducing fly ash, the calcium hydroxide and fly ash chemically combine in a process called pozzolanic action. It takes a relatively long time (compared with the cementitious action involved between cement and water), but the resulting concrete is stronger, less permeable, and more durable than before. And, if the designer can wait for the strength development, fly ash may be used to replace part of the portland cement, resulting in a lower-cost product.

In addition to strength and durability improvements, fly ash also improves the workability of concrete, primarily because of its spherical shape. It has been used to improve the pumpability of concrete mixes, and finishers report that fly ash concrete is easier to finish.

The previous discussion applies to both Class F and Class C fly ashes. However, *some* Class C fly ashes possess significant cementitious properties themselves, probably because of their relatively high calcium contents [14]. Thus, when used as an admixture, extremely high strengths can be obtained, as illustrated by the use of Class C fly ash to achieve 7,500 psi concrete in the Texas Commerce Tower in Houston [15]. These fly ashes can replace significant portions of portland cement with no detrimental effects, as demonstrated with 25 percent replacement percentages on several concrete pavements in Texas [16].

Fly ash poses two problems to the concrete designer and constructor. One is that while it has been shown to be a valuable addition to concrete and can result in significant cost savings, it is a by-product in the production of electric power. This means that it can vary in its properties from day to day, and thus a good-quality management program is needed to ensure that only high-quality fly ash is used. Furthermore, because fly ash produced by different plants will vary in quality, it is mandatory that each fly ash be tested prior to its use to assure that it has the desired quality. Second, adding fly ash means the concrete batch has five major ingredients rather than four. The chances for a mistake being made in the batching are increased, making it desirable to increase the quality-control efforts.

BLENDED CEMENTS

Blended cements are cements that have been made by intergrinding portland cement and fly ash at the plant. These cements are governed by ASTM C595 [15] and can be used anywhere type 1 portland cement can be used. Because of the energy savings resulting through the use of fly ash, the use of this type of cement will increase as more and more cement plants start producing it.

PLACING CONCRETE IN COLD WEATHER

When concrete is placed in cold weather, some provision must be made to keep the concrete above freezing during the first few days after it has been placed.

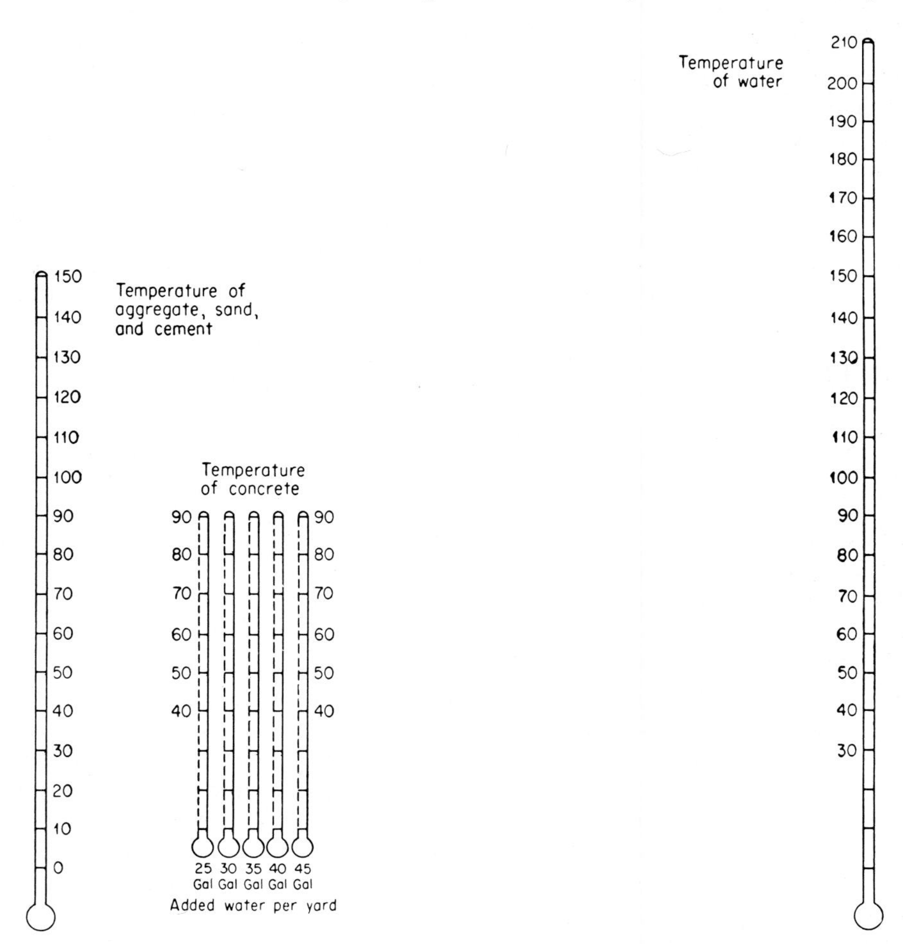

Figure 20-20 Chart for determining the temperature of concrete.

Specifications generally require that the concrete be kept at not less than 70°F for 3 days or not less than 50°F for 5 days after placement. ACI 306R-77 contains guidelines for concreting in cold weather [18]. Preheating the water is generally the most effective method of providing the necessary temperature for placement.

When the temperatures of the ingredients are known, the chart in Fig. 20-20 may be used to determine the temperature of concrete. A straight line across all three scales, passing through any two known temperatures, will permit the determination of the third temperature. If the sand is surface-dry, the solid lines of the scales giving the temperature of concrete should be used. However, if the sand contains about 3 percent moisture, the dotted lines should be used.

PLACING CONCRETE IN HOT WEATHER

When the temperature of fresh concrete exceeds around 85 to 90°F, the resulting strength and durability of the concrete can be reduced. Therefore most specifications require the concrete to be placed at a temperature less than 90°F. When concrete is placed in hot weather, the ingredients should be cooled before mixing. ACI 305R-77 contains guidelines for mixing and placing concrete in hot weather [19]. Methods of cooling include using ice instead of water in the mix and cooling the aggregate with liquid nitrogen.

PROBLEMS

20-1 A concrete batch calls for the following quantities per cubic yard of concrete, based on saturated surface-dry conditions of the aggregate. Determine the required weight of each solid ingredient and the number of gallons of water required for a 6.5-cu-yd batch. Also determine the wet unit weight of the concrete in pounds per cubic foot.

Cement, 5.6 bags
Fine aggregate, 1,420 lb
Coarse aggregate, 1,840 lb
Water, 34 gal

20-2 In Prob. 20-1, assume that the fine aggregate contains 8 percent free moisture, by weight, and the coarse aggregate contains 3 percent free moisture, by weight. Determine the required weights of cement, fine aggregate, coarse aggregate, and volume of added water for a 6.5-cu-yd batch.

20-3 A concrete retaining wall whose total volume will be 236 cu yd is to be constructed by using job-mixed concrete containing the following quantities per cubic yard, based on surface-dry sand and gravel:

Cement, 5.6 bags
Sand, 1,340 lb
Gravel, 1,864 lb
Water, 33 gal

The sand and gravel will be purchased by ton weight, including any moisture present at the time they are weighed. The gross weights, including the moisture present at the time of weighing, are as follows:

Item	Gross weight, lb per cu yd	Percent moisture by gross weight
Sand	2,945	6
Gravel	2,968	3

It is estimated that 7 percent of the sand and 6 percent of the gravel will be lost or not recovered in the stockpile at the job.

Determine the total number of tons each of sand and gravel required for the project.

20-4 The aggregate, sand, and cement used in a concrete have an initial temperature of 76°F. The sand contains 4 percent moisture. If 35 gal of water, at a temperature of 46°F, is used per cubic yard of concrete, determine the temperature of the concrete.

20-5 The aggregate, sand, and cement used in a concrete have an initial temperature of 44°F. The sand contains 4 percent moisture. If 30 gal of water is used per cubic yard of concrete, what should be the temperature of the water to produce a concrete having a temperature of 72°F?

REFERENCES

1. ACI Committee 304, "Recommended Practice for Measuring, Mixing, Transporting, and Placing Concrete," ACI 304-73, Reaffirmed 1978, *ACI Manual of Concrete Practice, Part 2,* American Concrete Institute, Detroit, MI, 1983.
2. Mindess, Sidney, and J. Francis Young: "Concrete," Prentice-Hall, Englewood Cliffs, NJ, 1981.
3. ASTM Committee C9, "Test for Slump of Portland Cement Concrete," *Annual Book of ASTM Standards,* vol. 04.02, 1983.
4. ACI Committee 211, "Standard Practice for Selecting Proportions for Normal, Heavyweight, and Mass Concrete," ACI 211.2-81, *ACI Manual of Concrete Practice, Part 1,* American Concrete Institute, Detroit, MI, 1983.
5. ASTM Committee C94, "Standard Specification for Ready-Mixed Concrete," *Annual Book of ASTM Standards,* vol. 04.02, 1983.
6. National Ready Mixed Concrete Association, 900 Spring Street, Silver Spring, MD 20910.
7. Ledbetter, Bonnie S., W. B. Ledbetter, and Eugene H. Boeke, Mixing, Moving, and Mashing Concrete—75 Years of Progress, *Concrete International,* pp. 69–76, November 1980.
8. ACI Committee 304, "Placing Concrete by Pumping Methods," ACI 304.2R-71, Revised 1982, *ACI Manual of Concrete Practice, Part 2,* American Concrete Institute, Detroit, MI, 1983.
9. ACI Committee 304, Pumped Concrete Climbs 75 Flights, *Engineering News Record,* pp. 28–29, March 5, 1981.
10. ACI Committee 309, "Standard Practice for Consolidation of Concrete," ACI 309-72, Revised 1982, *ACI Manual of Concrete Practice, Part 2,* American Concrete Institute, Detroit, MI, 1983.
11. ACI Committee 506, "Recommended Practice for Shotcreting," ACI 506-66, Reaffirmed 1978, *ACI Manual of Concrete Practice, Part 5,* American Concrete Institute, Detroit, MI, 1983.
12. Berry, E. E., and V. M. Malhotra. "Fly Ash for Use in Concrete—A Critical Review," *Journal of the ACI,* pp. 59–73, March–April 1980.
13. EPA, Guidelines for Federal Procurement of Cement and Concrete Containing Fly Ash, *Federal Register,* vol. 48, no. 20, pp. 4230–4253, January 28, 1983.

14. McKerall, W. C., and W. B. Ledbetter: Variability and Control of Class C Fly Ash, *Cement, Concrete, and Aggregates*, CCAGOP, vol. 14, no. 2, pp. 87–93. Winter 1982, American Society for Testing and Materials, Philadelphia, PA.
15. Cook, James E.: Research and Application of High Strength Concrete Using Class C Fly Ash, *Concrete International*, July 1982.
16. Ledbetter, W. B., D. J. Teague, R. L. Long, and B. N. Banister, "Construction of Fly Ash Test Sites and Guidelines for Construction," *Research Report* 240-2, Texas Transportation Institute, Texas A&M University, College Station, TX, October 1981, 111 p.
17. ASTM C9, "Specification for Blended Hydraulic Cements," *Annual Book of ASTM Standards*, vol. 04.02, 1983.
18. ACI Committee 306, "Cold Weather Concreting," ACI 306R-78, *ACI Manual of Concrete Practice, Part 2*, American Concrete Institute, Detroit, MI, 1983.
19. ACI Committee 305, "Hot Weather Concreting," ACI 305R-77, *ACI Manual of Concrete Practice, Part 2*, American Concrete Institute, Detroit, MI, 1983.
20. CMI Corporation, P.O. Box 1985, Oklahoma City, OK 73101.
21. Rexworks, Inc., P.O. Box 2037, Milwaukee, WI 53021.
22. Challenge-Cook Bros., Inc., P.O. Box 1272, Industry, CA 91749.
23. Allen Engineering Corp., P.O. Box 1058, Paragould, AK 72450.

APPENDIX

A

INTEREST TABLES

			$i = 0.5\%$			
n	P/F	P/A	A/F	A/P	F/P	F/A
1	0.9950249	0.9950249	1.0000000	1.0050000	1.005000	1.000000
2	0.9900745	1.9850990	0.4987531	0.5037531	1.010025	2.005000
3	0.9851488	2.9702480	0.3316722	0.3366722	1.015075	3.015025
4	0.9802475	3.9504960	0.2481328	0.2531328	1.020151	4.030100
5	0.9753707	4.9258660	0.1980100	0.2030100	1.025251	5.050251
6	0.9705181	5.8963850	0.1645955	0.1695955	1.030378	6.075502
7	0.9656896	6.8620740	0.1407285	0.1457285	1.035529	7.105880
8	0.9608852	7.8229590	0.1228289	0.1278289	1.040707	8.141409
9	0.9561047	8.7790640	0.1089074	0.1139074	1.045911	9.182116
10	0.9513479	9.7304120	0.0977706	0.1027706	1.051140	10.228030
11	0.9466149	10.6770300	0.0886590	0.0936590	1.056396	11.279170
12	0.9419053	11.6189300	0.8106640	0.0860664	1.061678	12.335560
13	0.9372192	12.5561500	0.0746422	0.0796422	1.066986	13.397240
14	0.9325565	13.4887100	0.0691361	0.0741361	1.072321	14.464230
15	0.9279169	14.4166200	0.0643644	0.0693644	1.077683	15.536550
16	0.9233004	15.3399300	0.0601894	0.0651894	1.083071	16.614230
17	0.9187068	16.2586300	0.0565058	0.0615058	1.088487	17.697300
18	0.9141362	17.1727700	0.0532317	0.0582317	1.093929	18.785790
19	0.9095882	18.0823600	0.0503025	0.0553025	1.099399	19.879720
20	0.9050629	18.9874200	0.0476665	0.0526665	1.104896	20.979120
21	0.9005601	19.8879800	0.0452816	0.0502816	1.110420	22.084010
22	0.8960797	20.7840600	0.0431138	0.0481138	1.115972	23.194430
23	0.8916216	21.6756800	0.0411347	0.0461347	1.121552	24.310400
24	0.8871857	22.5628700	0.0393206	0.0443206	1.127160	25.431960
25	0.8827718	23.4456400	0.0376519	0.0426519	1.132796	26.559120

			$i = 0.5\%$			
n	P/F	P/A	A/F	A/P	F/P	F/A
26	0.8783799	24.3240200	0.0361116	0.0411116	1.138460	27.691910
27	0.8740099	25.1980300	0.0346856	0.0396856	1.144152	28.830370
28	0.8696616	26.0676900	0.0333617	0.0383617	1.149873	29.974520
29	0.8653349	26.9330200	0.0321291	0.0371291	1.155622	31.124400
30	0.8610297	27.7940500	0.0309789	0.0359789	1.161400	32.280020
32	0.8524836	29.5032800	0.0288945	0.0338945	1.173043	34.608620
34	0.8440223	31.1955500	0.0270559	0.0320559	1.184803	36.960580
36	0.8356449	32.8710200	0.0254219	0.0304219	1.196681	39.336110
38	0.8273507	34.5298600	0.0239604	0.0289604	1.208677	41.735450
40	0.8191389	36.1722300	0.0226455	0.0276455	1.220794	44.158850
48	0.7870984	42.5803200	0.0184850	0.0234850	1.270489	54.097830
60	0.7413722	51.7255600	0.0143328	0.0193328	1.348850	69.770030
72	0.6983024	60.3395100	0.0115729	0.0165729	1.432044	86.408860
80	0.6709885	65.8023100	0.0101970	0.0151970	1.490339	98.067720
100	0.6072868	78.5426500	0.0077319	0.0127319	1.646669	129.333700
120	0.5496327	90.0734500	0.0061021	0.0111021	1.819397	163.879400
180	0.4074824	118.5035000	0.0034386	0.0084386	2.454094	290.818700
240	0.3020961	139.5808000	0.0021643	0.0071643	3.310205	462.040900

			$i = 0.75\%$			
n	P/F	P/A	A/F	A/P	F/P	F/A
1	0.9925558	0.992558	1.0000000	1.0075000	1.007500	1.000000
2	0.9851671	1.977723	0.4981320	0.5056320	1.015056	2.007500
3	0.9778333	2.955556	0.3308458	0.3383458	1.022669	3.022556
4	0.9705542	3.926110	0.2472050	0.2547050	1.030339	4.045225
5	0.9633292	4.889440	0.1970224	0.2045224	1.038067	5.075565
6	0.9561580	5.845598	0.1635689	0.1710689	1.045852	6.113631
7	0.9490402	6.794638	0.1396749	0.1471749	1.053696	7.159484
8	0.9419754	7.736613	0.1217555	0.1292555	1.061599	8.213180
9	0.9349632	8.671576	0.1078193	0.1153193	1.069561	9.274779
10	0.9280032	9.599580	0.0966712	0.1041712	1.077583	10.344340
11	0.9210949	10.520670	0.0875509	0.0950509	1.085664	11.421920
12	0.9142382	11.434910	0.0799515	0.0874515	1.093807	12.507590
13	0.9074324	12.342350	0.0735219	0.0810219	1.102010	13.601390
14	0.9006773	13.243020	0.0680115	0.0755115	1.110276	14.703400
15	0.8939725	14.136990	0.0632364	0.0707364	1.118603	15.813680
16	0.8873177	15.024310	0.0590588	0.0665588	1.126992	16.932280
17	0.8807123	15.905020	0.0553732	0.0628732	1.135445	18.059270
18	0.8741561	16.779180	0.0520977	0.0595977	1.143960	19.194720

			$i = 0.75\%$			
n	P/F	P/A	A/F	A/P	F/P	F/A
19	0.8676488	17.646830	0.0491674	0.0566674	1.152540	20.338680
20	0.8611899	18.508020	0.0465306	0.0540306	1.161184	21.491220
21	0.8547790	19.362800	0.0441454	0.0516454	1.169893	22.652400
22	0.8484159	20.211210	0.0419775	0.0494775	1.178667	23.822300
23	0.8421001	21.053310	0.0399985	0.0474985	1.187507	25.000960
24	0.8358314	21.889150	0.0381847	0.0456847	1.196414	26.188470
25	0.8296093	22.718760	0.0365165	0.0440165	1.205387	27.384880
26	0.8234336	23.542190	0.0349769	0.0424769	1.214427	28.590270
27	0.8173038	24.359490	0.0335518	0.0410518	1.223535	29.804700
28	0.8112197	25.170710	0.0322287	0.0397287	1.232712	31.028230
29	0.8051808	25.975890	0.0309972	0.0384972	1.241957	32.260940
30	0.7991869	26.775080	0.0298482	0.0373482	1.251272	33.502900
32	0.7873326	28.355650	0.0277663	0.0352663	1.270111	36.014830
34	0.7756542	29.912780	0.0259305	0.0334305	1.289234	38.564580
36	0.7641490	31.446810	0.0242997	0.0317997	1.308645	41.152720
38	0.7528144	32.958080	0.0228416	0.0303416	1.328349	43.779820
40	0.7416480	34.446940	0.0215302	0.0290302	1.348349	46.446480
48	0.6986141	40.184780	0.0173850	0.0248850	1.431405	57.520710
60	0.6386997	48.173370	0.0132584	0.0207584	1.565681	75.424140
72	0.5839236	55.476850	0.0105255	0.0180255	1.712553	95.007000
80	0.5500417	59.994440	0.0091682	0.0166682	1.818044	109.072500
100	0.4736903	70.174620	0.0067502	0.0142502	2.111084	148.144500
120	0.4079373	78.941690	0.0051676	0.0126676	2.451357	193.514300
180	0.2605494	98.593410	0.0026427	0.0101427	3.838043	378.405800
240	0.1664128	111.145000	0.0014973	0.0089973	6.009151	667.886900

			$i = 1\%$			
n	P/F	P/A	A/F	A/P	F/P	F/A
1	0.9900990	0.990099	1.0000000	1.0100000	1.010000	1.000000
2	0.9802960	1.970395	0.4975125	0.5075125	1.020100	2.010000
3	0.9705901	2.940985	0.3300221	0.3400221	1.030301	3.030100
4	0.9609803	3.901965	0.2462811	0.2562811	1.040604	4.060401
5	0.9514657	4.853431	0.1960398	0.2060398	1.051010	5.101005
6	0.9420452	5.795476	0.1625484	0.1725484	1.061520	6.152015
7	0.9327181	6.728194	0.1386283	0.1486283	1.072135	7.213535
8	0.9234832	7.651678	0.1206903	0.1306903	1.082857	8.285670
9	0.9143398	8.566017	0.1067404	0.1167404	1.093685	9.368527
10	0.9052870	9.471304	0.0955821	0.1055821	1.104622	10.462210

			i = 1%			
n	*P/F*	*P/A*	*A/F*	*A/P*	*F/P*	*F/A*
11	0.8963237	10.367630	0.0864541	0.0964541	1.115668	11.566830
12	0.8874492	11.255080	0.0788488	0.0888488	1.126825	12.682500
13	0.8786626	12.133740	0.0724148	0.0824148	1.138093	13.809330
14	0.8699630	13.003700	0.0669012	0.0769012	1.149474	14.947420
15	0.8613495	13.865050	0.0621238	0.0721238	1.160969	16.096900
16	0.8528213	14.717870	0.0579446	0.0679446	1.172579	17.257860
17	0.8443775	15.562250	0.0542581	0.0642581	1.184304	18.430440
18	0.8360173	16.398270	0.0509820	0.0609820	1.196147	19.614750
19	0.8277399	17.226010	0.0480518	0.0580518	1.208109	20.810890
20	0.8195445	18.045550	0.4541530	0.0554153	1.220190	22.019000
21	0.8114302	18.856980	0.0430308	0.0530308	1.232392	23.239190
22	0.8033962	19.660380	0.0408637	0.0508637	1.244716	24.471590
23	0.7954418	20.455820	0.0388858	0.0488858	1.257163	25.716300
24	0.7875661	21.243390	0.0370735	0.0470735	1.269735	26.973460
25	0.7797684	22.023160	0.0354068	0.0454068	1.282432	28.243200
26	0.7720480	22.795200	0.0338689	0.0438689	1.295256	29.525630
27	0.7644039	23.559610	0.0324455	0.0424455	1.308209	30.820890
28	0.7568356	24.316440	0.0311244	0.0411244	1.321291	32.129100
29	0.7493422	25.065780	0.0298950	0.0398950	1.334504	33.450390
30	0.7419229	25.807710	0.0287481	0.0387481	1.347849	34.784890
32	0.7273041	27.269590	0.0266709	0.0366709	1.374941	37.494070
34	0.7129733	28.702670	0.0248400	0.0348400	1.402577	40.257700
36	0.6989250	30.107500	0.0232143	0.0332143	1.430769	43.076880
38	0.6851534	31.484660	0.0217615	0.0317615	1.459527	45.952720
40	0.6716310	32.834690	0.0204556	0.3045560	1.488864	48.886370
48	0.6202604	37.973960	0.0163338	0.0263338	1.612226	61.222610
60	0.5504496	44.955040	0.0122444	0.0222444	1.816697	81.669670
72	0.4884961	51.150390	0.0095502	0.0195502	2.047099	104.709900
80	0.4511179	54.888210	0.0082189	0.0182189	2.216715	121.671500
100	0.3697112	63.028880	0.0058657	0.0158657	2.704814	170.481400
120	0.3029948	69.700520	0.0043471	0.0143471	3.300387	230.038700
180	0.1667834	83.321660	0.0020017	0.0120017	5.995802	499.580200
240	0.0918058	90.819420	0.0010109	0.0110109	10.892550	989.255300

			i = 1.25%			
n	*P/F*	*P/A*	*A/F*	*A/P*	*F/P*	*F/A*
1	0.9876543	0.9876543	1.0000000	1.0125000	1.012500	1.000000
2	0.9754611	1.963115	0.4968944	0.5093944	1.025156	2.012500

			$i = 1.25\%$			
n	P/F	P/A	A/F	A/P	F/P	F/A
3	0.9634183	2.926534	0.3292012	0.3417012	1.037971	3.037656
4	0.9515243	3.878058	0.2453610	0.2578610	1.050945	4.075627
5	0.9397771	4.817835	0.1950621	0.2075621	1.064082	5.126572
6	0.9281749	5.7460100	0.1615338	0.1740338	1.077383	6.190654
7	0.9167159	6.6627260	0.1375887	0.1500887	1.090850	7.268037
8	0.9053984	7.5681240	0.1196331	0.1321331	1.104486	8.358888
9	0.8942207	8.4623450	0.1056706	0.1181706	1.118292	9.463374
10	0.8831809	9.3455260	0.0945031	0.1070031	1.132271	10.581670
11	0.8722775	10.2178000	0.0853684	0.0978684	1.146424	11.713940
12	0.8615086	11.0793100	0.0777583	0.0902583	1.160755	12.860360
13	0.8508727	11.9301800	0.0713210	0.0838210	1.175264	14.021120
14	0.8403681	12.7705500	0.0658051	0.0783051	1.189955	15.196380
15	0.8299932	13.6005500	0.0610265	0.0735265	1.204829	16.386330
16	0.8197463	14.4202900	0.0568467	0.0693467	1.219890	17.591160
17	0.8096260	15.2299200	0.0531602	0.0656602	1.235138	18.811050
18	0.7996306	16.0295500	0.0498848	0.0623848	1.250577	20.046190
19	0.7897587	16.8193100	0.0469555	0.0594555	1.266210	21.296770
20	0.7800086	17.5993200	0.0443204	0.0568204	1.282037	22.562980
21	0.7703788	18.3696900	0.0419375	0.0544375	1.298063	23.845020
22	0.7608680	19.1305600	0.0397724	0.0522724	1.314288	25.143080
23	0.7514745	19.8820400	0.0377967	0.0502967	1.330717	26.457370
24	0.7421971	20.6242300	0.0359866	0.0484866	1.347351	27.788080
25	0.7330341	21.3572700	0.0343225	0.0468225	1.364193	29.135430
26	0.7239843	22.0812500	0.0327873	0.0452873	1.381245	30.499630
27	0.7150463	22.7963000	0.0313668	0.0438668	1.398511	31.880870
28	0.7062185	23.5025200	0.0300486	0.0425486	1.415992	33.279380
29	0.6974998	24.2000200	0.0288223	0.0413223	1.433692	34.695380
30	0.6888887	24.8889100	0.0276785	0.0401785	1.451613	36.129070
32	0.6719841	26.2412700	0.0256079	0.0381079	1.488130	39.050440
34	0.6554943	27.5604600	0.0237839	0.0362839	1.525566	42.045300
36	0.6394092	28.8472700	0.0221653	0.0346653	1.563944	45.115500
38	0.6237187	30.1025000	0.0207198	0.0332198	1.603287	48.262940
40	0.6084133	31.3269300	0.0194214	0.0319214	1.643619	51.489560
48	0.5508565	35.9314800	0.0153307	0.0278307	1.815355	65.228390
60	0.4745676	42.0345900	0.0112899	0.0237899	2.107181	88.574510
72	0.4088441	47.2924700	0.0086450	0.0211450	2.445920	115.673600
80	0.3701668	50.3866600	0.0073465	0.0198465	2.701485	136.118800
100	0.2887333	56.9013400	0.0050743	0.0175743	3.463404	197.072300
120	0.2252144	61.9828500	0.0036335	0.0161335	4.440213	275.217000
180	0.1068795	71.4496400	0.0014959	0.0139959	9.356334	668.506700
240	0.0507215	75.9422800	6.679E-40	0.0131679	19.715490	1,497.239000

	i = 1.5%					
n	P/F	P/A	A/F	A/P	F/P	F/A
1	0.9852217	0.9852217	1.0000000	1.0150000	1.000000	1.000000
2	0.9706617	1.9558830	0.4962779	0.5112779	1.030225	2.015000
3	0.9563170	2.9122000	0.3283830	0.3433830	1.045678	3.045225
4	0.9421842	3.8543850	0.2444448	0.2594448	1.061364	4.090903
5	0.9282603	4.7826450	0.1940893	0.2090893	1.077284	5.152267
6	0.9145422	5.6971870	0.1605252	0.1755252	1.093443	6.229551
7	0.9010268	6.5982140	0.1365562	0.1515562	1.109845	7.322994
8	0.8877111	7.4859250	0.01185840	0.1335840	1.126493	8.432839
9	0.8745922	8.3605170	0.1046098	0.1196098	1.143390	9.559332
10	0.8616672	9.2221850	0.0934342	0.1084342	1.160541	10.702720
11	0.8489332	10.0711200	0.0842938	0.0992938	1.177949	11.863260
12	0.8363874	10.9075100	0.0766800	0.0916800	1.195618	13.041210
13	0.8240270	11.7315300	0.0702404	0.0852404	1.213552	14.236830
14	0.8118493	12.5433800	0.0647233	0.0797233	1.231756	15.450380
15	0.7998515	13.3432300	0.0599444	0.0749444	1.250232	16.682140
16	0.7880310	14.1312600	0.0557651	0.0707651	1.268986	17.932370
17	0.7763853	14.9076500	0.0520797	0.0670797	1.288020	19.201360
18	0.7649116	15.6725600	0.0488058	0.0638058	1.307341	20.489380
19	0.7536075	16.4261700	0.0458785	0.0608785	1.326951	21.796720
20	0.7424740	17.1686400	0.0432457	0.0582457	1.346855	23.123670
21	0.7314979	17.9001400	0.0408655	0.0558655	1.367058	24.470520
22	0.7206876	18.6208200	0.0387033	0.0537033	1.387564	25.837580
23	0.7100371	19.3308600	0.0367308	0.0517308	1.408377	27.225140
24	0.6995439	20.0304100	0.0349241	0.0499241	1.429503	28.633520
25	0.6892058	20.7196100	0.0332635	0.0482635	1.450945	30.063020
26	0.6790205	21.3986300	0.0317320	0.0467320	1.472710	31.513970
27	0.6689857	22.0676200	0.0303153	0.0453153	1.494800	32.986680
28	0.6590992	22.7267200	0.0290011	0.0440011	1.517222	34.481480
29	0.6493589	23.3760800	0.0277788	0.0427788	1.539981	35.998700
30	0.6397624	24.0158400	0.0266392	0.0416392	1.563080	37.538680
32	0.6209929	25.2671400	0.0245771	0.0395771	1.610324	40.688290
34	0.6027741	26.4817300	0.0227619	0.0377619	1.658996	43.933090
36	0.5850897	27.6606800	0.0211524	0.0361524	1.709140	47.275970
38	0.5679242	28.8050500	0.0197161	0.0347161	1.760798	50.719890
40	0.5512623	29.9158500	0.0184271	0.0334271	1.814018	54.267890
48	0.4893617	34.0425500	0.0143750	0.0293750	2.043478	69.565220
60	0.4092960	39.3802700	0.0103934	0.0253934	2.443220	96.214650
72	0.3423300	43.8446700	0.0078078	0.0228078	2.921158	128.077200
80	0.3038901	46.4073200	0.0065483	0.0215483	3.290663	152.710900
100	0.2256294	51.6247000	0.0043706	0.0193706	4.432046	228.803000
120	0.1675232	55.4984500	0.0030185	0.0180185	5.969323	331.288200

			$i = 1.5\%$			
n	P/F	P/A	A/F	A/P	F/P	F/A
180	0.0685666	62.0955600	0.0011042	0.0161042	14.584370	905.624500
240	0.0280640	64.7957300	4.331E-40	0.0154331	35.632820	2,308.854000

			$i = 1.75\%$			
n	P/F	P/A	A/F	A/P	F/P	F/A
1	0.9828010	0.982801	1.0000000	1.0175000	1.017500	1.000000
2	0.9658978	1.948699	0.4956630	0.5131630	1.035306	2.017500
3	0.9492853	2.897984	0.3275675	0.3450675	1.053424	3.052806
4	0.9329585	3.830943	0.2435324	0.2610324	1.071859	4.106230
5	0.9169125	4.747855	0.1931214	0.2106214	1.090617	5.178089
6	0.9011425	5.648998	0.1595226	0.1770226	1.109702	6.268706
7	0.8856438	6.534641	0.1355306	0.1530306	1.129122	7.378408
8	0.8704116	7.405053	0.1175429	0.1350429	1.148882	8.507530
9	0.8554413	8.260494	0.1035581	0.1210581	1.168987	9.656412
10	0.8407286	9.101223	0.0923753	0.1098753	1.189444	10.825400
11	0.8262689	9.927492	0.0832304	0.1007304	1.210260	12.014840
12	0.8120579	10.739550	0.0756138	0.0931138	1.231439	13.225100
13	0.7980913	11.537640	0.0691728	0.0866728	1.252990	14.456540
14	0.7843649	12.322010	0.0636556	0.0811556	1.274917	15.709530
15	0.7708746	13.092880	0.0588774	0.0763774	1.297228	16.984450
16	0.7576163	13.850500	0.5469960	0.0721996	1.319929	18.281680
17	0.7445861	14.595080	0.0510162	0.0685162	1.343028	19.601610
18	0.7317799	15.326860	0.0477449	0.0652449	1.366531	20.944630
19	0.7191940	16.046060	0.0448206	0.0623206	1.390445	22.311170
20	0.7068246	16.752880	0.0421912	0.0596912	1.414778	23.701610
21	0.6946679	17.447550	0.0398146	0.0573146	1.439537	25.116390
22	0.6827203	18.130270	0.0376564	0.0551564	1.464729	26.555930
23	0.6709782	18.801250	0.0356880	0.0531880	1.490361	28.020650
24	0.6594380	19.460690	0.0338857	0.0513857	1.516443	29.511020
25	0.6480963	20.108780	0.0322295	0.0497295	1.542981	31.027460
26	0.6369497	20.745730	0.0307027	0.0482027	1.569983	32.570440
27	0.6259948	21.371730	0.0292908	0.0467908	1.597457	34.140420
28	0.6152283	21.986950	0.0279815	0.0454815	1.625413	35.737880
29	0.6046470	22.591600	0.0267642	0.0442642	1.653858	37.363290
30	0.5942476	23.185850	0.0256298	0.0431298	1.682800	39.017150
32	0.5739825	24.343860	0.0235781	0.0410781	1.742213	42.412200
34	0.5544084	25.462380	0.0217736	0.0392736	1.803725	45.927120
36	0.5355018	26.542750	0.0201751	0.0376751	1.867407	49.566130
38	0.5172400	27.586280	0.0187499	0.0362499	1.933338	53.333620

			i = 1.75%			
n	*P/F*	*P/A*	*A/F*	*A/P*	*F/P*	*F/A*
40	0.4996010	28.594230	0.0174721	0.0349721	2.001597	57.234130
48	0.4348585	32.293800	0.0134657	0.0309657	2.299599	74.262780
60	0.3531303	36.963990	0.0095534	0.0270534	2.831816	104.675200
72	0.2867622	40.756450	0.0070360	0.0245360	3.487210	142.126300
80	0.2496011	42.879930	0.0058209	0.0233209	4.006392	171.793800
100	0.1764242	47.061470	0.0037488	0.0212488	5.668156	266.751800
120	0.1247010	50.017090	0.0024932	0.0199932	8.019183	401.096200
180	0.0440357	54.626530	8.061E-40	0.0183061	22.70885	1,240.506000
240	0.0155503	56.254270	2.764E-40	0.0177764	64.30730	3,617.560000

			i = 2%			
n	*P/F*	*P/A*	*A/F*	*A/P*	*F/P*	*F/A*
1	0.9803922	0.9803922	1.0000000	1.0200000	1.020000	1.000000
2	0.9611688	1.9415610	0.4950495	0.5150495	1.040400	2.020000
3	0.9423223	2.8838830	0.3267547	0.3467547	1.061208	3.060400
4	0.9238454	3.8077290	0.2426238	0.2626238	1.082432	4.121608
5	0.9057308	4.7134600	0.1921584	0.2121584	1.104081	5.204040
6	0.8879714	5.6014310	0.1585258	0.1785258	1.126162	6.308121
7	0.8705602	6.4719910	0.1345120	0.1545120	1.148686	7.434283
8	0.8534904	7.3254810	0.1165098	0.1365098	1.171659	8.582969
9	0.8367553	8.1622370	0.1025154	0.1225154	1.195093	9.754628
10	0.8203483	8.9825850	0.0913265	0.1113265	1.218994	10.949720
11	0.8042630	9.7868400	0.0821779	0.1021779	1.243374	12.168720
12	0.7884932	10.5753400	0.0745596	0.0945596	1.268242	13.412090
13	0.7730325	11.3483700	0.0681184	0.0881184	1.293607	14.680330
14	0.7578750	12.1062500	0.0626020	0.0826020	1.319479	15.973940
15	0.7430147	12.8492600	0.0578255	0.0778255	1.345868	17.293420
16	0.7284458	13.5777100	0.0536501	0.0736501	1.372786	18.639290
17	0.7141626	14.2918700	0.0499698	0.0699698	1.400241	20.012070
18	0.7001594	14.9920300	0.0467021	0.0667021	1.428246	21.412310
19	0.6864308	15.6784600	0.0437818	0.0637818	1.456811	22.840560
20	0.6729713	16.3514300	0.0411567	0.0611567	1.485947	24.297370
21	0.6597758	17.0112100	0.0387848	0.0587848	1.515666	25.783320
22	0.6468390	17.6580500	0.0366314	0.0566314	1.545980	27.298980
23	0.6341559	18.2922000	0.0346681	0.0546681	1.576899	28.844960
24	0.6217215	18.9139300	0.0328711	0.0528711	1.608437	30.421860
25	0.6095309	19.5234600	0.0312204	0.0512204	1.640606	32.030300
26	0.5975793	20.1210400	0.0296992	0.0496992	1.673418	33.670910
27	0.5858620	20.7069000	0.0282931	0.0482931	1.706886	35.344320

			$i = 2\%$			
n	P/F	P/A	A/F	A/P	F/P	F/A
28	0.5743746	21.2812700	0.0269897	0.0469897	1.741024	37.051210
29	0.5631123	21.8443800	0.0257784	0.0457784	1.775845	38.792230
30	0.5520709	22.3964600	0.0246499	0.0446499	1.811362	40.568080
32	0.5306333	23.4683300	0.0226106	0.0426106	1.884541	44.227030
34	0.5100282	24.4985900	0.0208187	0.0408187	1.960676	48.033800
36	0.4902231	25.4888400	0.0192329	0.0392329	2.039887	51.994370
38	0.4711872	26.4406400	0.0178206	0.0378206	2.122299	56.114940
40	0.4528904	27.3554800	0.0165557	0.0365557	2.208040	60.401980
48	0.3865376	30.6731200	0.0126018	0.0326018	2.587070	79.353520
60	0.3047823	34.7608900	0.0087680	0.0287680	3.281031	114.051500
72	0.2403187	37.9840600	0.0063268	0.0263268	4.161140	158.057000
80	0.2051097	39.7445100	0.0051607	0.0251607	4.875439	193.772000
100	0.1380330	43.0983500	0.0032027	0.0232027	7.244646	312.232300
120	0.0928922	45.3553900	0.0020481	0.0220481	10.765160	488.258200
180	0.0283119	48.5844000	5.827E-40	0.0205827	35.320830	1,716.042000
240	0.0086290	49.5685500	1.741E-40	0.0201741	115.888700	5,744.437000

			$i = 2.5\%$			
n	P/F	P/A	A/F	A/P	F/P	F/A
1	0.9756098	0.9756097	1.0000000	1.0250000	1.025000	1.000000
2	0.9518144	1.9274240	0.4938272	0.5188272	1.050625	2.025000
3	0.9285994	2.8560240	0.3251372	0.3501372	1.076891	3.075625
4	0.9059506	3.7619740	0.2408179	0.2658179	1.103813	4.152516
5	0.8838543	4.6458280	0.1902469	0.2152469	1.131408	5.256328
6	0.8622969	5.5081250	0.1565500	0.1815500	1.159693	6.387737
7	0.8412652	6.3493910	0.1324954	0.1574954	1.188686	7.547430
8	0.8207466	7.1701370	0.1144673	0.1394673	1.218403	8.736116
9	0.8007284	7.9708650	0.1004569	0.1254569	1.248863	9.954519
10	0.7811984	8.7520640	0.0892588	0.1142588	1.280085	11.203380
11	0.7621448	9.5142090	0.0801060	0.1051060	1.312087	12.483470
12	0.7435559	10.2577600	0.0724871	0.0974871	1.344889	13.795550
13	0.7254204	10.9831800	0.0660483	0.0910483	1.378511	15.140440
14	0.7077272	11.6909100	0.0605365	0.0855365	1.412974	16.518950
15	0.6904656	12.3813800	0.0557665	0.0807665	1.448298	17.931930
16	0.6736249	13.0550000	0.0515990	0.0765990	1.484506	19.380220
17	0.6571951	13.7122000	0.0479278	0.0729278	1.521618	20.864730
18	0.6411659	14.3533600	0.0446701	0.0696701	1.559659	22.386350
19	0.6255277	14.9788900	0.0417606	0.0667606	1.598650	23.946010
20	0.6102709	15.5891600	0.0391471	0.0641471	1.638616	25.544660

$i = 2.5\%$

n	P/F	P/A	A/F	A/P	F/P	F/A
21	0.5953863	16.1845500	0.0367873	0.0617873	1.679582	27.183270
22	0.5808647	16.7654100	0.0346466	0.0596466	1.721571	28.862860
23	0.5666972	17.3321100	0.0326964	0.0576964	1.764611	30.584430
24	0.5528754	17.8849900	0.0309128	0.0559128	1.808726	32.349040
25	0.5393906	18.4243800	0.0292759	0.0542759	1.853944	34.157760
26	0.5262347	18.9506100	0.0277687	0.0527687	1.900293	36.011710
27	0.5133997	19.4640100	0.0263769	0.0513769	1.947800	37.912000
28	0.5008778	19.9648900	0.0250879	0.0500879	1.996495	39.859800
29	0.4886613	20.4535500	0.0238913	0.0488913	2.046407	41.856300
30	0.4767427	20.9302900	0.0227776	0.0477776	2.097568	43.902700
32	0.4537706	21.8491800	0.0207683	0.0457683	2.203757	48.150280
34	0.4319053	22.7237900	0.0190068	0.0440068	2.315322	52.612880
36	0.4110937	23.5562500	0.0174516	0.0424516	2.432535	57.301410
38	0.3912849	24.3486000	0.0160701	0.0410701	2.555682	62.227300
40	0.3724306	25.1027700	0.0148362	0.0398362	2.685064	67.402550
48	0.3056712	27.7731500	0.0110060	0.0360060	3.271490	90.859580
60	0.2272836	30.9086600	0.0073534	0.0323534	4.399790	135.991600
72	0.1689981	33.2400800	0.0050842	0.0300842	5.917228	196.689100
80	0.1387046	34.4518200	0.0040260	0.0290260	7.209568	248.382700
100	0.0846474	36.6141100	0.0023119	0.0273119	11.813720	432.548600
120	0.0516578	37.9336900	0.0013618	0.0263618	19.358150	734.326000
180	0.0117410	39.5303600	2.970E-40	0.0252970	85.171790	3,366.871000
240	0.0026685	39.8932600	6.689E-50	0.0250669	374.737900	14,949.520000

$i = 3\%$

n	P/F	P/A	A/F	A/P	F/P	F/A
1	0.9708738	0.9708738	1.0000000	1.0300000	1.030000	1.000000
2	0.9425959	1.9134700	0.4926108	0.5226108	1.060900	2.030000
3	0.9151417	2.8286110	0.3235304	0.3535304	1.092727	3.090900
4	0.8884870	3.7170980	0.2390270	0.2690270	1.125509	4.183627
5	0.8626088	4.5797070	0.1883546	0.2183546	1.159274	5.309136
6	0.8374843	5.4171910	0.1545975	0.1845975	1.194052	6.468410
7	0.8130915	6.2302830	0.1305064	0.1605064	1.229874	7.662462
8	0.7894092	7.0196920	0.1124564	0.1424564	1.266770	8.892336
9	0.7664167	7.7861090	0.0984339	0.1284339	1.304773	10.159110
10	0.7440939	8.5302030	0.0872305	0.1172305	1.343916	11.463880
11	0.7224213	9.2526240	0.0780774	0.1080774	1.384234	12.807800
12	0.7013799	9.9540040	0.0704621	0.1004621	1.425761	14.192030

			$i = 3\%$			
n	P/F	P/A	A/F	A/P	F/P	F/A
13	0.6809513	10.6349600	0.0640295	0.0940295	1.468534	15.617790
14	0.6611178	11.2960700	0.0585263	0.0885263	1.512590	17.086320
15	0.6418619	11.9379400	0.0537666	0.0837666	1.557967	18.598910
16	0.6231669	12.5611000	0.0496108	0.0796108	1.604706	20.156880
17	0.6050164	13.1661200	0.0459525	0.0759525	1.652848	21.761590
18	0.5873946	13.7535100	0.0427087	0.0727087	1.702433	23.414440
19	0.5702860	14.3238000	0.0398139	0.0698139	1.753506	25.116870
20	0.5536758	14.8774700	0.0372157	0.0672157	1.806111	26.870370
21	0.5375493	15.4150200	0.0348718	0.0648718	1.860295	28.676490
22	0.5218925	15.9369200	0.0327474	0.0627474	1.916103	30.536780
23	0.5066917	16.4436100	0.0308139	0.0608139	1.973587	32.452880
24	0.4919337	16.9355400	0.0290474	0.0590474	2.032794	34.426470
25	0.4776056	17.4131500	0.0274279	0.0574279	2.093778	36.459260
26	0.4636947	17.8768400	0.0259383	0.0559383	2.156591	38.553040
27	0.4501891	18.3270300	0.0245642	0.0545642	2.221289	40.709630
28	0.4370768	18.7641100	0.0232932	0.0532932	2.287928	42.930920
29	0.4243464	19.1884500	0.0221147	0.0521147	2.356566	45.218850
30	0.4119868	19.6004400	0.0210193	0.0510193	2.427262	47.575420
32	0.3883370	20.3887700	0.0190466	0.0490466	2.575083	52.502760
34	0.3660449	21.1318400	0.0173220	0.0473220	2.731905	57.730180
36	0.3450324	21.8322500	0.0158038	0.0458038	2.898278	63.275950
38	0.3252261	22.4924600	0.0144593	0.0444593	3.074784	69.159450
40	0.3065568	23.1147700	0.0132624	0.0432624	3.262038	75.401260
48	0.2419988	25.2667100	0.0095778	0.0395778	4.132252	104.408400
60	0.1697331	27.6755600	0.0061330	0.0361330	5.891603	163.053400
72	0.1190474	29.3650900	0.0040540	0.0340540	8.400017	246.667200
80	0.0939771	30.2007600	0.0031117	0.0331117	10.640890	321.363000
100	0.0520328	31.5989100	0.0016467	0.0316467	19.218630	607.287700
120	0.0288093	32.3730200	8.899E-40	0.0308899	34.710990	1,123.700000
180	0.0048899	33.1703400	1.474E-40	0.0301474	204.503400	6,783.446000
240	8.300E-40	33.3056700	2.492E-50	0.0300249	1,204.853000	40,128,420000

			$i = 4\%$			
n	P/F	P/A	A/F	A/P	F/P	F/A
1	0.9615385	0.9615385	1.0000000	1.0400000	1.040000	1.000000
2	0.9245562	1.8860950	0.4901961	0.5301961	1.081600	2.040000
3	0.8889964	2.7750910	0.3203485	0.3603485	1.124864	3.121600
4	0.8548042	3.6298950	0.2354900	0.2754900	1.169859	4.246464
5	0.8219271	4.4518220	0.1846271	0.2246271	1.216653	5.416323

			i = 4%			
n	P/F	P/A	A/F	A/P	F/P	F/A
6	0.7903145	5.2421370	0.1507619	0.1907619	1.265319	6.632975
7	0.7599178	6.0020550	0.1266096	0.1666096	1.315932	7.898294
8	0.7306902	6.7327450	0.1085278	0.1485278	1.368569	9.214226
9	0.7025867	7.4353320	0.0944930	0.1344930	1.423312	10.582800
10	0.6755642	8.1108960	0.0832909	0.1232909	1.480244	12.006110
11	0.6495809	8.7604770	0.0741490	0.1141490	1.539454	13.486350
12	0.6245970	9.3850740	0.0665522	0.1065522	1.601032	15.025810
13	0.6005741	9.9856480	0.0601437	0.1001437	1.665074	16.626840
14	0.5774751	10.5631200	0.0546690	0.0946690	1.731676	18.291910
15	0.5552645	11.1183900	0.0499411	0.0899411	1.800944	20.023590
16	0.5339082	11.6523000	0.0458200	0.0858200	1.872981	21.824530
17	0.5133732	12.1656700	0.0421985	0.0821985	1.947900	23.697510
18	0.4936281	12.6593000	0.0389933	0.0789933	2.025817	25.645410
19	0.4746424	13.1339400	0.0361386	0.0761386	2.106849	27.671230
20	0.4563869	13.5903300	0.0335818	0.0735818	2.191123	29.778080
21	0.4388336	14.0291600	0.0312801	0.0712801	2.278768	31.969200
22	0.4219554	14.4511200	0.0291988	0.0691988	2.369919	34.247970
23	0.4057263	14.8568400	0.0273091	0.0673091	2.464716	36.617890
24	0.3901215	15.2469600	0.0255868	0.0655868	2.563304	39.082600
25	0.3751168	15.6220800	0.0240120	0.0640120	2.665836	41.645910
26	0.3606892	15.9827700	0.0225674	0.0625674	2.772470	44.311740
27	0.3468166	16.3295900	0.0212385	0.0612385	2.883369	47.084210
28	0.3334775	16.6630600	0.0200130	0.0600130	2.998703	49.967580
29	0.3206514	16.9837100	0.0188799	0.0588799	3.118651	52.966290
30	0.3083187	17.2920300	0.0178301	0.0578301	3.243398	56.084940
35	0.2534155	18.6646100	0.0135773	0.0535773	3.946089	73.652220
40	0.2082890	19.7927700	0.0105235	0.0505235	4.801021	95.025520
45	0.1711984	20.7200400	0.0082625	0.0482625	5.841176	121.029400
50	0.1407126	21.4821800	0.0065502	0.0465502	7.106683	152.667100
60	0.0950604	22.6234900	0.0042018	0.0442018	10.519630	237.990700
70	0.0642194	23.3945100	0.0027451	0.0427451	15.571620	364.290500
80	0.0433843	23.9153900	0.0018141	0.0418141	23.049800	551.245000
90	0.0293089	24.2672800	0.0012078	0.0412078	34.119330	827.983300
100	0.0198000	24.5050000	8.080E-40	0.0408080	50.504950	1,237.624000

$i = 5\%$						
n	P/F	P/A	A/F	A/P	F/P	F/A
1	0.9523810	0.9523810	1.0000000	1.0500000	1.050000	1.000000
2	0.9070295	1.8594100	0.4878049	0.5378049	1.102500	2.050000
3	0.8638376	2.7232480	0.3172086	0.3672086	1.157625	3.152500
4	0.8227025	3.5459510	0.2320118	0.2820118	1.215506	4.310125
5	0.7835262	4.3294770	0.1809748	0.2309748	1.276282	5.525631
6	0.7462154	5.0756920	0.1470175	0.1970175	1.340096	6.801913
7	0.7106813	5.7863730	0.1228198	0.1728198	1.407100	8.142008
8	0.6768394	6.4632130	0.1047218	0.1547218	1.477455	9.549109
9	0.6446089	7.1078220	0.0906901	0.1406901	1.551328	11.026560
10	0.6139133	7.7217350	0.0795046	0.1295046	1.628895	12.577890
11	0.5846793	8.3064140	0.0703889	0.1203889	1.710339	14.206790
12	0.5568374	8.8632520	0.0628254	0.1128254	1.795856	15.917130
13	0.5303214	9.3935730	0.0564558	0.1064558	1.885649	17.712980
14	0.5050680	9.8986410	0.0510240	0.1010240	1.979932	19.598630
15	0.4810171	10.3796600	0.0463423	0.0963423	2.078928	21.578656
16	0.4581115	10.8377700	0.0422699	0.0922699	2.182875	23.657490
17	0.4362967	11.2740700	0.0386991	0.0886991	2.292018	25.840370
18	0.4155207	11.6895900	0.0355462	0.0855462	2.406619	28.132380
19	0.3957340	12.0853200	0.0327450	0.0827450	2.526950	30.539000
20	0.3768895	12.4622100	0.0302426	0.0802426	2.653298	33.065950
21	0.3589424	12.8211500	0.0279961	0.0779961	2.785963	35.719250
22	0.3418499	13.1630000	0.0259705	0.0759705	2.925261	38.505210
23	0.3255713	13.4885700	0.0241368	0.0741368	3.071524	41.430480
24	0.3100679	13.7986400	0.0224709	0.0724709	3.225100	44.502000
25	0.2953028	14.0939400	0.0209525	0.0709525	3.386355	47.727100
26	0.2812407	14.3751900	0.0195643	0.0695643	3.555673	51.113450
27	0.2678483	14.6430300	0.0182919	0.0682919	3.733456	54.669130
28	0.2550936	14.8981300	0.0171225	0.0671225	3.920129	58.402580
29	0.2429463	15.1410700	0.0160455	0.0660455	4.116136	62.322710
30	0.2313774	15.3724500	0.0150514	0.0650514	4.321942	66.438850
35	0.1812903	16.3741900	0.0110717	0.0610717	5.516015	90.320310
40	0.1420457	17.1590900	0.0082782	0.0582782	7.039989	120.799800
45	0.1112965	17.7740700	0.0062617	0.0562617	8.985008	159.700200
50	0.0872037	18.2559300	0.0047767	0.5477670	11.467400	209.348000
60	0.0535355	18.9292900	0.0028282	0.0528282	18.679190	353.583700
70	0.0328662	19.3426800	0.0016992	0.0516992	30.426430	588.528500
80	0.0201770	19.5964600	0.0010296	0.0510296	49.561440	971.228800
90	0.0123869	19.7522600	6.271E-40	0.0506271	80.730370	1,594.607000
100	0.0076045	19.8479100	3.831E-40	0.0503831	131.501300	2,610.025000

	$i = 6\%$					
n	P/F	P/A	A/F	A/P	F/P	F/A
1	0.9433962	0.9433962	1.0000000	1.0600000	1.060000	1.000000
2	0.8899964	1.8333930	0.4854369	0.5454369	1.123600	2.060000
3	0.8396193	2.6730120	0.3141098	0.3741098	1.191016	3.183600
4	0.7920937	3.4651060	0.2285915	0.2885915	1.262477	4.374616
5	0.7472582	4.2123640	0.1773964	0.2373964	1.338226	5.637093
6	0.7049605	4.9173240	0.1433626	0.2033626	1.418519	6.975319
7	0.6650571	5.5823810	0.1191350	0.1791350	1.503630	8.393838
8	0.6274124	6.2097940	0.1010359	0.1610359	1.593848	9.897468
9	0.5918985	6.8016920	0.0870222	0.1470222	1.689479	11.491320
10	0.5583948	7.3600870	0.0758680	0.1358680	1.790848	13.180790
11	0.5267875	7.8868750	0.0667929	0.1267929	1.898299	14.971640
12	0.4969694	8.3838440	0.0592770	0.1192770	2.012196	16.869940
13	0.4688390	8.8526830	0.0529601	0.1129601	2.132928	18.882140
14	0.4423010	9.2949840	0.0475849	0.1075849	2.260904	21.015070
15	0.4172651	9.7122490	0.0429628	0.1029628	2.396558	23.275970
16	0.3936463	10.1059000	0.0389521	0.0989521	2.540352	25.672530
17	0.3713644	10.4772600	0.0354448	0.0954448	2.692773	28.212880
18	0.3503438	10.8276000	0.0323565	0.0923565	2.854339	30.905650
19	0.3305130	11.1581200	0.0296209	0.0896209	3.025599	33.759990
20	0.3118047	11.4699200	0.0271846	0.0871846	3.207135	36.785590
21	0.2941554	11.7640800	0.0250045	0.0850045	3.399564	39.992730
22	0.2775051	12.0415800	0.0230456	0.0830456	3.603537	43.392290
23	0.2617973	12.3033800	0.0212785	0.0812785	3.819750	46.995830
24	0.2469785	12.5503600	0.0196790	0.0796790	4.048935	50.815580
25	0.2329986	12.7833600	0.0182267	0.0782267	4.291871	54.864510
26	0.2198100	13.0031700	0.0169043	0.0769043	4.549383	59.156380
27	0.2073680	13.2105300	0.0156972	0.0756972	4.822346	63.705770
28	0.1956301	13.4061600	0.0145926	0.0745926	5.111687	68.528110
29	0.1845567	13.5907200	0.0135796	0.0735796	5.418388	73.639800
30	0.1741101	13.7648300	0.0126489	0.0726489	5.743491	79.058190
35	0.1301052	14.4982500	0.0089739	0.0689739	7.686087	111.434800
40	0.0972222	15.0463000	0.0064615	0.0664615	10.285720	154.762000
45	0.0726501	15.4558300	0.0047005	0.0647005	13.764610	212.743500
50	0.0542884	15.7618600	0.0034443	0.0634443	18.420150	290.335900
60	0.0303143	16.1614300	0.0018757	0.0618757	32.987690	533.128200
70	0.0169274	16.3845400	0.0010331	0.0610331	59.075930	967.932200
80	0.0094522	16.5091300	5.725E-40	0.0605725	105.796000	1,746.600000
90	0.0052780	16.5787000	3.184E-40	0.0603184	189.464500	3,141.075000
100	0.0029472	16.6175500	1.774E-40	0.0601774	339.302100	5,638.368000

	$i = 7\%$					
n	P/F	P/A	A/F	A/P	F/P	F/A
1	0.9345794	0.9345794	1.0000000	1.0700000	1.070000	1.000000
2	0.8734387	1.8080180	0.4830918	0.5530918	1.144900	2.070000
3	0.8162979	2.6243160	0.3110517	0.3810517	1.225043	3.124900
4	0.7628952	3.3872110	0.2252281	0.2952281	1.310796	4.439943
5	0.7129862	4.1001970	0.1738907	0.2438907	1.402552	5.750739
6	0.6663422	4.7665400	0.1397958	0.2097958	1.500730	7.153291
7	0.6227497	5.3892890	0.1155532	0.1855532	1.605781	8.654021
8	0.5820091	5.9712990	0.0974678	0.1674678	1.718186	10.259800
9	0.5439337	6.5152320	0.0834865	0.1534865	1.838459	11.977990
10	0.5083493	7.0235820	0.0723775	0.1423775	1.967151	13.816450
11	0.4750928	7.4986740	0.0633569	0.1333569	2.104852	15.783600
12	0.4440120	7.9426860	0.0559020	0.1259020	2.252192	17.888450
13	0.4149644	8.3576510	0.0496508	0.1196508	2.409845	20.140640
14	0.3878172	8.7454680	0.0443449	0.1143449	2.578534	22.550490
15	0.3624460	9.1079140	0.0397946	0.1097946	2.759032	25.129020
16	0.3387346	9.4466490	0.0358576	0.1058576	2.952164	27.888050
17	0.3165744	9.7632230	0.0324252	0.1024252	3.158815	30.840220
18	0.2958639	10.0590900	0.0294126	0.0994126	3.379932	33.999030
19	0.2765083	10.3356000	0.0267530	0.0967530	3.616528	37.378960
20	0.2584190	10.5940100	0.0243929	0.0943929	3.869684	40.995490
21	0.2415131	10.8355300	0.0222890	0.0922890	4.140562	44.865180
22	0.2257132	11.0612400	0.0204058	0.0904058	4.430402	49.005740
23	0.2109469	11.2721900	0.0187139	0.0887139	4.740530	53.436140
24	0.1971466	11.4693300	0.0171890	0.0871890	5.072367	58.176670
25	0.1842492	11.6535800	0.0158105	0.0858105	5.427433	63.249040
26	0.1721955	11.8257800	0.0145610	0.0845610	5.807353	68.676470
27	0.1609304	11.9867100	0.0134257	0.0834257	6.213868	74.483820
28	0.1504022	12.1371100	0.0123919	0.0823919	6.648838	80.697690
29	0.1405628	12.2776700	0.0114487	0.0814487	7.114257	87.346530
30	0.1313671	12.4090400	0.0105864	0.0805864	7.612255	94.460790
35	0.0936629	12.9476700	0.0072340	0.0772340	10.676580	138.236900
40	0.0667804	13.3317100	0.0050091	0.0750091	14.974460	199.635100
45	0.0476135	13.6055200	0.0034996	0.0734996	21.002450	285.749300
50	0.0339478	13.8007500	0.0024598	0.0724598	29.457030	406.528900
60	0.0172573	14.0391800	0.0012292	0.0712292	57.946430	813.520400
70	0.0087727	14.1603900	6.195E-40	0.0706195	113.989400	1,614.134000
80	0.0044596	14.2220100	3.136E-40	0.0703136	224.234400	3,189.063000
90	0.0022670	14.2533300	1.591E-40	0.0701591	441.103000	6,287.185000
100	0.0011525	14.2692500	8.076E-50	0.0700808	867.716300	12,381.66000

	i = 8%					
n	*P/F*	*P/A*	*A/F*	*A/P*	*F/P*	*F/A*
1	0.9259259	0.9259259	1.0000000	1.0800000	1.080000	1.000000
2	0.8573388	1.7832650	0.4807692	0.5607692	1.166400	2.080000
3	0.7938322	2.5770970	0.3080335	0.3880335	1.259712	3.246400
4	0.7350299	3.3121270	0.2219208	0.3019208	1.360489	4.506112
5	0.6805832	3.9927100	0.1704565	0.2504565	1.469328	5.866601
6	0.6301696	4.6228800	0.1363154	0.2163154	1.586874	7.335929
7	0.5834904	5.2063700	0.1120724	0.1920724	1.713824	8.922803
8	0.5402689	5.7466390	0.0940148	0.1740148	1.850930	10.636630
9	0.5002490	6.2468880	0.0800797	0.1600797	1.999005	12.487560
10	0.4631935	6.7100810	0.0690295	0.1490295	2.158925	14.486560
11	0.4288829	7.1389640	0.0600763	0.1400763	2.331639	16.645490
12	0.3971138	7.5360780	0.0526950	0.1326950	2.518170	18.977130
13	0.3676979	7.9037760	0.0465218	0.1265218	2.719624	21.495300
14	0.3404610	8.2442370	0.0412969	0.1212969	2.937194	24.214920
15	0.3152417	8.5594790	0.0368295	0.1168295	3.172169	27.152110
16	0.2918905	8.8513690	0.0329769	0.1129769	3.425943	30.324280
17	0.2702690	9.1216380	0.0296294	0.1096294	3.700018	33.750230
18	0.2502490	9.3718870	0.0267021	0.1067021	3.996020	37.450240
19	0.2317121	9.6035990	0.0241276	0.1041276	4.315701	41.446260
20	0.2145482	9.8181470	0.0218522	0.1018522	4.660957	45.761960
21	0.1986557	10.0168000	0.0198323	0.0998323	5.033834	50.422920
22	0.1839405	10.2007400	0.0180321	0.0980321	5.436540	55.456760
23	0.1703153	10.3710600	0.0164222	0.0964222	5.871464	60.893300
24	0.1576993	10.5287600	0.0149780	0.0949780	6.341181	66.764760
25	0.1460179	10.6747800	0.0136788	0.0936788	6.848475	73.105940
26	0.1352018	10.8099800	0.0125071	0.0925071	7.396353	79.954420
27	0.1251868	10.9351600	0.0114481	0.0914481	7.988061	87.350770
28	0.1159137	11.0510800	0.0104889	0.0904889	8.627106	95.338830
29	0.1073275	11.1584100	0.0096185	0.0896185	9.317275	103.965900
30	0.0993773	11.2577800	0.0088274	0.0888274	10.062660	113.283200
35	0.0676345	11.6545700	0.0058033	0.0858033	14.785340	172.316800
40	0.0460309	11.9246100	0.0038602	0.0838602	21.724520	259.056500
45	0.0313279	12.1084000	0.0025873	0.0825873	31.920450	386.505600
50	0.0213212	12.2334800	0.0017429	0.0817429	46.901610	573.770200
60	0.0098759	12.3765500	7.979E-40	0.0807979	101.257100	1,253.213000
70	0.0045744	12.4428200	3.676E-40	0.0803676	218.606400	2,720.080000
80	0.0021188	12.4735100	1.699E-40	0.0801699	471.954800	5,886.935000
90	9.814E-40	12.4877300	7.859E-50	0.0800786	1,018.915000	12,723.940000
100	4.546E-40	12.4943200	3.638E-50	0.0800364	2,199.761000	27,484.520000

n	P/F	P/A	A/F	A/P	F/P	F/A
			$i = 9\%$			
1	0.9174312	0.9174312	1.0000000	1.0900000	1.090000	1.000000
2	0.8416800	1.7591110	0.4786890	0.5684689	1.188100	2.090000
3	0.7721835	2.5312950	0.3050548	0.3950548	1.295029	3.278100
4	0.7084252	3.2397200	0.2186687	0.3086687	1.411582	4.573129
5	0.6499314	3.8896510	0.1670925	0.2570925	1.538624	5.984711
6	0.5962673	4.4859190	0.1329198	0.2229198	1.677100	7.523335
7	0.5470342	5.0329530	0.1086905	0.1986905	1.828039	9.200435
8	0.5018663	5.5348190	0.0906744	0.1806744	1.992563	11.028470
9	0.4604278	5.9952470	0.0767988	0.1667988	2.171893	13.021040
10	0.4224108	6.4176580	0.0658201	0.1558201	2.367364	15.192930
11	0.3875329	6.8051910	0.0569467	0.1469467	2.580426	17.560290
12	0.3555347	7.1607250	0.0496507	0.1396507	2.812665	20.140720
13	0.3261786	7.4869040	0.0435666	0.1335666	3.065805	22.953380
14	0.2992465	7.7861500	0.0384332	0.1284332	3.341727	26.019190
15	0.2745380	8.0606880	0.0340589	0.1240589	3.642482	29.360920
16	0.2518698	8.3125580	0.0302999	0.1202999	3.970306	33.003400
17	0.2310732	8.5436310	0.0270462	0.1170462	4.327633	36.973700
18	0.2119937	8.7556250	0.0242123	0.1142123	4.717120	41.301340
19	0.1944897	8.9501150	0.0217304	0.1117304	5.141661	46.018460
20	0.1784309	9.1285460	0.0195465	0.1095465	5.604411	51.160120
21	0.1636981	9.2922440	0.0176166	0.1076166	6.108808	56.764530
22	0.1501817	9.4424250	0.0159050	0.1059050	6.658600	62.873340
23	0.1377814	9.5802070	0.0143819	0.1043819	7.257874	69.531940
24	0.1264049	9.7066120	0.0130226	0.1030226	7.911083	76.789810
25	0.1159678	9.8225800	0.0118063	0.1018063	8.623081	84.700900
26	0.1063925	9.9289720	0.0107154	0.1007154	9.399158	93.323980
27	0.0976078	10.0265800	0.0097349	0.0997349	10.245080	102.723100
28	0.0895484	10.1161300	0.0088520	0.0988520	11.167140	112.968200
29	0.0821545	10.1982800	0.0080557	0.0980557	12.172180	124.135400
30	0.0753711	10.2736500	0.0073364	0.0973364	13.267680	136.307500
35	0.0489861	10.5668200	0.0046358	0.0946358	20.413970	215.710800
40	0.0318376	10.7573600	0.0029596	0.0929596	31.409420	337.882400
45	0.0206922	10.8812000	0.0019017	0.0919017	48.327290	525.858700
50	0.0134485	10.9616800	0.0012269	0.0122690	74.357520	815.083600
60	0.0056808	11.0479900	5.142E-40	0.0905142	176.031300	1,944.792000
70	0.0023996	11.0844500	2.165E-40	0.0902165	416.730100	4,619.223000
80	0.0010136	11.0998500	9.132E-50	0.0900913	986.551700	10,950.570000
90	4.282E-40	11.1063500	3.855E-50	0.0900386	2,335.527000	25,939.180000
100	1.809E-40	11.1091000	1.628E-50	0.0900163	5,529.041000	61,422.670000

	i = 10%					
n	*P/F*	*P/A*	*A/F*	*A/P*	*F/P*	*F/A*
1	0.9090909	0.9090909	1.0000000	1.1000000	1.100000	1.000000
2	0.8264463	1.735537	0.4761905	0.5761905	1.210000	2.100000
3	0.7513148	2.486852	0.3021148	0.4021148	1.331000	3.310000
4	0.6830135	3.169865	0.2154708	0.3154708	1.464100	4.641000
5	0.6209213	3.790787	0.1637975	0.2637975	1.610510	6.105100
6	0.5644739	4.355261	0.1296074	0.2296074	1.771561	7.715610
7	0.5131581	4.868419	0.1054055	0.2054055	1.948717	9.487171
8	0.4665074	5.334926	0.0874440	0.1874440	2.143589	11.435890
9	0.4240976	5.759024	0.0736405	0.1736405	2.357948	13.579480
10	0.3855433	6.144567	0.0627454	0.1627454	2.593742	15.937420
11	0.3504939	6.495061	0.0539631	0.1539631	2.853117	18.531170
12	0.3186308	6.813692	0.0467633	0.1467633	3.138428	21.384280
13	0.2896644	7.103356	0.0407785	0.1407785	3.452271	24.522710
14	0.2633313	7.366687	0.0357462	0.1357462	3.797498	27.974980
15	0.2393920	7.606080	0.0314738	0.1314738	4.177248	31.772480
16	0.2176291	7.823709	0.0278166	0.1278166	4.594973	35.949730
17	0.1978447	8.021553	0.0246641	0.1246641	5.054470	40.544700
18	0.1798588	8.201412	0.0219302	0.1219302	5.559917	45.599170
19	0.1635080	8.364920	0.0195469	0.1195469	6.115909	51.159090
20	0.1486436	8.513564	0.0174596	0.1174596	6.727500	57.275000
21	0.1351306	8.648694	0.0156244	0.1156244	7.400250	64.002500
22	0.1228460	8.771540	0.0140051	0.1140051	8.140275	71.402750
23	0.1116782	8.883218	0.0125718	0.1125718	8.954302	79.543020
24	0.1015256	8.984744	0.0112998	0.1112998	9.849733	88.497330
25	0.0922960	9.077040	0.0101681	0.1101681	10.834710	98.347060
26	0.0839055	9.160945	0.0091590	0.1091590	11.91818	109.181800
27	0.0762777	9.237223	0.0082576	0.1082576	13.10999	121.099900
28	0.0693433	9.306567	0.0074510	0.1074510	14.42099	134.209900
29	0.0630394	9.369606	0.0067281	0.1067281	15.86309	148.630900
30	0.0573086	9.426914	0.0060792	0.1060792	17.44940	164.494000
35	0.0355841	9.644159	0.0036897	0.1036897	28.10244	271.024400
40	0.0220949	9.779051	0.0022594	0.1022594	45.25926	442.592600
45	0.0137192	9.862808	0.0013910	0.1013910	72.89048	718.904800
50	0.0085186	9.914814	8.592E-40	0.1008592	117.39090	1,163.909000
60	0.0032843	9.967157	3.295E-40	0.1003295	304.4816	3,034.816
70	0.0012662	9.987338	1.268E-40	0.1001268	789.7470	7,887.470
80	4.882E-40	9.9951180	4.884E-50	0.1000488	2,048.400	20,474.00
90	1.882E-40	9.998118	1.883E-50	0.1000188	5,313.023	53,120.23
100	7.257E-50	9.999274	7.257E-60	0.1000073	13,780.61	137,796.1

	i = 11%					
n	*P/F*	*P/A*	*A/F*	*A/P*	*F/P*	*F/A*
1	0.9009009	0.9009009	1.0000000	1.1100000	1.110000	1.000000
2	0.8116224	1.7125230	0.4739337	0.5839337	1.232100	2.110000
3	0.7311914	2.4437150	0.2992131	0.4092131	1.367631	3.342100
4	0.6587310	3.1024460	0.2123264	0.3223264	1.518070	4.709731
5	0.5934513	3.6958970	0.1605703	0.2705703	1.685058	6.227801
6	0.5346408	4.2305380	0.1263766	0.2363766	1.870415	7.912860
7	0.4816584	4.7121960	0.1022153	0.2122153	2.076160	9.783274
8	0.4339265	5.1461230	0.0843211	0.1943211	2.304538	11.859430
9	0.3909248	5.5370480	0.0706017	0.1806017	2.558037	14.163970
10	0.3521845	5.8892320	0.0598014	0.1698014	2.839421	16.722010
11	0.3172833	6.2065150	0.0511210	0.1611210	3.151757	19.561430
12	0.2858408	6.4923560	0.0440273	0.1540273	3.498451	22.713190
13	0.2575143	6.7498700	0.0381510	0.1481510	3.883280	26.211640
14	0.2319948	6.9818650	0.0332282	0.1432282	4.310441	30.094920
15	0.2090043	7.1908700	0.0290652	0.1390652	4.784589	34.405360
16	0.1882922	7.3791620	0.0255167	0.1355167	5.310894	39.189950
17	0.1696326	7.5487940	0.0224715	0.1324715	5.895093	44.500840
18	0.1528222	7.7016170	0.0198429	0.1298429	6.543553	50.395940
19	0.1376776	7.8392940	0.0175625	0.1275625	7.263344	56.939490
20	0.1240339	7.9633280	0.0155756	0.1255756	8.062311	64.202830
21	0.1117423	8.0750700	0.0138379	0.1238379	8.949166	72.265140
22	0.1006687	8.1757390	0.0123131	0.1223131	9.933574	81.214310
23	0.0906925	8.2664320	0.0109712	0.1209712	11.026270	91.147880
24	0.0817050	8.3481370	0.0097872	0.1197872	12.239160	102.174200
25	0.0736081	8.4217450	0.0087402	0.1187402	13.585460	114.413300
26	0.0663136	8.4880580	0.0078126	0.1178126	15.079860	127.998800
27	0.0597420	8.5478000	0.0069892	0.1169892	16.738650	143.078600
28	0.0538216	8.6016220	0.0062571	0.1162571	18.579900	159.817300
29	0.0484879	8.6501100	0.0056055	0.1156055	20.623690	178.397200
30	0.0436828	8.6937930	0.0050246	0.1150246	22.892300	199.020900
35	0.0259236	8.8552400	0.0029275	0.1129275	38.574850	341.589600
40	0.0153844	8.9510510	0.0017187	0.1117187	65.000870	581.826100
45	0.0091299	9.0079100	0.0010135	0.1110135	109.530200	986.638600
50	0.0054182	9.0416530	5.992E-40	0.1105992	184.564800	1,668.771000
60	0.0019082	9.0735620	2.103E-40	0.1102103	524.0572	4,755.066
70	6.720E-40	9.0848000	7.397E-50	0.1100740	1,488.019	13,518.36
80	2.367E-40	9.0887570	2.604E-50	0.1100260	4,225.113	38,401.02
90	8.336E-50	9.0901510	9.170E-60	0.1100092	11,996.87	109,053.4
100	2.936E-50	9.0906420	3.229E-60	0.1100032	34,064.17	309.665.2

	i = 12%					
n	*P/F*	*P/A*	*A/F*	*A/P*	*F/P*	*F/A*
1	0.8928571	0.8928571	1.0000000	1.1200000	1.120000	1.000000
2	0.7971939	1.6900510	0.4716981	0.5916981	1.254400	2.120000
3	0.7117802	2.4018310	0.2963490	0.4163490	1.404928	3.374400
4	0.6355181	3.0373490	0.2092344	0.3292344	1.573519	4.779328
5	0.5674269	3.6047760	0.1574097	0.2774097	1.762342	6.352847
6	0.5066311	4.1114070	0.1232257	0.2432257	1.973823	8.115189
7	0.4523492	4.5637570	0.0991177	0.2191177	2.210681	10.089010
8	0.4038832	4.9676400	0.0813028	0.2013028	2.475963	12.299690
9	0.3606100	5.3282500	0.0676789	0.1876789	2.773079	14.775660
10	0.3219732	5.6502230	0.0569842	0.1769842	3.105848	17.548740
11	0.2874761	5.9376990	0.0484154	0.1684154	3.478550	20.654580
12	0.2566751	6.1943740	0.0414368	0.1614368	3.895976	24.133130
13	0.2291742	6.4235480	0.0356772	0.1556772	4.363493	28.029110
14	0.2046198	6.6281680	0.0308712	0.1508712	4.887112	32.392600
15	0.1826963	6.8108640	0.0268242	0.1468242	5.473566	37.279710
16	0.1631217	6.9739860	0.0233900	0.1433900	6.130394	42.753280
17	0.1456443	7.1196300	0.0204567	0.1404567	6.866041	48.883670
18	0.1300396	7.2496700	0.0179373	0.1379373	7.689966	55.749710
19	0.1161068	7.3657770	0.0157630	0.1357630	8.612762	63.439680
20	0.1036668	7.4694440	0.0138788	0.1338788	9.646293	72.052440
21	0.0925596	7.5620030	0.0122401	0.1322401	10.803850	81.698740
22	0.0826425	7.6446460	0.0108105	0.1308105	12.100310	92.502580
23	0.0737880	7.7184340	0.0095600	0.1295600	13.552350	104.602900
24	0.0658821	7.7843160	0.0084634	0.1284634	15.178630	118.155200
25	0.0588233	7.8431390	0.0075000	0.1275000	17.000060	133.333900
26	0.0525208	7.8956600	0.0066519	0.1266519	19.040070	150.333900
27	0.0468936	7.9425540	0.0059041	0.1259041	21.324880	169.374000
28	0.0418693	7.9844230	0.0052439	0.1252439	23.883870	190.698900
29	0.0373833	8.0218060	0.0046602	0.1246602	26.749930	214.582800
30	0.0333779	8.0551840	0.0041437	0.1241437	29.959920	241.332700
35	0.0189395	8.1755040	0.0023166	0.1223166	52.799620	431.663500
40	0.0107468	8.2437770	0.0013036	0.1213036	93.050970	767.091400
45	0.0060980	8.2825160	7.363E-40	0.1207363	163.987600	1,358.230000
50	0.0034602	8.3044980	4.167E-40	0.1204167	289.002200	2,400.018000
60	0.0011141	8.3240490	1.338E-40	0.1201338	897.5969	7,471.641
70	3.587E-40	8.3303440	4.306E-50	0.1200431	2,787.800	23,223.33
80	1.155E-40	8.3323710	1.386E-50	0.1200139	8,658.483	72,145.69
90	3.719E-50	8.3330230	4.462E-60	0.1200045	26,891.93	224,091.1
100	1.197E-50	8.3332340	1.437E-60	0.1200014	83,522.27	696,010.6

			i = 13%			
n	*P/F*	*P/A*	*A/F*	*A/P*	*F/P*	*F/A*
1	0.8849558	0.8849558	1.0000000	1.1300000	1.130000	1.000000
2	0.7831467	1.6681020	0.4694836	0.5994836	1.276900	2.130000
3	0.6930502	2.3611530	0.2935220	0.4235220	1.442897	3.406900
4	0.6133187	2.9744710	0.2061942	0.3361942	1.630474	4.849797
5	0.5427599	3.5172310	0.1543145	0.2843145	1.842435	6.480271
6	0.4803185	3.9975500	0.1201532	0.2501532	2.081952	8.322706
7	0.4250606	4.4226100	0.0961108	0.2261108	2.352605	10.404660
8	0.3761599	4.7987700	0.0783867	0.2083867	2.658444	12.757260
9	0.3328848	5.1316550	0.0648689	0.1948689	3.004042	15.415710
10	0.2945883	5.4262430	0.0542896	0.1842896	3.394567	18.419750
11	0.2606977	5.6869410	0.0458415	0.1758415	3.835861	21.814320
12	0.2307059	5.9176470	0.0389861	0.1689861	4.334523	25.650180
13	0.2041645	6.1218120	0.0333503	0.1633503	4.898011	29.984700
14	0.1806766	6.3024880	0.0286675	0.1586675	5.534753	34.882710
15	0.1598908	6.4623790	0.0247418	0.1547418	6.254270	40.417460
16	0.1414962	6.6038750	0.0214262	0.1514262	7.067326	46.671730
17	0.1252179	6.7290930	0.0186084	0.1486084	7.986078	53.739060
18	0.1108123	6.8399050	0.0162009	0.1462009	9.024268	61.725140
19	0.0980640	6.9379690	0.0141344	0.1441344	10.197420	70.749410
20	0.0867823	7.0247520	0.0123538	0.1423538	11.523090	80.946830
21	0.0767985	7.1015500	0.0108143	0.1408143	13.021090	92.469920
22	0.0679633	7.1695130	0.0094795	0.1394795	14.713830	105.491000
23	0.0601445	7.2296580	0.0083191	0.1383191	16.626630	120.204800
24	0.0532252	7.2828830	0.0073083	0.1373083	18.788090	136.831500
25	0.0471020	7.3299850	0.0064259	0.1364259	21.230540	155.619600
26	0.0416831	7.3716680	0.0056545	0.1356545	23.990510	176.850100
27	0.0368877	7.4085560	0.0049791	0.1349791	27.109280	200.840600
28	0.0326440	7.4412000	0.0043869	0.1343869	30.633490	227.949900
29	0.0288885	7.4700880	0.0038672	0.1338672	34.615840	258.583400
30	0.0255651	7.4956530	0.0034107	0.1334107	39.115900	293.199200
35	0.0138757	7.5855720	0.0018292	0.1318292	72.068510	546.680800
40	0.0075312	7.6343760	9.865E-40	0.1309865	132.781600	1,013.704000
45	0.0040876	7.6608640	5.336E-40	0.1305336	244.641400	1,874.165000
50	0.0022186	7.6752420	2.891E-40	0.1302891	450.735900	3,459.507000
60	6.536E-40	7.6872800	8.502E-50	0.1300850	1,530.053	11,761.95
70	1.925E-40	7.6908270	2.503E-50	0.1300250	5,193.870	39,945.15
80	5.672E-50	7.6918710	7.374E-60	0.1300074	17,630.94	135,614.9
90	1.671E-50	7.6921790	2.172E-60	0.1300022	59,849.42	460,372.4
100	4.922E-60	7.6922700	6.399E-70	0.1300006	203,162.9	1,562,784.0

	$i = 14\%$					
n	P/F	P/A	A/F	A/P	F/P	F/A
1	0.8771930	0.8771930	1.0000000	1.1400000	1.140000	1.000000
2	0.7694675	1.646661	0.4672897	0.6072897	1.299600	2.140000
3	0.6749715	2.321632	0.2907315	0.4307315	1.481544	3.439600
4	0.5920803	2.913712	0.2032048	0.3432048	1.688960	4.921144
5	0.5193687	3.433081	0.1512835	0.2912835	1.925415	6.610104
6	0.4555865	3.888668	0.1171575	0.2571575	2.194973	8.535519
7	0.3996373	4.288305	0.0931924	0.2331924	2.502269	10.730490
8	0.3505591	4.638864	0.0755700	0.2155700	2.852586	13.232760
9	0.3075079	4.946372	0.0621684	0.2021684	3.251949	16.085350
10	0.2697438	5.216116	0.0517135	0.1917135	3.707221	19.337300
11	0.2366174	5.452733	0.0433943	0.1833943	4.226232	23.044520
12	0.2075591	5.660292	0.0366693	0.1766693	4.817905	27.270750
13	0.1820694	5.842362	0.0311637	0.1711637	5.492411	32.088650
14	0.1597100	6.002072	0.0266091	0.1666091	6.261349	37.581060
15	0.1400965	6.142168	0.0228090	0.1628090	7.137938	43.842410
16	0.1228917	6.265060	0.0196154	0.1596154	8.137249	50.980350
17	0.1077997	6.372859	0.0169154	0.1569154	9.276464	59.117600
18	0.0945611	6.467420	0.0146212	0.1546212	10.575170	68.394070
19	0.0829484	6.550369	0.0126632	0.1526632	12.055690	78.969230
20	0.0727617	6.623131	0.0109860	0.1509860	13.743490	91.024930
21	0.0638261	6.686957	0.0095449	0.1495449	15.667580	104.768400
22	0.0559878	6.742944	0.0083032	0.1483032	17.861040	120.436000
23	0.0491121	6.792056	0.0072308	0.1472308	20.361580	138.297000
24	0.0430808	6.835137	0.0063028	0.1463028	23.212210	158.658600
25	0.0377902	6.872927	0.0054984	0.1454984	26.461920	181.870800
26	0.0331493	6.906077	0.0048000	0.1448000	30.166580	208.332700
27	0.0290783	6.935155	0.0041929	0.1441929	34.389910	238.499300
28	0.0255073	6.960662	0.0036645	0.1436645	39.204490	272.889200
29	0.0223748	6.983037	0.0032042	0.1432042	44.693120	312.093700
30	0.0196270	7.002664	0.0028028	0.1428028	50.950160	356.786800
35	0.0101937	7.070045	0.0014418	0.1414418	98.100180	693.572700
40	0.0052943	7.105041	7.451E-40	0.1407451	188.883500	1,342.025000
45	0.0027497	7.123217	3.860E-40	0.1403860	363.679100	2,590.565000
50	0.0014281	7.132656	2.002E-40	0.1402002	700.233000	4,994.521000
60	3.852E-40	7.140106	5.395E-50	0.1400540	2,595.919	18,535.13
70	1.039E-40	7.142115	1.455E-50	0.1400145	9,623.645	68,733.18
80	2.803E-50	7.142657	3.924E-60	0.1400039	35,676.98	254,828.4
90	7.561E-60	7.142803	1.059E-60	0.1400011	132,262.5	944,724.8
100	2.039E-60	7.142843	2.855E-70	0.1400003	490,326.2	3,502,323.0

	$i = 15\%$					
n	P/F	P/A	A/F	A/P	F/P	F/A
1	0.8695652	0.8695652	1.0000000	1.1500000	1.150000	1.000000
2	0.7561437	1.6257090	0.4651163	0.6151163	1.322500	2.150000
3	0.6575162	2.2832250	0.2879770	0.4379770	1.520875	3.472500
4	0.5717532	2.8549780	0.2002654	0.3502654	1.749006	4.993375
5	0.4971767	3.3521550	0.1483156	0.2983156	2.011357	6.742381
6	0.4323276	3.7844830	0.1142369	0.2642369	2.313061	8.753738
7	0.3759370	4.1604200	0.0903604	0.2403604	2.660020	11.066800
8	0.3269018	4.4873220	0.0728501	0.2228501	3.059023	13.726820
9	0.2842624	4.7715840	0.0595740	0.2095740	3.517876	16.785840
10	0.2471847	5.0187690	0.0492521	0.1992521	4.045558	20.303720
11	0.2149432	5.2337120	0.0410690	0.1910690	4.652391	24.349280
12	0.1869072	5.4206190	0.0344808	0.1844808	5.350250	29.001670
13	0.1625280	5.5831470	0.0291105	0.1791105	6.152788	34.351920
14	0.1413287	5.7244760	0.0246885	0.1746885	7.075706	40.504710
15	0.1228945	5.8473700	0.0210171	0.1710171	8.137062	47.580410
16	0.1068648	5.9542350	0.0179477	0.1679477	9.357621	55.717470
17	0.0929259	6.0471610	0.0153669	0.1653669	10.761260	65.075090
18	0.0808051	6.1279660	0.0131863	0.1631863	12.375450	75.836360
19	0.0702653	6.1982310	0.0113364	0.1613364	14.231770	88.211810
20	0.0611003	6.2593310	0.0097615	0.1597615	16.366540	102.443600
21	0.0531307	6.3124620	0.0084168	0.1584186	18.821520	118.810100
22	0.0462006	6.3586630	0.0072658	0.1572658	21.644750	137.631600
23	0.0401744	6.3988370	0.0062784	0.1562784	24.891460	159.276400
24	0.0349343	6.4337710	0.0054298	0.1554298	28.625180	184.167800
25	0.0303776	6.4641490	0.0046994	0.1546994	32.918950	212.793000
26	0.0264153	6.4905640	0.0040698	0.1540698	37.856800	245.712000
27	0.0229699	6.5135340	0.0035265	0.1535265	43.535310	283.568800
28	0.0199738	6.5335080	0.0030571	0.1530571	50.065610	327.104100
29	0.0173685	6.5508770	0.0026513	0.1526513	57.575450	377.169700
30	0.0151031	6.5659800	0.0023002	0.1523002	66.211770	434.745100
35	0.0075089	6.6166070	0.0011349	0.1511349	133.17550	881.170200
40	0.0037332	6.6417780	5.621E-40	0.1505621	267.86350	1,779.090000
45	0.0018561	6.6542930	2.789E-40	0.1502789	538.76930	3,585.128000
50	9.228E-40	6.6605150	1.385E-40	0.1501385	1,083.65700	7,217.716000
60	2.281E-40	6.6651460	3.422E-50	0.1500342	4,383.999	29,219.99
70	5.638E-50	6.6662910	8.458E-60	0.1500085	17,735.72	118,231.5
80	1.394E-50	6.6665740	2.091E-60	0.1500021	71,750.88	478,332.5
90	3.445E-60	6.6666440	5.168E-70	0.1500005	290,272.3	1,935,142.0
100	8.516E-70	6.6666610	1.277E-70	0.1500001	1,174,313.0	7,828,750.0

			$i = 20\%$			
n	P/F	P/A	A/F	A/P	F/P	F/A
1	0.8333333	0.8333333	1.0000000	1.2000000	1.200000	1.000000
2	0.6944444	1.5277780	0.4545455	0.6545455	1.440000	2.200000
3	0.5787037	2.1064810	0.2747253	0.4747253	1.728000	3.640000
4	0.4822531	2.5887350	0.1862891	0.3862891	2.073600	5.368000
5	0.4018776	2.9906120	0.1343797	0.3343797	2.488320	7.441600
6	0.3348980	3.3255100	0.1007057	0.3007057	2.985984	9.929920
7	0.2790816	3.6045920	0.0774239	0.2774239	3.583181	12.915900
8	0.2325680	3.8371600	0.0606094	0.2606094	4.299817	16.499080
9	0.1938067	4.0309670	0.0480795	0.2480795	5.159780	20.798900
10	0.1615056	4.1924720	0.0385228	0.2385228	6.191736	25.958680
11	0.1345880	4.3270600	0.0311038	0.2311038	7.430084	32.150420
12	0.1121567	4.4392170	0.0252650	0.2252650	8.916100	39.580500
13	0.0934639	4.5326810	0.0206200	0.2206200	10.699320	48.496600
14	0.0778866	4.6105670	0.0168931	0.2168931	12.839180	59.195920
15	0.0649055	4.6754730	0.0138821	0.2138821	15.407020	72.035110
16	0.0540879	4.7295610	0.0114361	0.2114361	18.488430	87.442130
17	0.0450732	4.7746340	0.0094401	0.2094401	22.186110	105.930600
18	0.0375610	4.8121950	0.0078054	0.2078054	26.623330	128.167000
19	0.0313009	4.8434960	0.0064625	0.2064625	31.948000	154.740000
20	0.0260841	4.8695800	0.0053565	0.2053565	38.337600	186.688000
21	0.0217367	4.8913160	0.0044439	0.2044439	46.005120	225.025600
22	0.0181139	4.9094300	0.0036896	0.2036896	55.206140	271.030700
23	0.0150949	4.9245250	0.0030653	0.2030653	66.247370	326.236900
24	0.0125791	4.9371040	0.0025479	0.2025479	79.496850	392.484200
25	0.0104826	4.9475870	0.0021187	0.2021187	95.396220	471.981100
26	0.0087355	4.9563230	0.0017625	0.2017625	114.475500	567.377300
27	0.0072796	4.9636020	0.0014666	0.2014666	137.370600	681.852800
28	0.0060663	4.9696680	0.0012207	0.2012207	164.844700	819.223300
29	0.0050553	4.9747240	0.0010162	0.2010162	197.813600	984.068000
30	0.0042127	4.9789360	8.461E-40	0.2008461	237.376300	1,181.882000
35	0.0016930	4.9915350	3.392E-40	0.2003392	590.668200	2,948.341000
40	6.804E-40	4.9965980	1.362E-40	0.2001362	1,469.772	7,343.858000
45	2.734E-40	4.9986330	5.470E-50	0.2000547	3,657.262	18,281.310000
50	1.099E-40	4.9994510	2.198E-50	0.2000220	9,100.438	45,497.190000
60	1.775E-50	5.9999110	3.529E-60	0.2000035	56,347.51	281,732.6
70	2.866E-60	4.9999860	5.732E-70	0.2000006	348,889.0	1,744,440.0
80	4.629E-70	4.9999980	9.258E-80	0.2000001	2,160,228.0	10,801,137.0
90	7.476E-80	5.0000000	1.495E-80	0.2000000	13,375,565	66,877,822
100	1.207E-80	5.0000000	2.145E-90	0.2000000	82,817,975	4.1409E8

	i = 25%					
n	*P/F*	*P/A*	*A/F*	*A/P*	*F/P*	*F/A*
1	0.8000000	0.800000	1.0000000	1.2500000	1.250000	1.000000
2	0.6400000	1.440000	0.4444444	0.6944444	1.562500	2.250000
3	0.5120000	1.952000	0.2622951	0.5122951	1.953125	3.812500
4	0.4096000	2.361600	0.1734417	0.4234417	2.441406	5.765625
5	0.3276800	2.689280	0.1218467	0.3718467	3.051758	8.207031
6	0.2621440	2.951424	0.0888195	0.3388195	3.814697	11.258790
7	0.2097152	3.161139	0.0663417	0.3163417	4.768372	15.073490
8	0.1677722	3.328911	0.0503985	0.3003985	5.960464	19.841860
9	0.1342177	3.463129	0.0387562	0.2887562	7.450581	25.802320
10	0.1073742	3.570503	0.0300726	0.2800726	9.313226	33.252900
11	0.0858993	3.656403	0.0234929	0.2734929	11.64153	42.566130
12	0.0687195	3.725122	0.0184476	0.2684476	14.55192	54.207660
13	0.0549756	3.780098	0.0145434	0.2645434	18.18989	68.759580
14	0.0439805	3.824078	0.0115009	0.2615009	22.73737	86.949470
15	0.0351844	3.859263	0.0091169	0.2591169	28.42171	109.686800
16	0.0281475	3.887410	0.0072407	0.2572407	35.52714	138.108500
17	0.0225180	3.909928	0.0057592	0.2557592	44.40892	173.635700
18	0.0180144	3.927942	0.0045862	0.2545862	55.51115	218.044600
19	0.0144115	3.942354	0.0036556	0.2536556	69.38894	273.555800
20	0.0115292	3.953883	0.0029159	0.2529159	86.73617	342.944700
21	0.0092234	3.963107	0.0023273	0.2523273	108.42020	429.680900
22	0.0073787	3.970485	0.0018584	0.2518584	135.52530	538.101100
23	0.0059030	3.976388	0.0014845	0.2514845	169.40660	673.626400
24	0.0047224	3.981111	0.0011862	0.2511862	211.75820	843.032900
25	0.0037779	3.984888	9.481E-40	0.2509481	264.69780	1,054.791000
26	0.0030223	3.987911	7.579E-40	0.2507579	330.87220	1,319.489000
27	0.0024179	3.990329	6.059E-40	0.2506059	413.59030	1,650.361000
28	0.0019343	3.992263	4.845E-40	0.2504845	516.98790	2,063.952000
29	0.0015474	3.993810	3.875E-40	0.2503875	646.23490	2,580.939000
30	0.0012379	3.995048	3.099E-40	0.2503099	807.79360	3,227.174000
35	4.056E-40	3.998377	1.015E-40	0.2501015	2,465.19000	9,856.76100
40	1.329E-40	3.999468	3.324E-50	0.2500332	7,523.16400	30,088.660000
45	4.356E-50	3.999826	1.089E-50	0.2500109	22,958.87000	91,831.500000
50	1.427E-50	3.999943	3.568E-60	0.2500036	70,064.92000	280,255.700000
60	1.532E-60	3.999994	3.831E-70	0.2500004	652,530.4	2,610,118.0
70	1.646E-70	3.999999	4.114E-80	0.2500000	6,077,163.0	24,308,649
80	1.767E-80	4.000000	4.417E-90	0.2500000	56,597,994	2.2639E8
90	1.897E-90	4.000000	4.74E-100	0.2500000	5.2711E8	2.1084E9
100	2.04E-100	4.000000	5.09E-110	0.2500000	4.9091E9	1.964E10

	i = 30%					
n	P/F	P/A	A/F	A/P	F/P	F/A
1	0.7692308	0.7692308	1.0000000	1.3000000	1.300000	1.000000
2	0.5917160	1.3609470	0.4347826	0.7347826	1.690000	2.300000
3	0.4551661	1.8161130	0.2506266	0.5506266	2.197000	3.990000
4	0.3501278	2.1662410	0.1616292	0.4616292	2.856100	6.187000
5	0.2693291	2.4355700	0.1105815	0.4105815	3.712930	9.043100
6	0.2071762	2.6427460	0.0783943	0.3783943	4.826809	12.756030
7	0.1593663	2.8021120	0.0568736	0.3568736	6.274852	17.582840
8	0.1225895	2.9247020	0.0419152	0.3419152	8.157307	23.857690
9	0.0942996	3.0190010	0.0312354	0.3312354	10.604500	32.015000
10	0.0725382	3.0915390	0.0234634	0.3234634	13.785850	42.619500
11	0.0557986	3.1473380	0.0177288	0.3177288	17.921600	56.405350
12	0.0429220	3.1902600	0.0134541	0.3134541	23.298090	74.326950
13	0.0330169	3.2232770	0.0102433	0.3102433	30.287510	97.625040
14	0.0253976	3.2486750	0.0078178	0.3078178	39.373760	127.912500
15	0.0195366	3.2682110	0.0059778	0.3059778	51.185890	167.286300
16	0.0150282	3.2832390	0.0045772	0.3045772	66.541660	218.472200
17	0.0115601	3.2948000	0.0035086	0.3035086	86.504160	285.013900
18	0.0088924	3.3036920	0.0026917	0.3026917	112.455400	371.518000
19	0.0068403	3.3105320	0.0020662	0.3020662	146.192000	483.973400
20	0.0052618	3.3157940	0.0015869	0.3015869	190.049600	630.165500
21	0.0040475	3.3198420	0.0012192	0.3012192	247.064500	820.215100
22	0.0031135	3.3229550	9.370E-40	0.3009370	321.183900	1,067.280000
23	0.0023950	3.3253500	7.202E-40	0.3007202	417.539100	1,388.464000
24	0.0018423	3.3271920	5.537E-40	0.3005537	542.800800	1,806.003000
25	0.0014172	3.3286090	4.257E-40	0.3004257	705.641000	2,348.803000
26	0.0010901	3.3297000	3.274E-40	0.3003274	917.333300	3,054.444000
27	8.386E-40	3.3305380	2.518E-40	0.3002518	1,192.533000	3,971.778000
28	6.450E-40	3.3311830	1.936E-40	0.3001936	1,550.293000	5,164.311000
29	4.962E-40	3.3316790	1.489E-40	0.3001489	2,015.381000	6,714.604000
30	3.817E-40	3.3320610	1.145E-40	0.3001145	2,619.996000	8,729,985000
35	1.028E-40	3.3329910	3.084E-50	0.3000308	9,727.860	32,422.870000
40	2.769E-50	3.3332410	8.306E-60	0.3000083	36,118.86	120,393.900000
45	7.457E-60	3.3333080	2.237E-60	0.3000022	134,106.8	447,019.400000
50	2.008E-60	3.3333270	6.025E-70	0.3000006	497,929.2	1,659,761.000000
60	1.457E-70	3.3333330	4.370E-80	0.3000000	6,864,377.0	22,881,254
70	1.057E-80	3.3333330	3.170E-90	0.3000000	94,631,268	3.1544E8
80	7.67E-100	3.3333330	2.30E-100	0.3000000	1.3046E9	4.3486E9
90	5.56E-110	3.3333330	1.67E-110	0.3000000	1.798E10	5.995E10
100	4.03E-120	3.3333330	1.21E-120	0.3000000	2.479E11	8.264E11

	i = 35%					
n	*P/F*	*P/A*	*A/F*	*A/P*	*F/P*	*F/A*
1	0.7407407	0.7407407	1.000000	1.3500000	1.350000	1.000000
2	0.5486968	1.2894380	0.4255319	0.7755319	1.822500	2.350000
3	0.4064421	1.6958800	0.2396645	0.5896645	2.460375	4.172500
4	0.3010682	1.9969480	0.1507642	0.5007642	3.321506	6.632875
5	0.2230135	2.2199610	0.1004583	0.4504583	4.484033	9.954381
6	0.1651952	2.3851570	0.0692597	0.4192597	6.053445	14.438410
7	0.1223668	2.5075230	0.0487999	0.3987999	8.172151	20.491860
8	0.0906421	2.5981650	0.0348870	0.3848870	11.03240	28.664010
9	0.0671423	2.6653080	0.0251912	0.3751912	14.89375	39.696410
10	0.0497350	2.7150430	0.0183183	0.3683183	20.10656	54.590160
11	0.0368408	2.7518840	0.0133875	0.3633875	27.14385	74.696720
12	0.0272894	2.7791730	0.0098193	0.3598193	36.64420	101.840600
13	0.0202144	2.7993870	0.0072210	0.3572210	49.46967	138.484800
14	0.0149736	2.8143610	0.0053204	0.3553204	66.78405	187.954400
15	0.0110916	2.8254530	0.0039256	0.3539256	90.15847	254.738500
16	0.0082160	2.8336690	0.0028994	0.3528994	121.7139	344.897000
17	0.0060859	2.8397550	0.0021431	0.3521431	164.3138	466.610900
18	0.0045081	2.8442630	0.0015850	0.3515850	221.8236	630.924700
19	0.0033393	2.8476020	0.0011727	0.3511727	299.4619	852.748300
20	0.0024736	2.8500760	8.679E-40	0.3508679	404.2736	1,152.210000
21	0.0018323	2.8519080	6.425E-40	0.3506425	545.7693	1,556.484000
22	0.0013572	2.8532650	4.757E-40	0.3504757	736.7886	2,102.253000
23	0.0010054	2.8542700	3.522E-40	0.3503522	994.6646	2,839.042000
24	7.447E-40	2.8550150	2.608E-40	0.3502608	1,342.7970	3,833.706000
25	5.516E-40	2.8555670	1.932E-40	0.3501932	1,812.7760	5,176.504000
26	4.086E-40	2.8559750	1.431E-40	0.3501431	2,447.2480	6,989.280000
27	3.027E-40	2.8562780	1.060E-40	0.3501060	3,303.7850	9,436.528000
28	2.242E-40	2.8565020	7.849E-50	0.3500785	4,460.1090	12,740.310000
29	1.661E-40	2.8566680	5.814E-50	0.3500581	6,021.1480	17,200.4200000
30	1.230E-40	2.8567910	4.306E-50	0.3500431	8,128.5490	23,221.5700000
35	2.744E-50	2.8570640	9.603E-60	0.3500096	36,448.69	104,136.3000000
40	6.119E-60	2.8571250	2.142E-60	0.3500021	163,437.10	466,960.4000000
45	1.365E-60	2.8571390	4.776E-70	0.3500005	732,857.6	2093,876.0000000
50	3.043E-70	2.8571420	1.065E-70	0.3500001	3286,158.0	9389,020.0000000
60	1.513E-80	2.8571430	5.297E-90	0.3500000	66073317	1.8878E8
70	7.53E-100	2.8571430	2.63E-100	0.3500000	1.3285E9	3.7957E9
80	3.74E-110	2.8571430	1.31E-110	0.3500000	2.671E10	7.632E10
90	1.86E-120	2.8571430	6.52E-130	0.3500000	5.371E11	1.535E12
100	9.26E-140	2.8571430	3.24E-140	0.3500000	1.080E13	3.085E13

	i = 40%					
n	*P/F*	*P/A*	*A/F*	*A/P*	*F/P*	*F/A*
1	0.7142857	0.7142857	1.0000000	1.4000000	1.400000	1.000000
2	0.5102041	1.2244900	0.4166667	0.8166667	1.960000	2.400000
3	0.3644315	1.5889210	0.2293578	0.6293578	2.744000	4.360000
4	0.2603082	1.8492290	0.1407658	0.5407658	3.841600	7.104000
5	0.1859344	2.0351640	0.0913609	0.4913609	5.378240	10.945600
6	0.1328103	2.1679740	0.0612601	0.4612601	7.529536	16.323840
7	0.0948645	2.2628390	0.0419228	0.4419228	10.541350	23.853380
8	0.0677604	2.3305990	0.0290742	0.4290742	14.757890	34.394730
9	0.0484003	2.3789990	0.0203448	0.4203448	20.661050	49.152620
10	0.0345716	2.4135710	0.0143238	0.4143238	28.925470	69.813660
11	0.0246940	2.4382650	0.0101277	0.4101277	40.49565	98.739130
12	0.0176286	2.4559040	0.0071821	0.4071821	56.69391	139.234800
13	0.0125990	2.4685030	0.0051039	0.4051039	79.37148	195.928700
14	0.0089993	2.4775020	0.0036324	0.4036324	111.12010	275.300200
15	0.0064281	2.4839300	0.0025879	0.4025879	155.56810	386.420200
16	0.0045915	2.4885210	0.0018451	0.4018451	217.7953	541.988300
17	0.0032796	2.4918010	0.0013162	0.4013162	304.9135	759.783700
18	0.0023426	2.4941440	9.392E-40	0.4009392	426.8789	1064.697000
19	0.0016733	2.4958170	6.704E-40	0.4006704	597.6304	1491.576000
20	0.0011952	2.4970120	4.787E-40	0.4004787	836.6826	2089.206000
21	8.537E-40	2.4978660	3.418E-40	0.4003418	1,171.356	2,925.889000
22	6.098E-40	2.4984760	2.441E-40	0.4002441	1,639.898	4,097.245000
23	4.356E-40	2.4989110	1.743E-40	0.4001743	2,295.857	5,737.142000
24	3.111E-40	2.4992220	1.245E-40	0.4001245	3,214.200	8,032.999000
25	2.222E-40	2.4994440	8.891E-50	0.4000889	4,499.880	11,247.200000
26	1.587E-40	2.4996030	6.350E-50	0.4000635	6,299.831	15,747.080000
27	1.134E-40	2.4997170	4.536E-50	0.4000454	8,819.764	22,046.910000
28	8.099E-50	2.4997980	3.240E-50	0.4000324	12,347.670	30,866.670000
29	5.785E-50	2.4998550	2.314E-50	0.4000231	17,286.740	43,214.340000
30	4.132E-50	2.4998970	1.653E-50	0.4000165	24,201.430	60,510.080000
35	7.683E-60	2.4999810	3.073E-60	0.4000031	130,161.100	325,400.300000
40	1.428E-60	2.4999960	5.714E-70	0.4000006	700,037.700	1,750,092.000000
45	2.656E-70	2.4999990	1.062E-70	0.4000001	3,764,971.000	9,412,424.000000
50	4.939E-80	2.5000000	1.975E-80	0.4000000	20,248,916.000	50,622,288.000000
60	1.707E-90	2.5000000	6.83E-100	0.4000000	5.8571E8	1.4643E9
70	5.90E-110	2.5000000	2.36E-110	0.4000000	1.694E10	4.235E10
80	2.04E-120	2.5000000	8.16E-130	0.4000000	4.901E11	1.225E12
90	7.05E-140	2.5000000	2.82E-140	0.4000000	1.418E13	3.544E13
100	2.44E-150	2.5000000	9.76E-160	0.4000000	4.100E14	1.025E15

	$i = 45\%$					
n	P/F	P/A	A/F	A/P	F/P	F/A
1	0.6896552	0.6896552	1.000000	1.4500000	1.450000	1.000000
2	0.4756243	1.1652790	0.4081633	0.8581633	2.102500	2.450000
3	0.3280167	1.4932960	0.2196595	0.6696595	3.048625	4.552500
4	0.2262184	1.7195150	0.1315595	0.5815595	4.420506	7.601125
5	0.1560127	1.8755270	0.0831834	0.5331834	6.409734	12.021630
6	0.1075950	1.9831220	0.0542553	0.5042553	9.294114	18.431370
7	0.0742034	2.0573260	0.0360679	0.4860679	13.476470	27.725480
8	0.0511748	2.1085000	0.0242707	0.4742707	19.540880	41.201950
9	0.0352930	2.1437930	0.0164629	0.4664629	28.334270	60.742820
10	0.0243400	2.1681330	0.0112262	0.4612262	41.084690	89.077090
11	0.167862	2.1849200	0.0076827	0.4576827	59.572800	130.161800
12	0.115767	2.1964960	0.0052705	0.4552705	86.380560	189.734600
13	0.0079839	2.2044800	0.0036217	0.4536217	125.251800	276.115100
14	0.0055061	2.2099860	0.0024915	0.4524915	181.615100	401.367000
15	0.0037973	2.2137840	0.0017153	0.4517153	263.341900	582.982100
16	0.0026189	2.2164030	0.0011816	0.4511816	381.845800	846.324000
17	0.0018061	2.2182090	8.142E-40	0.4508142	553.676400	1,228.170000
18	0.0012456	2.2194540	5.612E-40	0.4505612	802.830800	1,781.846000
19	8.590E-40	2.2203130	3.869E-40	0.4503869	1,164.105000	2,584.677000
20	5.924E-40	2.2209060	2.668E-40	0.4502668	1,687.952000	3,748.782000
21	4.086E-40	2.2213140	1.839E-40	0.4501839	2,447.530000	5,436.734000
22	2.818E-40	2.2215960	1.268E-40	0.4501268	3,548.919000	7,884.264000
23	1.943E-40	2.2217900	8.746E-50	0.4500875	5,145.932000	11,433.180000
24	1.340E-40	2.2219240	6.032E-50	0.4500603	7,461.602000	16,579.110000
25	9.243E-50	2.2220170	4.160E-50	0.4500416	10,819.320000	24,040.720000
26	6.374E-50	2.2220810	2.869E-50	0.4500287	15,688.0200	34,860.040000
27	4.396E-50	2.2221250	1.978E-50	0.4500198	22,747.6300	50,548.060000
28	3.032E-50	2.2221550	1.364E-50	0.4500136	32,984.0600	73,295.680000
29	2.091E-50	2.2221760	9.409E-60	0.4500094	47,826.8800	106,279.700000
30	1.442E-50	2.2221900	6.489E-60	0.4500065	69,348.9800	154,106.600000
35	2.250E-60	2.2222170	1.012E-60	0.4500010	444,508.5	987,794.500000
40	3.510E-70	2.2222210	1.579E-70	0.4500002	2,849,181.0	6,331,512.000000
45	5.476E-80	2.2222220	2.464E-80	0.4500000	18,262,495	40,583,319.000000
50	8.543E-90	2.2222220	3.844E-90	0.4500000	1.1706E80	2.6013E8000000
60	2.08E-100	2.2222220	9.36E-110	0.4500000	4.8093E9	1.069E10000000
70	5.06E-120	2.2222220	2.28E-120	0.4500000	1.976E11	4.391E11000000
80	1.23E-130	2.2222220	5.54E-140	0.4500000	8.118E12	1.804E13000000
90	3.00E-150	2.2222220	1.35E-150	0.4500000	3.335E14	7.412E14000000
100	7.30E-170	2.2222220	3.28E-170	0.4500000	1.370E16	3.045E16000000

			i = 50%			
n	*P/F*	*P/A*	*A/F*	*A/P*	*F/P*	*F/A*
1	0.6666667	0.6666667	1.0000000	1.5000000	1.500000	1.000000
2	0.4444444	1.111111	0.4000000	0.9000000	2.250000	2.500000
3	0.2962963	1.407407	0.2105263	0.7105263	3.375000	4.750000
4	0.1975309	1.604938	0.1230769	0.6230769	5.062500	8.125000
5	0.1316872	1.736626	0.0758294	0.5758294	7.593750	13.187500
6	0.0877915	1.824417	0.0481203	0.5481203	11.390630	20.78125
7	0.0585277	1.882945	0.0310831	0.5310831	17.085940	32.17188
8	0.0390184	1.921963	0.0203013	0.5203013	25.628910	49.25781
9	0.0260123	1.947975	0.0133535	0.5133535	38.443360	74.88672
10	0.0173415	1.965317	0.0088238	0.5088238	57.665040	113.33010
11	0.0115610	1.976878	0.0058481	0.5058481	86.497560	170.99510
12	0.0077073	1.984585	0.0038836	0.5038836	129.746300	257.49270
13	0.0051382	1.989724	0.0025824	0.5025824	194.619500	387.23900
14	0.0034255	1.993149	0.0017186	0.5017186	291.929300	581.85850
15	0.0022837	1.995433	0.0011444	0.5011444	437.893900	873.78780
16	0.0015224	1.996955	7.624E-40	0.5007624	656.840800	1,311.68200
17	0.0010150	1.997970	5.080E-40	0.5005080	985.261200	1,968.52200
18	6.766E-40	1.998647	3.385E-40	0.5003385	1,477.892000	2,953.78400
19	4.511E-40	1.999098	2.256E-40	0.5002256	2,216.838000	4,431.67600
20	3.007E-40	1.999399	1.504E-40	0.5001504	3,325.257000	6,648.51300
21	2.005E-40	1.999599	1.003E-40	0.5001003	4,987.885	9,973.77000
22	1.337E-40	1.999733	6.684E-50	0.5000668	7,481.828	14,961.6600
23	8.910E-50	1.999822	4.456E-50	0.5000446	11,222.740	22,443.4800
24	5.940E-50	1.999881	2.970E-50	0.5000297	16,834.110	33,666.22
25	3.960E-50	1.999921	1.980E-50	0.5000198	25,251.170	50,500.34
26	2.640E-50	1.999947	1.320E-50	0.5000132	37,876.75	75,751.51
27	1.760E-50	1.999965	8.801E-60	0.5000088	56,815.13	113,628.3
28	1.173E-50	1.999977	5.867E-60	0.5000059	85,222.69	170,443.4
29	7.823E-60	1.999984	3.911E-60	0.5000039	127,834.00	255,666.1
30	5.215E-60	1.999990	2.608E-60	0.5000026	191,751.10	383,500.1
35	6.868E-70	1.999999	3.434E-70	0.5000003	1,456,110.0	2,912,217.0
40	9.044E-80	2.000000	4.522E-80	0.5000000	11,057,332	22,114,663
45	1.191E-80	2.000000	5.955E-90	0.5000000	83,966,617	1.6793E8
50	1.568E-90	2.000000	7.84E-100	0.5000000	6.3762E80	1.2752E9
60	2.72E-110	2,000000	1.36E-110	0.5000000	3.677E10	7.354E10
70	4.72E-130	2,000000	2.36E-130	0.5000000	2.120E12	4.241E12
80	8.18E-150	2.000000	4.09E-150	0.5000000	1.223E14	2.445E14
90	1.42E-160	2,000000	7.09E-170	0.5000000	7.050E15	1.410E16
100	2.46E-180	2,000000	1.23E-180	0.5000000	4.066E17	8.131E17

APPENDIX

B

DEFINITIONS OF CERTAIN SI UNITS

Name	Unit	Definition
Energy	joule	The joule is the work done when the point of application of a force of one newton is displaced a distance of one meter in the direction of the force.
Force	newton	The newton is that force which, when applied to a body having a mass of one kilogram, gives it an acceleration of one meter per second squared.
Frequency	hertz	The hertz is the frequency of a periodic phenomenon of which the period is one second.
Power	watt	The watt is the power which gives rise to the production of energy at the rate of one joule per second.
Pressure or stress	pascal	The pascal is the pressure or stress of one newton per square meter.
Temperature (thermodynamic)	kelvin	The kelvin is the unit of thermodynamic temperature measured from absolute zero; it is the same size as the degree Celsius.
Temperature (practical)	degree Celsius	The degree Celsius is the temperature in kelvins minus 273.15. The Celsius scale was formerly called centigrade.

Source: Standard for Metric Practice, ASTM E 380-76, IEEE 268-1976, American Society for Testing and Materials, 1916 Race Street, Philadelphia, PA 19103.

APPENDIX

C

ALPHABETICAL LIST OF UNITS WITH THEIR SI NAMES AND CONVERSION FACTORS

To convert from	to	Symbol	Multiply by
Acre (U.S. survey)	square meter	m^2	4.047×10^3
Acre-foot	cubic meter	m^3	1.233×10^3
Atmosphere (standard)	pascal	Pa	1.013×10^5
Board foot	cubic meter	m^3	$2.359 \div 10^3$
Degrees Celsius	kelvin	K	$t_K = t_{°C} + 273.15$
Degree Fahrenheit	Celsius degree	°C	$t_{°C} = (t_{°F} - 32)/1.8$
Degree Fahrenheit	kelvin	K	$t_K = (t_{°F} + 459.67)/1.8$
(Degree) Kelvin	Celsius degree	°C	$t_{°C} = t_K - 273.15$
Foot	meter	m	$3.048 \div 10$
Foot, square	square meter	m^2	$9.290 \div 10^2$
Foot, cubic	cubic meter	m^3	$2.831 \div 10^2$
Feet, cubic, per minute	cubic meters/per second	m^3/s	$4.917 \div 10^4$
Feet per second	meters/per second	m/s	$3.048 \div 10$
Foot-pound force	joule	J	1.355×1
Foot-pounds per minute	watt	W	$2.259 \div 10^2$
Foot-pounds per second	watt	W	1.355×1
Gallon (U.S. liquid)	cubic meter	m^3	$3.785 \div 10^3$
Gallons per minute	cubic meters per second	m^3/s	$6.309 \div 10^5$
Horsepower (550 ft-lb/sec)	watt	W	7.457×10^2
Horsepower	kilowatt	kW	$7.457 \div 10$
Inch	meter	m	$2.540 \div 10^2$
Inch, square	square meter	m^2	$6.452 \div 10^4$
Inch, cubic	cubic meter	m^3	$1.639 \div 10^5$
Inch	millimeter	mm	2.540×10
Kelvin	degree Celsius	°C	$t_{°C} = t_K - 273.15$
Mile	meter	m	1.609×10^3
Mile	kilometer	km	1.609×1
Miles per hour	kilometers per hour	km/h	1.609×1
Miles per minute	meters per second	m/s	2.682×10
Pound	kilogram	kg	$4.534 \div 10$
Pounds per cubic yard	kilograms per cubic meter	kg/m^3	$5.933 \div 10$
Pounds per cubic foot	kilograms per cubic meter	kg/m^3	1.602×10
Pounds per gallon (U.S.)	kilograms per cubic meter	kg/m^3	1.198×10^2
Pounds per square foot	kilograms per square meter	kg/m^2	4.882×1
Pounds per square inch (psi)	pascal	Pa	6.895×10^3
Ton (2,000 lb)	kilogram	kg	9.072×10^2
Ton (2,240 lb)	kilogram	kg	1.016×10^3
Ton (metric)	kilogram	kg	1.000×10^3
Tons (2,000 lb) per hour	kilograms per second	kg/s	$2.520 \div 10$
Yard, cubic	cubic meter	m^3	$7.646 \div 10$
Yards, cubic, per hour	cubic meter per hour	m^3/h	$7.646 \div 10$

Note: All SI symbols are expressed in lower-case letters except those that are used to designate a person, which are capitalized.

Sources: Standard for Metric Practice, ASTM E 380-76, IEEE 268-1976, American Society for Testing and Materials, 1916 Race Street, Philadelphia, PA 19103.

National Standard of Canada Metric Practice Guide, CAN-3-001-02-73/CSA Z 234.1-1973, Canadian Standards Association, 178 Rexdale Boulevard, Rexdale, Ontario, Canada M94 IRS.

APPENDIX

D

FACTORS FOR CONVERTING CERTAIN U.S. CUSTOMARY (ENGLISH) UNITS TO METRIC UNITS

In general, the units appearing in this list do not appear in the list of SI units but they are used frequently, and it is probable that they will continue to be used by the construction industry. The units meter and liter may be spelled metre and litre. Both spellings are acceptable.

Multiply USC (English) unit	by	To obtain metric unit
Acre	0.4047	Hectare
Cubic foot	0.0283	Cubic meter
Foot-pound	0.1383	Kilogram-meter
Gallon (U.S.)	0.833	Imperial gallon
Gallon (U.S.)	3.785	Liters
Horsepower	1.014	Metric horsepower
Cubic inch	0.016	Liter
Square inch	6.452	Square centimeter
Miles per hour	1.610	Kilometers per hour
Ounce	28.350	Grams
Pounds per square inch	0.0689	Bars
Pounds per square inch	0.0703	kilograms per square centimeter

APPENDIX

E

U.S. CUSTOMARY (ENGLISH) UNIT EQUIVALENTS

Unit	Equivalent
1 acre	43,560 square feet
1 atmosphere	14.7 lb per square inch
1 BTU	788 foot-pounds
1 BTU	0.000393 horsepower-hour
1 foot	12 inches
1 cubic foot	7.48 gallons liquid
1 square foot	144 square inches
1 gallon	231 cubic inches
1 gallon	4 quarts liquid
1 horsepower	550 foot-pounds per second
1 mile	5,280 feet
1 mile	1,760 yards
1 square mile	640 acres
1 pound	16 ounces avoirdupois
1 quart	32 fluid ounces
1 long ton	2,240 pounds
1 short ton	2,000 pounds

APPENDIX

F

METRIC UNIT EQUIVALENTS

Unit	Equivalent
1 centimeter	10 millimeters
1 square centimeter	100 square millimeters
1 hectare	10,000 square meters
1 kilogram	1,000 grams
1 liter	1,000 cubic centimeters
1 meter	100 centimeters
1 kilometer	1,000 meters
1 cubic meter	1,000 liters
1 square meter	10,000 square centimeters
1 square kilometer	100 hectares
1 kilogram per square meter	0.97 atmosphere
1 metric ton	1,000 kilograms

INDEX